D0394294

Women in the Biological Sciences

Women in the Biological Sciences

A Biobibliographic Sourcebook

Edited by

Louise S. Grinstein, Carol A. Biermann,
and
Rose K. Rose

With a Foreword by Martha Oakley Chiscon

NATIONAL UNIVERSITY LIBRARY SAN DIEGO

NON CIRCULATING

GREENWOOD PRESS
Westport, Connecticut • London

Library of Congress Cataloging-in-Publication Data

Women in the biological sciences : a biobibliographic sourcebook /
  edited by Louise S. Grinstein, Carol A. Biermann, and Rose K. Rose ;
  with a foreword by Martha Oakley Chiscon.
     p.   cm.
  Includes bibliographical references and index.
  ISBN 0–313–29180–2 (alk. paper)
  1. Women biologists—Biography.   I. Grinstein, Louise S.
II. Biermann, Carol A.   III. Rose, Rose K.
QH26.W66   1997
574'.082—dc20
  [B]   96–43783

British Library Cataloguing in Publication Data is available.

Copyright © 1997 by Louise S. Grinstein, Carol A. Biermann, and Rose K. Rose

All rights reserved. No portion of this book may be
reproduced, by any process or technique, without the
express written consent of the publisher.

Library of Congress Catalog Card Number: 96–43783
ISBN: 0–313–29180–2

First published in 1997

Greenwood Press, 88 Post Road West, Westport, CT 06881
An imprint of Greenwood Publishing Group, Inc.

Printed in the United States of America

The paper used in this book complies with the
Permanent Paper Standard issued by the National
Information Standards Organization (Z39.48–1984).

10 9 8 7 6 5 4 3 2 1

To Jack Richman
—*L.S.G.*

To Gregg Walter and Annice Rose Biermann
and
In memory of
Professor Priscilla Frew Pollister and Dr. Jean Broadhurst
—*C.A.B.*

To Esther H. Rose, M.D.
—*R.K.R.*

CONTENTS

FOREWORD

Martha Oakley Chiscon

Lillian Smith, author of such fine works as *Strange Fruit, Killers of the Dream* and *The Journey,* insightfully has written, "To believe in something not yet proved and to underwrite it with our lives is the only way we can leave the future open. Humans, surrounded by facts, permitting themselves no surmise, no intuitive flash, no great hypothesis, are in a locked cell. . . . To find the point where hypothesis and fact meet; the delicate equilibrium between dream and reality; to question, knowing that you may never find the full answer. . . . This is what our journey is about, I think."

I recently was reminded of this while rereading what I've read so many times before—a collection of unpublished letters in my possession from Nobel Laureate Barbara McClintock, written nearly 50 years ago to one of her graduate students. They are filled not just with science material, but also with the guiding, sensitive, caring, even compassionate thoughts underlying a mentor's warm regard and genuine respect for her mentee. Elsewhere are included intimate or even, at times, mundane moments which might have come from the lives of any of us: the illness and untimely death of a relative, disdain for the weather and growing conditions, the anniversary celebration of parents, concern for colleagues and their happiness and well-being. And, interspersed amongst it all, there are references to McClintock's very personal involvement with biology: the hypotheses; the immense work involved in the gathering and interpretation of data, coupled with the stress, the frustration and weariness; the inner belief of being able to see through the morass to solve a problem; and, ultimately, the *fun*, that short and simple word implying fulfillment from one's journey through life. McClintock recounts:

Beadle mentioned to you something about this problem I am working on. It is a fine one but really frightfully complex. . . . I told him how frustrated I had been over it but, since then, the picture has cleared in certain essential places. . . . It has necessitated a tremendous organization of the material in order to develop the needed lines of attack. . . . I have been working day and night on it for a month. . . . The night before last, I was so all-in and felt so hopelessly rushed. . . . Yesterday, I discussed the matter with Witkin and Caspari. Both of them are well acquainted with my problem and both feel it is a striking one. Caspari states it is the most interesting story he has heard of in years. This will give you some idea of how I would feel if I made a mess of it. It is terribly complex but it does make sense (or I think so now). I am frightfully tired from this continuous strain. I just can't seem to relax because of the built-up tensions. Have not had a real vacation for eight years and just ache for one now. . . .

The material is fascinating. I am itching to show it to you for it does get into one's brain, like it or not. But, it twists one's brain likewise and I often feel as if mine were cork-screwed. . . . I do feel I am on the right track although all of the facts are by no means clear. There is very great variation, but all of it has order. It is finding this order that is the difficulty. . . . I am working hard at it much of the time, with all its complexities that are frustrating. . . . The problem is such a brain twister and has had me somewhat tied up in knots. . . . I think I know something about it but still not much. It is a real brain twister. . . . If I have 20 more years, maybe I will see some light. . . . Even so, I do like to do it and am having a good time. [It] really is swell and wonderful fun.

You will find in the pages of this biobibliographic sourcebook what to me are fascinating accounts of the journeys of a goodly number of such remarkable humans, all of them women. Each chose the biological sciences as one way of seeing their world. They believed in something not yet proved and underwrote it with their lives. They looked to their future as open, permitting themselves that intuitive flash, that working hypothesis. Whatever the answers that would come, they knew they had questioned. Many, in their determination to allow the cells of their own lives to remain unlocked, in turn unlocked many of the myriad of doors leading to the better comprehension of the cellularity of life and of life itself.

Much has been written about the difficulties many women have encountered in a scientific world not always amenable to women. Some women in science possibly were less historically documented or cited. Overlooked, their work was incorporated into the work of those who politically and financially dominated the laboratories of the time. Other journeys fared better, the travelers benefiting from the paths forged previously. And others, of course, were indeed fortunate enough to be welcomed and nurtured by many colleagues in the establishment. You will find some of each in the pages which follow. One hopes that this volume will instruct and hopefully inspire not only those that are, but also those wondering whether they should be contributing to the biological sciences. Historically or contemporarily, readers will find themselves in good company.

I personally hope this volume will be used extensively by teachers as well as students. These accounts still are not found as frequently as merit might demand within the lectures, discussions, texts, and learning or teaching aids of our time. There are still messages to be told by teachers who themselves often are undervalued, underpaid, undertenured, and underpromoted. As my husband succinctly has put it, "You are the means by which your own teachers and theirs before them, your present and past colleagues and the researchers in whose laboratories they once worked, continue to have a voice in the shaping of a new generation. It becomes your privilege to bring together these many accomplishments, in a way hopefully destined to demonstrate possible interrelationships, raise new questions to be answered in the future, and place old questions within the reach of this newly assembled knowledge, in hopes of their earlier solution. ... Students must realize that what you are saying as a creative teacher is in many ways unique, having never been brought together, interpreted and taught quite this way before—and that what *they* now will make of and do with this part of their total life's education may prove to be hybrid vigor at its finest.''

I urge that everyone interested in any aspect of the sciences contemplate a scene from Anton Chekhov's *Three Sisters*, in which Masha is found looking wistfully out a window, examining in her mind what types of things humans like her must question to keep their lives from being relegated to simple existence. She then turns to her male colleagues on stage with her and says, ''How can you live and not know why the cranes fly, why children are born, why the stars shine in the sky. ... You must either know why you live, or else . . . nothing else matters either . . . everything's just wild grass, all nonsense and waste.''

In the pages which follow, you will discover productive women who once looked through their own windows and discovered the biological sciences. Here are women who, in the course of their lives, did want to know why they lived. *It did matter,* as hopefully it shall to those who utilize this volume.

PREFACE

Louise S. Grinstein, Carol A. Biermann,
and Rose K. Rose

The recent expansion of women's consciousness-raising associations and women's studies programs has heightened queries concerning the participation of women in the various fields of the biological sciences. History of science publications, including existing survey books devoted exclusively to women in science, and general biographical dictionaries portray an extremely limited picture of the roles played by women in the growth and development of these sciences.

Generally, biology textbooks provide only a cursory look at the accomplishments of women in the biological disciplines. Only a handful of women biologists are likely to be noted, including Nobel laureates such as Gerty Cori, Gertrude Elion, Barbara McClintock, and Rosalyn Yalow.

This volume aims to present a definitive archival collection of original entries on a larger group of individuals, including their work as well as their lives. An attempt was made to survey as many of the branches of biological investigation as possible.

Sixty-five representative women from different countries and eras were chosen for inclusion in this work. These women may have gained recognition through (1) attainment of advanced degrees despite extensive familial and societal pressures; (2) innovative research results in some aspect of the biological sciences; (3) influence exerted in teaching and guidance of students at both the undergraduate and graduate levels; (4) active participation and leadership in professional societies; (5) extensive scholarly publications; (6) participation on journal editorial boards; (7) extensive field experiences; and (8) influence exerted upon public and political scientific policy making of their time. A woman was deemed eligible for inclusion if her work satisfied several of these criteria, even though in terms of any one she may not have been particularly outstanding. In order to provide a historical perspective, the scope of this volume was limited to those deceased or else born on or before 1930.

Entries are arranged in alphabetical order. Where a woman has been known by more than one name, we have tried to use the name under which much of her professional work has been published.

For purposes of stylistic uniformity and consistency in the volume as a whole, contributors were asked to follow a set format. Each chapter has three sections: biography, work, and bibliography. Cross-references to other women discussed in the volume are given by an asterisk following first mention in an entry. Since space limitations have forced us to shorten some contributors' entries, the reader is cautioned not to interpret the length of any contribution (or part of it) as a measure of the fame of the scientist or the importance of her work.

The biography section includes known information about the subject's family background, education, and career. Particular attention is given to any circumstances and influences that affected her profession.

In the work section, an estimate is made of the individual's career and its significance. Every attempt has been made to present the relevant material in language as nontechnical as possible without destroying its meaning.

The bibliography section is divided into listings of "works by" and "works about" the subject. The listing of "works by" is chronological. Space limitations forced us to abridge the lists supplied by the contributors. If a substantial bibliography of the individual's work is easily available elsewhere, we refer the reader to that source. In addition, we omit such works as abstracts of lectures and unpublished papers, book reviews, patents, and unpublished technical reports. The listing of "works about" is alphabetical by author.

Omitted from the individual listings but collected in an appendix at the end of the volume are citations to standard reference works and other compilations. Other appendixes highlight and summarize information presented. One appendix lists the subjects in chronological order by birthdate and depicts the time span of their lives. Another appendix summarizes in tabular form the data on place of birth, highest education attained, place of work, and specific field of interest. Subject and name indexes are also included.

It is our objective that this volume will help encourage young women to seek careers in the biological sciences. We also hope that it will serve as a springboard for further research on the past, present, and future role of women in these sciences.

This volume could not have been brought to fruition without the efforts of our contributors. We wish to express to them our thanks and appreciation for their suggestions, advice, and participation. In many cases their work required deep commitment and compassion.

We wish to acknowledge the assistance of the librarians at Kingsborough Community College of City University of New York (CUNY) in the task of cross-checking and indexing citations. Professors Jeanne Galvin, Barbara Goldman, Florence Houser, Allan Mirwis, Coleridge Orr, Roberta Pike, and Angelo Tripicchio were especially helpful. Our thanks and appreciation are extended also to Audrey Bernfield, director, Office Enrichment Programs of Harvard Medical School, and the librarians at the American Museum of Natural History (New

York City), Columbia University, the Mt. Sinai Medical Library, the New York Academy of Medicine, the New York Botanical Garden Library, New York University, and the Library of Science and Medicine of Rutgers University. In addition, special thanks are extended to Peg Henderson at the Diatom Library of the Philadelphia Academy of Natural Sciences; reference archivist Beverly B. Allen of Emory University's Woodruff Library; Dorothy E. Mosakowski, coordinator of archives and special collections at Clark University; librarians at the Woman's Medical College of Pennsylvania; Debbie Stern and N. Nelson of the Albert Einstein Medical Library; Irene Ferguson of the Edinburgh University Library; M. J. Hughes, registrary's clerk, Cambridge University; Steven P. Johnson, Wildlife Conservation Society; Peter Carini, Archivist, Mt. Holyoke College; Julie Moring at the University of Oklahoma Library; and Scott Russell of the Oklahoma Academy of Science. They offered much-needed help in answering questions regarding bibliographic entries.

We also wish to thank Cynthia Harris and the other members of the editorial staff at Greenwood Press for their guidance, encouragement, and patience in producing this work.

Finally, we wish to thank Helen, Serge, and Masha Kovarsky for their skilled typing and formatting.

INTRODUCTION

Jane Butler Kahle

It is difficult to determine when and how most girls become turned off to science. Does the influence of children's television or clearly marked boys' and girls' aisles in toy stores play a role? When does a girl first learn to smile prettily or to act demurely? When does a boy learn that risk taking (on the playground and in the classroom) is rewarded? Although we do not know when active and passive roles are learned and internalized, we do know that acceptance of such different roles affects a young person's confidence in, and attitudes about, science. Recent studies suggest that appropriate behavior for girls and boys is learned by the age of 24 to 26 *months*! At that early age, female and male behavior stereotypes are set, and boys, more rigidly than girls, define what they will and will not do. Acceptance of different behavior roles affects the attitudes of girls and boys in a variety of ways. Interviews of kindergarten children three weeks after they enter school suggest that they bring to school different ideas about science. At that age, neither boys nor girls can define science, but—even without that knowledge—more boys than girls reply that they want to be scientists, that they are good in science, and that they have done science (Campbell).

One of the challenges of education—and of this book—is to change the negative beliefs and attitudes that girls, on the whole, hold about science. In the third grade equal percentages of girls (67%) and boys (66%) respond that what they learn in science classes is useful in everyday life. By seventh grade both girls' and boys' responses have declined by about 10 percent. However, the percentage of boys with positive attitudes is stable through high school, while the percentage of girls who respond positively falls by another 11 percent. The same is true of students' responses concerning a career in science; that is, girls and boys respond similarly in seventh grade, but girls' responses decline precipitously by high school graduation (Mullis and Jenkens).

Recently, several summaries have explicated the types of research needed to help us understand gender differences in attitudes, beliefs, and participation in science and to ameliorate them. One area identified is pertinent to this book;

that is, research is needed that analyzes the effects of differential representations of various groups in school texts, children's literature, and content curricula. This book provides positive representation of women in the biological sciences and is an excellent addition to the scarce literature available. It is readable and, hopefully, will be widely accessible.

How often have all of us who are concerned about the participation and achievement of women in science heard our colleagues say, ''I would include information about women, if I knew where to find it. Can you give me a list of famous (or competent) women in science?'' This book provides not only a list but interesting biobibliographical essays for 65 outstanding women scientists. The essays have wide appeal and use because each contains a personal description of the scientist's life as well as details of her scientific work and accomplishments. Supplementing the essays and making the book useful for neophytes as well as experts is the bibliography that is part of each description. The bibliographies are divided into two sections: one is a list of the scientist's own writings; the other is a listing of articles and/or books about her. The latter makes the book very useful for teachers and for young women who want to explore further the conditions and successes surrounding the work of women scientists.

One entry, Gerty Cori, presents several themes that are common to many of the women's lives. Dr. Cori overcame great odds to receive a scientific education. She married a scientist who was very supportive of her work, but it was not until she received the Nobel Prize that she was accorded equal opportunities in the workplace. For years she labored in the shadow of her husband. Over and over again those themes are repeated. The reader will find many fascinating stories concerning both well known women scientists—Rachel Carson and Barbara McClintock, for example—and lesser known or unknown women.

It is my hope that this book will become a common reference for textbook authors and publishers, for teachers and students, and for every girl who dares to dream about becoming a scientist. It fulfills a great need and it may be instrumental in changing the attitudes of a generation of girls who, armed with knowledge of the odds, may be less afraid of them.

REFERENCES

Pat Campbell (President, Campbell-Kibler Associates, 80 Lakeside Drive, Groton, MA 01450) has provided the information about kindergarten students' attitudes (Personal communication, 1993).

Mullis, I. V., and L. B. Jenkens. *The Science Report Card: Elements of Risk and Recovery.* Princeton, NJ: Educational Testing System, 1988.

Women in the
Biological Sciences

ELIZABETH CABOT CARY AGASSIZ
(1822–1907)

Marilyn Bailey Ogilvie

BIOGRAPHY

Elizabeth Cabot Cary was born on December 5, 1822, in Boston, Massachusetts, at the house of her grandfather, Colonel Perkins. She was the second of seven children of Thomas Graves and Mary Ann Cushing (Perkins) Cary. Both the Cary and the Perkins families had come to Massachusetts in the early seventeenth century and had become wealthy; however, Thomas Cary's father lost his fortune, and Thomas himself was unsuccessful in his law practice in Brattleboro, Vermont. To support his family he moved to Boston and joined his father-in-law's firm, Perkins and Company, which had been successful in the China trade. Six years later the company closed, and Thomas Cary became treasurer of the Hamilton and Appleton Mills of Lowell, Massachusetts, a position he held until his death in 1859. Elizabeth (also known as Lizzie) spent most of her childhood in her grandfather Perkins's large house at Temple Place, Boston. Her childhood was one of controlled confusion, for, in addition to the Cary children, the house was host to innumerable cousins. Elizabeth Cary's sister, Caroline (Mrs. Charles P. Curtis), noted in her *Memories*, an unpublished account, that there were 21 children in all: eight Cabots, six Gardiners, and seven Carys. Cultural activities were plentiful in the extended family, and Elizabeth grew up with a love for music. The Carys' policy toward their children's education was to provide them with a governess to teach both boys and girls until they were 14, at which point they went to school. Elizabeth, however, never attended school because of her delicate health. She had additional lessons at home in languages, drawing, and music. Nothing in her education presaged an interest in natural science.

In 1846 Louis Agassiz (1807–73), professor at the University of Neuchatel, Switzerland, arrived in Boston to study North American natural history. He had three children and was married at this time. During his stay in Boston he presented a series of popular lectures at the Lowell Institute and made a great impression on Cary's mother. After church one Sunday, Mary Ann Cary reputedly remarked to her daughter that the person who sat in the Lowell's pew was the first man whom she would want Elizabeth to marry. Louis Agassiz immigrated to the United States and accepted the chair of natural history at the Lawrence Scientific School, a newly acquired branch of Harvard University. His wife had died, and his children remained in Europe to attend school.

Meanwhile, Elizabeth Cary had been introduced to Bostonian intellectual circles by her sister, Mary, who had married Cornelius Conway Felton, professor of Greek at Harvard. Through the Feltons, Elizabeth Cary met Agassiz, whom she married on April 25, 1850. Shortly after their marriage, in 1851, Louis Agassiz accepted a position at the Medical School in Charleston, South Carolina, and was accompanied by Elizabeth. This lectureship, during the winter term, did not interfere with his duties at the Lawrence Scientific School, where he lectured in the fall and spring. His three children came to the United States, and a happy relationship developed between them and their stepmother. She was particularly close to Alexander Agassiz, who was later to become a famous marine zoologist. He wrote that she was mother, sister, companion, and friend to him (1913).

The strategy of marital collaboration in science was evident in their careers. In 1856 Elizabeth Agassiz helped out with the finances by setting up a girls' school in her Cambridge home with the help of her two older stepchildren. Louis helped her with this project by giving lectures and planning the curriculum. Her duties did not involve teaching, but she served as head of the school and generally oversaw the pupils in the various branches. In 1863 they closed the school partially because of the uncertainties generated by the Civil War and partially because Louis Agassiz's income had increased sufficiently so that additional money was no longer necessary. The school experiment was the first time that Elizabeth Agassiz showed an interest in natural history. She attended Louis's lectures and took notes avidly. When she showed him her notes, he commented that while they were most gracefully expressed, they were nonsense from a scientific point of view.

In 1865 Elizabeth Agassiz accompanied her husband to Brazil to study the fauna of that region for the benefit of the fledgling Museum of Comparative Zoology (Agassiz Museum) at Harvard. They remained in Brazil until August 1866. Elizabeth Agassiz was the self-appointed clerk of the expedition. She kept a detailed journal, including anecdotes about their companions, one of whom was the youthful psychologist William James. She sent this journal as letters to her family. Later, Louis Agassiz's lectures and Elizabeth's own personal experiences were published under both their names with the title, *A Journey in Brazil* (1868).

Elizabeth and Louis Agassiz collaborated on another project, the Hassler Expedition, which involved deep-sea dredging along the Atlantic and Pacific coasts of the Americas from December 1871 until August 1872. Elizabeth's duties again consisted of taking detailed notes. This journal was not ready for the press at the time of Louis Agassiz's death, and it has never been published as a whole. Portions were published in three articles that appeared in the *Atlantic Monthly*.

Their last collaborative project occurred in 1873, when Elizabeth aided Louis in the planning and administration of the coeducational Anderson School of Natural History, both a summer school and a marine laboratory, on Penikese Island in Buzzard's Bay, Massachusetts. In 1873 Louis's death of a cerebral hemorrhage brought an end to their collaborative projects. His death also opened a new chapter in Elizabeth's life. Her natural history days were over. She spent her time caring for her stepson Alexander's three children (his young wife had died eight days after the death of Louis) and working on a biography of her husband. She was active in the establishment of the Harvard Annex for women (later Radcliffe College) and served as its first president (1894–1903). She suffered a cerebral hemorrhage in 1904 and died of a second one in 1907.

WORK

Elizabeth Agassiz's interest in science derived from her husband's. She was totally without scientific training and received all her information from her association with Louis. She was essential in preserving, elucidating, and popularizing his ideas, due to her ability with words. Her first book, *A First Lesson in Natural History* (1859), was prepared under Louis's direction. Its revision, *Seaside Studies*, was published in 1865 in collaboration with Alexander Agassiz. It is a well-written textbook and field guide on marine zoology. In addition to drawings of specimens (made by Alexander), with descriptions and accounts of the animals' geographical distribution, the book includes information on the best mode of catching jellyfish, a consideration of the embryology of echinoderms, a discussion of the distribution of life in the ocean, and a general description of the radiates. Elizabeth's preface states that she hoped to remedy a deficiency by supplying a popular book describing marine animals.

In addition to the two books, Elizabeth produced a short account for the *Atlantic Monthly* from her diary of the journey to Brazil (1866) and a long account, *A Journey in Brazil* (1868), written in collaboration with Louis. Her record-keeping abilities were important again when she accompanied Louis on the Hassler Expedition. Louis Agassiz had theorized that the entire South American landmass had once been covered by a vast ice sheet. His idea of a continuous former glacial chain extending from south to north was supported by the evidence of past glaciation encountered around the Strait of Magellan. The only published reports of these findings were Elizabeth Agassiz's three articles in the *Atlantic Monthly* in 1872 and 1873.

The two-volume biography of Louis that Elizabeth compiled after his death is an important source of his life (1885). Little of the personality of the author and few details of her life appear in the narrative.

Elizabeth Agassiz was thrust into the world of science. Like Caroline Herschel (1750–1848), who became an astronomer because she worshiped her brother,

William, and Elizabeth Campbell (1868–1961), who learned astronomy to please her husband, Wallace, Elizabeth Agassiz's interest in natural history was spawned by her love for Louis and her desire for his approbation. These women, who were without formal training in the sciences but were schooled by husbands or other relatives, made contributions to science because of their association with a beloved one. They either collaborated with their mentors or, as in the case of Caroline Herschel, became so interested in the subject itself that they worked on their own. Elizabeth Agassiz's work in natural history did not outlive Louis. Nevertheless, her role as scribe and popularizer made her important in the history of science. She is a prime example of a woman who found a genuine interest in a subject through collaboration with a spouse.

BIBLIOGRAPHY

Works by Elizabeth Cabot Cary Agassiz

Scientific Works

[Actea, pseud.]. *A First Lesson in Natural History*. Boston: Little, Brown, 1859.
(with Alexander Agassiz) *Seaside Studies in Natural History. Marine Animals of Massachusetts Bay. Radiates*. Boston: Ticknor and Fields, 1865.
"An Amazonian picnic." *Atlantic Monthly* 17 (Mar. 1866): 313–323.
(with Louis Agassiz) *A Journey in Brazil*. Boston: Ticknor and Fields, 1868.
"The Hassler Glacier in the Straits of Magellan." *Atlantic Monthly* 30 (Oct. 1872): 472–478.
"In the Straits of Magellan." *Atlantic Monthly* 31 (Jan. 1873): 89–95.
"A cruise through the Galapagos." *Atlantic Monthly* 31 (May 1873): 579–584.
Louis Agassiz: His Life and Correspondence. 2 vols. London: Macmillan, 1885.

Works about Elizabeth Cabot Cary Agassiz

Agassiz, Alexander. *Letters and Recollections of Alexander Agassiz with a Sketch of His Life and Work*. Edited by G. R. Agassiz. Boston: Houghton Mifflin, 1913.
    Contains some biographical material on Elizabeth Agassiz.
Lurie, Edward. *Louis Agassiz, a Life in Science*. Chicago: University of Chicago Press, 1960.
    This biography of Louis Agassiz contains information about Elizabeth.
Paton, Lucy Allen. *Elizabeth Cary Agassiz. A Biography*. Boston: Houghton Mifflin, 1919.
    This biography is a major source for Elizabeth Agassiz, but its inadequacies indicate a definite need for a new biography using available primary materials, many of which are at the Schlesinger Library of Radcliffe College.
Tharp, Louise Hall. *Adventurous Alliance. The Story of the Agassiz Family of Boston*. Boston: Little, Brown, 1959.
    Information on the principal families involved and the relationships between people. Elizabeth Agassiz plays an important role in this book.

HATTIE ELIZABETH ALEXANDER (1901–1968)

Soraya Ghayourmanesh-Svoronos

BIOGRAPHY

Hattie Elizabeth Alexander, the second of eight children, was born to William Bain, a merchant, and Elsie May (Townsend) Alexander on April 5, 1901, in Baltimore, Maryland, where she grew up and received her formal education. Her first paternal American ancestor was Joseph Alexander, a native of Scotland, who came to the United States from Ireland in 1714 and settled in Cecil County, Maryland. Her immediate grandparents were John McKnitt and Mary Elizabeth Henderson.

Following graduation from the Western High School for Girls, Alexander attended Goucher College, where she earned her A.B. degree in 1923. There she excelled as an enthusiastic athlete and was a rather unambitious, average student. Nevertheless, she was always regarded as a young lady of great potential, and the college yearbook recognized that "ambition fires her; hygiene claims her; kindness portrays her." In fact, her interest in hygiene led her to choose courses in bacteriology and physiology.

After her graduation, she worked as a bacteriologist at the U.S. Public Health Service in Washington, D.C. (1923–24), and the branch laboratory of the Maryland Public Health Service in Washington (1924–26) in order to save money to help her pay for medical school. She attended Johns Hopkins University Medical School, where she earned her M.D. degree in 1930. She interned at the Harriet Lane Home of Johns Hopkins Hospital, Baltimore (1930–31), where she began searching for the cure of a common and always fatal disease of babies, *Hemophilus influenzae* meningitis. She then received an appointment at Babies Hospital of the Columbia Presbyterian Medical Center in New York City. For the rest of her life she continued her association with Babies Hospital and the Vanderbilt Clinic of Presbyterian Hospital at the Medical Center. She served as adjunct assistant physician (1933–38), assistant attending pediatrician (1938–42), associate attending pediatrician (1942–51), attending pediatrician (1951–66), and consultant for life (from 1966).

Alexander's first academic appointment was the Holt Fellowship (1932–34). At the College of Physicians and Surgeons of Columbia University she was promoted to an instructor of diseases of children (1935) and through the various ranks to professor in 1958 and professor emeritus (1966). She also served the college as acting executive officer of the Department of Pediatrics from January to June 1956 and as a member of the medical board of the Presbyterian Hospital from 1959 to 1966. Alexander served as a consultant to the North County Community Hospital, Glen Cove, New York (1953–55), as well as its successor, Community Hospital (1955–66). In 1954 she became honorary consultant to the North Shore Hospital, Manhasset, New York, and the New York Infirmary.

Alexander's efforts and ideas earned her numerous grant awards from several sources, such as the National Science Foundation (1955), the March of Dimes (1956), and the National Foundation for Infantile Paralysis (1958).

On March 4, 1944, Alexander won the prize given annually by the Mead Johnson Company for "outstanding scientific contributions to pediatrics in the United States." On April 3, 1954, Alexander, along with 18 other Goucher graduates, was honored by the college administration for accomplishments in science. For her work, the New York Infirmary conferred upon Alexander the Elizabeth Blackwell Award (1956). On June 23, 1961, she was the first woman ever to be presented the Oscar B. Hunter annual award of the American Therapeutics Society. In 1964, she was the first woman to be elected the president of the American Pediatric Society. During her year as president she helped found a new journal, *Pediatric Research*. She was presented medals by Columbia University for her participation in war research work (1945), at the 75th anniversary of Babies Hospital (1963), and at the ceremonies commemorating the 200th anniversary of the medical school. Alexander was also the recipient of the Heart Award of the Variety Club of Philadelphia (1966) and of an honorary doctoral degree in science from Wheaton College (1967).

Alexander's religious affiliation was with the Congregational Church, and she was an independent Democrat. She never married or had children. She shared a house in Port Washington, New York, with Dr. Elizabeth Ufford. On June 24, 1968, Alexander died at the Harkness Pavilion of Columbia Presbyterian, the same hospital where she spent her entire professional career. For almost two years she had lived with the awareness that she had carcinoma. Nevertheless, she was actively pursuing her studies on bacterial genetics until a few weeks before her death. A funeral service was held at the Congregational Church in Manhasset, Long Island. At the time of her death Alexander was survived by four sisters, Elsie M. Norfolk, Mary L. Wilkins, Dorothy B. Gibson, and Mildred M. Ridings, and a brother, William B. Alexander. She was fond of music, travel, her speedboat, and growing exotic flowers.

WORK

Although Alexander started her college career interested more in sports than in science, she quickly developed a great interest in science and medicine, primarily due to her obsession with hygiene. She worked for three years to save enough money to be able to enroll at Johns Hopkins Medical School.

Her internship at the Harriet Lane Home influenced her future lifelong interest in influenzal meningitis, a disease caused by *Hemophilus influenzae*, which is unrelated to the influenza virus. This was the turning point in her career that led to her rise in the ranks of elite pediatric microbiologists.

At Babies Hospital, in the pediatric service of Rustin McIntosh, Alexander was given full charge of the microbiology laboratory. The laboratory's functions were to serve the community and develop research projects. She quickly established rigid standards, and her laboratory soon earned the highest acclaim as a model of excellence. Although Alexander was a rather reluctant lecturer, she undertook a heavy load of clinical teaching. She always insisted that her students develop critical thinking skills to support their preliminary diagnoses with objective evidence.

Initial research studies by Alexander involved the diagnosis and treatment of bacterial meningitis, especially that caused by *Hemophilus influenzae*. At the time, treating this disease with an anti-influenzal serum, prepared in horses, was known to be a failure. Scientists at Rockefeller Institute (now Rockefeller University) had developed a serum in rabbits that was very successful against pneumonia. Alexander decided to use such a serum to solve her problem. In collaboration with the immunochemist Michael Heidelberger, she immunized rabbits with large doses of influenza bacilli, which she had extracted from the spinal fluid of stricken chickens, and then tried to determine the potency of the developed serum. In a publication ("Type 'b' . . . " 1939), she reported that this serum, for the first time in history, led to the complete cure of infants who were critically ill with influenzal meningitis. The procedure quickly became standard, and within two years fatalities from the disease dropped by 80%, which led to Alexander's name becoming world renowned. With Heidelberger she developed a method that could measure the strength of the serum, thus ensuring the prescription of correct amounts. Of equal importance was the fact that ways were developed to determine the severity of the illness. Thus, the determination of the treatment's effectiveness was feasible.

In the early 1940s Alexander's work involved the use of sulfa drugs and other antibiotics to combat influenzal meningitis. She was the first to realize that the resistance to antibiotics, often developed by cultures of influenzal bacilli, was the direct result of genetic mutation. In a series of experiments she demonstrated the transformation of *Hemophilus influenzae* and showed that its activity is due to deoxyribonucleic acid (DNA). As a result, she was involved with the field of microbial genetics, which had just emerged as one of the most promising fields in the biological sciences. Her treatment methods led to the reduction of children's fatalities to less than 10%.

In 1944 Rockefeller Institute scientists reported that the genetic constituent DNA could cause changes in the hereditary characteristics of pneumococci. This suggestion was received with great skepticism. Alexander, in collaboration with Grace Leidy, attempted to confirm their research by producing hereditary changes in *Hemophilus influenzae* with DNA obtained from this organism. Her newly developed techniques led to successful results (Alexander and Leidy 1950). Later she applied her procedures to other procaryotic organisms and also to various viruses.

The successful incorporation of Alexander's results into theoretical biology, with direct applications to clinical medicine, created a great impact. While she relentlessly continued her research work, she also remained active on the wards and published prolifically. Her precise, well-organized writing earned her the 1954 Stevens Triennial Prize for the best essay on a medical subject. Apart from her research activities at Columbia, she delivered the Benjamin Knox Rachford Lectures at the Children's Hospital of the University of Cincinnati under the title "Treatment of *Haemophilus influenzae* Infections" (1942). She delivered the Alpha Omega Lectures at Boston University under the title "The Genetic Control of Heritable Traits in Microorganisms" (1957). Alexander gave the Felton Lectures Series at the Fairfield Hospital for Infectious Diseases in Melbourne, Australia (1960). During the same year she gave several Brennemann Memorial Lectures in Los Angeles on the insight into inheritance, the nature of the *Staphylococcus* problem, the guiding principles in antibiotic therapy, and the perception of virus infections as an invasion by nucleic acid. She was a visiting professor at the University of Pittsburgh (1963). In 1964 Alexander participated in the distinguished symposium on "The Child" at the dedication of the New Children's Medical and Surgical Center and the celebration of the 75th anniversary of the Johns Hopkins Hospital.

Alexander was a member of the American Pediatric Society since 1951, serving as chairman of its council (1956–57) and vice president of the society (1959–60). She served as the secretary of the New York Academy of Medicine (1953–54) and chairman of its pediatric section (1954–55). She was also a diplomate of the American Board of Pediatrics, a fellow of the American Association for the Advancement of Science, and a member of the Society for Pediatric Research, the Harvey Society, the Society for Experimental Biology and Medicine, and the American Academy of Pediatrics. Serving the public, Alexander became consultant to the U. S. Secretary of War on the influenza commission (1941–45) and later became a member of the pediatric advisory committee of the New York City Department of Health (1958–60).

NOTE

The author wishes to express deep appreciation to Dr. LeLeng To, chair of the Department of Biological Sciences at Goucher College, for valuable information used in preparing this article.

BIBLIOGRAPHY

Works by Hattie Elizabeth Alexander

Scientific Works

"Prognostic value of precipitin test in meningococcus meningitis." *Journal of Clinical Investigation* 16 (1937): 207–211.

(with G. Rake) "Studies on meningococcus infection: A further note on presence of meningococcus precipitinogens in cerebrospinal fluid." *Journal of Experimental Medicine* 65 (1937): 317–321.

"Influenza bacillus infections." In *Holt's Diseases of Infancy and Childhood*, 11th ed., edited by L. E. Holt et al., 1114–1117. New York: Holt and McIntosh (Appleton Century), 1939.

"Response to antiserums in meningococcic infections of human beings and mice. A comparative study." *American Journal of Diseases of Children* 58 (1939): 746–752.

"Type 'b' anti-influenzal rabbit serum for therapeutic purposes." *Proceedings of the Society for Experimental Biology and Medicine* 40 (1939): 313–314.

(with M. Heidelberger) "Chemical studies on bacterial agglutination. Agglutinin and precipitin content of antisera to *Haemophilus influenzae*, type b." *Journal of Experimental Medicine* 71 (1940): 1–11.

"Treatment of bacterial meningitis." *Bulletin of the New York Academy of Medicine* 17 (1941): 100–115.

(with H. R. Craig, R. G. Shirley, et al.) "Validity of etiological diagnosis of pneumonia in children by rapid nasopharyngeal mucus." *Journal of Pediatrics* 18 (1941): 31–35.

"Treatment of influenzal meningitis." *Connecticut State Medical Journal* 6 (1942): 167–173.

(with C. Ellis and G. Leidy) "Treatment of type specific *Hemophilus influenzae* infections in infancy and childhood." *Journal of Pediatrics* 20 (1942): 673–698.

"Experimental basis for treatment of *Haemophilus influenzae* infections." *American Journal of Diseases of Children* 66 (1943): 160–171.

"Treatment of *Haemophilus influenzae* infections and of meningococcic and pneumococcic meningitis." *American Journal of Diseases of Children* 66 (1943): 172–187.

(with G. Leidy) "Experimental investigations as basis for treatment of type b *Haemophilus influenzae* meningitis in infants and children." *Journal of Pediatrics* 23 (1943): 640–655.

"Treatment of type b *Haemophilus influenzae* meningitis." *Journal of Pediatrics* 25 (1944): 517–532.

(with M. Heidelberger and G. Leidy) "Protective or curative element in type b *H. influenzae* rabbit serum." *Yale Journal of Biology and Medicine* 16 (1944): 425–434.

"Streptomycin pediatrics." *Journal of Pediatrics* 29 (1946): 192–198.

(with G. Leidy) "Influence of streptomycin on type b *Haemophilus influenzae*." *Science* 104 (1946): 101–102.

(———— and C. MacPherson) "Production of types a, b, c, d, e and f *H. influenzae* antibody for diagnostic and therapeutic purposes.'' *Journal of Immunology* 54 (1946): 207–211.

(with G. Leidy, G. Rake, et al.) ''*Hemophilus influenzae* meningitis treated with streptomycin.'' *Journal of the American Medical Association* 136 (1946): 434–440.

(with C.F.C. MacPherson, M. Heidelberger, et al.) ''Specific polysaccharides of types a, b, c, d, e and f *Hemophilus influenzae*.'' *Journal of Immunology* 52 (1946): 207–219.

''Treatment of *H. influenzae* meningitis.'' *Journal of Michigan State Medical Society* 46 (1947): 193–198.

''The treatment of purulent meningitides.'' In *Advances in Pediatrics*, vol. 2, edited by S. Z. Levine et al., 121–150. New York: Interscience, 1947.

(with G. Leidy) ''Mode of action of streptomycin on type b *Hemophilus influenzae*. I. Origin of resistant organisms.'' *Journal of Experimental Medicine* 85 (1947): 329–338.

(———) ''Mode of action of streptomycin on type b *Hemophilus influenzae*. II. Nature of resistant variants.'' *Journal of Experimental Medicine* 85 (1947): 607–621.

(———) ''The present status of treatment for influenzal meningitis.'' *American Journal of Medicine* 2 (1947): 457–466.

''The Hemophilus group.'' In *Textbook of Bacterial and Mycotic Infections of Man*, edited by R. J. Dubos, 472–492. Philadelphia: Lippincott and Sons, 1948.

''Origin of resistance of bacteria and therapeutic implications.'' *Pediatrics* 1 (1948): 273–277.

''*H. influenzae* infections.'' In *Nelson Loose Leaf Medicine Textbook*, 691–714. New York: Thomas Nelson, 1949.

''Meningitis, non-tuberculous.'' In *Streptomycin: Nature and Practical Applications*, edited by S. A. Waksman, 360–380. Baltimore: William and Wilkins, 1949.

''The problem of microbial resistance to chemotherapeutic agents.'' In *Evaluation of Chemotherapeutic Agents. Symposium II of the Section on Microbiology*, edited by C. M. MacLeod, 65–80. New York: Columbia University Press, 1949.

(with G. Leidy) ''Influenzae infections in children.'' *Connecticut State Medical Journal* 13 (1949): 713–721.

(———) ''Mechanism of emergence of resistance to streptomycin in 5 species of gram negative bacilli.'' *Pediatrics* 4 (1949): 214–221.

(———) ''Mode of action of streptomycin on *H. influenzae* type b. III. Nature of streptomycin action on sensitive *H. influenzae*.'' *Pediatrics* 3 (1949): 277–285.

(——— and W. Redman) ''Comparison of the action of streptomycin, polymyxin B, aureomycin and chloromycetin on *H. pertussis*, *H. parapertussis*, *H. influenzae* and 5 enteric strains of gram negative bacilli.'' *Journal of Clinical Investigation* 28 (1949): 867–870.

(with C.F.C. MacPherson and G. Leidy) ''Quantitative determination in type-specific antisera to *Hemophilus influenzae*, of the antibody that cross-reacts with encapsulated pneumococci.'' *Journal of Bacteriology* 57 (1949): 443–446.

(with W. Redman) ''Mechanism of emergence of resistance to streptomycin of *H. pertussis* and *H. parapertussis* during treatment with this antibiotic.'' *Pediatrics* 4 (1949): 461–467.

(with G. Leidy) ''Transformation type specificity of *H. influenzae*.'' *Proceedings of the Society for Experimental Biology and Medicine* 73 (1950): 485–487.

(———, W. Redman, et al.) ''Experimental basis for prediction of therapeutic efficacy of streptomycin in infections caused by gram negative bacilli." *Pediatrics* 5 (1950): 78–89.

(with C. MacPherson and W. Redman) "A quantitative method for measuring *H. pertussis* antibody." *Pediatrics* 5 (1950): 443– 447.

(with G. Leidy) "Determination of inherited traits of *H. influenzae* by desoxyribonucleic acid fractions isolated from type specific cells." *Journal of Experimental Medicine* (1951): 345–359.

(————) "Induction of heritable new type in type specific strains of *H. influenzae*." *Proceedings of the Society for Experimental Biology and Medicine* 78 (1951): 625–626.

(with C. MacPherson, P. H. Maurer, et al.) "A method for the quantitative measurement of agglutinin mitogen in antisera to *Hemophilus pertussis*, phase I." *Canadian Journal of Medicinal Science* 30 (1952): 284–293.

(with S. Zamenhof, G. Leidy, et al.) "Purification of the desoxypentose nucleic acid of *Hemophilus influenzae* having transforming activity." *Archives of Biochemistry and Biophysics* 40 (1952): 50–55.

"Guides to optimal therapy in bacterial meningitis." *Journal of the American Medical Association* 152 (1953): 662–666.

(with R. G. Ames, S. M. Cohen, et al.) "Comparison of the therapeutic efficacy of four agents in pertussis." *Pediatrics* 11 (1953): 323–337.

(with E. Hahn) "*In vitro* production of new types of *Hemophilus influenzae*." *Journal of Experimental Medicine* 97 (1953): 467–482.

(with G. Leidy) "Induction of streptomycin resistance in sensitive *Hemophilus influenzae* by extracts containing desoxyribonucleic acid from resistant *Hemophilus influenzae*." *Journal of Experimental Medicine* 97 (1953): 17–31.

(with W. Redman) "Transformation of type specificity of meningococci." *Journal of Experimental Medicine* 97 (1953): 797–806.

(with S. Zamenhof and G. Leidy) "Biological activity of the nucleic acids." *Canadian Journal of Medicinal Science* 31 (1953): 252–262.

(————) "Studies on the chemistry of the transforming activity. I. Resistance to physical and chemical agents." *Journal of Experimental Medicine* 98 (1953): 373–397.

(————, et al.) "Polyribophosphate, the type-specific substance of *Hemophilus influenzae*, type b." *Journal of Biological Chemistry* 203 (1953): 695–704.

(with G. Leidy and E. Hahn) "Studies on the nature of *Hemophilus influenzae* cells susceptible to heritable changes by desoxyribonucleic acids." *Journal of Experimental Medicine* 99 (1954): 505–533.

"*H. ducreyi* infections." In *Textbook of Medicine*, 9th ed., edited by R. L. Cecil et al., 204. Philadelphia: Saunders, 1955.

"*H. influenzae* infections." In *Textbook of Medicine*, 9th ed., edited by R. L. Cecil et al., 202–204. Philadelphia: Saunders, 1955.

"Poliomyelitis virus variation." *Annals of the New York Academy of Sciences* 61 (1955): 940–942.

(with K. Sprunt, I. M. Mountain, et al.) "Production of poliomyelitis virus with combined antigenic characteristics of type 1 and type 2." *Virology* 1 (1955): 236–249.

"Treatment of pyogenic meningitis." In *Neurology and Psychiatry in Childhood*, edited by R. McIntosh and C. Hare, chapter 1. Baltimore: Williams and Wilkins, 1956.

(with D. S. Damrosch and C. Ellis) "Correlation of *in vitro* susceptibility of tubercle bacilli in patients with miliary and meningeal tuberculosis with their response to therapy.'' *American Review of Tuberculosis and Pulmonary Diseases* 74 (1956): 232–240.

(with G. Leidy and E. Hahn) ''On the specificity of the desoxyribonucleic acid which induces streptomycin resistance in *Hemophilus.*'' *Journal of Experimental Medicine* 104 (1956): 305–320.

(with S. Zamenhof, G. Leidy, et al.) ''Inactivation and instabilization of the transforming principle by mutagenic agents.'' *Journal of Bacteriology* 72 (1956): 1–11.

''Prophylactic effects of isoniazid on primary tuberculosis in children. A preliminary report.'' *American Review of Tuberculosis and Pulmonary Diseases* 76 (1957): 942–963.

''Treatment of bacterial infections of the central nervous system.'' In *The Medical Clinics of North America*, edited by H. H. Merritt, 575–586. Philadelphia: Saunders, 1958.

(with G. Koch, I. M. Mountain, et al.) ''Infectivity of ribonucleic acid of poliovirus on HeLa cell monolayers.'' *Virology* 5 (1958): 172–173.

(with G. Koch, I. M. Mountain, and O. Van Damme) ''Infectivity of ribonucleic acid from poliovirus in human cell monolayers.'' *Journal of Experimental Medicine* 108 (1958): 493–503.

(with G. Leidy and E. Hahn) ''Interspecific transformation in *Hemophilus*: A possible index of relationship between *H. influenzae* and *H. aegyptius.*'' *Proceedings of the Society for Experimental Biology and Medicine* 102 (1959): 86–88.

(with G. Leidy, K. Sprunt, et al.) ''Sensitivity of populations of clonal lines of HeLa cells to polioviruses.'' *Proceedings of the Society for Experimental Biology and Medicine* 102 (1959): 81–85.

(with I. M. Mountain) ''Infectivity of ribonucleic acid (RNA) from type 1 poliovirus in embryonated egg.'' *Proceedings of the Society for Experimental Biology and Medicine* 101 (1959): 527–532.

(with K. Sprunt and W. M. Redman) ''Combination of antigenic traits of type 1 and type 2 poliovirus.'' *Journal of Immunology* 82 (1959): 232–240.

(———) ''Infectious ribonucleic acid derived from enteroviruses.'' *Proceedings of the Society for Experimental Biology and Medicine* 101 (1959): 604–608.

''Infectivity of ribonucleic acid of poliovirus on HeLa cell monolayers.'' In *Viral Infections of Infancy and Childhood*, edited by H. M. Rose, 1–9. New York: Hoeber, 1960.

''An insight into inheritance.'' *Medical Journal of Australia* 2 (1960): 761–765.

(with G. Koch and S. Koenig) ''Quantitative studies on the infectivity of ribonucleic acid from partially purified and highly purified poliovirus preparations.'' *Virology* 10 (1960): 329–343.

(with G. Leidy, E. Hahn, et al.) ''Biochemical aspects of virulence of *Hemophilus influenzae.*'' *Annals of the New York Academy of Sciences* 88 (1960): 1195–1201.

(with K. Sprunt, W. M. Redman, et al.) ''Factors influencing degree of infectivity of enterovirus ribonucleic acid.'' *Proceedings of the Society for Experimental Biology and Medicine* 103 (1960): 306–309.

(with K. Sprunt) ''Enterovirus ribonucleic acid in tissue culture.'' *National Cancer Institute Monograph*, no. 7, 1961.

(——— and S. Koenig) ''Factors influencing degree of infectivity of enterovirus ribonucleic acid (RNA). II. Role of host cell RNA inactivators in the low degree of infectivity of poliovirus RNA in low salt concentration." *Virology* 13 (1961): 135–138.

(———) "Factors influencing the infectivity of poliovirus ribonucleic acid." *Proceedings of the Society for Experimental Biology and Medicine* 108 (1961): 755–760.

"Bacterial meningitis." "Influenza bacillus infections." "Meningococcal infections." "Tularemia." In *Pediatrics*, 13th ed., by L. E. Holt, Jr., et al., 1021–1029, 1108–1111, 1129–1131, 1250–1252. New York: Appleton-Century-Crofts, 1962.

"Disease caused by *Hemophilus*." In *Cecil-Loeb Textbook of Medicine*, edited by P. B. Beeson and W. McDermott, 212–218. Philadelphia: Saunders, 1963.

(with G. Leidy and I. Jaffee) "Further evidence of a high degree of genetic homology between *H. influenzae* and *H. aegyptius*." *Proceedings of the Society for Experimental Biology and Medicine* 118 (1965): 671–679.

(———) "Genetic modifiers of the phenotypic level of deoxyribonucleic acid-conferred novobiocin resistance in *Haemophilus*." *Journal of Bacteriology* 92 (1966): 1464–1468.

Works about Hattie Elizabeth Alexander

"Columbia gets polio grant." *New York Times* (Apr. 2, 1958): 16.

"Dr. Hattie Alexander, 67, dies: Columbia research pediatrician." *New York Times* (June 25, 1968): 41.

Hogue, J. "The contribution of Goucher women to the biological sciences." *Goucher Alumnae Quarterly* (Summer 1951): 21–22.

"Influenzal meningitis." *New York Times* (Mar. 5, 1944): 9.

McIntosh, Rustin. "Hattie Alexander." *Pediatrics* 42 (1968): 544.

"Polio grant given Columbia." *New York Times* (Jan. 11, 1956): 63.

"Polio research grant." *New York Times* (Jan. 2, 1955): 65.

"Serum that kills cancer cells in test tube developed in horse." *New York Times* (May 11, 1955): 25.

"Therapy award for pediatrician." *New York Times* (June 23, 1961): 8.

Turner, L. "From C student to winning scientist." *Goucher Alumnae Quarterly* (Winter 1952): 18–20.

"Woman wins science prize." *New York Times* (Mar. 5, 1944): 26.

"Women in science cited." *New York Times* (Apr. 4, 1954): 82.

AGNES ROBERTSON ARBER
(1879–1960)

Maura C. Flannery

BIOGRAPHY

Though she never held an academic or professional position, and though she worked alone in a laboratory she had fitted up in a small bedroom in her home, Agnes Robertson Arber nonetheless had a distinguished career in botany. As Harry Godwin (1970, 206) has noted, "She was in the vanguard of the movement of women into scientific research." She published seven books and over 80 articles on botany, its history, and its philosophical underpinnings. Her accomplishments were so distinguished that she was the first woman botanist to be made a fellow of the Royal Society.

Agnes Robertson was born in London on February 23, 1879, into a family that had a long tradition of interest in both art and science, the two fields in which she was involved throughout her life. Her father, Henry Robert Robertson, was a Scottish artist; his father directed a private school and had a considerable interest in art and botany, interests that he shared with his granddaughter. Her mother, Agnes Lucy Turner, was a descendant of Robert Chamberlain, founder of the china manufacturing company, Chamberlain and Sons of Worcester, England.

Agnes Robertson was the oldest of four children. Her brother, Donald Straun Robertson, was Regius Professor of Greek at the University of Cambridge, and her sister, Janet Robertson, was a portrait painter. In the introductions to a number of her books, Robertson thanks both these siblings for their assistance in her work and, in particular, expresses appreciation to her brother for discussions that broadened her understanding of the world of literature and were important to the development of her more philosophical works.

Robertson's interests in botany and art were encouraged early in her life by her mother and father. Her mother shared with her daughter a deep curiosity about plants, and her father gave her regular drawing lessons from the time she was three years old until she went off to school. When she was eight years old, she was enrolled in the North London Collegiate School for Girls, which provided excellent instruction in the sciences. Here Robertson's enthusiasm for botany developed. Her botany teacher was Edith Aitken, who had attended Girton College of Cambridge University. Aitken introduced Robertson to the work of Johann Goethe, and this began a lifetime interest in his botanical and philosophical work.

During her years at this school Robertson studied *Lyte's Herbal* of 1578. This illustrated work sparked her interest in herbals and ultimately led to her writing a book on the subject. Also at the school she met the botanist Ethel Sargant, who spoke to the Science Club. Sargant, who was to be the major influence on Robertson's scientific career, appreciated Robertson's talents and invited the student to work in her private laboratory in her home. Like many women of that day who were interested in science, Sargant found it easier to pursue her studies independently rather than attempt to obtain any of the few professional positions open to women.

Robertson worked in Sargant's laboratory during one of her university vacations and for a year in 1902–3, aiding the older woman with her work on the anatomy of seedlings. She learned many techniques involved in the study of plant morphology or form, including making and interpreting serial sections. In 1897 Robertson entered University College of London and received her B.Sc. degree in 1899. During her years there she received several medals and prizes and gained first-class honors in all the examinations she took. Among her teachers was the distinguished botanist and ecologist Arthur Tansley.

After graduation from University College, she enrolled in Newnham College of Cambridge University, having received an entrance scholarship. She studied chemistry, physics, botany, and geology for Part I of the Natural Sciences Tripos at Cambridge and botany and geology for Part II, again gaining first-class honors. Her instructors included botanist A. C. Seward, biologist Francis Darwin, and geneticist William Bateson. It was the latter's lectures on the rediscovery of Gregor Mendel's work that Robertson found most stimulating.

When she left Newnham College in 1902, she worked for a year in Ethel Sargant's private laboratory. From 1903 to 1908, Robertson held the Quain Studentship in Biology at University College, where she studied gymnosperms, including fossil forms, and where she received her Doctor of Science degree in 1905. She remained at the college until 1909, holding a lectureship in botany in her last year.

In 1909 Robertson left London and moved back to Cambridge as the wife of Edward Alexander Newell Arber (1870–1918). They were married on August 5, 1909, and had a happy life together, a life that was infused with a mutual interest in botany. Newell Arber was university demonstrator in paleobotany at Cambridge. Agnes Robertson had known him since her student days at Cambridge, when he was one of her botany instructors. He was a member of Trinity College, Cambridge, where he was responsible for the Sedgwick Museum's collection of fossil plants, to which he added over 5,000 specimens. He also worked in the Natural History Department of the British Museum, arranging fossils and preparing a catalog of botanical fossil specimens. He lectured on both elementary and advanced paleobotany, was also interested in present-day species, and wrote a book on the Alpine plants of Switzerland.

Newell Arber assisted his wife in her botanical work with helpful criticism on ways to improve her papers and with suggestions on the techniques of research and publication. His ideas even helped spark her research interests. In her book on *Water Plants* (1920, 5), Agnes Arber notes that her husband suggested that she study these plants, since they were so varied and abundant in the waterways of Cambridge. This book was dedicated to him and was published two years after his untimely death in 1918. The Arbers had one child, Muriel, who was always close to her mother and whom her mother thanks in the foreward to *The Manifold and the One* (1957) for the many discussions on philosophy that they had enjoyed over the years.

After her marriage and move to Cambridge, Agnes Arber did botanical research at Newnham College for a number of years. She held a research fellowship there in 1912–13 and again in 1918–20. She continued to work at the Balfour Laboratory, which then belonged to Newnham College, until 1927, when she converted one of the bedrooms in her home into a laboratory. In a talk she gave to students at Girton College of Cambridge University about this time, Arber described the advantages of private research. Working in a large laboratory facility means that there are usually assistants around who can help solve the technical problems that so often arise in using or constructing equipment or in other aspects of research. While Arber recognized this advantage (i.e., less time is wasted in such technicalities), she saw a further advantage in the mental effort spent on dealing with these difficulties when working alone. She thought it meant putting more brain work into her technique and thus into the project as a whole. She also thought that working alone was less likely to lead to doing mechanistic work. Since the lone researcher has to handle all tasks, there is no fear that a worker may specialize in a particular task and lose the view of the research as a whole.

In deciding to work in her own laboratory, Arber was obviously following the lead of her mentor and teacher, Ethel Sargant, who, after studying at the Jodrell Laboratory of the Royal Botanic Gardens, Kew, set up her own laboratory in her home. Both Sargant and Arber used their own money to build their laboratories, but Arber also received grants-in-aid from the Royal Society, and she held a Leverhulme Research Fellowship from 1936 to 1938. She found working by herself much more fruitful than working in a busy laboratory with a great deal of bustling activity to distract her. She held to Sargant's view that independence is the essence of research and found that solitude allowed her the concentration of mind necessary for independent and original thought. Perhaps this solitude led Arber in later years to more philosophical considerations of botany and of biology in general.

Throughout her career Agnes Arber was a prolific writer. Her first paper, on the anatomy of the gymnosperm *Macrozamia heteromera*, was published in

1902 and was an outgrowth of her work with her instructor, A. C. Seward. Her last book came out in 1957, three years before her death. Entitled *The Manifold and the One*, it was a detailed study of the philosophical aspects of oneness in diversity. These first and last publications seem very different from each other. Besides revealing the breadth of her studies, these two publications also indicate two areas that were of interest to her throughout her life. She was always fascinated by plant form, but she also saw beyond specific morphological problems to underlying philosophical questions.

Agnes Arber died on March 22, 1960, at a nursing home in Cambridge. She was 81 years old and was survived by her daughter, Muriel, a geologist and teacher (Arber 1988). Arber was buried in the churchyard at Girton College in the same grave as her husband. Arber was thought of as a recluse by many, because she loved to work in solitude, but those who knew her well attested to her warmth and friendliness and her enjoyment of conversation. As William T. Stearn noted in his obituary, "She was a genial and gracious person whose kindness and friendship will be gratefully remembered by the students of botany she helped and encouraged" (1960, 263). She also had the respect of her scientific peers. Stearn quotes A. G. Tansley, her teacher and fellow botanist, who noted in 1952 that Arber was "the most distinguished as well as the most erudite British plant morphologist" (Stearn 1960, 263). In 1946 Arber became the third woman elected to the Royal Society and the first woman botanist so honored. In 1948 the Linnaean Society presented her with the Linnaean Gold Medal, another prestigious award.

WORK

Arber's first book was on a botanical topic, but it involved the history of botany rather than the latest research. In 1912 Arber published *Herbals: Their Origin and Evolution*, in which she looked at the development of the printed herbal from 1470 to 1670. Arber's background in art is manifest here. The book has a large number of illustrations, and Arber comments on the artistic merit of the herbals as well as on their scientific significance. She also does not miss an opportunity to examine the philosophical aspects of her topic. She notes that, from the beginning, the study of plants has been approached from two different standpoints, the philosophical and the utilitarian. These differences are mirrored in the herbals, some of which were written to enlarge on philosophical aspects of botany and some for the very practical purpose of allowing for the identification of plants to be used in medicine and cooking.

Long before the art historian William Ivins (1953, 42) made the point, Arber noted that the illustrations in the herbals did not improve progressively over the years, as might have been expected. Instead, the opposite frequently occurred. The illustrations in later books were often inferior to those of earlier ones. The main reason for this was that successive copying led to poorer and poorer drawings, to the point that it was often difficult to identify the plant from the illustration.

Arber greatly revised her book on herbals for a second edition in 1938. This book is probably her best-known work. It is cited in many subsequent histories of botanical illustration, including the work of R. Anderson and W. Blunt. Anderson calls it an admirable book (1977, 32), and Blunt cites it several times to bolster his points (1950, 41, 48, 51). Arber never lost her interest in scientific illustration and its importance to the advancement of biology and of botany in particular.

While Arber remained interested in the historical and philosophical aspects of botany, she also made many original contributions to botanical knowledge. Her work dealt primarily with the anatomy and morphology of monocotyledonous plants. This work was done over a period of about 50 years and can be divided into four parts. During each of these phases, the work was published in journal articles and then later used as the basis for a book, in which the general principles involved were developed. The first such book is *Water Plants: A Study of Aquatic Angiosperms* (1920), in which Arber discusses a group of plants that are united by lifestyle rather than by evolutionary relationship. She describes the life histories of a number of these plants and then discusses physiological and ecological considerations. There are also a number of chapters on the evolution of these plants. This book, as is the case with Arber's other botanical studies, has a large number of illustrations, most of which were done by her. Most of the illustrations are composed of several different views of the same plant or several examples from different species of the structure of interest. Each illustration is hand-lettered. A great deal of time must have gone into these illustrations.

In 1925, in the second phase of her career, Arber published *Monocotyledons: A Morphological Study*, as a volume in the series of *Cambridge Botanical Handbooks*. In 1910 Ethel Sargant had accepted an invitation from the editors of this series to write such a volume, but she was plagued by ill health. Shortly before her death in 1918, Sargant suggested that Arber should take up the work on the book. This is a general survey of the monocotyledons, with many illustrations, including 140 by Arber. She begins the book with a chapter on the principles of morphology and a discussion of the philosophy of plant form. These two subjects appear again and again in her books, though Arber never repeats herself. She uses the particular subject under study to illustrate her points, and through the years, her analysis deepens. In her early books her attention to general questions is brief and peripheral, but in her later work these questions loom larger.

During her work in preparation for this book on monocotyledons, Arber came upon a number of issues of morphology that she thought deserved further study in a more limited group of organisms. This led to the third phase of her research and to her book *The Gramineae: A Study of Cereal, Bamboo, and Grass*, which was published in 1934, again as one of the *Cambridge Botanical Handbooks*.

This book involved ten years of research. During this time, Arber published ten articles on this group of plants in *The Annals of Botany*, which dealt with a number of complex and little-studied aspects of these species. In these papers and in the book, Arber relied heavily on illustrations to make her points. Many of these contain a series of drawings showing the same structure at different stages of growth or at different levels from bottom to top.

But this book is more than a morphological study. Arber admitted in the introduction that she had not used any one organizing principle for her material. Though she had attempted a number of these approaches, she found them all too restrictive and decided instead to cover the topics she found most relevant and fascinating. This approach led to a work that is also most interesting to the reader. Arber begins with an historical overview of the Gramineae, emphasizing their importance in human history. She then goes into a thorough examination of the morphology of these plants and ends with an investigation of their ecological and economic importance. As always in her botanical studies, the emphasis is on morphology and its significance.

After the completion of this work on the Gramineae, Arber published a series of studies on flower structure. In 1937 she wrote a review paper on floral anatomy, which included criticisms of several of the then-current views on how structures, such as the carpel, had developed. Arber found it impossible to maintain her laboratory and continue her morphological studies during World War II. But this just changed the focus of her work rather than suspending it. She turned to an area that had always interested her: the philosophical aspects of the morphology of flowering plants. This investigation marked the beginning of the last of the four phases of her life's work and resulted in the publication of *The Natural Philosophy of Plant Form* in 1950. In this book she investigates the mental processes involved in developing ideas about morphology and discusses how different botanists have viewed the underlying ideas of morphology. H. Hamshaw Thomas ("Agnes Arber, 1879–1960" 1960, 7) considers this "perhaps her most important book." In it she goes beyond the investigation of specific questions of plant morphology and inquires into the mental processes involved in forming a unified conception of a plant and into how these modes of thought have developed historically.

In the years after the publication of this book, Arber continued to explore philosophical questions related to biological inquiry. The fruit of this work was *The Mind and the Eye*, published in 1954. This book has become a classic in the literature on the philosophy of biology. In it Arber dissects the process of inquiry and looks at each stage of the process individually. She begins with the first step, which is defining a problem for investigation. The second stage is the search for relevant data, either by observation or experimentation, and then its organization. Then comes the process of attempting to interpret the data, either by fitting it into the existing framework of scientific explanation or by reconstructing that framework. The fourth stage in the process of inquiry is the attempt to test the validity of the interpretation or solution. If this attempt is successful, the next stage involves convincing others of the solution.

This analysis is not much different from that presented by many other scientists and philosophers of science, but two things set Arber's work apart from that of others. First, Arber adds a sixth step to her description of biological inquiry, which is pursued by some biologists more than others: putting the specific work in the context of science in general. This involves criticizing presuppositions and discovering how the intellectual and sensory elements are interconnected.

What also sets Arber's work apart is that, while many philosophers of science use physics as the paradigmatic science and draw their examples exclusively from this science, Arber uses examples from biological inquiry. The subtitle of *The Mind and the Eye* is *A Study of the Biologist's Standpoint*, and it is just that. In describing how biology differs from physics, Arber notes that biological hypotheses represent ways of looking at nature that, like artistic creations, cannot be assessed by "any simple and easily defined procedure, any mathematical-experimental system" (*The Mind and the Eye* 1954, 29).

In 1957 Arber published her last book, *The Manifold and the One*, which is very different from all her other writings. This book has little to do with science or with scientific inquiry. It is, instead, an investigation of the idea that a unity underlies the diversity found in the world, a one beneath the manifold. She traces this idea through the thought of the ancient Greek and Eastern philosophers. The range of her knowledge and her understanding of the writings by a large number of thinkers are truly impressive.

Through her writings Agnes Arber made a unique contribution to the development of the science of biology. She not only made significant discoveries in plant morphology but also helped broaden our understanding of biology as a field of inquiry through her work in the philosophy of science. Making all her writings noteworthy was her lucid and forthright writing style. She describes even the most complex of structures clearly, and she always puts specific findings in context by showing how they relate to larger issues in botany. For example, in *Water Plants*, after a thorough review of the adaptations of vascular plants to aquatic environments, she notes that one of the "unfortunate results" of the publication of Charles Darwin's *The Origin of Species* (1859) was the teleological undercurrent that appeared in the writings of biologists: "On the theory that every existing organ and structure either has, or has had in the past, a special adaptive purpose and 'survival value,' it readily becomes a recognized habit to draw deductions as a function from structure, without checking such deductions experimentally" (*Water Plants* 1920, 260).

Such critical comments pepper all of Arber's scientific writings. Even in her earliest publications, she writes with the surety evident in the preceding quote. In one sense, her writing style never matures, because it is so mature from the beginning. But her work does evolve in that it becomes more philosophical over time. Each of her books successively devotes more space to the sixth stage of inquiry, putting scientific work in context and criticizing it.

Arber's work is important in the history of plant morphology because she attempted to unite two very different scientific traditions. Many English botanists saw morphological studies as valuable only to the extent that they revealed phylogenetic or evolutionary relationships. These workers paid a great deal of attention to comparative morphology, since they saw this approach as a way to discover relationships. On the other hand, the botanists working in France and Germany followed a more ancient tradition that existed long before the publication of *The Origin of Species*. The botanists working in this tradition saw the study of form and structure as valuable for its own sake. Many of these biologists had paid little attention to comparative morphology. Arber sought to unite these views in *Monocotyledons*, by first differentiating between the pure morphology practiced by the French and Germans and morphology applied to evolutionary questions as practiced by the English. She then went on to make use of both methods in her investigation of this group of plants.

Another noteworthy aspect of her writing is not only the depth of her knowledge of plant morphology, which is very evident in her botanical papers and books, but the breadth of her knowledge as well. In *Monocotyledons* there is an illustration of an Assyrian bas-relief of an eagle-headed figure pollinating a date-palm, one of many such illustrations of historical interest found in a number of her books. In *The Gramineae* Arber reveals a vast knowledge of the history of agriculture. This depth of background makes Arber's work such a joy to read; she is always including items that obviously fascinated her and that she trusts her readers will also find interesting.

Arber's early training in art served her in good stead throughout her career. She believed, with Goethe, that drawing is essential to careful observation. What is amazing about Arber's work is that she managed not only to make so many careful morphological studies but also to create hundreds of illustrations, each carefully hand-lettered, to accompany the text.

Obviously, the importance of illustrations to the science of botany was apparent to her even in her earliest work, since her first book was on the history of herbals. At the other end of her career, the very title of her book on biological inquiry, *The Mind and the Eye*, indicates the importance she placed on observation and on the communication of those observations to others. In this book she discusses the significance not only of illustrations but of mental images as well. She argues that the use of pictorial imagery in thinking is a fundamental need of the human mind, and like the psychologist of art Rudolf Arnheim (1969) she contends that there is a fusion of mental and visual thinking.

She was ahead of her time in her interest in analogical thinking in science. Now it is common for linguists and philosophers to see metaphor and analogy as crucial to thought processes, learning, and discovery, but when Arber argued for this position in the 1950s, it was much less popular. In this and in her conviction of the importance of visual thinking, Arber was espousing a philosophical position very far from that of logical positivism, which still held sway in science in the 1950s. Her thought is closer to that found in many recent works on the psychological facets of scientific inquiry.

Several factors may have played a role in the development of Arber's novel viewpoint. First, Arber's training was in biology rather than in physics, the science in which many of the logical positivists were trained. Since biology is the most visual of the sciences (Ritterbush 1968), it is not surprising that she valued visual thinking so highly. Arber's interest in analogical thinking may be attributed to her involvement with the sixth stage of scientific inquiry, in seeing how individual facts and hypotheses fit into the larger picture, in seeing how ideas are related. Finally, feminist critics of science might argue that this aspect of Arber's work is directly related to the fact that she was a woman, since women tend to think more holistically than men.

The feminist view can also be used to account for the strong aesthetic aspect of Arber's work. This would include not only her heavy reliance on visual imagery in her scientific papers but also her interest in form and in unifying types underlying the diversity of plant form. In this, Arber's work is directly related to the botanical work of Goethe. While he saw all plant forms as based on the leaf, she saw at least some of them as based on the leaf and shoot. In *The Natural Philosophy of Plant Form*, she develops this idea. In this book she also uses the type concept, the idea that plants are related to each other by being variations on particular types or archetypal forms. But Arber does not carry the type concept as far as Goethe did. While he saw it as an end in itself, she saw it as congruous with evolution, with forms related to a particular type by means of evolutionary descent.

Obviously, the aesthetic quality of unity was important to Arber not only in terms of her work on botanical types as a way of unifying the great diversity of plant form but also in terms of her interest in meditation as a way of experiencing the underlying unity of all existence, a concept she explored in *The Manifold and the One*. This book deals with Eastern as well as Western views on how to achieve a meditative state of unity with nature. While Arber seems to have dealt with many different areas in her work, from morphology through history of science to Eastern philosophy, there is an underlying theme in all her work. That theme is the desire to find foundations, whether they be the historical foundations of modern botany, the types forming the foundation of morphological diversity, or the fundamental unifying foundation beneath the multiplicity of everyday experiences. In light of this quest for unity, it is not surprising that her last book would be a philosophical, rather than a scientific, work. In a sense, her whole life's work was moving toward such a climax.

NOTE

The author wishes to thank Muriel A. Arber for her gracious assistance in the preparation of this article about her mother.

BIBLIOGRAPHY

Works by Agnes Robertson Arber

Scientific Works

Space does not permit the listing of the complete works of Agnes Robertson Arber. Listed here are all works by Arber except those cited in Thomas ("Agnes Arber 1879–1960" 1960). Included here are all references cited in the text.

"Notes on the anatomy of *Macrozamia heteromera,* Moore." *Proceedings of the Cambridge Philosophical Society* 12 (1902): 1.

Herbals: Their Origin and Evolution. Cambridge, Great Britain: Cambridge University Press, 1912.

Water Plants: A Study of Aquatic Angiosperms. Cambridge, Great Britain: Cambridge University Press, 1920.

Monocotyledons: A Morphological Study. Cambridge, Great Britain: Cambridge University Press, 1925.

The Gramineae: A Study of Cereal, Bamboo, and Grass. Cambridge, Great Britain: Cambridge University Press, 1934.

The Natural Philosophy of Plant Form. Cambridge, Great Britain: Cambridge University Press, 1950.

The Mind and the Eye. A Study of the Biologist's Standpoint. Cambridge, Great Britain: Cambridge University Press, 1954.

Other Works

The Manifold and the One. London: John Murray, 1957.

Works about Agnes Robertson Arber

"Agnes Arber." *Phytomorphology* 11 (1961): 197–198.

"Agnes Arber." *Proceedings of the Linnaean Society of London* 172 (1961): 128.

Arber, M. A. "List of published works of Agnes Arber, E.A.N. Arber and Ethel Sargant." *Journal of the Society for the Bibliography of Natural History* 4 (1968): 370–384.

————. *Lyme Landscape with Figures.* Exeter, Great Britain: Dorset Books, 1988.

Flannery, M. C. "Goethe and Arber: Unity in diversity." *American Biology Teacher* 57 (8) (1995): 544–547.

Godwin, H. "Agnes Robertson Arber." *Dictionary of Scientific Biography,* vol. I. New York: Scribner's, 1970.

Stearn, W. T. "Mrs. Agnes Arber." *Taxon* 6 (1960): 261–263.

Thomas, H. H. "Agnes Arber." *Nature, London* 186 (1960): 847.

————. "Agnes Arber, 1879–1960." *Biographical Memoirs of Fellows of the Royal Society* 6 (1960): 1–11.

Other References

Anderson, R. *An Illustrated History of the Herbals.* New York: Columbia University Press, 1977.

Arnheim, R. *Visual Thinking*. Berkeley: University of California Press, 1969.

Blunt, W. *The Art of Botanical Illustration*. London: Collins, 1950.

Darwin, C. *The Origin of Species*. London: Murray, 1859.

Ivins, W. *Prints and Visual Communication*. Cambridge, MA: Harvard University Press, 1953.

Lyte, H. *A Nievve Herball, or Historie of Plants*. Antwerp: Henry Loë Bookeprinter, 1578.

Ritterbush, P. "The biological muse." *Natural History* 77 (9) (1968): 26–31.

CHARLOTTE AUERBACH (1899–1994)

Linda E. Roach and Scott S. Roach

BIOGRAPHY

Charlotte Auerbach, a third-generation German-Jewish scientist, was born in Krefeld, Germany, on May 14, 1899. She was influenced by her father, a chemist; an uncle, a physicist; and her grandfather, an anatomist, for whom Auerbach's plexus in the human intestine was named (McGraw-Hill 1980).

With degrees from the universities of Würzburg, Freiburg, and Berlin, Auerbach fled Germany and Nazism in 1933 and settled in Edinburgh, Scotland. Here, at the Institute of Animal Genetics, she earned her Ph.D. degree in 1935. She remained affiliated with the Institute during her entire career.

Her first position was that of assistant instructor of animal genetics. She was promoted to lecturer (1947) and then through the ranks to professor of genetics (1967). In 1969 she was awarded the title of professor emeritus. Her long and fruitful career at the Institute of Animal Genetics was punctuated with sabbaticals. In 1958 Auerbach was a visiting professor at Oakridge National Laboratories in the United States. Auerbach was a member of a number of professional societies, including the Royal Society of London and the Royal Society of Edinburgh. She held an honorary membership in the Genetical Society of Japan (1966) and was a foreign associate of the U. S. National Academy of Sciences (1970) and the Danish Academy of Sciences (1968).

Auerbach received honorary doctorates from several universities, including Leiden in the Netherlands (1975), Cambridge (1977), Dublin (1977), and the University of Indiana at Bloomington (1985). In 1947 the University of Edinburgh awarded her a D.Sc. degree and in 1967 honored her with a chair. She received the Keith Medal from the Royal Society of Edinburgh (1947) and the Darwin Medal from the Royal Society of London (1977). Environmental mutagen societies in both the United States (1972) and Europe (1974) recognized her research.

Interested in the public's understanding of science and mutations, Auerbach wrote two books for the general public, commenting on the effects of mutagens on humans as well as other animals. She served on several committees with environmental agendas (McGraw-Hill 1980).

Auerbach never married. She took care of her aging, invalid mother. Towards the end of her life she lived in a nursing home in Scotland and died on March 17, 1994.

WORK

Auerbach's dissertation involved determining the effects of mutant genes on the development of the fruit fly, *Drosophila melanogaster*. Her goal was to determine how genes operated. When Hermann Joseph Muller (Nobel laureate, 1946), a mutation researcher, joined the Institute of Animal Genetics, where she held a personal chair, he introduced her to mutation research.

I well remember the day when Muller came to the laboratory in which I worked, sat down beside me, and asked me what general purpose I was pursuing with my research. I answered that I wanted to find out something about the way genes act. He replied that the morphological and histological effects of mutant genes which I was studying were still far removed from the primary action of the gene and that the best way to find out something about the gene itself was to see by which means it could be made to mutate. His enthusiasm for mutation research was infectious and from that day on I switched to mutation research. I have never regretted it. (Auerbach 1978, 319–320)

Muller had demonstrated (1927–28) that X-rays were powerful mutagenic agents. At the beginning of World War II, Edinburgh pharmacologists Alfred Joseph Clark and John Michael Robson noted that the properties of mustard gas produced effects similar to those produced by X-rays.

While many researchers had attempted to induce mutations by chemical means, Auerbach and Robson were the first to discover and report that mustard gas (dichloroethyl sulphide) was a highly effective chemical mutagen that produced visible mutations in mice and *Drosophila melanogaster*. This was the first reported success with chemical mutagens. "This stretch of my pilgrim's progress was the stoniest and most demanding one but also the most thrilling," stated Auerbach (Auerbach 1978, 328).

The work with mustard gas was dangerous, because testing apparatus was, indeed, crude. Auerbach and her coworkers were plagued with rashes and burns from their exposure. Despite warnings from her dermatologist, Auerbach was spurred to further research.

The inhibitory action of mustard gas on *Drosophila* cell mitosis was confirmed. Later a higher mutation rate (equivalent to 2,000–3,000 rads of X-rays) was demonstrated. Auerbach noted that semilethal mutations were frequent after treatment with mustard gas and that some mutations were delayed. Other mutations identified included dominant and recessive lethals, dominant and recessive mutations, chromosome breaks, and chromosomal rearrangement.

"Because of the nature of these chemicals, all our results were classified; the first short note on them appeared in *Nature* in 1946, and the first full papers came out in 1947," recalled Auerbach (Auerbach 1978, 328). Secrecy surrounding the project provided time for synthesis of ideas without pressure to publish or compete with other researchers. Her hope was that knowledge gained from mutation research would shed light on the nature of the gene. "Moreover, it could be hoped that among chemical mutagens there might be some with particular affinities for individual genes. Detection of such substances not only would be of high theoretical interest but would open up the long sought-for way to the production of directed mutations" (Auerbach 1967, 1141).

Auerbach suggested that chemical mutations may have played a role in evolution. She studied the effects of many chemical mutagens, such as sulfur mustard, nitrogen mustard, formaldehyde, chlorethyl methanesulfonate, and diepoxybutane (DEB), and compared their effects to those of X-rays and ultraviolet light. Yet, her primary interest was in the process of mutation, not the mutagenic abilities or properties of these chemicals. These chemicals are considered radiomimetic substances because they mimic the effects of radiation. The lesions caused by chemical mutagens resemble those caused by radiation, but in chemical induction the lesions may be delayed for several cell divisions.

While she studied chemical mutagenic effects on mice, most of her research concentrated on *Drosophila melanogaster* and *Neurospora crassa*. Because formaldehyde is present in the environment, albeit in small amounts, Auerbach studied and reported its effects on various animals.

Auerbach has been nicknamed the "mother of chemical mutagenesis," but she considered her major contribution to genetics research an in-depth study of mutagenesis focused on the selective effects of the chemical mutagens on particular genes (McGraw-Hill 1980). She reported on the biological processes of mutagenesis involved when chemical mutagens were introduced to an organism. Of interest to her were the sites of mutations, delayed actions of chemical mutagens, and specificities of the chemical mutagens.

Contrary to the views of molecular biologists at the time, Auerbach became convinced that mutagenesis must be directed by cellular enzymes, complex biochemical pathways, membrane structures, and regulatory capabilities. Her ideas were not encouraged, merely tolerated. "Indeed," she said, "coming from the rank undergrowth of biological phenomena, with its vines and thorns stuck all over us, we did not fit into this tidy picture" (Auerbach 1978, 332). Further investigations showed that Auerbach was correct in her assumptions.

Two books on genetics, *Genetics in the Atomic Age* (1956) and *The Science of Genetics* (1961), were written by Auerbach for the general population in such a manner as to inform the public about genetics and to popularize science. Concerned about the effects of mutagenic substances on humans, Auerbach reported on the genetic effects of radiation and chemicals. She believed that exposure to these substances by humans should be limited until further research could be completed.

Charlotte Auerbach loved teaching. She wrote two textbooks, *Mutation: An Introduction to Research on Mutagenesis. Part 1: Methods* (1962) and *Mutation Research: Problems, Results and Perspectives* (1976), to help senior genetics students plan and carry out research. She also produced a handbook, *Notes for Introductory Courses in Genetics* (1965), to assist students in their studies.

BIBLIOGRAPHY

Works by Charlotte Auerbach

Scientific Works

Space does not permit the listing of the complete works of Charlotte Auerbach. Listed here are all works by Auerbach except those cited in Auerbach et al. (1977). Included here are her dissertation and all references cited in the text.

"The Development of the Legs, Wings, and Halteres in Wild-type and Certain Mutant Strains of *Drosophila melanogaster.*" Ph.D. diss., University of Edinburgh, 1935.

"Chemically induced mosaicism in *Drosophila melanogaster.*" *Proceedings of the Royal Society of Edinburgh* 62b (1946): 211–222.

"The induction of mustard gas of chromosomal instabilities of *Drosophila melanogaster.*" *Proceedings of the Royal Society of Edinburgh* 62b (1947): 307–320.

"Nuclear effects of chemical substances." *Proceedings of the 6th International Congress of Cytology, Experimental Cell Research* (Suppl. 1) (1947): 93–96.

(with J. M. Robson) "The production of mutations by chemical substances." *Proceedings of the Royal Society of Edinburgh* 62b (1947): 271–283.

(———) "Tests of chemical substances for mutagenic action." *Proceedings of the Royal Society of Edinburgh* 62b (1947): 284–291.

"Chemical induction of mutations." *Proceedings of the 8th International Congress of Genetics. Hereditas Suppl.* (1948): 128–147.

"Chemical mutagenesis." *Biological Reviews of the Cambridge Philosophical Society* 24 (1949): 355–391.

(with D. S. Falconer) "A new mutant in the progeny of mice treated with nitrogen mustard." *Nature* 163 (1949): 678–679.

"Differences between effects of chemical and physical mutagens." *Pubblicazioni della Stazione Zooligica di Napoli* (Suppl.) 22 (1950): 1–21.

"S-H poisoning and mutation." *Experientia* 6 (1950): 17.

"Problems in chemical mutagenesis." *Cold Spring Harbor Symposium on Quantitative Biology* 76 (1951): 199–213.

Genetics in the Atomic Age. Edinburgh: Oliver and Boyd, 1956.

"Genetical effects of radiation and chemicals." *Experientia* 13 (1957): 217–224.

"Radiomimetic substances." *Radiation Research* 9 (1958): 33–47.

"Spontaneous mutations in dry spores of *Neurospora crassa.*" *Zeitschrift für induktive Abstammungs-und Vererbungs Lehre* 90 (1959): 335–346.

(with E. M. Sonbati) "Sensitivity of the *Drosophila*: Tests to the mutagenic action of mustard gas." *Zeitschrift für induktive Abstammungs-und Vererbungs Lehre* 91 (1960): 237–252.

(with B. Woolf) "Alpha and beta loci in *Drosophila.*" *Genetics* 45 (1960): 1691–1703.

"Chemicals and their effects." In *Symposium on Mutations and Plant Breeding*, National Research Council Publication 891, 120–144. Washington, DC: National Academy of Sciences, 1961.

"Hazards of radiation." *Nature* 189 (1961): 629.

The Science of Genetics. New York: Harper and Row, 1961, 1964.

Mutation: An Introduction to Research on Mutagenesis. Part 1. Methods. Edinburgh: Oliver and Boyd, 1962.

"The production of visible mutations in *Drosophila* by chlorethyl methanesulfonate." *Genetical Research* 3 (1962): 461–466.

(with D. S. Falconer and J. A. Isaacson) "Test for sex-linked lethals in irradiated mice." *Genetical Research* 3 (1962): 444–447.

"Stages in the cell cycle and germ cell development." In *Radiation Effects in Physics, Chemistry and Biology*, edited by M. Ebert and A. Howard, 152–168. Chicago: Year Book Medical, 1963.

Notes for Introductory Courses in Genetics. Edinburgh: Kallman, 1965.

"Chemical induction of recessive lethals in *Neurospora crassa*." *Microbial Genetics Bulletin* 17 (1966): 5.

"*Drosophila* tests in pharmacology." *Nature* 210 (1966): 104.

"The chemical production of mutations." *Science* 158 (1967): 1141–1147.

(with D. Ramsay) "Differential effect of incubation temperature on nitrous acid-induced reversion frequencies at two loci in *Neurospora*." *Mutation Research* 4 (1967): 508–510.

"Remark on the 'Tables for determining the statistical significance of mutation frequencies.' " *Mutation Research* 10 (1970): 256.

(with D. Ramsay) "Analysis of a case of mutagen specificity in *Neurospora crassa*. II. Interaction between treatments with diepoxybutane (DEB) and ultraviolet light." *Molecular and General Genetics* 109 (1970): 1–17.

(———) "Analysis of a case of mutagen specificity in *Neurospora crassa*. III. Fractionated treatment with diepoxybutane (DEB)." *Molecular and General Genetics* 109 (1970): 285–291.

(with B. J. Kilbey) "Mutation in eukaryotes." *Annual Review of Genetics* 5 (1971): 163–218.

(with D. Ramsay) "The problem of viability estimates in tests for reverse mutations." *Mutation Research* 11 (1971): 353–360.

(———) "Analysis of the storage effect of diepoxybutane (DEB)." *Mutation Research* 18 (1973): 129–141.

Mutation Research: Problems, Results and Perspectives. London: Chapman and Hall, 1976.

(with M. Moutschen-Dahmen and J. Moutschen) "Genetic and cytogenetical effects of formaldehyde and related compounds." *Mutation Research* 39 (1977): 317–362.

"A pilgrim's progress through mutation research." *Perspectives in Biology and Medicine* 21 (1978): 319–334.

Works about Charlotte Auerbach

McGraw-Hill Modern Scientists and Engineers. Vol.1, 32–33. New York: McGraw-Hill, 1980.

FLORENCE AUGUSTA MERRIAM BAILEY (1863–1948)

Harriet Kofalk

BIOGRAPHY

Locust Grove, near Leyden, New York, was named by Florence Augusta Merriam's grandparents for the grove of trees that surrounded their homestead. She was born in the large house called "Homewood," which her father built before the Civil War on a hill above his parents' home. She had two older brothers, Charles Collins and Clinton Hart (called C. Hart to differentiate him from his father, Clinton), and a sister, Gertrude, who died at age five, the day before Florence was born. Her father, Clinton Merriam, was a businessman who had a mercantile house and then founded a brokerage firm on Wall Street in New York City. He later served two terms as a Republican representative to Congress. The Merriam ("merry home") family origins were in England, and the family history has been traced back to the 1300s by other relatives.

Her mother, Caroline (Hart) Merriam, was a graduate of Rutgers College who loved astronomy and often took her young daughter up to the glassed-in cupola built on the top of the house to see the constellations. The family was very religious, especially Mrs. Merriam. The mother carefully tended flower gardens, which were another source of inspiration to young Florence Merriam. In memoirs to her nieces years later, she wrote, "Especially dear to my child heart were the snow-drops, babies' breath, the sweet-faced little violets and the small sweet-smelling pinks. The tall, handsome tiger lilies belonged to a different clan" ("Pages . . ." unpublished). Already her connections and friendship with the natural world were evident.

The Merriams were well-to-do, as their large house and land holdings indicate. In the world of finance, the father did well until the Civil War, when he decided to leave the New York world of business and retire where he could hear the robins sing in the morning, near his parents' home. Homewood had already been built for that eventual move, and the timing allowed them to live comfortably, unlike those whose businesses went broke because of that war. At age 40, her father then managed the family farm, with the help of a resident farmer and his family. He also cared for his aging parents, who continued to live on the homestead nearby. Hearing her father tell the story of the robins was Florence Merriam's earliest connection with birds and inspired her life's course. Later, her father's interest in natural history led him to the writings of John Muir, and as soon as the transcontinental railroad was complete, he ventured to California to meet him.

Merriam was educated primarily at home in her early years, as were her brothers. This was partly dictated by her delicate health, probably tuberculosis, although this word was never used, due to the prevalent fear of the disease in those times. In her memoirs, she described this early education as "coming mainly from the woods and fields" ("Pages . . ." unpublished). Merriam attended public school half-days one year in Syracuse, New York, while undergoing medical treatment at the insistence of her physician/brother. She stayed with the family of the physician who was treating her and attended school with his daughter. When the Merriam family wintered in New York City, as snow made travel in Upstate New York difficult, she briefly attended public school there as well. In order to prepare her for upper-level education, she was sent to Mrs. Piatt's School in Utica, New York.

Her brother, Collins, 13 years older, was gone from home during her childhood years, but Hart Merriam, 8 years her elder, was her constant companion and mentor. By the age of 12 he was already collecting mammals and birds and beginning a lifelong interest in taxidermy, which eventually led his father to build a three-story museum to house the results. This convinced his young budding-naturalist sister that she preferred *live* birds over stuffed specimens.

Her father and brother took her on walks through the woods when she was a young child, pointing out tracks and evidence of wildlife and teaching her through their own enthusiasm for nature. Plants, too, were special cause for note. At Christmas one year, when each member took a role in a family play, Florence played the spring beauty, a flower associated with her sister, who had died. She entered Smith College as a "special student," with nondegree status, which allowed her to take whatever classes she chose. Miss Jordan, her favorite teacher and a cousin to naturalist David Starr Jordan, was instrumental in furthering her already deep love for nature in all its forms. She left college as she had come, as a special student, without being awarded a degree with her classmates in 1886. Thirty-five years later, in 1921, Smith College conferred upon her the degree of Bachelor of Arts in recognition of her life's work. In 1933 she was awarded an honorary LL.D. degree by the University of New Mexico for her major work, *Birds of New Mexico*.

Her brother, C. Hart Merriam, moved to Washington, D.C., to head the new U. S. Biological Survey, the federal government's first entry into the field of natural sciences. His first assistant, Vernon Bailey, became the Survey's chief field naturalist, a 46-year career. Bailey married his boss's sister, Florence Merriam, on December 16, 1899, after a ten-year friendship. They remained married until his death in 1942. Although she never bore children, she not only was

"Auntie Florence" to her nieces and nephews but also mothered many of her husband's protegés and their wives.

Neither of the Baileys during their married years professed any traditional religious affiliation, although both held deep convictions about the goodness of humankind. Mrs. Bailey had long correspondence with her brother in their early years about the positive meanings of life, and this philosophy stayed with her throughout her life. Her husband's family was deeply religious in traditional ways, and he followed those traditions at least until his marriage. While in the field, he visited any church that was nearby, curious about all forms of belief. Content in their private beliefs, the Baileys did not continue this practice after they married.

Florence Merriam Bailey chose to stay out of the political area, although no doubt nominally Republican, based on her family's tradition. She spoke little of politics, except in relation to birds and the need for legislation to protect them from extinction, primarily due to their use in decorating women's hats. Her husband and close brother both died in 1942. She went into seclusion in their home in Washington, D.C., where she lived with two women friends who took care of her. She took daily walks, experimented with the newest vitamins, and passed away in 1948.

During her lifetime, Bailey always worked and published under her own name, unusual in her time. She was considered one of the four best women nature writers of the late nineteenth century. All of her writing was done voluntarily, simply to express her own connection with, and concern for, the birds. She held no official position in any traditional work sense but continually worked for the conservation of her bird friends. Author of ten books and more than 100 magazine articles, mostly for lay readers, she received reviews in scientific as well as nonscientific publications and was respected by scientific ornithologists (when they came into being) as well as nonscientific enthusiasts. Although ever a critic of unscrupulous hunters, she admired sportsmen who established societies for game protection and occasionally wrote articles for their magazines, despite protests from her purist friends.

Because of her upper-class upbringing, Bailey was comfortable in the Washington, D.C., entertainment scene and gave dinners weekly during the winters for scientific colleagues. It has been said that a guest book from the Baileys' home would have included all the scientific luminaries of the day. They built a home soon after they were married, in the typical three-story fashion of the time and place. The living room was created around a mural of two tigers, painted by their friend, noted wildlife artist Charles R. Knight, on a large piece of mahogany that formed the backdrop for the main fireplace mantel. It can be seen today in the archives of the Smithsonian Institution.

Bailey was always encouraged by her family in her endeavors. She was her brother's close ally until he married, when she realized that her own life, not his, was her pursuit. With his constant help, she became increasingly expert in her work with live birds, in contrast to the scientific approach of shooting them first and asking questions later. She spent days and weeks sitting quietly with her opera glasses, observing birds in their homes, and writing about what she observed. Her aunt was a botanist, her mother an astronomy advocate, her brother a physician, one uncle a paleontologist, and another a geologist. They shared a common definition of a naturalist as one who knows all the wonders of nature, not a specialist in just one of them. Each of the children was tutored in the specialty of others; each had ready access to assistance in many fields. Notables often visited the family home, and the children were included in walks with them through the countryside. During their adult years, the Baileys did the same for visitors to Washington and in the field wherever they were.

For Bailey, public and private life intertwined. Birds were her friends; humans were also. All were to be nurtured and cared for. During World War I she extended her writing to the young soldiers she observed on the train as she crossed the country for a summer field season in the West. Her profession was her life. She pioneered new ways of seeing not only birds but life around her and describing them so that others might share her vantage point. Such writing has become common in today's world, but the mix of personal and scientific was new when Bailey was writing. She knew her audience of laypeople and wrote for them. With a strong scientific background, tutored by her brother, she had the knowledge necessary but chose to help others get outdoors rather than write to colleagues in ornithology. Her articles in professional journals always had the personal touch that endears her to some of us and yet leaves her outside the scientific realm today. As with many pioneers, her seminal work is now forgotten as we move onward, but the evidence of human–bird connections is even now being rediscovered and revalued by many readers and writers.

Writing about birds has changed greatly, as has their scientific study. The poetic essence that pervades writings like Bailey's came into disrepute as we devoted more energy to quantifying rather than qualifying the natural sciences. By her focus on both, she brings a wholeness sorely lacking in much ornithological writing today.

Bailey fits well Deborah Strom's comment in her book on *Birdwatching with American Women*: "Some women writers found in bird study a model for tranquility within a turbulent world." Bailey describes this to her Smith College alumnae in a letter written during World War II: "And at night, the peace of the star-filled canopy of the heavens. Is that not what we need to remember above all in these terrible days of man-made disasters? Let us raise our eyes to the heavens above us." This statement speaks of many dimensions of Bailey's life—carrying on her mother's love for astronomy, her sense of spiritual connection with the larger harmony of the universe, and her lifelong passion for watching the birds in their realm.

WORK

During her last year at college (1885–86), she heard that George Bird Grinnell was founding an Audubon Society. He was a friend of her brother, Hart Merriam, and she wrote him immediately to ask how she and her friends might help. Before long, a third of the Smith College campus was involved in the very first chapter. Their enthusiasm was aided by another family friend, naturalist John Burroughs, then retired, who agreed to come and take the girls for bird walks one week. This early society later dissipated; it was restarted in Massachusetts a few years later, which spawned the group that has now become the National Audubon Society.

Bailey's contribution to ornithological research was significant in bringing scientific understanding and heartfelt living into public awareness. She spent more than a half century watching the birds across the country and cataloging their movements, their nesting, and simply their presence. She shared her experiences not only with scientists who did not have the opportunity that life afforded her but also with women and young people whom she encouraged to get outdoors. Her aim was to help them appreciate living birds so they would not use bird feathers and bodies as mere decorations.

Bailey's research led naturally to teaching. She started classes for schoolteachers through the Audubon Society in Washington, D.C., and led bird walks and classes for young people there.

Fieldwork was her specialty, as it was her husband's. At first on her own, she traveled west for her health, with her physician/brother's encouragement. Having trained her himself, he also saw her travels as a way to enhance his own understanding of the West, and he kept her well supplied with forms and literature so she could record her findings for the Biological Survey records. Women at the turn of the century were finding the freedom to travel a new advantage, which Bailey enjoyed fully. Trains made the West increasingly accessible, and she explored well beyond their bounds by wagon and by horseback for more than a half century.

After she and Vernon Bailey married, they traveled together most years until the late 1930s, spending the summer (and occasionally winter) field season wherever his work took them. They both wanted children, but no pregnancy came to term. Probably, some summers that she remained in Washington were times of miscarriage. When well enough to travel, she was constantly by her husband's side. He focused on the mammals; she on the birds. They collaborated on two books but generally each kept individual field notes and wrote separately. They edited for one another, and her brother also kept a sharp eye on his sister's work, especially in her younger years.

Bailey was the first woman to become an associate member of the American Ornithologists' Union (AOU), in 1885, as well as its first woman fellow (1929). She was also the first woman to receive the AOU's coveted Brewster Medal (1932). Both the fellowship and the medal were in recognition of her major work, *Birds of New Mexico*. With 800 pages, it was not a book to carry in the field! Her earlier *Handbook of Birds of the Western United States* complemented Frank Chapman's handbook on Eastern birds and provided an impetus for Roger Tory Peterson later to create his own series of handbooks.

Bailey claimed no scientific breakthroughs and sought none. In retrospect, however, she certainly helped to achieve scientific status for the study of *live* birds through her untiring insistence on "knowing the birds at home." Her published observations represented new knowledge at that time of bird behavior and distribution, although her role and her intent were simply to share her observations so others might be inspired to observe more closely. She spoke out for the benefits of living outdoors, based on her own life. She lived by her principles of moving humanity forward, not as we usually think of progress but as "upliftment"—one of her favorite words—which is the most powerful way one can speak out for life (Kofalk 1989).

Career was not important to Bailey; birds were. She focused her attention on them wherever she traveled, and she went many places throughout the American West at a time when relatively few people had done so. What she experienced she knew was valuable, and she wanted to share it. Writing was one of the few media open to her, given her lifestyle, and she took full advantage of it. She published virtually every year of her adult life. Through her inspiration, many people could go into the field with her, at least through her eyes. Her intent to inspire this was evident from her earliest writings. In the preface to *Birds through an Opera Glass* (1890), she wrote, "[I]t is above all the careworn indoor workers to whom I would bring a breath of the woods, pictures of sunlit fields, and a hint of the simple, childlike gladness, the peace and comfort that is offered us every day by these blessed winged messengers of nature." She followed this by very practical "hints" for those willing to go into the field and discover these joys for themselves. Guided by a devoted brother and later a loving husband, she fulfilled her intent and much more. Unlike her contemporary Grace Seton-Thompson, the first wife of naturalist Ernest Thompson Seton, who followed her husband into the field to be with him and wrote *Woman Tenderfoot in the Rockies*, Florence Merriam Bailey was *No Woman Tenderfoot*, the title of her biography (Kofalk 1989). She was, as its subtitle indicates, a "pioneer naturalist" in the fullest sense of both terms.

BIBLIOGRAPHY

Works by Florence Augusta Merriam Bailey

Scientific Works

Space does not permit listing the complete works of Florence Merriam Bailey. They are cited in Kofalk (1989). Included here are all references cited in the text of this article.
Birds Through an Opera Glass. Boston: Houghton-Mifflin, 1890.
Handbook of Birds of the Western United States. Boston: Houghton Mifflin, 1902.
Birds of New Mexico. Santa Fe: New Mexico Department of Game and Fish, 1928.

Other Works

Letters, Class of 1886, 1886–1947. Smith College, Northhampton, MA (unpublished).
"Pages from the Merriam Family History for the Children of Lyman Lyon Merriam" (unpublished).

Works about Florence Augusta Merriam Bailey

Kofalk, Harriet. *No Woman Tenderfoot—Florence Merriam Bailey, Pioneer Naturalist.* College Station: Texas A&M University Press, 1989.

Other References

Chapman, Frank. *A Handbook of Birds of Eastern North America.* New York: Appleton, 1895.
Seton-Thompson, Grace. *Woman Tenderfoot in the Rockies.* New York: Doubleday, Page, 1900.
Strom, Deborah (ed). *Birdwatching with American Women.* New York: W. W. Norton, 1986.

RACHEL LITTLER BODLEY (1831–1888)

Ronald L. Stuckey

BIOGRAPHY

Rachel Littler Bodley, a pioneer in the professional education of women, taught botany, chemistry, and toxicology and prepared published studies showing the success of women college graduates in the medical profession. She encouraged women to enter medical missionary work. She herself wanted to be a part of this work early in life but could not do so because of delicate health. Born in Cincinnati, Ohio, December 7, 1831, Bodley was the eldest daughter and third of five children of Anthony Prichard Bodley, a carpenter and later a pattern maker, and Rebecca Wilson (Talbott) Bodley, a teacher who operated a private school in Cincinnati (Alsop 1950, 1971).

The father was of Scotch-Irish descent. His great grandfather, Thomas Bodley, emigrated from the north of Ireland in the early 1700s and settled in what is now Montgomery County, Pennsylvania. There he married Mrs. Eliza (McIntosh) Knox from Edinburgh, Scotland. Rachel Bodley's father left Montgomery County at age 21, crossed the Allegheny Mountains on foot to Pittsburgh, and from there descended the Ohio River in a canoe, arriving in Cincinnati during 1817 (Bolton 1888).

The mother was descended from John Talbott, a Quaker who emigrated from England and settled near Winchester, Virginia. In 1806 his descendants, including Samuel and Rachel (Littler) Talbott, and their only daughter, Rachel Bodley's mother, Rebecca Wilson, moved by wagon to the banks of the Monongahela River in western Pennsylvania. The family descended the Ohio River in 1817 and also established a residence at Cincinnati (Bolton 1888).

Rachel Bodley had two older brothers and two younger sisters, all raised in the Presbyterian faith. The education of the Bodley children was conducted by their mother, and Rachel Bodley was a pupil in her mother's school until her 12th year. In 1844 she entered the Wesleyan Female College of Cincinnati, the first chartered college for women in the United States, founded two years earlier for the purpose of giving women a higher education. Many of the school's graduates were remarkable for their Christian labors in churches and charitable work in their communities (Bolton 1888). At the college she distinguished herself in writing for the literary society and graduated in 1849, prior to her 18th birthday.

Following graduation she was appointed an assistant teacher at the Wesleyan Female College, where she advanced to preceptress in the higher collegiate studies. Although her teaching skills were praised as outstanding and successful, Bodley was not satisfied with her attainments and was unhappy with work there.

In the fall of 1860, at age 29, Bodley left Cincinnati for Philadelphia and became a special student in advanced chemistry and physics at the Polytechnic College of Pennsylvania, then the leading institution in the country for instruction in the applied sciences. She also studied practical anatomy and physiology in the Woman's Medical College of Pennsylvania (now Medical College of Pennsylvania). After nearly two years of classwork, she returned to her teaching career in Cincinnati (Bolton 1888). While there Bodley continued her studies through private lessons in higher mathematics, music, French, German, elocution, drawing, microscopy, and phonography. Later, in Philadelphia, she completed her regular course of medical study begun in 1860 at the Woman's Medical College (Bolton 1888).

Following her studies in Philadelphia, Rachel Bodley returned to Cincinnati in February 1862 to begin an appointment as professor of natural sciences in the Cincinnati Female Seminary (sometimes referred to as the Ohio Female College). She held this teaching position for three years and then in 1865 was invited to the chair of chemistry and toxicology in the Woman's Medical College. Bodley became the first woman to hold the title of professor of chemistry in a medical school and the first to excel as a teacher of that science (Miles 1976). In January 1874 she was elected Dean of the Faculty, a position she held until her death.

Bodley did not confine her educational and administrative responsibilities to the Woman's Medical College but also accepted invitations to teach and lecture for short periods at other institutions. She was school director of the 29th School Section of Philadelphia (1882–85, 1887–88) and one of seven women to inspect institutions in Philadelphia County for the Board of Public Charities of the State of Pennsylvania (1883).

Bodley received many honors and recognitions for her contributions to science and literature. Among these were elections to the State Historical Society of Wisconsin (1864), the Academy of Natural Sciences of Philadelphia (1871), the Cincinnati Society of Natural History (1873), the New York Academy of Sciences (1876), the American Chemical Society (charter member and first woman member, 1876), the Franklin Institute of Philadelphia (1880), and the Public Educational Society of Philadelphia (1882). At the Franklin Institute in 1880, Bodley delivered, by invitation, six popular lectures on chemistry (Alsop 1950, 1971; Bolton 1888). The honorary degree A.M. was conferred on Bodley in 1871 by the Wesleyan Female College, her alma mater. She was the first of three graduates to be awarded this honor. In 1879 she was awarded an honorary M.D. degree by the Woman's Medical College (Alsop 1950, 1971). A century later Bodley has continued to be recognized for her achievements and was included in biographical dictionaries of notable American women (Alsop 1971), American educators (Dirkes 1978), scientists (Elliott 1979), physicians (Levin 1980), and chemists (Miles 1976).

Rachel L. Bodley, who never married, lived in Philadelphia. Death came following heart failure at age 56, on June 15, 1888, while at her home. She was buried at Spring Grove Cemetery, Cincinnati. The presentations delivered at her memorial service, held at the Woman's College that fall, on October 13, 1888, are preserved in the ceremony's published papers (1888).

WORK

Upon Bodley's arrival as a faculty member in the natural sciences at the Cincinnati Female Seminary, she "found chaos reigning in the domain of science" and selected the Herbarium as her first task (Bodley 1865). She organized the plant collections of Joseph Clark, a native of Scotland and resident of Cincinnati (1823–58), which had been deposited with the Seminary by a relative following Clark's death. Bodley's extensive efforts in arranging Clark's herbarium were described in great detail by her in the Preface of the resultant printed *Catalogue of Plants* (Bodley 1865). This work, completed in leisure hours over a period of three years, was designed as a reference to assist the student and traveler in studying the plants of the Cincinnati neighborhood.

Bodley's preparation of this material, based on Clark's collections, is the first printed local flora of Ohio prepared by a woman and is her most significant contribution to botany (Stuckey 1992). This work was later recognized by Asa Gray, America's foremost botanist, as a "very satisfactory indeed" contribution to science (Bolton 1888).

Throughout her life Bodley maintained an interest in botany, and during many summers she traveled to natural scenic and historic places. During her trips Bodley always carried equipment for collecting and preparing herbarium specimens of plants. She studied many unusual plants, among them the Venus flytrap, the lily of the valley, the snowdrops, the dwarf horse chestnut, and the Alpine sandwort (Bolton 1888).

Bodley's extensive herbarium was presented by Dr. Henry Leffmann to the Wagner Free Institute of Science in Philadelphia. Leffmann was a professor in the Institute's Department of Chemistry and member of its Board of Trustees (Leffmann et al. 1909). In June 1973, the Institute's Herbarium was deposited with the Academy of Natural Sciences of Philadelphia, where the specimens were integrated with the Academy's Herbarium. Bodley had collected plants at identified localities in at least 18 states. A sample specimen label is shown in Stuckey (1992). Her written papers in botany were mainly contributed to the *Philadelphia Ledger and Transcript*, where a series on seaweeds, collected at

Longport, New Jersey, attracted considerable and favorable attention (Harsh-berger 1899).

When Bodley joined the Woman's Medical College, she was not only the first woman chemist on the faculty but also the first faculty member appointed from outside the city. She represented a new kind of personality, motivated by a quest for knowledge in science.

The scientific approach was new and unexpected at the College.

It was an intellectual concern, which grew out of an individual bent, and did not spring out of any "womanly" intuitive, maternal or sympathetic qualities. The feminine qual-ities of meticulous attention to detail and patient observation were useful to science. . . . Rachel Bodley's students were no longer to win their way by "womanliness" but by learning. Knowledge itself was the keystone to achievement. Health would . . . rest on facts; disease was to be cured by facts. (Alsop 1950, 105–106)

Bodley's scientific knowledge became linked with a practical purpose. She was teaching the science of medicine, which had a strong foundation, rather than the art of medicine. Her thoroughness and enthusiasm for scientific study opened doors to membership in scientific societies, which increased her stature in the College (Alsop 1950).

During her 14 years as Dean of the College, Bodley established the School in its first newly constructed building (1874). She lengthened the course of instruction to three years and introduced a progressive curriculum with an in-crease in the amount of demonstration and practical instruction. She expanded opportunities for clinical training that included the construction of a surgical amphitheater and clinical hall. She also appointed new faculty members in the disciplines of *materia medica*, obstetrics, gynecology, and experimental physi-ology. Bodley developed a cosmopolitan student body, and the graduates oc-cupied many diverse positions throughout the country, including several who became medical missionaries in eastern hemisphere countries (Alsop 1950, 110–126).

By means of a questionnaire sent to 244 known living graduates, Bodley reported on an extensive statistical study of the college's graduates. One of the concerns most often asked by the severest of critics concerning women in med-icine is the relationship between marriage and the practice of medicine. Of the 52 married doctors who responded, 45 reported that marriage was favorable and did not affect their practice. Many of the graduates "have achieved brilliant success since assuming the duties and responsibilities of married life" (Bodley 1881).

This study, the first of its kind, served as her address to the graduates at commencement on March 17, 1881. Published as "The college story" (1881), her report attracted considerable attention (Alsop 1950, 1971). Thomas Went-worth Higginson, in the *Woman's Journal*, characterized her address as "the first really good and careful collection of facts . . . bearing on the professional life of women'' in the medical practice. "The care with which the facts were obtained, and the clearness with which they are stated, give them a value almost unique" (Bolton 1888). Bodley's report provides the baseline data to which subsequent surveys were compared. Drachman (1986, 60–61) noted that these later surveys provided factual, quantifiable evidence to assist in refuting the traditional belief of inherent differences between men and women physicians and to support instead the concept of a growing acceptance of their equality.

A particular interest of Bodley's was medical missionary work. She encouraged students to enter this field and followed with pleasure the careers of her graduates. She invited returning missionaries to stay with her and her mother at their home. It became a center of entertainment for the many visitors to the college with whom she had contact in the religious, medical, and educational worlds (Alsop 1950, 135–144; 1971).

A memorable event occurred at the 1886 commencement exercises. One of the graduates was Dr. Anandibai Joshee, a Brahmin lady from India. Bodley extended an invitation to Joshee's distant cousin, a distinguished kinswoman in England, Pundita Ramabai Sarasvati, to be her guest for this event. The Pundita came and witnessed the conferring of the degree of Doctor of Medicine upon her Hindu cousin. The next evening a formal reception was held for the two distinguished ladies. A large audience of both women and men listened to the Pundita's address, "The Women of India" (Alsop 1950; Lovejoy 1957).

In April 1886 Bodley prepared a little pamphlet entitled, "The Welcome to Pundita Ramabai," entailing a complete record of the two events, which was widely distributed. Acknowledgment by Queen Victoria of England was significant in that it could have a considerable bearing on her subjects in India who were working for the elevation of women in society (Bolton 1888). Bodley also undertook the business affairs connected with the publication of Pundita Ramabai's book, *The High-Caste Hindu Woman* (1887), and wrote an introduction to it (Bodley 1887). Her devotion to Pundita Ramabai's book and the enterprise of publishing it helped form circles of support for child-widows in India (Withington 1928).

Bodley helped organize the centennial celebration in honor of Joseph Priestley, the discoverer of oxygen. She was a first vice president of the Priestley Centennial Association and the only woman upon whom such an honor was conferred (Bolton 1888).

Among the attributes that Rachel Bodley cultivated to make her career successful, as given by her close friend Sarah K. Bolton (1888), were "good health, acute powers of observation, a refined and modest manner, carefulness in details, a systematic division of time, and an orderly arrangement of material." Bodley's later biographer and college historian, Gulielma F. Alsop (1950, 111), described her as a "warm . . . [and] dignified person, dressed in the sedate black of the period."

Bodley gave her full time and talents to the Woman's Medical College, promoting its interests and, at the same time, striving to elevate and secure respect and recognition for women and their work. Her voluminous correspondence with women throughout the world and her literary contributions to periodicals in many states brought wide recognition to the College and to herself as a professional woman ("Obituary" 1888).

BIBLIOGRAPHY

Works by Rachel Littler Bodley

Scientific Works

Catalogue of Plants, Contained in [the] Herbarium of Joseph Clark. Cincinnati, OH: R. P. Thompson, 1865.

Other Works

Introductory Lecture to the Class of the Woman's Medical College of Pennsylvania, Delivered at the Opening of the Nineteenth Annual Session, October 15, 1868. Philadelphia: Merrihew and Son, 1868.
"Letter to determine the location of the Centennial of Chemistry, 1774–1874, at the grave site of Joseph Priestley, in Northumberland, Pennsylvania." *The American Chemist* 5 (1874): 35–36.
Valedictory Address to the Twenty-second Graduating Class of the Woman's Medical College of Pennsylvania, March 13, 1874. Philadelphia: Rodgers, 1874.
Introductory Lecture Delivered at the Opening of the Twenty-sixth Annual Session of the Woman's Medical College of Pennsylvania, October 7, 1875. Philadelphia: Grant, Faires, and Rodgers, 1875.
"The college story." *Valedictory Address to the Twenty-ninth Graduating Class of the Woman's Medical College of Pennsylvania, March 17th, 1881.* Philadelphia: Grant, Faires, and Rodgers, 1881.
"Introduction." In *The High-Caste Hindu Woman*, 2d ed., by Pundita Ramabai Sarasvati, i–xxiv. Philadelphia: J. B. Rodgers, 1887.

Works about Rachel Littler Bodley

Abram, Ruth J. "Documenting our past: Daughters of Aesculapius." *Journal of the American Medical Women's Association* 38 (1983): 71–72, 82.
———. "This promised land: Women doctors one hundred years later." In *Send Us a Lady Physician: Women Doctors in America, 1835–1920*, edited by Ruth J. Abram, 247–248. New York: W. W. Norton, 1985.
Alsop, Gulielma Fell. "Rachel Bodley: Professor of Chemistry (1865–1874)"; "Rachel Bodley: Dean (1874–1888)"; "The college story"; "Pioneer women missionaries of the Woman's Medical College." In *History of the Woman's Medical College, Philadelphia, Pennsylvania 1850–1950*, 101–109, 110–126, 127–134, 135–144. Philadelphia: J. B. Lippincott, 1950.
———. "Rachel Littler Bodley." In *Notable American Women: A Biographical Dictionary 1607–1950*, vol. 1, edited by Edward T. James, 186–187. Cambridge, MA: Belknap Press of Harvard University Press, 1971.

————. "Women in medicine, Rachel Bodley, 1831–1888, chemist-scientist, third woman dean of the Woman's Medical College." *Journal of the American Medical Women's Association* 4 (1949): 534–536.

Bolton, Sarah K. "Rachel L. Bodley." In *Successful Women*, 149–174. Boston, MA: Lothrop, 1888. Reprinted. Plainview, NY: Books for Libraries Press, 1974.

Dirkes, M. Ann. "Rachel Littler Bodley." In *Biographical Dictionary of American Educators*, vol. 1, edited by John F. Ohles, 145. Westport, CT: Greenwood Press, 1978.

Drachman, Virginia G. "The limits of progress: The professional lives of women doctors, 1881–1926." *Bulletin of the History of Medicine* 60 (1986): 58–72.

Elliott, Clark A. "Bodley, Rachel Littler." In *Biographical Dictionary of American Science: The Seventeenth through the Nineteenth Centuries*, 33. Westport, CT: Greenwood Press, 1979.

Harshberger, John W. "Rachel L. Bodley." In *The Botanists of Philadelphia and Their Work*, 283–285. Philadelphia: privately published, 1899.

Leffmann, Henry, et al. "The Wagner Free Institute of Science of Philadelphia." In *Founders' Week Memorial Volume etc.*, edited by Frederick P. Henry, 206–213. Philadelphia: F. A. Davis, 1909.

Levin, Beatrice S. *Women and Medicine*. Metuchen, NJ: Scarecrow Press, 1980.

Lovejoy, Esther Pohl. *Women Doctors of the World*. New York: Macmillan, 1957.

Miles, Wyndham D. "Rachel L. Bodley." In *American Chemists and Chemical Engineers*, 38. Washington, DC: American Chemical Society, 1976.

Morantz-Sanchez, Regina Markell. "Women and the profession: The doctor as a lady." In *Sympathy and Science: Women Physicians in American Medicine*, 90–143. New York: Oxford University Press, 1985.

"Obituary: Dr. Rachel L. Bodley." *Philadelphia Ledger and Transcript* (June 10, 1888): 8.

"The Ohio Female College." *The Cincinnatus* 2 (1857): 204–206.

Papers Read at the Memorial Hour: Commemorative of the Late Prof. Rachel L. Bodley, M.D. Philadelphia: Woman's Medical College of Pennsylvania, 1888.

[Rachel L. Bodley elected a permanent vice president]. *The American Chemist* 5 (1874): 38.

Shotwell, John B. "Institutions now closed: Ohio Female College." In *A History of the Schools of Cincinnati*, 544. Cincinnati: School Life, 1902.

Stuckey, Ronald L. "Rachel Littler Bodley." In *Women Botanists of Ohio Born before 1900 with Reference Calendars from 1776 to 2028*, 5–6, 8. Columbus, OH: RLS Creations, 1992.

Willard, Frances E., and Mary A. Livermore. "Miss Rachel L. Bodley." In *American Women: Fifteen Hundred Biographies with over 1,400 Portraits*, vol. 1, edited by Frances E. Willard and Mary A. Livermore, 100–101. New York: Mast, Crowell, and Kirkpatrick, 1897. Reprinted. Detroit: Gale Research, 1973.

Withington, Alfreda B. "Rachel L. Bodley (1831–1888)." In *Dictionary of American Medical Biography*, edited by Howard A. Kelly and Walter L. Burrage, 116–117. New York: Appleton, 1928. Reprinted. Boston: Milford House, 1971.

EMMA LUCY BRAUN (1889–1971)

Ronald L. Stuckey

BIOGRAPHY

Emma Lucy Braun, one of the truly dedicated pioneer ecologists of the first half of the twentieth century, was an original thinker in the disciplines of plant ecology, vascular plant taxonomy, plant geography, and land conservation. Born in Cincinnati, Ohio, on April 19, 1889, Emma Lucy Braun was the younger of two daughters of George Frederick and Emma Moriah (Wright) Braun. She was educated in the Cincinnati schools, where her father was a principal, and her mother was a teacher. Her paternal ancestors were of German-French descent; her maternal ancestors of English origin. Braun's early interest in the natural world, like that of her older sister, Annette Frances (1884–1978), was fostered by their parents, who took the girls to the woods and identified wildflowers. Their mother was especially interested in botany and had prepared a small herbarium collection of dried, pressed plants for study. While in high school, Braun began collecting and drying plants, which was the beginning of her extensive personal herbarium of 11,891 specimens that was assembled throughout her life and now resides in the herbarium at the U.S. National Museum, Smithsonian Institution, Washington, D.C.

Braun's collegiate education was completed at the University of Cincinnati, where she earned the B.A. degree (1910), the M.A. degree in geology (1912), and the Ph.D. degree in botany (1914). During the summer of 1912 she studied under Professor Henry C. Cowles, the eminent plant ecologist at the University of Chicago. Although her botanical adviser was Professor Harris M. Benedict, she received considerable guidance from the University's geologist, Professor Nevin M. Fenneman, one of the masters in this country on geomorphology, physiography, and geological history. Braun also gained much from her contemporaries Merritt L. Fernald and Henry A. Gleason on floristics and phytogeography. Gleason, Edgar N. Transeau, and Frederick L. Clements refined many of her ecological ideas.

Braun never married; she lived with her older unmarried sister, Annette, an entomologist who became internationally known as an authority on *Microlepidoptera* (moths) and the first woman to obtain a Ph.D. degree from the University of Cincinnati. She assisted and shared in all of Lucy Braun's work (Stein 1988). The two sisters lived together in suburban Cincinnati, and they maintained a strict Victorian lifestyle, as exemplified by their parents. Their home and garden, surrounded by mostly undisturbed woods, contained an area, both inside and outside, that they called the "science wing," which served as their laboratory. Their garden was, in fact, an experimental garden, where many rare and unusual plants were grown for closer observation and study.

Lucy Braun's entire teaching and research career was in botany at the University of Cincinnati. Initially, she was an assistant in geology (1910–13) and then became an assistant in botany (1914–17) and advanced to instructor in botany (1917–23), assistant professor of botany (1923–27), associate professor of botany (1927–46), professor of plant ecology (1946–48), and professor emeritus of plant ecology (1948–71). Her three advanced courses were World Botany, a geographical approach to the different kinds of plants that grow on the various continents; Geographical Botany, a study of the vegetational types throughout North America; and Plant Succession, an ecological approach toward an understanding of the different plant communities in the area about Cincinnati. Early retirement from teaching allowed her to conduct research in areas of special interest, including fieldwork and writing that continued nearly to the end of her life.

For her achievements in the field of botany, Braun was awarded many honors. She received the Mary Soper Pope Medal (1952), an award of merit in botany for her intimate knowledge of physiographic processes and a thorough familiarity with the eastern North American flora and its geological history. This award was given by the Cranbrook Institute of Science (Bloomfield Hills, Michigan). She held an honorary life membership in the Cincinnati Wild Flower Preservation Society (1953) and a Certificate of Merit for her contribution to the knowledge of the origin and structure of the eastern North American deciduous forest (1956), presented by the Botanical Society of America. She was cited as one of 69 distinguished American botanists by the Botanical Society of America (1961). She held an honorary life membership in the Ohio Academy of Science (1963). She was awarded an honorary Doctor of Science degree from the University of Cincinnati (1964) and the Eloise Payne Luquer Medal for special achievement in botany (1966), presented by the Garden Clubs of America. Braun received an Ohio Citation for contributions to the understanding of the vascular flora of Ohio, for active efforts in acquiring and preserving natural areas in Ohio, and for authoring three important botanical books (1967), awarded by the Martha Kinney Cooper Ohioana Library Association of Columbus, Ohio.

As a member of the Ohio Academy of Science, she was the first woman elected its president (1933–34), and she was the first woman elected president of the Ecological Society of America (1950) and the first woman listed in the Ohio Conservation Hall of Fame (1971). Braun is commemorated in the names of four plants: *Erigeron pulchellus* var. *brauniae* and *Eupatorium luciabrauniae*, both by Merrit L. Fernald, *Silphium terebinthinaceum* var. *lucy-brauniae* by Julian A. Steyermark, and *Viola x brauniae* by Tom S. Cooperrider.

E. Lucy Braun died of congestive heart failure in her home at the age of 81. She is buried with her parents and sister in Spring Grove Cemetery, Cincinnati. A flat, simple tombstone with the inscription "E. Lucy Braun 1889–1971" marks her burial site.

WORK

The scientific contributions that Braun made were in the young and developing field of plant ecology. Her first published ecological studies were on the area in the vicinity of Cincinnati. These included her Ph.D. dissertation (1914) on the physiographic ecology (published in 1916) and the vegetation of conglomerate rocks (1917), as well as the composition and geographical sources of the local flora (1921). She was the first to describe the vegetation segregate, which she named the "mixed mesophytic forest" ("The physiographic . . ." 1916). Braun was also particularly interested in the unique vegetation of the unglaciated dolomite region in Adams County, Ohio, and published in detail on the subject (1928).

She also wrote on the glacial and postglacial plant migrations, as indicated by relic colonies of southern Ohio (1928). Among these were the prairie communities of Adams County, which she suggested were pre-Illinoian in age. One source of evidence was the drainage pattern in southern Ohio, where the preglacial, mostly north-flowing streams and the preglacial divides were developed previous to Illinoian glaciation. Braun was impressed with the distribution of the prairie communities, as they were most numerous in the uplands and/or dry areas between these pre-Illinoian divides.

Then there followed publications on the floristic affinities and forest composition of the Illinoian Till Plain of southwestern Ohio. She also conducted ecological studies on the forests of the Cumberland Plateau and Mountains of Kentucky, resulting in further publications. Her studies provided the stimulus for writing a monumental book, *Deciduous Forests of Eastern North America* (1950). Her book gives a comprehensive and coordinated account of the entire deciduous forest. It lays the foundation for the measurement and evaluation of all future ecological changes in the hardwood forest. Reprinted three times, this book remains her most remembered and lasting scholarly achievement.

Braun also contributed extensively to the study of floristics and taxonomy of vascular plants. In the 1920s and 1930s she cataloged the flora of the Cincinnati area and compared it with the flora of the same region 100 years earlier. This study (1934), one of the first of its type in the United States, provided a model for comparing the changes in a given flora over a span of time. Her other writings included a key to the deciduous trees of Ohio, articles on selected plants of southwestern Ohio and eastern Kentucky, descriptions of nine plants new to science, and *An Annotated Catalog of the Spermatophytes of Kentucky* (1943). With these contributions as a foundation and the desire to continue advancing knowledge of the flora, she organized (1951) an Ohio Flora Committee within the Ohio Academy of Science. Its objective was to prepare a comprehensive study of the vascular flora of Ohio. Braun chaired the committee and moved the project forward by writing two authoritative books, *The Woody Plants of Ohio* (1961) and *The Monocotyledoneae* [*of Ohio*] (1967).

Knowledge of plant distribution naturally follows from extensive ecological and floristic studies. In her writings, Braun believed that during continental glaciation, the deciduous forest was maintained adjacent to the south of the glacial boundary, rather than being a zone of tundra and boreal forest. She expanded on the theory that the southern Appalachians were the center of the survival of plants during glaciation, and from there the forest communities spread. Her ideas on the origin of the prairie elements within the forest constitute an often overlooked, but innovative, viewpoint. She contended that two eastward migrations of prairie flora occurred during two different xerothermic periods. The first of these was the pre-Illinoian, during which time the dry prairie flora developed in unglaciated southern Ohio (especially Adams County). The second was the Wisconsinan and early post-Wisconsinan, during which time the meso-phytic prairie flora developed in Wisconsinan glacial drift of western Ohio. This phytogeographic knowledge culminated in an extensive analytical summary, "The phytogeography of unglaciated eastern United States and its interpretation" (1955). Although she drew freely from the work of others, her botanical and ecological studies diverged from those of her contemporaries, exhibiting originality in methodology and philosophy.

Braun's research was conducted primarily in the field, where she traveled over 65,000 miles during 25 years of her investigations. She was active in field studies nearly to the very end of her life, and her research miles increased substantially. Beginning in 1915, she made 13 trips to the western United States, five of them to the Pacific Coast. From 1934 to 1963, she drove her own car, and in those times, when she was much younger, no road was too difficult for her to travel. On all of her trips, both eastern and western, she took color slides, made notes on the vegetation, and prepared herbarium specimens. Both her university class lectures and popular talks were of great interest because of the exquisitely colored slides that were shown.

Braun exerted leadership roles in many conservation efforts to save remnants of natural areas, particularly in Adams County, Ohio. She founded the Cincinnati chapter of the Wild Flower Preservation Society (1924); published the first inventory of Ohio's natural areas (1926); contributed articles on native plants and conservation to the magazine *Wild Flower* (1924–36); and edited the Society's national magazine, *Wild Flower* (1928–33).

During the 1940s Braun described four species and four varieties of vascular plants as new to science, all from localities in Kentucky, and one hybrid, a fern, from Adams County, Ohio. At least seven women students are known to have received advanced degrees in botany under the direction of E. Lucy Braun at the University of Cincinnati from 1923 to 1938.

Braun was one of the most original thinkers in the developing field of plant ecology during the first half of the twentieth century in North America. In the field of conservation and the preservation of natural areas, she was a faithful and dedicated worker. As an independent research investigator, she held to firm and strong beliefs, which allowed little deviation for argument and disagreement. Braun was a perfectionist, striving for thoroughness and accuracy with meticulous attention to detail in all of her scientific work. Her book *Deciduous Forests* (1950) was acclaimed by F. R. Fosberg (1951) as a definitive work that "reached a level of excellence seldom or never before attained in American ecology or vegetation science, at least in any work of comparable importance." It remains her most remembered and lasting achievement.

BIOGRAPHY

Works by Emma Lucy Braun

Scientific Works

Space does not permit the listing of the complete works of Emma Lucy Braun. The references given below are only those cited in the text, as well as her dissertation. A complete list of Braun's published papers, both scientific and popular, is provided in Stuckey (1973).

"The Physiographic Ecology of the Cincinnati Region." Ph.D. diss., University of Cincinnati, 1914.

"A northern occurrence of *Dentaria multifida Muhl.*" *Journal of the Cincinnati Society of Natural History* 22 (1916): 16–17.

"The physiographic ecology of the Cincinnati region." *Ohio Biological Survey* 2 (7) (1916): 115–211.

"The vegetation of conglomerate rocks of the Cincinnati region." *Plant World* 20 (1917): 380–392.

"Composition and source of the flora of the Cincinnati (Ohio) region." *Ecology* 2 (1921): 161–180.

"Glacial and post-glacial plant migrations indicated by relic colonies of southern Ohio." *Ecology* 9 (1928): 284–302.

"The vegetation of the Mineral Springs region of Adams County, Ohio." *Ohio Biological Survey* 3 (15) (1928): 375–517.

> An unpublished manuscript, "Vascular plants of the Mineral Springs region of Adams County," is in the library of the Gray Herbarium of Harvard University.

"The Lea Herbarium and the flora of Cincinnati." *American Midland Naturalist* 15 (1934): 1–75.

An Annotated Catalog of Spermatophytes of Kentucky. Cincinnati, OH: privately published. Planographed by John S. Swift, 1943.

> Review: M.L. Fernald. *Rhodora* 45 (1943): 277–278.

Deciduous Forests of Eastern North America. Philadelphia, PA: Blakeston, 1950. Facsimile eds. New York: Hafner, 1964, 1967, 1972.
>   Reviews: V.M. Conway. *Journal of Ecology* 39 (1951): 424–425; F.R. Fosberg. *Scientific Monthly* 73 (1951): 66–67.
"The phytogeography of unglaciated eastern United States and its interpretation." *Botanical Review* 21 (1955): 297–375.
The Woody Plants of Ohio: Trees, Shrubs, and Woody Climbers Native, Naturalized, and Escaped. Columbus: Ohio State University Press, 1961. Facsimile eds., New York: Hafner, 1969; Columbus: Ohio State University Press, 1991.
The Monocotyledoneae [of Ohio]: Cat-tails to Orchids with Grameneae by Clara G. Weishaupt. Columbus: Ohio State University Press, 1967.

Works about Emma Lucy Braun

Cooperrider, Tom S. "Ohio's herbaria and the Ohio Flora Project." *Ohio Journal of Science* 84 (1984): 189–196.
———. *"Viola x brauniae (Viola rostrata x V. striata)." Michigan Botanist* 25 (1986): 107–109.
Cusick, Allison W. "The prairies of Adams County, Ohio: 50 years after the studies of E. Lucy Braun." In *Proceedings of the Sixth North American Prairie Conference*, edited by Ronald L. Stuckey and Karen J. Reese, 56–57. *Ohio Biological Survey, Biological Notes*, no. 15. Columbus: Ohio State University Press, 1981.
Durrell, Lucille. "Memories of E. Lucy Braun." In *Proceedings of the Sixth North American Prairie Conference*, edited by Ronald L. Stuckey and Karen J. Reese, 37–39. *Ohio Biological Survey, Biological Notes*, no. 15. Columbus: Ohio State University Press, 1981.
Lampe, Lois. "E(mma) Lucy Braun." *Ohio Journal of Science* 71 (1971): 247–248. Obituary.
Langenheim, Jean H. "Genealogy of some women plant ecologists, in the path and progress of American women ecologists." *Bulletin of the Ecological Society of America* 69 (4) (1988): 184–197.
Peskin, Perry K. "A walk through Lucy Braun's prairie." *Explorer* 20 (4) (1978): 15–21.
Stein, Lisa K. "The sisters Braun: Uncommon dedication." *Cincinnati Museum of Natural History* 21 (2) (1988): 9–13.
Stuckey, Ronald L. "E. Lucy Braun (1889–1971). Outstanding botanist and conservationist: A biographical sketch, with bibliography." *Michigan Botanist* 12 (1973): 83–106.
———. "Contributors to the development of the Prairie Peninsula concept." In *Proceedings of the Sixth North American Prairie Conference*, edited by Ronald L. Stuckey and Karen J. Reese, 24–36. *Ohio Biological Survey, Biological Notes*, no. 15. Columbus: Ohio State University Press, 1981.
———. "Origin and development of the concept of the Prairie Peninsula." In *Proceedings of the Sixth North American Prairie Conference*, edited by Ronald L. Stuckey and Karen J. Reese, 4–23, including contribution of Emma Lucy Braun, 17–20. *Ohio Biological Survey, Biological Notes*, no. 15. Columbus: Ohio State University Press, 1981.
———. *E. Lucy Braun: Ohio's Foremost Woman Botanist. A Collection of Biographical*

Accounts, Maps and Photos. Columbus, OH: Ronald L. Stuckey, RLS Creations, 1990. Reprinted, 1994.

————. "E. Lucy Braun, foremost woman botanist of Ohio." In *Women Botanists of Ohio: Born before 1900, with Reference Calendars from 1776 to 2028*, 34–39. Columbus, OH: Ronald L. Stuckey, RLS Creations, 1992.

Ware, S. "Polar ordination of Braun's mixed mesophytic forest." *Castanea* 47 (1982): 403–407.

ELIZABETH GERTRUDE KNIGHT BRITTON (1858–1934)

Lee B. Kass

BIOGRAPHY

Elizabeth Gertrude Knight was born on January 9, 1858, in New York City. She was one of five daughters of Sophie Ann (Compton) and James Knight, of Welsh and Scottish ancestry. She attended private elementary school on the island of Cuba, where her family spent time with her grandfather Knight, who ran a furniture factory and sugar estate near Matanzas. While in Cuba, she gained a fluency in Spanish and developed an interest in natural history, encouraged by her father's interests in the plants, animals, and geology of the island (Howe, "Elizabeth . . ." 1934). Beginning at the age of 11, she resided with her maternal grandmother, Compton, in New York City and attended Dr. Benedict's private school. Knight then entered Normal (now Hunter) College, graduating in 1875 at the age of 17. In that same year she was employed by her alma mater as a critic teacher in its Training Department. She held that position until 1883, when she was appointed teacher in botany at the Normal College. After her marriage to Dr. Nathaniel Lord Britton, she became honorary associate instructor of cryptogamic botany at Columbia and Barnard Colleges (becoming part of Columbia University in 1912) and the New York Botanical Garden.

As a teacher, her interest in botany was solidified when she took a field trip to Nova Scotia in 1879, and she discovered the rare curly grass fern, *Schizaea pusilla* Pursh., growing along with the small shoreweed *Littorella lacustris* Linn., on the shore of Grand Lake. The fern had been known in Newfoundland only from an herbarium specimen of Auguste Jean Marie Bachelot De la Pylaie, collected 60 years previously, and its rediscovery confirmed the original collection site (Eaton 1879). Knight's report about the shoreweed was the third finding for that plant in the Americas (Gray 1880). The significance of her findings is evident in the promptly published announcements by D. C. Eaton (1879), America's first pteridologist (fern biologist), and by Asa Gray (1880), foremost American botanist of the time.

Knight's botanical reputation soon led to her election to the Torrey Botanical Club on December 9, 1879. Founded in 1870, the organization was devoted to furthering the botanical knowledge of New York City (Leggett 1870). This group of professional and amateur botanists met regularly at the Columbia College Herbarium. Knight was curator of the club's collections from 1884 to 1885 (Barnhart 1940).

In November 1881, Knight published her first scientific report. Two years later (1883) her interest in mosses (bryology), which began as a young child in Cuba, was professionally recognized following publication of her first bryological contribution. The care that she gave to her research and publications is probably what precipitated the naturalist Dr. John Strong Newberry, professor of geology and mineralogy at Columbia College, to remark that she was "the best botanist in the City" and that her scientific method and logic were beyond those of any other woman he knew (Bonta 1991, 125). That scientific reputation was probably what led to her appointment as teacher of botany at the Normal College in 1883.

On August 27, 1885, at the age of 27, Knight married Dr. Nathaniel Lord Britton, an assistant in geology (soon to become instructor in geology and botany) at Columbia College. He had been a member of the Torrey Botanical Club since 1877, and they may have been introduced via this organization. One of Knight's colleagues at Normal College, William Wood, who considered her the second best botanist in the United States, believed that marriage would end her career (Bonta 1991, 126). Fortunately, his prediction was not fulfilled, and Mrs. Britton, as she was known professionally, proceeded to be scientifically active for the next 45 years. During her lifetime she authored over 300 publications, comprising scientific research on mosses, ferns, flowering plants, and wild flower preservation. Her works included biographies, book reviews, and editorials (Barnhart 1935).

It is unclear if Knight's marriage precluded her continued employment at the Normal College. Rossiter (1982, 15, 93) noted that, at the time, it was well known that "married women were not considered for employment at the early women's colleges . . . ; and any women already on the college faculty who did marry, . . . resigned immediately."

Soon after she married, Mrs. Britton became the unofficial curator of the moss collection at Columbia College, and she worked in the College's Herbarium without remuneration. From 1886 through 1888 she was editor of the *Bulletin of the Torrey Botanical Club*. During that time she also published results of her own research on flowering plants, mosses, and ferns and wrote two book reviews for that journal (Barnhart 1935). In 1888 she became one of three original members appointed to a standing committee of the Torrey Botanical Club to study the local cryptogamia (nonflowering plants), and she remained an active member of that committee for the rest of her life (Barnhart 1940). In 1897, as chair of the division of bryophyta for the (Women's) National Science Club (NWSC), she encouraged women scientists to study mosses (Rossiter 1982, 96).

Mrs. Britton held no official position at Columbia College; however, she acted as "major professor" to Ph.D. degree students in her field (Steere, "North American . . ." 1977). One of her students, Able Joel Grout, who graduated in 1897, became a noted American bryologist. Although Mrs. Britton is credited with the idea of establishing a society for the study of mosses, the society became a reality when Grout proposed "a Chapter for the study of mosses" in the third issue of *The Bryologist, a Department of the Fern Bulletin.* In 1899 the Sullivant Moss Chapter had 34 charter members, 16 of whom were women (Smith 1917).

By 1900 *The Bryologist* was being published independently of the *Fern Bulletin* and was edited, managed, and financed by Mrs. Annie Morrill Smith, secretary-treasurer of the Sullivant Moss Society (Smith 1917). Although Mrs. Britton was not pleased with Grout's initiative, she contributed six papers and one review to that third volume (Barnhart 1935). She served as president of the society from 1916 to 1919.

Another of Mrs. Britton's ideas that Grout pursued was her plan to publish a handbook of mosses, a proposal that she first publicly announced in 1890. In 1900 Grout published his own manual, *Mosses with a Hand-Lens,* followed in 1903 by *Mosses with a Hand-Lens and Microscope.* After her death, Grout admitted that Mrs. Britton "seemed to think [that he] was usurping her prerogatives" (Grout 1935).

It appears that the Brittons supported each other in their botanical endeavors. Probably because they had no children, Mrs. Britton was free to go on botanical expeditions with her husband. Together they visited herbaria and botanic gardens and attended scientific conferences and meetings, where she presented papers on her own research. Her research was primarily independent of her husband's. The few papers that are published in collaboration with others do not include her husband as a coauthor. Additionally, she personally collected, described, and published accounts of the mosses of many of the Antillean islands that they explored together.

While studying at the Royal Botanic Gardens at Kew, England, in 1888, Mrs. Britton proposed having a similar botanical garden in New York City. Upon returning to New York, she described the British Gardens at a meeting of the Torrey Botanical Club. This report prompted the members to organize a committee to consider the establishment of a botanical garden. With the aid of many prominent citizens of New York, the Torrey Botanical Club was responsible for the incorporation of the New York Botanical Garden (NYBG) in 1891. Dr. Britton was appointed director in chief in 1896, and she worked closely with him in designing the gardens and museum building. She worked at his side throughout his 33 years as director of the Garden, while continuing her own independent research.

Once established at the NYBG, she pursued her research on bryophytes and ferns, and for the next three decades she was considered the foremost American bryologist. Some of her colleagues believed that her interests in wild flower preservation, initiated in 1901, may have detracted from her research in bryology (Steere 1971). However, an examination of her publications in that area reveals over 170 research contributions on mosses alone, along with valuable treatments on the ferns (Barnhart 1935).

When the Columbia College Herbarium was transferred to, and incorporated into, the herbarium of the NYBG in 1899, Mrs. Britton continued as unofficial curator of its moss collection. Her efforts were finally recognized in 1912, when she was officially appointed honorary curator of mosses, a position she held without salary until her retirement.

In 1893 Mrs. Britton's scientific abilities were rewarded when she was the only woman nominated to be one of 25 charter members of the Botanical Society of America (BSA). This organization's membership was "restricted to those who have published worthy work and are actively engaged in botanical investigation" (Tippo 1956). To comply with those guidelines, at the first annual meeting of the BSA in 1895, Mrs. Britton was one of eight elected members who "presented papers." Although Mrs. Britton was very visible with her botanical investigations and continued to present and publish her work, Rossiter (1982) concluded that she was "conditioned enough to know her place and be voluntarily invisible" at the BSA "professional" banquets. In 1906 Britton advised her good friend Annie Morrill Smith of the group's unwritten prohibition on female attendance at their dinners, and by 1913 she wrote that she had still not attended a BSA dinner (Rossiter 1982, 85–86).

In 1905, while attending the Botanical Congress of Vienna, she was appointed to a committee to decide on moss nomenclature for the next Botanical Congress in 1910 (Bonta 1991, 129). She received additional recognition in 1906, when she was one of 19 women "starred" in the first edition of *American Men of Science* (AMS). This starred recognition, chosen by a select group of scientists, was given to the top 1,000 scientists listed in that volume. Mrs. Britton was starred in all five editions of AMS (1906–33).

Mrs. Britton is considered the pioneer in the movement for the protection of America's native plant life. Soon after the NYBG opened, she proposed establishing a fund for the preservation of native plants. She interested Olivia and Caroline Phelps Stokes in her idea, and in 1901, the Stokes sisters presented $3,000 to the Garden to be "used for the investigation and preservation of native plants, or for bringing the need of such preservation before the public." It was Mrs. Britton's idea to apply some of those funds toward essay competition prizes on the preservation of wild plants and for publishing and distributing those essays. The distribution of the first winning essay influenced the establishment of the Wild Flower Preservation Society of America, organized in 1902 by Mrs. Britton and other distinguished American botanists. Mrs. Britton, who was considered the motivating force behind that organization, was elected to the Board of Managers and soon became secretary, then secretary-treasurer. She contributed to the public's awareness of endangered wild flowers through her many published papers, lectures to schools and clubs, and correspondences. As a result of her activities, branch societies for wild flower preservation were established, conservation activities in schools and garden clubs were developed, and various states passed laws for the protection of native flora (Gager 1940).

Elizabeth Britton died of a stroke on February 25, 1934, and was buried in Moravian Cemetery on Staten Island, New York (Steere 1971). She was awarded many honors during her lifetime and posthumously. Many species of plants were named in her honor. In 1934 a double peak at the La Mina recreational area in Puerto Rico was named Mt. Britton in honor of the Brittons. Shortly after her death, the Board of Managers of the NYBG designated the entire moss collection of the garden as the Elizabeth Gertrude Britton Moss Herbarium (Barnhart 1940). Her husband died four months after the death of Mrs. Britton.

In May 1940, six years after her death, the New York Bird and Tree Club honored her memory with an all-day program devoted to the subject of wild flower preservation. During that event they dedicated a plaque to Mrs. Britton in the newly established Wild Flower Garden and presented $1,800 to establish the Elizabeth Gertrude Britton Fund for the "installation and maintenance of a Wild Flower Area . . . and for the development of interest in wild flowers and their preservation" ("Mrs. Britton . . ." 1940). To commemorate the 50th anniversary of their death, the April–June 1984 issue of *Brittonia* was dedicated to the Brittons, and a summary of Mrs. Britton's major contributions, including her influence on the establishment of the NYBG, was described (Slavick 1984).

Elizabeth Gertrude Knight Britton was a critically thinking, independent person. Perhaps her experience as a critic teacher qualified her for the careful and scrupulous reviews she gave to her own work and that of others. Unfortunately, her rigorous standards may have been what prompted some of her male colleagues to attack her in patronizing terms that are generally used to describe children and women: "spoiled," "over-privileged," "outspoken," "strong minded" (Steere, "North American . . ." 1977).

Enlightened scholars may interpret her as an enthusiastic person with ideas that catalyzed others. Her legacy remains in the establishment of the American Bryological and Lichenological Society, the NYBG, the Olivia and Caroline Phelps Stokes Fund for the preservation of native plants, and the Wild Flower Preservation Society of America Inc. and in her many publications about nonflowering and flowering plants.

At the request of N. L. Britton, John Hendley Barnhart compiled a list of 346 publications identified as Elizabeth Britton's work, which he published in January 1935.

WORK

In November 1881, Elizabeth Knight published her first botanical paper, a brief note entitled "Albinism." There she reported finding atypical plants with white flowers on *Pontederia cordata* L. (pickerelweed) among the plant's ordinarily blue flowers. Two years later (1883) Knight published her initial paper on mosses, which was the beginning of a series of contributions on them that were published from 1883 to 1930 (Barnhart 1935). Her report was significant because it was the first description of the reproductive capsule (fruit) of the sword moss, *Eustichium Norvegicum* Br. Eu. [=*Eustichium norvegicum* Brid.], which she had collected in Wisconsin and which had been previously known only in its sterile condition. She presented six illustrations with her description and carefully cited all known references to the plant. Her finding was the first of only two instances where the moss had been reported in its reproductive state (Lange 1981).

After moving to the herbarium at Columbia College in 1885 and becoming its unofficial curator of mosses, she worked actively to expand significantly its holdings and its reputation, by exchanges, purchases, and her personal field collections (Merrill 1934). She continued her research on mosses, and while she was editor of the *Bulletin of the Torrey Botanical Club* (1886–88), she published two original descriptions and wrote one review on mosses. One year later she began her series of 11 systematic papers on mosses titled "Contributions to American Bryology," which appeared between 1889 and 1895. The first two of these papers (1889, 1891) were enumerations of mosses collected by J. B. Leiberg. Mrs. Britton prepared and distributed sets of Leiberg's specimens (exsiccatae) from the Columbia College Herbarium beginning in 1890, although her name does not appear on the labels (Sayre 1971). Her other "Contributions" from that series included systematic revisions of numerous moss genera and descriptions of many new species of mosses (Barnhart 1935).

While at Columbia College in 1890, Mrs. Britton published a preliminary list of the mosses she had collected on Staten Island and a list of mosses she had collected in southwestern Virginia. In 1890 she announced her proposal to write a *Handbook of Mosses of Eastern America* in the *Bulletin of the Torrey Botanical Club*, as well as at the 1892 annual meeting of the American Association for the Advancement of Science (AAAS). To that organization she presented the general plan of her work and exhibited drawings that had been prepared to illustrate the book. Six years later, for the first issue of the "Moss Department," she gave Grout (1898) permission to publish one drawing from her manuscript. In 1892 she also issued her list of the mosses of West Virginia (Barnhart 1935).

The following year, at the meeting of the Botanical Club of the AAAS, Britton gave a presentation describing her success in obtaining the August Jaeger collection of mosses from Geneva, Switzerland, for the Columbia College Herbarium. That collection was a valuable addition to the bryological resources of America. Mrs. Britton also published a description of the Jaeger moss herbarium in August 1893 (Barnhart 1935).

Elizabeth Britton's moss book was never published. From 1894 through 1895, however, she did publish eight illustrated articles on "How to study the mosses" in *The Observer*. These were followed in 1896 by five additional popular articles on mosses in that same journal (Barnhart 1935). The articles popularized the study of mosses and stimulated serious amateurs to collect and study bryophytes.

In addition to encouraging amateurs in the study of mosses, Mrs. Britton also worked with, and guided, botanical scholars. Her most noted students from her association with Columbia College were J. F. Collins, A. J. Grout, and Marshall A. Howe. After her death the latter two wrote accounts of her life and work (Grout 1934; Howe, "Elizabeth Gertrude Britton" 1934). Regarding their interests in bryophytes, Britton advised and corresponded with many noted botanists.

Between 1898 and 1925, Mrs. Britton offered moss specimens for distribution, wrote many book reviews, and published original observations on mosses in the *Bryologist*. Her reputation as a bryophyte taxonomist was earned, in part, by her 12 publications on moss nomenclature appearing in the *Bryologist* between 1903 and 1914 (Barnhart 1935).

In 1906, she arranged for the moss and liverwort collection of the English bryologist William Mitten to be purchased for the NYBG herbarium. The significance and importance of the collection were discussed anonymously by her in 1907 (Barnhart 1935). Her studies on moss systematics often led to her borrowing specimens, including types, from other herbaria. She frequently took pieces of these specimens for her collections and left notes in the borrowed packets indicating that she had done so (Crum 1994). Mrs. Britton's separated pieces, known as "kleptotypes," are very valuable since many of the herbaria from which she took specimen pieces have been destroyed; therefore her segregates are the only known surviving type material for these taxa (Crum 1994).

Britton continued her research on mosses and contributed systematic treatments of eight families of mosses to the *North American Flora*, published in 1913. Six of these studies were published independently, and two were published with coauthors. Between 1914 and 1915 she collected and described the mosses of the Danish West Indies (now U.S. Virgin Islands), the Virgin Islands (British), and Bermuda. In 1918 her treatment of the Bermuda mosses appeared in N. L. Britton's *Flora of Bermuda*, and in 1920 her collections and descriptions of the Bahama mosses were published in N. L. Britton and C. F. Millspaugh's *The Bahama Flora*. She published additional accounts of Bahama mosses in 1921, of Cuban mosses in 1922, and of Puerto Rican mosses in 1924. The last account of her moss work was the "Report of the Honorary Curator of Mosses," published in 1930, just four years before her death (Barnhart 1935).

While working at the Columbia College herbarium, Elizabeth Britton and her husband identified and described the South American collections of Henry Hurd Rusby, who had been employed by the pharmaceutical firm of Parke, Davis and Company to collect drug plants there between 1885 and 1887 (Rossi-Wilcox 1993). Mrs. Britton studied the cryptogams (mosses and ferns), and Dr. Britton investigated the phanerogams (flowering plants). To aid in their identifications of Rusby's specimens, the Brittons both worked at the Royal Botanical Gardens at Kew, England (July through September 1888), and Mrs. Britton studied mosses at the Linnaean Society of London, where she was restricted to working upstairs, since women were not admitted to the main floor (Bonta 1991, 126).

On returning to New York, she published an enumeration of Rusby's South American ferns (1888). In that paper she described a new species of fern, *Acrosticum Eatonianum*, naming it after Daniel Cady Eaton, who had identified it for her as a new species and who had been the first person to announce Elizabeth Knight's finding of *Schizaea* when she was 17 years old (Eaton 1879). She delayed publication of an enumeration of Rusby's South American mosses until 1896 because of the difficulty she had with the material.

In 1895 she reported on the ferns collected in Bolivia by Miguel Bang, and continuing her research on ferns, she published a revision of the Adder's-tongue ferns (*Ophioglossum*) in 1897. The latter was the fifth paper she published out of 16 investigations on ferns that appeared between 1888 and 1923 (Barnhart 1935).

Mrs. Britton's love of flowering plants is not only evident in her publications on wild flower preservation but also demonstrated by the many research articles she published on higher plants and the plant lists and herbarium specimens she helped to prepare (Vail 1890). As indicated earlier, her first scientific publication described albinism in flowers on pickerelweed, while her other studies ranged from embryo development to relations of plants and animals. Her first articles on wild flower preservation appeared in 1901, soon after the opening of the NYBG, and her publications continued until ill health prevented her from working. Between 1912 and 1929 she published a series of 14 articles that included color illustrations on "Wild plants needing protecting" in the *Journal of the New York Botanical Garden*. The plants she described in those articles ranged from herbaceous spring flowers to flowering shrubs and trees. These articles represented the most noted of her more than 45 papers published on the subject between 1901 and 1929. The papers Elizabeth Britton published between 1923 and 1925 on protecting wild plants used as Christmas decorations culminated in her successful national boycott against using wild American holly (*Ilex opaca*) for such purposes (Barnhart 1935; Gager 1940; Steere 1971).

Since Barnhart's compilation of Britton's work in 1935, additional reports of her anonymous contributions have been published (Sayre 1971, 1975; Slack 1987). Many authors have referred to the work of Elizabeth Britton as being accomplished in her "leisure" or "spare time," possibly because they believed that as a woman she could not have wanted to engage in science as a career.

BIBLIOGRAPHY

Works by Elizabeth Gertrude Knight Britton

Scientific Works

Space does not permit the listing of the complete works of Elizabeth Gertrude Knight Britton. This list excludes all works of Britton that are cited in Barnhart (1935) with the exception of those cited in the text. Additional contributions not listed by Barnhart appear in the following section as references by Grout ("Moss . . ." 1898); Peck (1899); Sayre (1971, 1975); and Vail (1890).

"Albinism." *Bulletin of the Torrey Botanical Club* 8 (1881): 125.

"On the fruit of *Eustichium Norvegicum* Br. Eu." *Bulletin of the Torrey Botanical Club* 10 (1883): 99–100.

"An enumeration of the plants collected by Dr. H. H. Rusby in South America. 1885–1886.—III. Pteridophyta." *Bulletin of the Torrey Botanical Club* 15 (1888): 247–253.

"Contributions to American bryology.—I. An enumeration of mosses collected by Mr. John B. Leiberg in Kootenai [*sic*] Co., Idaho." *Bulletin of the Torrey Botanical Club* 16 (1889): 106–112.

"Contributions to American bryology.—II. A supplementary enumeration of the mosses collected by Mr. John B. Leiberg in Idaho, with descriptions of two new species." *Bulletin of the Torrey Botanical Club* 18 (1891): 49–56.

"An enumeration of the plants collected by Dr. H. H. Rusby in Bolivia, 1885–1886.—II. Musci." *Bulletin of the Torrey Botanical Club* 23 (1896): 471–499.

Works about Elizabeth Gertrude Knight Britton

Bailey, L. H. "Species Batorum. The Genus *Rubrus* in North America. V. Flagellares." *Gentes Herbarum* 5 (1943): 231–422.

Barnhart, J. H. "Elizabeth Gertrude Knight Britton as a scientist." *Journal of the New York Botanical Garden* 41 (1940): 142–143.

———. "The published work of Elizabeth Gertrude Britton." *Bulletin of the Torrey Botanical Club* 62 (1935): 1–17.

    A chronological and indexed list of the 346 signed and anonymously published, identifiable works of Elizabeth Gertrude Knight Britton.

Bonta, M. M. "Elizabeth Gertrude Knight Britton, mother of American bryology." In *Women in the Field*, 124–131. College Station: Texas A & M University Press, 1991.

"Botanical Society of America." *Botanical Gazette* 20 (1895) : 403–405.

Cheney, L. S. "Fruiting *Eustichia Norvegica* Brid." *Botanical Gazette* 19 (1894): 384.

Crum, R. A. Personal communications, 1994.

Dirig, R. Personal communications, 1994.

E[aton], D. C. "*Schizaea pusilla*, Pursh." *Bulletin of the Torrey Botanical Club* 6 (1879): 361.

    First report of Elizabeth Knight's finding of this rare fern.

Gager, C. S. "Elizabeth G. Britton and the movement for the preservation of native American wild flowers." *Journal of the New York Botanical Garden* 41 (1940): 137–142.

Gray, A. "*Littorella* and *Schizaea* in Nova Scotia." *Botanical Gazette* 5 (1880): 4.

Grout, A. J. "The Bryologist." *Fern Bulletin* 6 (1898): 85–92.

———. "Elizabeth Gertrude (Knight) Britton." *Bryologist* 38 (1935): 1–3.

———. "Moss Department." *Fern Bulletin* 6 (1898): 17–20.

Hooker, B. F., and H. M. Williams. "Memorial and resolution." *Journal of the New York Botanical Garden* 35 (1934): 103–104.

Howe, M. A. "Elizabeth Gertrude Britton." *Journal of the New York Botanical Garden* 35 (1934): 97–103.

————. "Nathaniel Lord Britton 1859–1934." *Journal of the New York Botanical Garden* 35 (1934): 169–180.

Humphrey, H. B. "Elizabeth Gertrude Knight Britton, 1858–1934." In *Makers of North American Botany*, 38–39. New York: Ronald Press, 1961.

Lange, K. I. "Botanists and naturalists at Wisconsin Dells in the 19th century." *Michigan Botanist* 20 (1981): 37–43.

Macoun, J. *Autobiography of John Macoun, M. A. Canadian Explorer and Naturalist.* Ottawa: Ottawa Field-Naturalist's Club, 1922. (Memorial vol., 1979).

Merrill, E. D. "The Elizabeth Gertrude Britton Moss Herbarium is established." *Journal of the New York Botanical Garden* 35 (1934): 210–211.

"Mrs. Britton honored in dedication of plaque by New York Bird and Tree Club in Wild Flower Garden." *Journal of the New York Botanical Garden* 41 (1940): 129–137.

"News and Notes. The organization of the Botanical Society of America . . ." *Botanical Gazette* 19 (1894): 338.

"NYBG's visionary team: The Brittons." *The New York Botanical Garden* [Newsletter] 18 (1985): 5.

"Obituary notice, Elizabeth G. Knight Britton." *Bryologist* 37 (1934): 64.

"Papers presented to the Botanical Club of the AAAS." *Botanical Gazette* 17 (1892): 291–295.

Peck, C. H. "Plants of North Elba." *Bulletin of the New York State Museum* 6 (1899): 65–226.

"Porto [*sic*] Rico honors memory of Dr. Britton." *Journal of the New York Botanical Garden* 36 (1935): 16–17.

"Proceedings of the Botanical Club, AAAS, Madison meeting." *Botanical Gazette* 18 (1893): 342–349, 368.

Rossiter, M. W. *Women Scientists in America, Struggles and Strategies to 1940.* Baltimore: Johns Hopkins University Press, 1982.

Rossi-Wilcox, S. M. "Henry Hurd Rusby: A biographical sketch and selectively annotated bibliography." *Harvard Papers in Botany* 4 (1993): 1–30.

Rudolph, E. D. "Bryology and lichenology." In *A Short History of Botany in the United States*, edited by J. Ewan, 89–96. New York: Hafner, 1969.

Sayre, G. "Cryptogamae exsiccatae: An annotated bibliography of published exsiccatae of algae, lichens, hepaticae, and musci. IV. Bryophyta." *Memoirs of the New York Botanical Garden* 19 (1971): 175–276.

————. "Cryptogamae exsiccatae: An annotated bibliography of exsiccatae of algae, lichens, hepaticae, and musci. V. Collectors." *Memoirs of the New York Botanical Garden* 19 (1975): 277–423.

"Section G (Botany) AAAS, Columbus meeting." *Botanical Gazette* 28 (1899): 207–210.

Slack, N. G. "Charles Horton Peck, bryologist, and legitimation of botany in New York State." *Memoirs of the New York Botanical Garden* 45 (1987): 28–45. Quotes letters between E. G. Britton and C. H. Peck.

Slavick, A. "Elizabeth Gertrude Knight Britton (1858–1934)." *Brittonia* 36 (1984): 96–97.

Smith, A. M. "The early history of the *Bryologist* and the Sullivant Moss Society." *Bryologist* 20 (1917): 1–8.

Stafleau, F. A., and R. S. Cowan. *Taxonomic Literature: A Selective Guide to Botanical*

Publications and Collections with Dates, Commentaries and Types, vol. 1: A-G. 2d ed. Utrecht: Bohn, Scheltema and Holkema, 1976.

Many references are incorrectly cited, making them difficult to locate.

Steere, W. C. "Britton, Elizabeth Gertrude Knight." In *Notable American Women, 1607–1950: A Biographical Dictionary*, vol. 1: *A-F*, edited by E. T. James, 243–244. Cambridge, MA: Belknap Press of Harvard University Press, 1971.

———. "Introduction." In "New combinations and new taxa of mosses proposed by Nills Conrad Kindberg" by W. C. Steere and H. A. Crum, *Memoirs of the New York Botanical Garden* 28 (1977): 1–220.

———. "North American muscology and muscologists, a brief history." *Botanical Review* 43 (1977): 285–343.

Extensive bibliographical and biographical list of references.

Tippo, O. "The early history of the Botanical Society of America." *American Journal of Botany* 43 (1956): 852–858.

Vail, A. M. "Notes on the spring flora of southwestern Virginia." *Memoirs of the Torrey Botanical Club* 2 (1890): 27–53.

Includes a list of "Virginian Plants. Collected by others and Mrs. N. L. Britton [E. G. Britton], May 30th-June 9th, 1890," not listed by Barnhart 1935.

Other References

Grout, A. J. "The Bryologist." *Fern Bulletin* 6 (1898): 61–68.

———. *Mosses with a Hand-Lens*. New York: privately published, 1900.

———. *Mosses with a Hand-Lens and Microscope*. New York City: privately published, 1903.

———. "[Preface]." *Bryologist* 3 (1900): 1.

Leggett, W. H. (ed.). "Prefatory." *Bulletin of the Torrey Botanical Club* 1 (1870): 1.

Patterson, P. M. "Our society's new name: The American Bryological Society." *Bryologist* 52 (1949): 28.

RACHEL LOUISE CARSON (1907–1964)

Randy Moore

BIOGRAPHY

Rachel Louise Carson was born on May 27, 1907, in the coal-mining town of Springdale, Pennsylvania. Her parents were Robert Warden and Maria Frazier (McLean) Carson. Rachel Carson was the youngest of three children, ten years younger than her sister, Marian, and eight years younger than her brother, Robert.

Carson's mother was the daughter of a Presbyterian minister. She graduated from Washington Female Seminary in Washington, Pennsylvania, where she had studied the classics. She loved music and books, appreciated nature, and had a brief career as a schoolteacher before marrying in 1894. Her mother was 37 years old when Rachel Carson was born. She lived with her daughter for nearly 50 years and was the strongest influence on her life. Rachel Carson's father, in addition to helping run the small family farm, sold insurance and real estate.

A shy girl, Carson spent much of her time studying nature in the woods near her house. She missed many days of school at Springdale Grammar School because of illness, but her mother tutored her at home so she could keep up with her work. The Carson family operated a small farm, and it was her responsibility to feed the chickens. As a child, Carson displayed great affection for animals. That affection remained intense for the rest of her life. Carson was especially fond of animal stories, such as *The Tale of Peter Rabbit*, written and illustrated by Beatrix Potter.*

When she was ten years old, Rachel Carson wrote a story entitled "A Battle in the Clouds" (about a World War I airplane battle). It was accepted for publication in *St. Nicholas*, a popular magazine for children. The article won a prize, the Silver Badge, which Carson said brought her more joy than all the checks she later received for her best-selling books. Two other stories were bought for publication by *St. Nicholas* for just over three dollars (at the rate of a penny a word). Carson also wrote poetry, but her poems were never published.

In 1921 Carson enrolled in Springdale High School. Because Springdale High ended after the tenth grade, Carson completed grades 11 and 12 at nearby Parnassus High School in New Kensington. There she was admired by her classmates.

Encouraged by her publications in *St. Nicholas*, Carson decided to study writing in college. Aided by an annual scholarship applied to her tuition, she enrolled in the Pennsylvania State College for Women (later called Chatham College) in Pittsburgh in 1925. There she became an English major, joined the Omega literary club, and became a member of the staff of *The Arrow*, a student paper. At the end of her sophomore year, Carson took a required course in biology taught by Mary Scott Skinker. Skinker became a significant influence on Carson, who changed her major to biology in the middle of her junior year. In her senior year Carson was elected president of the science club. She was awarded a B.A. degree magna cum laude in 1929.

The following summer Carson worked at the Marine Biological Laboratory at Woods Hole, Massachusetts, where she saw the ocean for the first time at age 22. She then enrolled at Johns Hopkins University, where she worked part-time as an assistant in the laboratory of geneticist Raymond Pearl. Carson studied under the direction of Rheinart P. Cowles of the zoology department. (Cowles later helped Carson get a teaching job at the University of Maryland.) The subject of her research was kidney development in channel catfish. In 1931 Carson earned an M.A. degree in marine zoology from Johns Hopkins University. Her 108-page thesis was titled "The Development of the Pronephros during the Embryonic and Early Larval Life of the Catfish (*Ictalurus punctatus*)." Carson never married.

From 1931 to 1936 Carson taught at Johns Hopkins and the University of Maryland. During this time, she published articles in the *Baltimore Sun*. When her father died in 1935, leaving little money to support her mother and herself, Carson accepted a part-time job with the U.S. Bureau of Fisheries (now the U.S. Fish and Wildlife Service). There she wrote short radio scripts called "Romance under the Waters" (the staff called these scripts "Seven-Minute Fish Tales"), for which she was paid $19.25 per week.

In 1936 Carson's sister died, leaving two school-age children, Marjorie and Virginia. The children came to live with Carson. Fortunately, an opportunity arose for a full-time junior aquatic biologist at the U.S. Bureau of Fisheries. Carson, who was the only woman competing for the position, obtained the highest score on the civil service examination and was hired in August 1936.

Carson's shyness did not hinder her upward mobility in the U.S. Fish and Wildlife Service. She started as an assistant aquatic biologist in 1942 and was promoted through the ranks to biologist and chief editor (1949). Because she was awarded a Guggenheim Fellowship (1951), she was able to obtain a year's leave of absence. Financially secure because of her book sales, Carson left her job (1952) and built a one-story summer cottage by the sea at West Southport, Maine. She worked as a professional writer for the rest of her life. When her niece died in 1957, Carson adopted her nephew Roger Christie and thus became mother to a five-year-old.

Carson won many awards, including the Audubon Medal of the National Audubon Society (the first to a woman), the conservation award of the Izaak Walton League of America, the Cullum Medal of the American Geographical Society, and the Conservationist of the Year Award of the National Wildlife Federation. She was elected to membership in the American Academy of Arts and Sciences and also won the Schweitzer Medal of the Animal Welfare Institute (1963). She was especially proud of this award, because she had dedicated *Silent Spring* to Albert Schweitzer, the French theologian, medical missionary to Africa, and Nobel laureate who pioneered the "reverence of life" concept that Carson embraced. In 1952 Pennsylvania College for Women awarded Carson the degree of Doctor of Literature; Oberlin gave her a Doctor of Science degreee that same year.

The success and controversy generated by *Silent Spring* and her other books produced many invitations for Carson to speak before various groups. However, by 1961 she was becoming increasingly concerned about her health. She had been diagnosed with cancer in 1957, was treated again in 1958, and had a radical mastectomy in 1960. However, she was not told that her condition was malignant. Upon discovering this in 1960, Carson began receiving radiation treatment that often left her too sick to accept invitations to speak.

In 1963 Carson testified before the Senate Committee on Commerce concerning the regulation of pesticides. Her recommendation for an independent agency at the executive level, free from political control or influence by the chemical industry, was the cornerstone for the creation of the Environmental Protection Agency in December 1970. After testifying, Carson left Washington for what would be her last summer in Maine, near the sea she loved so deeply. Carson died on April 14, 1964, at the age of 56, in Silver Spring, Maryland.

After her death, Carson's work continued to impress and influence people. For example, in 1969 the Coastal Maine Wildlife Refuge was renamed the Rachel Carson Natural Wildlife Refuge. In 1981 the U. S. Postal Service issued a stamp bearing her name. Sixteen years after her death (1980) Carson was awarded the nation's highest civilian honor, the Presidential Medal of Freedom. That medal was presented by President Jimmy Carter to Roger, her adopted son.

WORK

Soon after accepting the job at the U.S. Bureau of Fisheries, Carson was asked by her supervisor to write something general about the sea. Her supervisor thought that her essay was not appropriate for the Bureau of Fisheries but that the *Atlantic Monthly* might be interested. Carson's article, entitled "Undersea," described how animals live in the sea (Carson 1937). Soon after its publication, Carson began working on *Under the Sea Wind: A Naturalist's Picture of Ocean Life,* her first book. This work, derived from "Undersea," was dedicated to her mother. Carson viewed the writing as a perfect marriage between her interests in biology and writing. She wrote virtually all of *Under the Sea Wind* at night after work. It took her three years to finish, and it was published on November 1, 1941. In contrast to the factual monographs common at the time, the book was a highly romanticized, anthropomorphic portrayal of the lives and habits of many aquatic species. Most important, Carson showed how these organisms are interrelated and dependent on the sea around them. Although critics and scientists praised *Under the Sea Wind* (it was a selection of the Scientific Book Club), sales were disappointing because World War II took up more interest.

During World War II Carson worked in Washington and Chicago. There she wrote pamphlets to help people plan wartime menus using fish, while at the same time preventing overconsumption of certain edible fish. These pamphlets included detailed descriptions of the behavior, anatomy, and habitat of 26 varieties of shellfish and fish. Carson soon became editor and then editor in chief of the U.S. Fisheries and Wildlife Service publications, including a series of booklets called *Conservation in Action*. Carson also considered writing a book about evolution, but she never did. In 1945 her suggestion to *Reader's Digest* that they publish an article about the effects of DDT (dichlorodiphenyltrichloroethane) on the natural world was rejected.

In 1949 Carson won a Eugene F. Saxton Memorial Fellowship, which enabled her to work on her second book, *The Sea around Us*. While writing that book, Carson went on a ten-day voyage aboard the *Albatross III*, a fishing trawler converted to a research vessel. That voyage was to map the ocean floor and do a census of fish off George's Bank, a commercial fishing area south of Nova Scotia. Carson was the first woman to go on that voyage; each of the 50 men who worked aboard the ship had to sign an agreement that enabled Carson to accompany them on their voyage. While on the cruise, Carson was impressed by the contours and canyons of the ocean floor and realized that the Earth was largely a water world dominated by oceans.

The Sea around Us, which was typed by Carson's mother, was published in 1951. Throughout the book Carson portrays Earth as a balanced planet where all is interconnected. She tells the ocean's life story as if the sea were a person. The tone throughout *The Sea around Us* illustrates her great reverence for the ocean. In contrast to *Under the Sea Wind*, which tells of fishermen intruding on the lives of fish and shore-dwelling birds, *The Sea around Us* concentrates on how scientists developed an understanding of the ocean.

The Sea around Us was extremely successful, both critically and commercially. It arrived in bookstores in July and was on the best-seller list by August. In its first year, *The Sea around Us* had 11 printings. *Yale Review* bought and published the chapter entitled "The birth of an island." In 1950 that article won the George Westinghouse Science Writing Award for the best science writing in a magazine. *The Sea around Us* also won the John Burroughs Medal for natural history writing (given by scientists) and the National Book Award (given by publishers). Although Carson did not approve a script for a movie made about *The Sea around Us* because of its scientific errors, the movie did win an Oscar in 1953 as the best full-length documentary film. *The Sea around Us* was a Book-of-the-Month alternate and was also condensed in *Reader's Digest*. Eventually, *The Sea around Us* was published in 33 languages and remained on the best-seller list for 86 weeks. The success of *The Sea around Us* prompted a reissue in 1952 of *Under the Sea Wind*. That book quickly joined *The Sea around Us* on the best-seller list, thereby ending Carson's financial worries.

In 1955 Carson published *The Edge of the Sea*. Although this book was meant to be a guide to seashore life on the Atlantic Coast, it was actually a personal essay that helped readers understand the lives of the coastline's inhabitants. *The Edge of the Sea*, like *The Sea around Us*, portrayed a vivid account of the biology, chemistry, and physics of the ocean and its shores. Its lyrical and precise language contained none of the melodrama of *Under the Sea Wind*.

The Edge of the Sea was another best-seller; parts were published in the *New Yorker*. Carson wrote a script for a television show based on the book but steadfastly resisted lucrative offers to endorse a variety of products. She loved the sea because of its mystery and beauty. As it had for Herman Melville and Joseph Conrad, the sea became Carson's focus and ultimately her greatest symbol.

In 1956 Carson wrote an article entitled "Help your child to wonder." This article showed the importance of encouraging awareness of nature in children. It emphasized that one does not need the knowledge of a scientist, but simply the eyes of an observer, to look for beauty. "Help your child to wonder" resulted from Carson's experiences with her nephew, who visited her at Southport each summer. She had wanted to expand the article into a book but never did. A year after her death, in 1965, the text was reprinted in book form as *The Sense of Wonder*, containing handsome photographs.

Carson's final book was the monumental *Silent Spring*, published in 1962. She had worked on *Silent Spring* since 1958, the year that her mother died. Many people had warned Carson against writing the book; few thought that anyone would be interested in a book about a topic as dreary as pesticides. However, they all proved to be wrong. Few books have had a greater impact than *Silent Spring*. It is one of the most important books of the twentieth century.

Carson wrote *Silent Spring* in response to a letter that she received in January 1958 from her friend Olga Huckins, describing how a small part of the world had been made lifeless by pesticides. Carson responded, "There would be no peace for me if I kept silent." She wrote in *Silent Spring*, "As cruel a weapon as the cave man's club, the chemical barrage has been hurled against the fabric of life."

The book, containing 350 pages (including 55 pages of references), showed readers that the indiscriminate and unrestricted use of pesticides (such as DDT) could cause widespread biological destruction. Today, the public is so well aware of these facts that it is hard to remember that we were once ignorant of them. Carson was the first person to teach us an ecology lesson.

As expected, Carson's ideas were attacked by those who would profit from the continued use of pesticides, including leaders from the chemical industry, agriculture groups, the Nutrition Foundation, and the American Medical Association. Most magazines, fearing lost advertising income, refused to publish excerpts of the book, despite Carson's fame. Farmers were furious, and chemical companies spent large amounts of money to attempt to discredit the author. One manufacturer of canned baby food even claimed that her ideas would cause "unwarranted fear" to mothers who used their product. Carson and her book were attacked as biased, overemphatic, scientifically unsound, and emotionally motivated.

The assaults on Carson were often inaccurate, disparaging, unscientific, distorted, and personal. Because she was female, many of the criticisms directed at her utilized stereotypes of her sex (e.g., "priestess of nature," an eccentric spinster who liked birds better than children). However, her testimony before congressional committees and her eloquent, fact-based prose were convincing. Indeed, her conclusions were supported by many experts, as well as a committee of the Office of Science and Technology, a special panel of President Kennedy's Science Advisory Committee. That panel's report vindicated Carson and stimulated the government to act against pollution.

Silent Spring, which became another best-seller, had literary interest as well as social impact. Parts of the book were published in three issues of the *New Yorker*. When Carson's interview was televised on "CBS Reports" (narrated by Eric Sevareid), two major corporations withdrew their sponsorship.

Silent Spring changed history by forcing readers to reconsider their ideas about the world. It helped show that humans must act responsibly with pesticides and awakened people to the danger of poisoning the Earth. *Silent Spring* soon led to a ban on the use of DDT in the United States and also produced revolutionary changes in laws affecting the environment.

More than any book published in the twentieth century, *Silent Spring* showed that an individual can change the course of history by changing how one views the world. *Silent Spring* helped launch the movement that made ecology a household word. In 1994, *Silent Spring* was reissued with a preface by Vice President Al Gore. At the request of *Newsweek*, Bruce Peterjohn of the North American Breeding Bird Survey reported that of the 40 species described as near extinction by Carson, 19 have had stable populations since 1966 (Easterbrook 1994). Included in the list of survivors were the cardinal, the bald eagle, various woodpeckers, and other endangered species. The fact that Carson's predictions encouraged environmental protection, so that her fears were not realized, is one of her greatest achievements.

BIBLIOGRAPHY

Works by Rachel Louise Carson

Scientific Works

Space does not permit the listing of the complete works of Rachel Louise Carson. Listed here are all works by Carson except those cited in Brooks (1989) and Gartner (1983). Included here is her master's thesis as well as all references cited in the text.

"The Development of the Pronephros during the Embryonic and Early Larval Life of the Catfish (*Ictalurus punctatus*)." Master's thesis, Johns Hopkins University, 1931.

"Undersea." *Atlantic Monthly* (Sept. 1937): 322–325.

Under the Sea Wind: A Naturalist's Picture of Ocean Life. New York: Simon and Schuster, 1941. Reissued, New York: Oxford University Press, 1952.

Chincoteague: A National Wildlife Refuge. Conservation in Action, no. 1. Washington, DC: Government Printing Office, 1947.

Mattamuskeet: A National Wildlife Refuge. Conservation in Action, no. 4. Washington, DC: Government Printing Office, 1947.

Parker River: A National Wildlife Refuge. Conservation in Action, no. 2. Washington, DC: Government Printing Office, 1947.

Guarding Our Wildlife Resources. Conservation in Action, no. 5. Washington, DC: Government Printing Office, 1948.

(with Vanez T. Wilson) *Bear River: A National Wildlife Refuge. Conservation in Action*, no. 8. Washington, DC: Government Printing Office, 1950.

The Sea around Us. New York: Oxford University Press, 1951.

The Edge of the Sea. Boston: Houghton Mifflin, 1955.

"Help your child to wonder." *Woman's Home Companion* (July 1956): 35–39. Republished as *The Sense of Wonder*. New York: Harper and Row, 1965.

Silent Spring. Boston: Houghton Mifflin, 1962.

"Rachel Carson answers her critics." *Audubon Magazine* (Sept. 1963): 262–265, 313–315.

Works about Rachel Louise Carson

Anticaglia, E. *Twelve American Women*. Chicago: Nelson-Hall, 1975.

Briggs, S. A. *Silent Spring: A View from 1987*. Chevy Chase, MD: Rachel Carson Council, 1987.

Brooks, P. "The courage of Rachel Carson." *Audubon Magazine* 89 (Jan. 1987): 12, 14–15.

———. *The House of Life: Rachel Carson at Work*. Boston: Houghton Mifflin, 1972. Reissued, 1989.

Includes a bibliography compiled by Carson's former editor.

———. *Speaking for Nature: How Literary Naturalists from Henry Thoreau to Rachel Carson Have Shaped America*. Boston: Houghton Mifflin, 1980.

Coates, R. A. *Great American Naturalists*. Minneapolis: Lerner, 1974.

Easterbrook, G. "Averting a death foretold." *Newsweek* (Nov. 28, 1994): 72–73.

Freeman, Martha (ed.). *Always, Rachel*. Boston: Beacon Press, 1995.

Gartner, C. B. *Rachel Carson*. New York: Ungar, 1983.

Hines, B. "Remembering Rachel." *Yankee Magazine* (June 1991): 62–66.

Hynes, H. P. "Catalysts of the American Environmental Movement." *Women of Power* 9 (Spring 1988): 37–41, 78–80.

———. *The Recurring Silent Spring*. New York: Pergamon Press, 1989.

Jezer, M. *Rachel Carson*. New York: Chelsea House, 1988.

King, Y. *What Is Ecofeminism?* New York: Ecofeminist Resources, 1990.

McCay, M. A. *Rachel Carson*. New York: Twayne, 1993.

McCurdy, P. P. "Twenty-five years after Rachel Carson." *Chemical Week* 139 (Aug. 20, 1986): 3.

Merchant, C. *The Death of Nature*. San Francisco: Harper and Row, 1989.

Norman, G. "The flight of Rachel Carson." *Recorder: The Chatham Alumnae Magazine* 54 (2) (Spring 1985): 4–8.

Norwood, V. "The nature of knowing: Rachel Carson and the American environment." *Signs* 12 (Summer 1987): 740–760.

Obituary. *New York Times* (Apr. 15, 1964).

Seif, D. T. "How I remember Rachel." *Recorder: The Chatham Alumnae Magazine* 54 (2) (Spring 1985): 9.

Sterling, P. *Sea and Earth: The Life of Rachel Carson*. New York: Crowell, 1970.

Other References

Potter, Beatrix. *The Tale of Peter Rabbit*. Private edition, 1901. London: Frederick Warne, 1902.

MARY AGNES MEARA CHASE (1869–1963)

Lesta J. Cooper-Freytag

BIOGRAPHY

Mary Agnes Chase was born Mary Agnes Meara (also spelled Mera) in Iroquois County, Illinois, on April 29, 1869. Her father, Martin John Meara, was a railroad blacksmith from Tipperary, Ireland. Her mother, Mary (Cassidy) Brannick Meara, was from Louisville, Kentucky. Mary Agnes was the third daughter and the second youngest of six children (the youngest child died in infancy). Her father died when she was barely two (in 1871), and the widowed mother and children moved to Chicago, where the family name was changed to Merrill. Here her mother and maternal grandmother worked to raise the fatherless children.

The children attended grammar school in the Chicago area and were expected to help out with the family's expenses. She never had the chance to attend college. "Girls did not get to go to college when I was young. I had to pick up my education as I went along" (*New York Times* 1963). She showed an early interest in plants and often told the story of bringing a bouquet of grass flowers to her grandmother in Chicago, who told her that grass did not have flowers. "But I was right and my grandmother was wrong," she said (*New York Times* 1963).

One of her first jobs was proofreading and setting type for a periodical called the *School Herald*; the editor of this journal was William Ingraham Chase. The two married on January 21, 1888, when she was 19, and he was 34 years old. William Chase was already seriously ill with tuberculosis at the time of their marriage, and he died within one year (January 3, 1889). There were no children of this union. Upon William Chase's death, Mary Agnes Chase was faced with serious debts related to the past publication of the journal. She did not turn to either family for help but lived frugally and worked diligently to pay off the debts on her own. During this time she worked primarily as a proofreader on the *Inter-Ocean* newspaper. She worked nights on the paper so that she could take extension courses at the Lewis Institute in Chicago and at the University of Chicago during the day.

Chase worked briefly in Wady Petra, Illinois, in her brother-in-law's store and developed a close relationship with her brother-in-law and his family. One nephew (Virginius) was especially close to Chase, and he boasted that he had rekindled the love of plants in Chase. Reportedly, he asked her to help him identify some of the local plants. In 1890 Chase returned to Chicago and her job on the *Inter-Ocean* newspaper, and there she and Virginius were further inspired in their botanical interests by their visit to the 1893 Columbian Exposition plant-collecting exhibit. Thus, while working at the *Inter-Ocean* newspaper, Chase began her informal and personal study of the flora of northern Illinois.

In 1897 she began recording her collections, and in 1898, while collecting around the Des Plaines River, she met Reverend Ellsworth Jerome Hill, a retired minister and amateur bryologist. A shared love for botany brought these two close together, and Chase became Reverend Hill's protegée. From 1898 to 1903 Reverend Hill employed Chase (gratis) to draw many of the new species that he described. During this time Chase's talent for illustration came to the attention of Charles Frederick Millspaugh, curator of botany of the Field Museum of Natural History, who also employed her (gratis) to illustrate two museum publications, ''Plantae Utowanae'' (1900) and ''Plantae Yucatanae'' (1903–4). Reverend Hill also taught Chase how to use a microscope and encouraged her to apply for the job of meat inspector for the U.S. Department of Agriculture (USDA) at the Chicago stockyards, a job she held from 1901 to 1903. At the same time she was also working (gratis) for both Reverend Hill and Millspaugh, honing her illustrative botanical skills and learning more and more of the native American flora. Under the urging of Reverend Hill, Chase applied for a job and was hired as a botanical artist for the USDA Bureau of Plant Industry in Washington, D.C. She moved to Washington, D.C., and on November 1, 1903, began what was to be a lifetime of dedicated service to the USDA.

Chase worked as an illustrator in the Division of Forage Plants until 1905. During this period, outside employment hours, she worked in the grass herbarium and began a series of papers on the genera of the Paniceae. In 1905 she began working as an illustrator with Albert Spear Hitchcock, the principal scientist in charge of systematic agrostology (the study of grasses). In 1907 she was appointed Hitchcock's scientific assistant in systematic agrostology, and was promoted to assistant botanist (1923) and then to associate botanist (1925). In 1936, after Hitchcock's death, she became senior botanist in charge of all systematic agrostology and, in 1937, the custodian of the Section of Grasses of the U.S. National Museum with responsibility for the grass herbarium.

Throughout her life, Chase was a reformer. Raised as a Roman Catholic, she later chose socialism as her guiding principle and was a fervent suffragette and prohibitionist. She marched with the noted suffragette Alice Paul and was jailed in January 1915 for helping maintain a fire fed by President Woodrow Wilson's speeches. She had vowed to burn any of Wilson's speeches that contained the words "freedom" and "liberty" as long as women did not have the right to vote. Chase was jailed again in August 1918 for picketing the White House. She was outspoken and dedicated to her causes and contributed to the Women's International League for Peace and Freedom, the National Woman's Party, the National Association for the Advancement of Colored People, and the Fellowship of Reconciliation.

Chase often opened her doors to students, and many of them (especially students from South America) were supported by Chase during their studies in the United States. Chase was equally faithful to her friends and family. She cared for her sister, Rose, from 1936 until her death in 1954 and shared her home with her longtime friend Florence Van Eseltine from 1953 until Chase's move to a nursing home in 1963.

Mary Agnes Chase officially retired from the USDA in April 1939 but remained scientifically active, serving as a research associate in the Division of Plants of the National Museum and custodian of grasses for the National Herbarium. In addition, in 1940, she visited Venezuela at the invitation of that government to carry out a survey and make recommendations on an agricultural program. In 1956 Chase was awarded a certificate of merit from the Botanical Society of America (one of only 50 recipients). The statement made at the time of presentation called her one of the world's outstanding agrostologists and preeminent among American students in this field. Chase received her one and only college degree (an honorary D.Sc.) from the University of Illinois in 1958 and, in the same year, received a medal for her service to the botany of Brazil. She was made the eighth honorary fellow of the Smithsonian Institution. In 1961 she was named a fellow of the Linnaean Society of London. Chase continued her work of describing and classifying grasses, completing a revised annotated index to grass species five months before her death. On her first day in a nursing home, she died of congestive heart failure (September 24, 1963) and, despite the brevity of her marriage to William Chase, 74 years earlier, requested that she be interred with him.

WORK

In the 36 years of official work for the USDA and the 24 unofficial years of work during her "retirement," Chase became one of the world's most renowned agrostologists. In those 60 years, Chase's contributions to the knowledge of grasses (especially of the Western Hemisphere) were outstanding. Especially important was her work in classifying and adding to the collection of grasses in the National Herbarium. She began her extensive collecting trips for the USDA in 1906. Over the years she visited a total of 19 states and Mexico. These trips were usually financed with her own money, and the specimens collected were given to the National Herbarium. The information accumulated on these trips proved invaluable in the preparation of the publication (with A. S. Hitchcock) of the *Manual of the Grasses of the United States*.

Her first long-term foreign expedition was to Puerto Rico in 1913, where she spent two months collecting grasses, bamboo, and ferns, including one new species: *Botrychium jenmani*. She went on her first trip to visit European herbaria in 1922–23. The primary site of her visit was to the Hackel Herbarium in the Natural History Museum of Vienna, but she also worked at the herbaria in Florence, Pisa, Geneva, Leiden, Brussels, and Paris. Although she preferred working in the field collecting her own specimens, her trip to the European herbaria allowed her to crystallize taxonomic concepts of several species she had studied.

In 1924 Chase received grants (from the USDA, the New York Botanical Garden, the Field Museum, the Missouri Botanical Garden, and the Gray Herbarium) to continue her fieldwork, this time in Brazil. There she collected over 20,000 plant specimens (500 specimens were of grasses). This trip was accomplished by car, horse, and foot over difficult and often hazardous terrain, scaling steep cliffs and traversing marshes and sandy savannas. Four years later (1929), at the age of 60, Chase returned to Brazil and, braving once again the difficult terrain, returned with more specimens. This second trip to Brazil yielded approximately 30 grass species not known to grow in Brazil and at least ten new varieties. These two Brazilian trips earned Chase the label of Uncle Sam's chief woman explorer of the USDA.

She returned to the European herbaria in 1935 to work at Montpellier, Caen, and Paris. One year after her official retirement (1940), Chase traveled to Venezuela at the behest of the Venezuelan Ministry of Agriculture. The government wanted her to conduct a survey of local forage grasses and recommend a range management program based on local flora. Chase visited several ecozones during this visit, including the Andes, the savanna, and the cloud forest. She collected as she traveled and, despite drought conditions, amassed over 400 types of grasses (including 11 species not known to be native to Venezuela).

The trip to Venezuela was to be Chase's last major collecting trip. She continued to do consulting work for the USDA. Later she dedicated herself to revisions of the *Manual of the Grasses of the United States* and *The First Book of Grasses* (1922, 1937, 1959). She cataloged and sorted the 10,031 grass-type specimens that she had been the first to describe. Her culminating work, according to many of her colleagues, was the annotated index to grass species (over 80,000 cards) published in 1962 with C. D. Niles. Mary Agnes Chase was indeed one of America's outstanding agrostologists. Perhaps her dedication to this area is best summed up by the philosophy she often expressed, that civilizations were based on grasses, and grasses held the world together.

BIBLIOGRAPHY

Works by Mary Agnes Meara Chase

Scientific Works

Space does not permit the listing of the complete works of Mary Agnes Meara Chase. This list includes all works by Chase with the exception of those cited in Fosberg and Swallen (1959). All references cited in the text are also included.

(with C. F. Millspaugh) "Plantae Utowanae." *Field Columbian Museum. Botanical series*, no. 50 (1900): 113–124.
Illustrations by Chase.
(———)"Plantae Yucatanae." *Field Museum of Natural History. Botanical series*, no. 69 (1903): 15–84; no. 92 (1904): 85–151.
Illustrations by Chase.
The First Book of Grasses. The Structure of Grasses Explained for Beginners. New York: Macmillan Press, 1922. Rev. eds. San Antonio, TX: W. A. Silvius, 1937; Washington, DC: Smithsonian Institution, 1959; reissued, 1964, 1977.
The Genera of Grasses of the United States, with Special Reference to the Economic Species. Rev. ed. Washington, DC: Government Printing Office, 1936.
(with A. S. Hitchcock) *Manual of the Grasses of the United States.* Rev. ed. Washington, DC: Government Printing Office, 1950.
Chase revised the original 1935 edition written by Hitchcock.
(with C. D. Niles) *Index to Grass Species.* Boston: G. K. Hall, 1962.

Works about Mary Agnes Meara Chase

Baker, Gladys. "Women in the United States Department of Agriculture." *Agricultural History* (Jan. 1976): 190–201.
Fosberg, R. F., and Jason R. Swallen. "Agnes Chase." *Taxon* 8 (June 1959): 145–151.
Furman, Bess. "Grass is her life-root." *New York Times* (June 11, 1958): 37.
Harney, Thomas. *The Magnificent Foragers.* Washington, DC: National Museum of Natural History, Smithsonian Institution, 1978.
Hillenbrand, Liz. "87-year-old grass expert still happy with subject." *Washington Post and Times Herald* (Apr. 30, 1956): 7.
Obituary. *New York Times* (Sept. 26, 1963): 35.
Stieber, M. T. "Manuscripts written and /or annotated by Agnes Chase pertinent to the grass collections at the Smithsonian Institution." *Huntia* 3 (2) (1979): 117–125.

Sources

Chase's field books (seven vols., 1897–1959). Botany branch, Smithsonian Institution Libraries.
Unpublished manuscripts.
Hitchcock-Chase correspondence. Chicago Field Museum of Natural History.
Hitchcock and Chase papers. Hunt Institute for Botanical Documentation, Carnegie-Mellon University.
Hitchcock-Chase Library. Smithsonian Institution Archives.

EUGENIE CLARK (1922–   )

Virginia L. Buckner

BIOGRAPHY

Eugenie Clark, born May 4, 1922, in New York City, was the only child of Charles and Yumico (Mitomi) Clark. Her father, a barber by profession, died when Eugenie was two years old, so she never knew her father's family. She was close to her mother's family. Her mother had been born in Japan of a mother who was Japanese and Scottish and a father who was a Japanese doctor. When she was two years old, Clark went to the beach with her family: her mother, her uncle, and her grandmother. The whole family loved to swim. At one time, her mother had been a swimming instructor, and her uncle performed difficult dives. By the age of nine Clark spent many Saturday mornings alone at the old Aquarium in Battery Park, New York City, while her mother worked at the cigar counter in the lobby of the Downtown Athletic Club. Then they often went for lunch at a tiny Japanese restaurant owned by "Nobusan," Masatomo Nobu, who married Clark's mother much later when Clark graduated from college.

As a young child Clark coaxed her mother into buying a 15-gallon aquarium for Christmas, and she started her fish collection with some tropical fish, including guppies and swordtails. The aquarium was soon followed by terraria, as she collected some salamanders, a horned toad, and an alligator. Clark became the youngest member of the Queens County Aquarium Society. Her interest in collecting fish increased. She began to keep records of her fish, their scientific names, when she obtained them, and what happened to them. Although she had watched guppies mate, it was especially exciting when the male Siamese fighting fish began to build his bubble nest and then mated. As the baby fish hatched, Clark had visions of making a fortune by selling fish. Biology was her favorite class at Bryant High School in Queens, New York, and she managed to find ways to work her interest in fish into every assignment possible.

In the fall of 1939, when Clark entered Hunter College (now part of the City University of New York), she no longer had as much time to work with her animals. Nor was there as much available space, for she and her mother were living in an apartment together with her grandmother. Her mother suggested that she take practical classes in college, such as typing and shorthand, but there never seemed to be time for those classes, only courses in zoology. Clark had decided to become an ichthyologist, so she took every class in zoology that she could. During the summers of 1940 and 1941, Clark went to the University of Michigan's Biological Station in northern Michigan. Here she collected fish and snakes, but she also kept in training for the swim team at Hunter College by swimming in Douglas Lake.

When Clark finished Hunter College in 1942, the Second World War was in progress, and there were no jobs for young, inexperienced zoologists. Clark got a job as a chemist in plastic research at the Celanese Corporation in Newark, New Jersey. She worked there for four years (1942–46) while she took night classes at New York University (NYU) to earn a master's degree. Her favorite class was a course in ichthyology, taught by Dr. Charles M. Breder, formerly the director of the Aquarium at Battery Park. He was now the curator of the Department of Fishes at the American Museum of Natural History. Breder sponsored Clark's research and acquainted her with a group of fish called the plectognaths. He encouraged Clark to study the anatomy and mechanism by which some plectognaths, the blowfish, inflate their bodies. This work was the basis of her master's thesis, and the master's degree was awarded in 1946. This material, coauthored by Breder, became her first publication. Clark married Hideo Umaki, a pilot, in 1942, and they were divorced in 1949.

Clark was introduced to Dr. Carl L. Hubbs at the annual convention of the Society of Ichthyologists and Herpetologists held in 1946. He offered her a job as his part-time research assistant at Scripps Institute of Oceanography at the University of California in La Jolla. Here Clark attended lectures and went to sea on the Institute's ship. At Scripps Clark first used a diving helmet and walked on the bottom of the ocean. She also obtained a research problem when Dr. Hubbs gave her a swell shark that had puffed up. She discovered that the mechanism used by the shark was similar to that used by the plectognaths she had studied earlier.

In 1947 Clark had the opportunity to go to the Philippines with the U.S. Fish and Wildlife Service to survey the possibility of fisheries around the Philippine Islands. Her work as a chemist at Celanese Corporation and her courses at Scripps made her uniquely qualified for the position. However, she could not leave Hawaii because her visa had been held up—supposedly because the Federal Bureau of Investigation (FBI) was checking her Japanese background. After waiting for her visa for weeks, she gave up. A man was hired for the job. She felt she did not get the position because she would have been the only woman on the expedition. While she was waiting for her visa, Clark studied the puffing apparatus of some Hawaiian puffer fish and bought the first of many face masks needed to watch the fish on the coral reefs.

Clark returned to NYU to work on her Ph.D. degree in zoology, which was granted in 1950. She became a research assistant to Dr. Myron Gordon at the

American Museum of Natural History. From 1948 to 1949 she was an associate in the Department of Animal Behavior at the museum, remaining in that position until 1980. Her project with Dr. Gordon concerned the mating behavior of platies and swordtails. Clark was the first American to inseminate fish artificially. During this period she studied at the Marine Biological Laboratory (MBL) at Woods Hole, Massachusetts (1948) and at the Lerner Marine Laboratory on the island of Bimini in the West Indies, where Breder was the director. In June 1949 Clark received a Pacific Science Board fellowship and was sent to investigate the fish of Micronesia. This trip included visits to Kwajalein, Guam, Saipan, and the Palauan Islands and was sponsored by the Office of Naval Research. In the Pacific islands she swam with native fishermen and learned the techniques of underwater spearfishing. She collected and identified many poisonous fish.

In 1950 Clark was awarded a Fulbright Scholarship to go to the Marine Biological Station of Fouad University at Ghardaqa, Egypt, a small desert station on the Red Sea. Here she was to study the plectognaths and other poisonous fish. Clark arrived in Egypt on Christmas Eve 1950 and immediately went to Ghardaqa to work with Dr. H.A.F. Gohar. At the marine station she lived in luxury, as she had a cook and a houseboy, who took care of all of the household chores. Although the Arab women around the marine station were heavily veiled, they accepted the American woman in a bathing suit who worked with their men. Before Clark had left for Egypt, she had met Ilias Konstantinu, a young doctor who had been raised in Greece, went to medical school in Austria, and was an intern in the United States. Konstantinu arrived at the marine station in June, and he and Clark were married in a Greek Orthodox church in Cairo. They honeymooned at the Red Sea, where they swam, and Konstantinu became an expert spear fisherman. Konstantinu returned to the United States and to Buffalo General Hospital, where he was an orthopedic surgeon. Clark remained in Egypt for several months finishing up her work before she joined her husband in Buffalo, New York.

In February 1952 Clark won the Eugene F. Saxton Memorial Trust fellowship, sponsored by Harper and Brothers. The same year she was awarded the Breadloaf Writer's fellowship, and the following year the Merit Award in Science from *Mademoiselle* magazine. Supported by these fellowships, she wrote of her experiences in the Red Sea in *Lady with a Spear* (1953). This very popular book has been translated into eight languages as well as Braille.

Clark received an invitation to lecture in Englewood, Florida, from Anne Vanderbilt, who had read *Lady with a Spear*. After the lecture she was approached by Anne and William Vanderbilt and asked to start a marine laboratory, a place where people could learn more about the sea. At that time there were no experts and no marine laboratories in that part of Florida. The marine laboratory was built on land owned by William Vanderbilt and his brother, Alfred. Clark, her husband, and two children, Hera (b. 1952) and Iris (b. 1954), moved to the west coast of Florida. Early in January 1955 Clark opened Cape Haze Marine Laboratory in Placida, Florida. She remained the executive director

of this marine laboratory until 1966. One of the first requests that came the day
after her arrival in Florida was from Dr. John Heller, director of the New England Institute for Medical Research. He needed shark livers for cancer research.
These were quickly procured. As Clark wanted to study the behavior of sharks,
shark pens were built, as well as tanks for other fish. She discovered how to
maintain sharks in captivity and observed their behavior. The marine laboratory
became world-famous, and other scientists came to study there. These were
happy years for her, and during this time two more children were born, Them-
istokles (b. 1956) and Nikolas (b. 1958). Her mother and stepfather moved to
Florida and opened a Japanese restaurant in Sarasota, and her mother often
looked after Clark's children. At this time Clark started scuba diving so she
could go deeper and stay down longer. In 1960 the marine laboratory was moved
to Siesta Key, Sarasota, and was renamed Mote Marine Laboratory of Sarasota.
William Mote had become the primary supporter of the laboratory, and the
Vanderbilts had moved farther south to be with their family.

In 1962 Clark was the leader of the American delegation for the Israel Red
Sea Expedition to Ethiopia, and in 1969 she became the founding member of
the international advisory committee of the Hebrew University of Jerusalem.

In the fall of 1965, the Crown Prince of Japan invited Clark to come there.
He had read her book *Lady with a Spear*, which had been translated into Japanese, and wanted to meet her. As a gift, she took with her a two-foot-long
nurse shark that had been trained to discriminate between a lighted and a dark-
ened target. Two years later, when the Crown Prince was in the United States,
he asked Clark to meet him in Miami. After an all-night discussion about fish,
he asked her to help him learn to skin-dive and collect fish.

As Clark's fame grew, she received recognition for her work: the Hadassah
Myrtle Wreath Award in Science (1964), the Nogi Award in Arts from the
Underwater Society of America (1965), the Golden Plate Award in Science from
the American Academy of Achievement (1965), and the Dugan Award in
aquatic sciences from the American Littoral Society (1969).

In 1966 Clark resigned from the marine laboratory in Florida. Her mother
had died (1959), and she had divorced Ilias Konstantinu (April 1966). Clark and
her four children moved north, where she became an instructor of biology at
Hunter College (1953–54). She married Chandler Brossard, a writer, and they
were divorced in 1968. After a year of teaching at Hunter, Clark took some
time off to write *The Lady and the Sharks* about her experiences at Cape Haze
Marine Laboratory. From 1966 to 1967 she was associate professor of zoology
at City University of New York. In 1968 Clark joined the faculty at the University of Maryland, where she became a full professor of zoology (1973). In
1969 she married Igor Klatzo, a neuropathologist, whom she divorced in 1974.

While a professor at the University of Maryland, Clark had the opportunity
to travel and do research throughout the world. The Red Sea remains her favorite
place. She became interested in ''sleeping'' sharks, which were first found in

caves in Mexico. In 1976 she took her stepfather to Japan, where she observed two other kinds of ''sleeping'' sharks.

Her many awards include the gold medal of the Society of Women Geographers (1975), the John Stoneman Marine Environmental Award (1982), the Governor of Sinai Medal (1984), the Lowell Thomas Award from the Explorers Club (1986), the Medal of the Governor of the Red Sea Egypt (1988), and the Nogi Award in Science (1988). She was awarded honorary Doctor of Science degrees by the University of Massachusetts at Dartmouth (1992), the University of Guelph, Ontario, Canada (1995), and Long Island University (1995).

Clark has an office in the Department of Zoology at the University of Maryland, College Park, Maryland, where she is professor emeritus and senior research scientist. She lives in Bethesda, Maryland. Clark remained close to her stepfather, who never learned to speak English and lived with her until his death.

WORK

Eugenie Clark is an ecologist and ichthyologist who is best known for her work with poisonous fish of the tropics and her studies of the behavior of sharks. Early in her research she learned how to inseminate viviparous fish artificially (guppies and swordtails) and was the first American to do this. She discovered the puffing mechanism used by various fish to inflate themselves (blowfish). Her lifetime research deals with the sexual behavior, morphology, and systematics of fish from Micronesia, the Red Sea, the Caribbean Sea, the Gulf of Mexico, and the Mediterranean Sea. During her research she collected many rare and poisonous fish. She discovered new species of fish, and she learned how to maintain sharks in captivity. She trained sharks to respond to various kinds of targets. In the later stages of her research she has examined sharks that appeared to be sleeping and discovered a fish, the Moses sole, that gives off a substance that is a very effective shark repellant. At the Mote Marine Laboratory, which she founded, she combined laboratory experiments with field observations. Through her research she has made many valuable contributions to the science of ichthyology.

Clark is highly respected by the scientific community. She has written numerous articles on her research, which have been published in many journals. She has served as an officer of scientific organizations and has garnered many prizes and awards. She has been active in professional associations and is a life member of the American Society of Ichthyologists and Herpetologists, the International Association of Professional Diving Scientists, the National Parks and Conservation Association (member of the Board of Trustees, 1969–75, and vice chairman, 1976–82), the American Littoral Society (vice president, 1960–91), and the American Elasmobranch Society. She also belongs to the Confederation Mondiale des Activities Subaquatiques and is an honorary life member of the Gesellschaft für Biologische Aquarian und Terrarienkund. Between 1955 and 1974, Clark received 15 research grants from the National Science Foundation, the Office of Naval Research, and the National Geographic Society. She is recognized as an authority on ichthyology by the U.S. government, as well as by other scientists. She is one of the few women to have been with Jacques Cousteau on his ship, the *Calypso*.

Always interested in teaching and young people, Clark began summer science training programs for high school students as soon as the Cape Haze Marine Laboratory was in operation. She remained the executive director of these programs while she was in Florida (1955–65). One of the young people who worked during the summer at the biological laboratory was Freddie Aronson, a 16-year-old boy who had trained the nurse shark that Clark took to Japan.

Clark has stimulated popular interest in sharks and in the life beneath the sea by her books, especially *Lady with a Spear*, and the many articles she has written for the *National Geographic*. The general public has been able to enjoy her adventures vicariously and to learn about sharks through her various television specials. Clark feels that her most important contribution is knowledge about sharks, especially that they are intelligent creatures. They have a memory, they can be taught, and they do not normally attack people. "I think I have tried to give a better reputation to sharks, and I feel this is my most worthwhile contribution" (Personal communications 1994–95).

Eugenie Clark's family is very important to her. The children often accompanied her on her travels and became expert swimmers. She states in *Lady with a Spear*, "Sharing the fun of fishing turns strangers into friends in a few hours."

BIBLIOGRAPHY

Works by Eugenie Clark

Scientific Works

"Specialization of the Plectognath Digestive Tract." M.S. thesis, New York University, 1946.

"Notes on the inflating power of the swell shark, *Cephaloscyllium uter.*" *Copeia* (4) (1947): 278–280.

(with C. M. Breder, Jr.) "A contribution to the visceral anatomy, development, and relationships of the Plectognathi." *Bulletin of the American Museum of Natural History* 88 (1947): 287–320.

"Notes on some Hawaiian plectognath fishes, including a key to the species." *American Museum Novitates* (1397) (1949): 1–22.

(with J. M. Moulton) "Embryological notes on Menidia." *Copeia* (2) (1949): 152–154.

(with R. W. Sperry) "Interocular transfer of visual discrimination habits in a teleost fish." *Physiological Zoology* 22 (1949): 372–378.

"A method for artificial insemination in viviparous fishes." *Science* 112 (1950): 722–723.

"Notes on the behavior and morphology of some West Indian plectognath fishes." *Zoologica* 35 (1950): 159–168.

"The Sexual Behavior of Two Sympatric Species of Poeciliid Fishes and Their Laboratory Induced Hybrids with an Analysis of the Factors Involved in the Isolating Mechanism." Ph.D. diss., New York University, 1950.

(with L. R. Aronson) "Sexual behavior in the guppy, *Lebistes reticulatus* (Peters)." *Zoologica* 36 (1951): 49–66.

(with R. P. Kamrin) "The role of the pelvic fins in the copulatory act of certain poeciliid fishes." *American Museum Novitates* (1509) (1951): 1–14.

"A scientific journey to the Red Sea." *Natural History* 61 (1952): 344–349, 414–419.

(with L. R. Aronson) "Evidences of ambidexterity and laterality in the sexual behavior of certain poeciliid fishes." *American Naturalist* 86 (1952): 161–171.

Lady with a Spear. New York: Harper and Brothers, 1953.

"Siakong, spear-fisherman preeminent." *Natural History* 62 (1953): 227–234.

(with H.A.F. Gohar) *The Fishes of the Red Sea.* Cairo, Egypt: Fouad I University Press, 1953.

(———) "The fishes of the Red Sea: Order Plectognathi." *Marine Biological Station: Ghardaqa* (8) (1953): 1–80.

(with L. R. Aronson and M. Gordon) "Mating behavior patterns in two sympatric species of xiphophorin fishes: Their inheritance and significance in sexual isolation." *Bulletin of the American Museum of Natural History* 103 (1954): 135–226.

(with J. H. Heller, M. S. Heller, et al.) "Squalene content of various shark livers." *Nature* 179 (1957): 919–920.

"Functional hermaphroditism and self-fertilization in a serranid fish." *Science* 129 (1959): 215–216.

"Instrumental conditioning of lemon sharks." *Science* 130 (1959): 217–218.

"Four shark attacks in the west coast of Florida, summer 1958." *Copeia* (1) (1960): 63–67.

(with W. Royal) "Natural preservation of human brain, Warm Mineral Springs, Florida." *American Antiquity* 26 (1960): 285–287.

"Maintenance of sharks in captivity, Part I. General." *Bulletin: Institute Oceanographique: Monaco* (sp. no. 1 A) (1962): 7–13.

"Maintenance of sharks in captivity, Part II. Experimental work on shark behavior." *Bulletin: Institute Oceanographique: Monaco* (sp. no. 1 D) (1963): 1–10.

"The maintenance of sharks in captivity, with a report on their instrumental conditioning." In *Sharks and Survival,* edited by P. W. Gilbert, 115–149. Boston: Heath, 1963.

"Massive aggregations of large rays and sharks in and near Sarasota, Florida." *Zoologica* 48 (1963): 61–66.

(with O. David, A. Tester, et al.) "Facilities for the experimental investigation of sharks." In *Sharks and Survival,* edited by P. W. Gilbert, 150–162. Boston: Heath, 1963.

(with J. C. Moore) "Discovery of right whales in the Gulf of Mexico." *Science* 141 (1963): 269.

"Mating of groupers." *Natural History* 74 (6) (1965): 22–25.

(with K. von Schmidt) "Sharks of the Central Gulf Coast of Florida." *Bulletin of Marine Science of the Gulf and the Caribbean* 15 (1965): 13–83.

"Pipefishes of the genus *Siokunichthys* Herald in the Red Sea with a description of a new species." *Bulletin of Sea Fisheries Research Station: Israel* (40) (1966): 3–6.

(with L. R. Aronson and F. R. Aronson) "Instrumental conditioning and light-dark discrimination in young nurse sharks." *Bulletin of Marine Science of the Gulf and the Caribbean* 17 (1966): 249–256.

(with K. von Schmidt) "A new species of *Trichonotus* (Pisces: Trichonotidae) from the Red Sea." *Bulletin of Sea Fisheries Research Station: Haifa* (42) (1966): 29–36.

"The need for conservation in the sea." *Oryx* 9 (1967): 151–153.

(with A. Ben-Tuvia and H. Steinitz) "Analysis of a coral reef fish population, Dahlak Archipelago, Red Sea." *Bulletin of Sea Fisheries Research Station: Haifa* (49) (1967): 15–31.

"Eleotid gobies collected during the Israel South Red Sea Expedition (1962) with a key to Red Sea species." *Bulletin of Sea Fisheries Research Station: Haifa* (49) (1968): 3–7.

The Lady and the Sharks. New York: Harper and Row, 1969.

"The Red Sea garden eel." *Underwater Naturalist* 7 (1971): 4–10.

"The Red Sea's garden of eels." *National Geographic* 142 (5) (1972): 724–735.

"Steinitz, father and son: Their influence on inter-oceanic waterway research." *Proceedings of the International Congress of Zoology* 17 (1972): 1–6.

(with W. Aron) "Heinz Steinitz: To realize a dream." *Israel Journal of Zoology* 21 (1972): 131–134.

" 'Sleeping' sharks in Mexico." *Underwater Naturalist* 8 (1973): 4–7.

(with A. Ben-Tuvia) "Red Sea fishes of the family Branchiostegidae with a description of a new genus and species *Asymmetrurus oreni*." *Bulletin of Sea Fisheries Research Station: Haifa* (60) (1973): 63–74.

(with S. Chao) "A toxic secretion from the Red Sea flatfish *Paradachirus marmoratus* (Lecepede)." *Bulletin of Sea Fisheries Research Station: Haifa* (60) (1973): 53–56.

"The Red Sea's sharkproof fish." *National Geographic* 146 (5) (1974): 718–727.

"Into the lairs of 'sleeping' sharks." *National Geographic* 147 (4) (1975): 570–584.

"The strangest sea." *National Geographic* 148 (3) (1975): 338–343.

(with A. George) "A comparison of the toxic soles *Pardachirus marmoratus* of the Red Sea and *Pardachirus pavoninus* from Japan." *Revue des Travaux de l'Institut des Peches Maritimes* 40 (1976): 545–546.

"Synagogues and sea fans." *National Parks and Conservation Magazine* 51 (1977): 13–20.

"Flashlight fish of the Red Sea." *National Geographic* 154 (5) (1978): 718–728.

"Red Sea fishes of the family Trypterygiidae with descriptions of eight new species." *Israel Journal of Zoology* 28 (1979): 65–113.

(with A. George) "Toxic soles *Pardachirus marmoratus* from the Red Sea and *Pardachirus pavoninus* from Japan, with notes on other species." *Environmental Biology of Fishes* 4 (1979): 103–123.

"Sharks: Magnificent and misunderstood." *National Geographic* 160 (2) (1981): 138–187.

"Hidden life of an undersea desert." *National Geographic* 164 (1) (1983): 128–144.

"Sand-diving behavior and territoriality of the Red Sea razorfish *Xyrichtys pentadactylus*." *Bulletin of the Institute of Oceanography and Fisheries*, Cairo 9 (1983): 225–242.

"Japan's Izu Oceanic Park." *National Geographic* 165 (4) (1984): 465–491.

(with P. Pemberton and R. Leen) "Population density of a colony of bluespotted jawfish off Baja California." *Underwater Naturalist* 15 (1985): 3–7.

(with E. Kristof and D. Lee) "New eyes for the dark reveal the world of sharks at 2,000 feet." *National Geographic* 170 (5) (1986): 680–691.

"Down the Cayman Wall." *National Geographic* 174 (5) (1988): 712–731.

(with J. S. Rabin and S. Holderman) "Reproductive behavior and social organization in the sand tilefish, *Malacanthus plumieri.*" *Environmental Biology of Fishes* 22 (1988): 273–286.

(with J. S. Rabin, E. Bunyan, et al.) "Social behavior of Caribbean tilefish." *Underwater Naturalist* 18 (1989): 20–23.

(with E. Kristof) "Deep-sea elasmobranchs observed from submersibles off Bermuda, Grand Cayman, and Freeport, Bahamas." In *Elasmobranchs as Living Resources, NOAA Technical Report* no. 90, 275–290. Washington, DC: National Maritime Fish Service, 1990.

(with J. F. Pohle and D. C. Shen) "Ecology and population dynamics of garden eels at Ras Mohammed, Red Sea." *National Geographic Research* 6 (1990): 306–318.

(with E. Kristof) "How deep do sharks go? Reflections on deep sea sharks." *Underwater Naturalist* 19 (1991): 79–84.

(with A. McGovern) *The Desert beneath the Sea.* New York: Scholastic, 1991.

(with M. Pohle and J. S. Rabin) "Spotted sandperch dynamics." *National Geographic Research and Exploration* 7 (1991): 138–155.

"Whale sharks: Gentle monsters of the deep." *National Geographic* 180 (6) (1992): 120–139.

(with J. F. Pohle) "Monogamy in the tile fish, *Malacanthus latovittatus*, compared with polygyny in related species." *National Geographic Research and Exploration* 8 (1992): 276–295.

(with D. Herold) "Monogamy, spawning and ecodynamics of the sea moth, *Eurypegasus dracones* (Pisces: Pegasidae)." *Environmental Biology of Fishes* 37 (1993): 219–236.

(with J. E. Randall) "*Helcogramma vulcana*, a new triplefin fish (Blenniodei: Tripteryiidae) from the Banda Sea, Indonesia." *Revue Francaise d'Aquariologie* 20 (1993): 27–32.

Other Works

"Field trip to the South Seas." *Natural History* 60 (1951): 8–15, 46.

"The lost quarry." *Natural History* 61 (1952): 258–263.

Personal communications to the author, 1994–95.

Works about Eugenie Clark

McGovern, A. *Shark Lady.* New York: Four Winds Press, 1979.

JEWEL ISADORA PLUMMER COBB
(1924–   )

Beatriz Chu Clewell

BIOGRAPHY

Jewel Isadora Plummer was born on January 17, 1924, in Chicago, Illinois, the only child in an upper-middle-class African-American family. Her parents, Frank V. and Carriebel (Cole) Plummer, were originally from Washington, D.C., although her mother was born in Augusta, Georgia.

Her father was a graduate of Cornell University and completed medical training at Rush Medical College in Chicago. He began practicing medicine in that city after his graduation in 1923, serving the newly arrived African-American migrants from the South. Her mother had studied interpretive dancing at Sargeants, a physical education college affiliated with Harvard University. In Chicago she worked as a teacher of dance in both the public schools and Works Public Administration (WPA) projects.

The accomplishments of African-American people were noted with pride, and racial matters were discussed often in the Plummer home. Many of the Plummer family friends and neighbors were successful, talented African-Americans. Plummer's uncle, Bob Cole, was a well-known producer of musicals in New York. In addition to the stimulating environment provided by the Plummer household, she had the use of her father's library, which contained a comprehensive collection of material about African-Americans, scientific journals and magazines, and periodicals of current events. Even more important, she was encouraged in her schoolwork by her parents. Her father took an interest in her science courses, and the two often talked about scientific papers and topics related to science. Plummer says: "My parents played a significant role in my education. . . . I was given a clear sense that education was a priority and I was expected to go as far as my potential allowed me to go" (Personal communications 1994).

During the first years of her education, Plummer attended Sexton Elementary School in a predominantly white neighborhood. She was later required to transfer to the overcrowded, dilapidated Betsy Ross Elementary School, because the city of Chicago had redrawn school district lines to ensure that fewer African-American families would be eligible to attend Sexton. She attended Englewood High School, which had a special honors or college-bound track that offered students the opportunity to take five years of science. Plummer was enrolled in this track. In writing about the early educational experiences that interested her in science, she says, "It was in Ms. Hyman's sophomore year biology class that I was given a microscope to view an entirely new world beyond my normal viewing capacity" (Cobb 1989, 39). The book *Microbe Hunters* by Paul De Kruif (1926) further stimulated her interest in biology. After two years of high school biology courses, Plummer decided that she wanted to be a biology teacher and to major in that subject in college.

As a high school graduate with advanced honors classes in English and five years of science, Plummer chose the University of Michigan at Ann Arbor, where she enrolled in 1941. In explaining her choice, she writes, "This decision was based on the long relationships I'd had with my teen-age friends from Michigan while spending summers in Idlewild, Michigan, a community of homes owned by Black folks from the Middle West" (Cobb 1989, 40). She describes Michigan as a good choice in terms of science courses and social opportunities (because of the relatively large African-American enrollment) but a "disaster" for African-American students in terms of dormitory living arrangements (Cobb 1989), because "the dormitories were segregated and the environment was hostile" (Personal communications 1994). Furthermore, social life on the mainstream campus was curtailed by the racial barriers that existed at that time. The popular hangouts for college students in the area did not welcome students of color, and no African-American students belonged to the big fraternities and sororities. Because of this situation, Plummer began to consider leaving the University of Michigan.

In 1942 Plummer and her parents were convinced by the Dean of Women at Talladega College, a historically black college in Alabama, to transfer to Talladega. The College had a strong science program, and there the young student found "encouraging professors and a good mentor [her bacteriology professor] who helped me to choose a graduate school" (Personal communications 1994). After three and a half years of college, Plummer graduated from Talladega College. Her mentor suggested that she apply to New York University's graduate program in biology, which she did, and she was admitted in 1944.

Plummer spent six years at New York University (NYU), during which she earned a master's degree and a doctorate in cell physiology. After attaining the master's degree, she abandoned her ambition to become a teacher, because of the discipline problems she encountered while working in New York City schools as a substitute teacher. She states that she was "pleased with the experience at NYU" (Personal communications 1994) and describes her six years there as "long but fulfilling" (Cobb 1989, 40). Her master's thesis was on a series of organic molecules, aromatic amidines, and their effect on the respiration of yeast cells. Her doctoral dissertation concerned the manner in which melanin pigment granules could be formed *in vitro* using the enzyme tyrosinase, which is needed for melanin pigment synthesis. Tyrosine and various substrates were tested as model systems for pigment formations; the results, methodology, and discussion then made up the dissertation, "Mechanisms of Pigment Formation." Concerning the process of completing her work toward the Ph.D. degree, Plummer writes:

My advisor and I worked together for months so that we both were satisfied that I had a "perfect" dissertation, capable of passing muster by the severest critic. Then finally, it was scrutinized by several readers and I came before the committee for my oral defense of my dissertation. I was, of course, extremely nervous. Despite that, I was comforted deep down inside with the awareness that I knew more about my subject . . . than anyone else in the world. (Cobb 1989, 41)

Immediately following the receipt of her Ph.D. degree in 1950, Plummer began a postdoctoral fellowship awarded by the National Cancer Institute to conduct research at the Harlem Hospital Cancer Research Foundation. As part of her first independent research activity, the young researcher undertook *in vitro* tissue culture studies of the tumors of cancer patients. During this time she also spent some time at the Columbia College of Physicians and Surgeons, where she learned new techniques for growing and analyzing nerve cells *in vitro*.

After her appointment in 1952 to the anatomy department faculty of the University of Illinois Medical School, she established the first tissue culture research laboratory and course in the field of cell biology. She extended the research she had begun as a graduate student. This research involved cytological studies of normal and malignant pigment cells and continued for 24 years until 1976, during which period she supported her research with external grants.

In 1954 she married and returned to New York and the Harlem Hospital Cancer Research Foundation. (Her marriage to Roy Raul Cobb produced a son, Roy Jonathan, born in 1957, and ended in divorce in 1967.) Soon after Plummer Cobb returned to the Harlem Hospital Cancer Research Foundation, the entire laboratory joined the NYU Post-Graduate Medical School, and at NYU's Bellevue Hospital Medical Center Cobb designed and established the new Tissue Culture Research Laboratory. At this point, she entered what she characterizes as "a most exciting phase of basic cell research" (Cobb 1989, 41). Her close collaborators were Dorothy G. Walker, her assistant, and Dr. Jane C. Wright.

Cobb left NYU in 1960 to assume a full-time teaching position at Sarah Lawrence College in the biology department, where she continued to conduct tissue culture studies. In 1969 she became Dean of the College and professor of zoology at Connecticut College. She remained at the College for seven years, during which time she arranged her schedule to spend the early morning hours in her research laboratory before going on to administrative or teaching duties. While Dean at Connecticut, Cobb established a program, the Postgraduate Premedical and Predental Program for Minority Students, which enrolled recent minority college graduates who had decided late in college that they wished to enter a health field. The one-year program prepared them to enter medical or dental schools. In the six years of the program, 90% of its 40 students entered professional schools of their choice.

From 1976 to 1981 Cobb was Dean of Douglass College of Rutgers University in New Jersey. After three years as Dean, she had to give up her research activities, but she continued to teach a course about tissue culture. She was appointed to the presidency of California State University at Fullerton. Upon leaving the presidency of Fullerton in 1990, she was named Trustee Professor at California State University at Los Angeles. She is currently the principal investigator of the Southern California Science and Engineering ACCESS Center and Network, a regional center in the Los Angeles basin, funded by the National Science Foundation (NSF) to motivate underrepresented minorities to study science and engineering. Since 1991 she has also been a member of the National Research Council's Office of Scientific and Engineering Personnel, Committee on Women and Minorities in Science (CWSE). This committee conducts studies to increase the participation of women in the scientific, educational, and governmental labor force and industry.

Cobb has been the recipient of myriad awards and honors, so numerous that it is impossible to list them all here. In 1952, the same year that she received the Key Pin Award for scholarship as outstanding woman alumna of 1952 from the Graduate School of Arts and Science at NYU, she was also elected to Sigma Xi, National Honorary Science Society. She is a Fellow of the New York Academy of Sciences and the American Association for the Advancement of Science. One residence hall for women majoring in mathematics and science has been named for her and Mary Bunting, a former dean at Douglass College at Rutgers. A second residence hall was named for Cobb at California State University at Fullerton. In 1993 she received the Lifetime Achievement Award from the NSF for Contributions to the Advancement of Women and Underrepresented Minorities. She has been the recipient of 21 honorary doctorates, including degrees from the Medical College of Pennsylvania, Rutgers, Bowdoin College, the City College of the City University of New York, Tuskegee University, and Rensselaer Polytechnic Institute.

In looking back at the influences on her choice of science as a career, Cobb feels that she received support and encouragement throughout her education. "No one ever said to me, 'You can't be a scientist,' " she says. "Classmates were supportive and, in general, I had encouraging people around me all the time. When I went to graduate school I had a lot of women friends" (Personal communications 1994). The fact that she was fortunate in having support for her scientific career aspirations does not blind Cobb to the reality that many barriers exist for women and minorities who wish to enter science. In an article that she wrote, "Filters for women in science" (Cobb 1979) she speaks of women in science as the filtrate that must pass through filters that have smaller pores than those for men. In this article, she suggests a number of remedies for

the lack of women in science fields. These remedies span the educational pipe-
line, from support for girls in elementary and high school to increase their
interest in science, to exposure to established women scientists for women undergraduate and graduate students in science.

In speaking of how the experiences of minority students enrolled in science
programs may have changed from the time that she attended graduate school,
Cobb states that she did not think that ''things have changed that much.'' She
believes, however, that there have been expanded opportunities for minority
students after the Martin Luther King assassination and that African-American
students can go into other areas. Furthermore, she continues, ''[T]here has been
little to encourage Black students to go into science. If they are not hooked into
a system where they have members of the family who are scientists and see
them as being successful in science, then it's unlikely that they'll choose science.'' She goes on to praise predominantly black colleges for their success in
producing graduate-level scientists and says that this is due to the mentors and
support networks that they provide (Personal communications 1994).

In telling about the way in which she was able to reconcile dual claims of
family and professional life, Cobb states: ''I managed just like men manage a
career and a private life. If one marries, it has to do with the kind of person
one marries. These days most women work, so it's not much of a surprise when
a woman has a career. It requires some kind of financial base, however'' (Personal communications 1994).

WORK

The major part of Jewel Plummer Cobb's research has been in cancer cell
biology, with human tumors and cancer chemotherapeutic agents. Her studies
have concentrated on human and other mammalian melanomas, as well as on
the effects of cancer chemotherapeutic agents and hormones on human and other
mammalian melanomas in tissue cultures.

Her postdoctoral research began at the Harlem Hospital Cancer Research
Foundation and involved tissue culture studies of the tumors of cancer patients,
as well as clinical studies of human patients treated *in vivo* with one of three
compounds: triethylene melamine (TEM), a radiomimetic compound; aureomycin, an antibiotic; and some 4–amino derivatives of folic acid. She and other
researchers at the Foundation wanted to find the mode of action of these compounds on cancer cells *in vitro* in order to discover if certain types of cells were
damaged more than others. They found that TEM *in vitro* at lower doses prevented human tumor cells from migrating along the flask surfaces. A-
Methopterin at significantly lower doses caused cell nuclei to enlarge, while
aureomycin at similar concentrations had no effect. Further studies comparing
the *in vivo* effects of chemotherapeutic agents with *in vitro* effects of the same
tissue from the same patient revealed a meaningful relationship between the *in
vitro* and *in vivo* effects of a chemotherapeutic agent used clinically in the same

patient with advanced neoplastic disease. She then undertook extensive research identifying a spectrum of *in vitro* cellular changes in tissue culture from a variety of sources.

When she moved to the University of Illinois Medical School in 1952, she began a new program that extended the graduate research she had previously undertaken. The work involved cytological studies of normal and malignant pigment cells. She also published data on human bladder cancer cell growth in tissue culture. When she returned to New York City and joined the Fifth Division Surgery of NYU-Bellevue Hospital Medical Center in 1955, she entered a most exciting phase of basic cell research undertaken in close coordination with the clinical cancer chemotherapy program directed by Louis T. Wright. In an early research experiment using mouse melanomas, Cobb wished to investigate whether the resistance to radium and X-ray therapy observed in patients with melanoma was a function of pigment density. Using tissue slices of densely pigmented and very pale areas of the same mouse tumor specimen, she exposed them to varying doses of X-rays and implanted them immediately in untreated host mice or grew them *in vitro*. The results showed that melanin protected the cells from X-ray damage.

In another significant study done at NYU, Cobb observed severe cellular changes following direct exposure of living cells *in vitro* to the antibiotic actinomycin D. These cellular changes were described *(in vitro)* and published for the first time. Later, other researchers reinforced these findings and ultimately explained the changes at the molecular level. Actinomycin D is now used as a laboratory tool to inhibit ribonucleic acid (RNA) synthesis. Subsequent research at NYU with Cobb's research assistant, Dorothy G. Walker, was often undertaken in concert with clinical studies on breast cancer and other tumors treated *in vivo* or *in vitro* with the nitrogen mustard derivatives. These included Thio-Tepa and Chlorambucil or Leukeran, the antibiotic puromycin derivative, the antibiotic actinomycin D or the folic acid antagonist A-Methopterin, now called Methotrexate.

What Cobb terms "my most interesting work" (Cobb 1989, 42) was an in-depth analysis of Thio-Tepa, a promising anticancer drug. This involved a cinematographic analysis of cell division movement behavior, photographed through a special phase contrast microscope. The paper ("Studies on human melanoma cells . . ." 1964) that emerged from this research was presented at the Eighth International Cancer Congress in Moscow in 1962. Another paper ("The comparative cytological effects . . ." 1960) describing the comparative cytological effects of several alkylating agents on human normal and neoplastic cells in tissue culture was published by the New York Academy of Sciences and was reprinted in several editions.

By 1960 Cobb had made numerous observations on the direct behavior of human normal and malignant cells, either as comparative studies in growth media or as part of a comparative chemotherapy study. The results of a five-year cytological study were published. After leaving NYU and assuming a full-time teaching position at Sarah Lawrence College in 1960, she continued her tissue culture studies. During this period the laboratory she established at Sarah Lawrence worked to perfect new techniques for organ cultures in sealed chambers placed in the peritoneal cavity of mice or in minidishes incubated at 37 degrees centigrade. At Connecticut College, where Cobb relocated in 1969, she built a new laboratory and, with two assistants, developed melanotic and amelanotic strains of the S91 mouse melanoma in cell culture. The work conducted during this period described significant changes in melanin intensity while reducing cell division in melanocyte stimulating hormone (MSH)-treated cells. When combined with Cytochalasin B, giant, multinucleated pigment cells formed. These data gave Cobb new insight into the dynamics of the pigment cell division cycle and the relationship of that cycle to the time of pigment formation.

The results of Cobb's research on the effects of newly discovered cancer chemotherapy drugs on human cancer cells are still valued today in medical research. Her pioneering laboratory work on the study of drugs used with cancer treatment has been a valuable contribution to the field.

In addition to being a productive and renowned research scientist, Cobb has been an effective teacher, administrator, and public servant. Her work with students at Sarah Lawrence College has produced scientists in the forefront of research in the respective fields of biology. She has been the recipient of five awards for her contribution to the field of education. She has been active in promoting programs to increase the number of women and underrepresented minorities in science. Her public service contributions include membership on various advisory boards and committees.

After a brilliant career as a scientist and educator, Jewel Plummer Cobb is uniquely qualified to give advice to women of color who wish to be successful in science careers: "I would tell [a woman of color who wanted to pursue a scientific career] to work very hard and have a [high] degree of perseverance and commitment. I would tell her to ask professors for assistance readily and freely all the time because that is what they are there for. I would advise her to take advantage of any summer programs, awards and [other support] that she can apply for" (Personal communications 1994).

BIBLIOGRAPHY

Works by Jewel Isadora Plummer Cobb

Scientific Works

"Effect of Several Aromatic Amidines on the Respiration and Aerobic Metabolism of Yeast Cells." Master's thesis, New York University, 1947 (as J. I. Plummer).
"Mechanisms of Pigment Formation." Ph.D. diss., New York University, 1950 (as J. I. Plummer).

(with G. Antikajian, L. T. Wright, et al.) "The effect of triethylene melamine, aureomycin, and some 4-amino derivatives of folic acid on tissues *in vitro.*" *Journal of the National Cancer Institute* 12 (1951): 269–274 (as J. I. Plummer).

"The *in vitro* effects of A-methopterin." *Proceedings of Second Conference on Folic Acid Antagonists in Leukemia Treatment, Blood VII* (Suppl.) (Jan. 1952): 152–190 (as J. I. Plummer).

(with L. T. Wright, G. Antikajian, et al.) "Triethylene melamine *in vitro* studies. I. Mitotic alterations produced in chick fibroblast tissue cultures." *Cancer Research* 12 (11) (1952): 796–800 (as J. I. Plummer).

(with M. J. Kopac) "The *in vitro* production of pigment granules." In *Pigment Cell Growth*, edited by M. Gordon, 305–317. New York: Academic Press, 1953 (as J. I. Plummer).

(with J. C. Wright, R. S. Coidan, et al.) "The *in vivo* and *in vitro* effects of chemotherapeutic agents on human neoplastic diseases." *Harlem Hospital Bulletin* 6 (1953): 58–63 (as J. I. Plummer).

"Tissue culture observations of the effects of chemotherapeutic agents on human tumors." *Transactions of the New York Academy of Sciences* 17 (1955): 237–249.

(with J. H. Keifer and H. Woods) "Human bladder neoplastic cells in tissue culture." *Journal of Urology* 73 (1955): 1039–1044.

"Effects of *in vitro* X-irradiation on pigmented and pale slices of Cloudman S91 mouse melanoma as measured by subsequent proliferation *in vivo.*" *Journal of the National Cancer Institute* 17 (1956): 657–666.

(with P. Foster, B. Billow, et al.) "The effect of triethylene thiophosphoramide on fifty patients with incurable neoplastic diseases." *Cancer* 10 (1957): 239–245.

(with J. C. Wright, S. L. Gumport, et al.) "Investigation of the relationship between clinical and tissue response to chemotherapeutic agents on human cancer." *Journal of Medicine* 257 (1957): 1207–1211.

(with D. G. Walker) "Effects of actinomycin D on tissue cultures of normal and neoplastic cells." *Journal of the National Cancer Institute* 21 (1958): 263–277.

(with J. C. Wright) "Studies on a craniopharyngioma in tissue culture." *Journal of Neuropathology and Experimental Neurology* 18 (1959): 563–568.

(———, F. M. Golomb, et al.) "Chemotherapy of disseminated carcinoma of the breast." *Annals of Surgery* 150 (2) (1959): 221–240.

"The comparative cytological effects of several alkylating agents on human normal and neoplastic cells in tissue culture." *Annals of the New York Academy of Sciences* 84 (1960): 513–542.

(with D. G. Walker) "Studies on human melanoma cells in tissue cultures. I. Growth characteristics and cytology." *Cancer Research* 20 (1960): 858–867.

(——— and J. C. Wright) "Observations on the action of triethylene thiophosphoramide (TSPA) within individual cells." *Acta Union International Contra Cancrum* 16 (1960): 567–583.

"Cells tell a story." *Sarah Lawrence Alumnae Magazine* (Winter 1961): 10–16.

(with D. G. Walker) "Comparative chemotherapy studies on primary short-term cultures of human normal, benign, and malignant tumor tissues—a five year study." *Cancer Research* 21 (1961): 583–590.

(———) "Effect on heterologous, homologous, and autologous serums on human normal and malignant cells *in vitro.*" *Journal of the National Cancer Institute* 27 (1961): 1–15.

(with F. M. Golomb, D. G. Walker, et al.) *"In vitro* selection of chemotherapeutic agents for perfusion therapy of human cancer." *Surgery* 51 (1962): 639–644.

(with J. C. Wright, S. L. Gumport, et al.) "Further investigation of the relation between the clinical and tissue culture response to chemotherapy agents on human cancer." *Cancer* 15 (1962): 284–293.

(with D. G. Walker) "Studies on human melanoma cells in tissue culture. II. Effects of several cancer chemotherapeutic agents on cytology and growth." *Acta Union International Contra Cancrum* 20 (1964): 206–208.

(———) "Cytologic studies on human melanoma cells in tissue culture after exposure to five chemotherapeutic agents." *Cancer Chemotherapy Reports* 52 (1968): 543–552.

"Cancer—a solution?" *Connecticut College Alumni Magazine* (Winter 1972): 10–11.

(with A. McGrath) "S91 mouse melanoma sublimes following total *in vitro* versus alternate *in vivo* passages." *Journal of the National Cancer Institute* 48 (1972): 885–891.

(———) *"In vitro* effects of melanocyte-stimulating hormone, adrenocorticotropic hormone, 17B-estradiol, or testosterone propionate on Cloudman S91 mouse melanoma cells." *Journal of the National Cancer Institute* 52 (1974): 567–570.

(with A. McGrath and N. Willetts) "Brief communication: Responses of Cloudman S91 melanoma cells to melanocyte-stimulating hormone: Enhancement by Cytochalsin B." *Journal of the National Cancer Institute* 56 (1976): 1079–1081.

"Toward licking cancer." *Rutgers University, Douglass College Alumnae Bulletin* (Spring 1977): 1–5.

Other Works

"The impact of the black experience on higher education in New England." In *An Occasional Paper of the School of Education, University of Connecticut*, edited by G. C. Atkyns, 100–115. Storrs, CT: School of Education, University of Connecticut, 1970.

"I am woman, black, educated." *Hartford Courant Sunday Supplement* (Feb. 4, 1973).

"Black women and higher education: A brief history." In *The Black Woman: Myths and Realities. A Symposium*, edited by D. J. Mitchell and J. H. Bell, 114–119. Cambridge, MA: Radcliffe College, 1975.

"Graduate study." *Connecticut College Alumni Magazine* (Spring 1976).

"Postbaccalaureate premedical programs for minority students." In *Minorities in Science: The Challenge for Change in Biomedicine*, edited by V. L. Melnick and F. D. Hamilton, 237–240. New York: Plenum/Rosetta Edition, 1977.

"Breaking down barriers to women entering science." *Physics Today* 32 (8) (1979): 78.

"Filters for women in science." *Annals of the New York Academy of Sciences* 323 (1979): 236–248.

(with Louis Benezet, Joel Conarroe, et al.) "Issues and problems: A debate. A colloquium." In *The Great Core Curriculum Debate: Education as Mirror of Culture*, 51–78. New Rochelle, NY: Change Magazine Press, 1979.

"Planning strategies for women in scientific professions." In *Women in Scientific and Engineering Professions*, edited by Violet B. Haas and Carolyn C. Perrucci, 75–85. Ann Arbor, MI: University of Michigan Press, 1984.

"A life in science: Research and service." *Sage* 6 (2) (1989): 39–43.

"Planning the academic future for women and minorities." *American Association of*

State Colleges and Universities, (A.A.S.C.U.), Memo to the President 30 (2) (May 18, 1990): 4.

"The role of women presidents/chancellors in intercollegiate athletics." In *Women at the Helm*, edited by J. A. Sturnick et al., 42–50. Washington, DC: AASCU Press, 1991.

"Societal barriers and strategies for succeeding in the technical sciences." In *Women in Engineering Conference, A National Initiative. Conference Proceedings*, edited by J. Z. Daniels, 3–7. West Lafayette, IN: Women in Engineering Program Advocates Network, 1991.

Personal communications to the author, 1994.

Works about Jewel Isadora Plummer Cobb

Irvin, Dona L. "Jewell [*sic*] Plummer Cobb." In *Notable Black American Women*, edited by J. C. Smith, 195–198. Detroit: Gale Research, 1992.

Kessler, J. H., et al. *Distinguished African American Scientists of the 20th Century*. Phoenix, AZ: Oryx Press, 1996.

Other References

De Kruif, P. H. *Microbe Hunters*. New York: Harcourt Brace Jovanovich, 1926.

JANE COLDEN (1724–1766)

Katalin Harkányi

BIOGRAPHY

Jane Colden was born on March 27, 1724, in the province of New York and died there on March 10, 1766. America's first woman botanist was the second daughter and fifth child of Cadwallader and Alice (Christy) Colden.

Although his parents resided in Dunse, Scotland, Cadwallader Colden was born in Ireland on February 17, 1688, where his mother was visiting relatives. He was raised in Berwickshire, Scotland, in a Presbyterian environment, since his father, Alexander, was a minister. Cadwallader Colden emigrated to the American colonies in 1710. After a long, successful life, he died near Flushing, New York, on September 28, 1776.

Alice Christy (whose maiden name is often spelled Christie in older documents) was the daughter of a Scottish minister. She was born on January 5, 1690, and died in March 1762. She married Cadwallader Colden in 1715, and in 1716 they sailed to the colonies and settled in Philadelphia.

Jane Colden's parents had ten children: Alexander, David (died in infancy), Elizabeth, Cadwallader, Jane, Alice, Sarah, John, Catharine, and David.

Her father was educated at the University of Edinburgh with the intention of becoming a minister and following in his father's footsteps. He received his A.B. degree in 1705 from the University. Because he was more interested in the sciences, he moved to London, where from 1705 to 1708 he studied medicine, anatomy, chemistry, and mathematics. It is not known at which university he was enrolled and from which one he received his M.D. degree.

Since establishing a medical practice in Scotland or in England was very expensive, and his family did not have the means to help him, Cadwallader Colden decided to relocate to America. He moved to Philadelphia in 1710, where one of his aunts lived. In 1715 Colden visited England and married Alice Christy. They returned to Philadelphia in 1716, where he planned to continue his medical career, but actually he became Surveyor General in the province of New York. Colden bought land in the town of Montgomery, Orange County

and in 1727 built his aristocratic colonial home, Coldengham (renamed Colden-
ham in the late 1700s and demolished in 1845). The Colden family relocated
to New York City in 1762, and their new home, Spring Hill, was built near
Flushing, Long Island. Cadwallader Colden was a Loyalist and remained one
until his death.

Colden was a political and historical writer. In addition to being a physician
and politician, he was a colonial scholar interested in physics, astronomy, and
botany. Some of his scientific writings were in physics, *An Explication of the
First Causes of Action in Matter*; in botany, "Plantae Coldenghamiae"; and in
mathematics, *An Introduction to the Doctrine of Fluxions, or the Arithmetic of
Infinites*.

Jane Colden's mother, Alice Christy, was born and raised in a Presbyterian
minister's home in Kelso, Scotland. She received the education of the gentlewomen of the era. Her warm, practical nature and intellect helped her run the
family and the estate during her husband's frequent travels. To a great degree
she was responsible for her children's education.

The Coldens had an aristocratic, highly civilized home. Both of Jane Colden's
parents were intellectuals and entertained distinguished visitors and scientific
men, such as John Bartram, Alexander Garden, and Peter Kalm. The family
lived in the wilderness, far from good schools, but had a valuable personal
library.

Jane Colden, just like her brothers and sisters, was educated at home by her
mother and father (Smith 1988, 1090). Her brother, Cadwallader, in his letters
to family members, testifies to the excellence of their mother's teachings (Eager
1846–47, 245). Colden's father recognized early her ability, reliability, and fondness for reading. Therefore, he took special care to educate her and provide an
intellectually supportive environment for her scholarly curiosity. He also encouraged her interest in research. Jane Colden mastered and used Carolus Linnaeus's classification system, which her father had translated from Latin to
English. Since she lived in the wilderness and because of her gender, Colden
could not freely travel and visit botanical gardens. Her father supplied her with
the next best thing: books on botany and gardens (Bonta 1991, 7). He also
taught her the art of spatter prints. For many years until her marriage, she was
actively engaged in finding, discovering, and classifying plants. There is no
evidence of her continued botanical activities after her marriage.

Jane Colden married when she was 35 years old, late in life according to
eighteenth-century standards. She became the wife of William Farquhar, a widower, on March 12, 1759, in the Trinity Church Parish, New York. Farquhar
was a Scotsman and a physician. He studied in Edinburgh, emigrated to the
colonies, and practiced medicine in New York City as well as in the surrounding
area. He was a founder of the St. Andrew's Society in New York and held
several offices of the Society. Farquhar died in 1787.

Colden and Farquhar had a happy marriage for seven years. Then their only
child was born, but mother and child died within a few weeks of each other.

Jane Colden Farquhar died on March 10, 1766, possibly in childbirth or as a result of it. The cause of her death is only alluded to in the various sources. The location of her grave is not known.

Through a chain of events, Jane Colden became well known in botanical circles in both America and Europe. Her father's letters contain references to his convictions concerning women's education and especially to their suitability to learn botany. He discussed his effort in educating his daughter. He sent examples of her descriptions of plants to European scientists. Visitors to Coldengham had seen her work. Finally, her own correspondence with scientists helped in recognizing her as a botanist.

WORK

Jane Colden was recognized and acknowledged by Linnaeus. It is known that one of her descriptions of plants was included in Linnaeus's *Species Plantarum* (Colden, "Description of Fibraurea . . ." 1821, 94–95). "If so, she was probably the only woman among the many naturalists around the world who contributed to that great work" (Smith 1988, 1091). According to some scholars, Colden was on equal footing with Peter Collinson, John Ellis, and J. F. Gronovius (Hollingsworth 1962, 32). Her work was well regarded by these men of science. Evidently, she was secure in her knowledge and scientific work, because if her observations differed from those of others, she was not afraid to state it. Writing about the *Clematis virginiana*, she states, "Neither Linnaeus take notice [*sic*] that there are some Plants of the Clematis that bear only Male flowers, but this I have observed with such care, that there can be no doubt of it" (Britten 1895, 15).

Unfortunately, to the regret of her contemporaries, no plant was named after her. Both Ellis and Collinson suggested it to Linnaeus, but he ignored their recommendations (Bonta 1991, 7). Ellis wrote to Linnaeus:

This young lady merits your esteem, and does honour to your system. She has drawn and described 400 plants in your method only; she uses English terms. Her father has a plant called after him Coldenia, suppose you should call this [Gold thread] Coldenella or any other name that might distinguish her among your Genera. (Denny 1948, 18)

Collinson asked Linnaeus to name a plant in honor of "the first lady that has so perfectly studied your system" (Bonta 1991, 7). She found the northern gold thread and called it *Fibraurea*. Today it is known as *Coptis groenlandica*, actually named by Richard A. Salisbury, who probably did not know Jane Colden's work concerning this plant (Eifert 1965, 54). She discovered another plant, which she named *Gardenia* in honor of Alexander Garden. It was unknown to her that this species was already classified and named by Linnaeus as *Hypericum virginicum*.

In an intellectual atmosphere and scientific environment, Colden became

America's first woman botanist. Her father's status as a botanist had helped pave the way for her. His scientific library and the books presented to her also augmented her chances of becoming a fine botanist. But achieving that status was entirely her own accomplishment.

By 1757 Jane Colden had described and sketched more than 300 local plants. She was an expert at identifying and classifying plant species indigenous to New York and vicinity. Her manuscript of New York flora contains detailed, careful morphological descriptions of plants; this indicates that her observations were taken from actual specimens. She provided folk names for the plants when known and listed their domestic and medicinal uses. Thus, we learn from her manuscript that the *Aclepias tuberosa* was used to alleviate colic (Britten 1895, 15). This work, after her death, came into the possession of Captain Frederick von Wangenheim. Later it was owned by a professor at the University of Goettingen, and finally by Sir Joseph Banks (Britten 1895, 13). Colden's manuscript found its way to the British Museum after Banks's death.

Jane Colden's cheese book (Cadwallader Colden 1918–37, 4: 220; 5: 85) is considered one of the first American manuals on making cheeses. In it she recorded the ingredients, exact amounts used, and circumstances under which she prepared her cheeses (Cadwallader Colden 1918–37, 5: 5–63). This book is another evidence of Colden's scientifically oriented mind and activities.

Colden corresponded with a number of great botanists and plant collectors of her time. References in contemporary records showed that she was well known for her botanical accomplishments and as a collector of plants and seeds. She could count Bartram, Garden, Ellis, and Robert Whytt as her friends.

Very few of Jane Colden's writings survived. Her manuscript of the New York flora is in the British Museum's Natural History collection; some correspondence is in Edinburgh, and other manuscripts with family papers are in the New York Historical Society. The latter were published in the Society's *Collections*.

BIBLIOGRAPHY

Works by Jane Colden

Scientific Works

"Flora Nov.—Eboracensis." British Museum (Natural History) Catalog. No. 26, c. 19. Manuscript. Ca. 1753–58.

"The description of a new plant: By Dr. Alexander Garden." *Essays and Observations, Physical and Literary* (Edinburgh Philosophical Society) 2 (1756): 1–7.

"Jane Colden to Charles Alston, New York, May 1, 1756." University of Edinburgh. Letter no. 99. Manuscript.

"Description." *Essays and Observations, Physical and Literary* (Edinburgh Philosophical Society) 2 (1770): 5–7.

"Description of *Fibraurea*, or *Gold Thread*." In *A Selection of the Correspondence of*

Linnaeus, edited by James Edward Smith, vol. 1, 94–95. London: Longmans, Hurst, Rees, Orme, and Brown, 1821. Reprinted, New York: Arno Press, 1978.

Botanical Manuscript of Jane Colden, 1724–1766. Edited by Harold Ricket and Elizabeth C. Hall. New York: Garden Club of Orange and Dutchess Counties, 1963.

Works about Jane Colden

Bonta, Marcia Myers. *Women in the Field: America's Pioneering Women Naturalists*. College Station: Texas A&M University Press, 1991.

Britten, James. "Bibliographical notes, viii—Jane Colden and the flora of New York." *Journal of Botany, British and Foreign* 33 (1895): 12–15.

Colden, Cadwallader. *The Letters and Papers of Cadwallader Colden*. Collections of the New York Historical Society, vols. 50–56; 67–68. 1917–35. Reprinted, New York: AMS Press, 1973.

Darlington, William. *Memorials of John Bartram and Humphry Marshall*. Philadelphia: Lindsay and Blakiston, 1849.

Denny, Margaret. "Naming the gardenia." *Scientific Monthly* 67 (1948): 17–22.

Eager, Samuel W. "The Colden family." *Newburgh Telegraph* (Apr. 25, 1861).

———. *An Outline History of Orange County*. Newburgh, NY: S. T. Callahan, 1846–47.

Eifert, Virginia S. *Tall Trees and Far Horizons: Adventures and Discoveries of Early Botanists in America*, chap. 4. New York: Dodd, Mead, 1965.

Gray, Asa. "Selections from the scientific correspondence of Cadwallader Colden with Gronovius, Linnaeus, Collinson, and other naturalists." *American Journal of Science and Arts* 44 (1843): 85–133. Also in *Selections from the Scientific Correspondence of Cadwallader Colden with Gronovious, Linnaeus, Collinson, and Other Naturalists*, arranged by Asa Gray. New Haven, CT: B. L. Hamlen, 1843.

Hindle, Brooke. "A colonial governor's family: The Coldens of Coldengham." *New York Historical Society Quarterly* 45 (1961): 233–250.

———. *The Pursuit of Science in Revolutionary America, 1735–1789*. Chapel Hill: University of North Carolina Press, 1956.

Hoermann, Alfred R. "A savant in the wilderness: Cadwallader Colden of New York." *New York Historical Society Quarterly* 62 (1978): 271–288.

Hollingsworth, Buckner. *Her Garden Was Her Delight*. New York: Macmillan, 1962.

Humphrey, Harry Baker. *Makers of North American Botany*. Chronica Botanica, no. 21. New York: Ronald Press, 1961.

Jarcho, Saul. "Biographical and bibliographical notes on Cadwallader Colden." *Bulletin of the History of Medicine* 32 (1958): 322–334.

Keys, Alice Mapeldsen. *Cadwallader Colden: A Representative Eighteenth Century Official*. New York: Columbia University Press, 1906.

Lokken, Roy N. "Cadwallader Colden's attempt to advance natural philosophy beyond the eighteenth-century mechanistic paradigm." *Proceedings of the American Philosophical Society* 122 (1978): 365–376.

Purple, Edwin Ruthven. *Genealogical Notes of the Colden Family in America*. New York: privately printed, 1873.

———. "Notes, biographical and genealogical, of the Colden family, and some of its collateral branches in America." *New York Genealogical and Biographical Record* 4 (1873): 161–183.

Reed, Elizabeth Wagner. *American Women in Science before the Civil War.* Minneapolis: University of Minnesota, 1992.

Schrader, H. A. "Flora Nov.—Eboracensis." *Journal fuer die Botanik* 2 (1801): 468–471.
    An account of Jane Colden's manuscript.

Smallwood, William Martin. *Natural History and the American Mind.* New York: AMS Press, 1967.

Smith, Beatrice Sheer. "Jane Colden (1724–1766) and her botanic manuscript." *American Journal of Botany* 75 (1988): 1090–1096.

Smith, James E. (ed.). *A Selection of the Correspondence of Linnaeus.* 2 vols. London: Longman, Hurst, Rees, Orme, and Brown, 1821.

Steacy, Stephen Charles. "Cadwallader Colden: Statesman and Savant of Colonial New York." 2 vols. Ph.D. diss., University of Kansas, 1987.

Stearns, Raymond Phineas. *Science in the British Colonies of America.* Urbana: University of Illinois Press, 1970.

Vail, Anna Murray. "Jane Colden, an early New York botanist." *Torreya* 7 (1907): 21–34; *Contributions from the New York Botanical Garden* 4 (1905–7): 21–34.

Wilson, Joan Hoff. "Dancing dogs of the colonial period: Women scientists." *Early American Literature* 7 (1973): 225–235.

LAURA NORTH HUNTER COLWIN
(1911–   )

Marjorie McCann Collier

BIOGRAPHY

Laura North Hunter was born on July 5, 1911, in Philadelphia, Pennsylvania. Her parents were Robert John and Helen Virginia (North) Hunter. She has a younger sister. Her ancestors from Holland first settled in what was then New Utrecht, Long Island, while others came from Scotland, Ireland, and England. Her more recent ancestors, most of whom were Presbyterian, lived in Pennsylvania.

Hunter's early environment consisted of winters in Philadelphia, where her father, a physician, practiced his specialty of otolaryngology, and her mother, who had majored in German and Latin at Bryn Mawr College, was a housewife. Her aunt, a schoolteacher, lived with them. During her childhood, summers were spent at the seashore of southern New Jersey. Later the family bought a summer home in Chester County, Pennsylvania. This early exposure to a variety of locales and the regard of her family for education provided many and diverse experiences for a child who was interested in nearly everything. Significant weekly events for the entire family were the Saturday lectures and exhibits at the University of Pennsylvania Museum.

By the time Hunter entered kindergarten, she knew that she wanted to go to Bryn Mawr. Her mother had sung "Pallas Athene," the college's anthem, as a lullaby and had entertained her daughters with many accounts of her days there. Hunter attended public school until she was 13, then transferred to the Phoebe Anna Thorne School, a private school for girls that was closely associated with Bryn Mawr. At Thorne, where she had a number of instructors from Bryn Mawr, Hunter received a sound secondary education and reinforcement of the philosophy that women are not inferior to men.

After Thorne, Hunter went to Bryn Mawr College. In her first year there, a required science course stimulated her interest in biology, which became her major subject. Her experience at Bryn Mawr College "marked [her] for life." She has described her time there as "lively, austere, merry, . . . irritating, invigorating," including memorable sensations such as the "pure, tickling delight" she felt at Professor Sally Schrader's lectures in first-year biology. The courses, the professors, and extracurricular programs, such as the one on the Summer School for Working Girls, all contributed to what she has called the "Continuing Effect." The benefits included not only preparation for graduate school but a way of dealing with onerous or boring tasks and a lasting joy in applying the mind to problems and new directions not necessarily restricted to science (Colwin 1982, 26–28).

After receiving the B.A. degree from Bryn Mawr College in 1932, Hunter entered the University of Pennsylvania. Her original intention was to study embryology, but the embryology professor was on sabbatical leave that year. Having read of the work of the protozoologist D. H. Wenrich, she decided to take a course with him. She became interested in protozoology and, under the supervision of Wenrich, worked on a ciliate found in the digestive tracts of sea cucumbers. She was awarded the M.A. degree by the University of Pennsylvania (1934) and immediately entered the doctoral program there. She continued to study the ciliate from her master's research, under the direction of Wenrich. Her doctoral research centered on the ciliate's reproduction, with emphasis on nuclear phenomena. Hunter received the Ph.D. degree from the University of Pennsylvania in 1938.

While a graduate student (1933) Hunter was introduced to the Marine Biological Laboratory (MBL) in Woods Hole, Massachusetts, an institution that originated, in part, from the Women's Education Association of Boston. From its beginning in 1888, it served as a summer laboratory for both female and male students and researchers. Here she collected and studied the marine organisms, attended lectures on a wide variety of subjects, and met biologists from all over the world. During the summer of 1937, in a communal laboratory for beginning investigators, she met embryologist Arthur Lentz Colwin, who was a recent graduate of the doctoral program at McGill University and a postdoctoral fellow at Yale University.

In 1936 Hunter accepted her first academic position as an instructor of biology at Pennsylvania College for Women (now Chatham College). The next year she was promoted to assistant professor and held that position until 1940, when she and Arthur Colwin were married. That same year she was hired as an instructor of zoology by Vassar College, where she remained until 1943. During the years of the Second World War, Arthur Colwin served in the Pacific. When he returned to civilian life and his position at Queens College (now part of City University of New York), they resumed their studies and began collaborating on problems in embryology. In 1945 Laura Colwin returned to Chatham College as an instructor of biology for one semester.

Laura Colwin was hired as a temporary instructor at Queens College (1947) to help with the influx of students after the war. She was rehired repeatedly as an instructor but was always assigned one hour less than the minimum for a full-time position, so that she would remain ineligible for tenure. After a number of years there, despite a very productive research collaboration with her husband and favorable evaluations of her teaching, she was denied a tenure-track position because the dean did not approve hiring both a husband and a wife in the same department. She felt hurt by this but did not protest because she thought it a small price to pay for having research space with her husband and collaborator. Several years later, the chairman of the department found that he could pay her more as a lecturer. She became and remained a lecturer while her husband was promoted to a professorship.

They continued with their research and teaching. In the summers they returned to the MBL as investigators, members of the corporation, and, eventually, trustees. Arthur Colwin taught the MBL embryology course, which is an advanced summer course for graduate students and postdoctoral fellows; Laura Colwin participated in an unofficial capacity. In 1953 Laura Colwin was awarded the Morrison Fellowship of the American Association of University Women, and Arthur Colwin was awarded a Fulbright Fellowship. This support allowed them a year of research at the Misaki Marine Biological Station of the University of Tokyo.

Other embryologists recognized that in their research Laura Colwin and Arthur Colwin were equal partners. They alternated as first author on the publications of their papers, which by the early 1960s had received worldwide acclaim. On their research grants, Laura Colwin was always coprincipal investigator. They alternated as speakers at the many conferences and symposia to which they were invited. From Japan to Europe they were referred to familiarly and respectfully as ''the Colwins.''

More important, Arthur and Laura Colwin saw themselves as equals. Their partnership was built on respect for each other, including the right to have differences of opinion about the interpretation of data. They collaborated so thoroughly and completely that the results (papers, presentations) were seamless and single-voiced despite being the work of two very different, strong-minded individuals.

Finally, in 1966, the efforts of the chairman of the Biology Department and other colleagues in the science departments of Queens College were successful in having Laura Colwin appointed a full professor, much to her surprise and pleasure.

Although she was more widely recognized for her research and preferred research to teaching, Colwin spent as much time and effort on teaching as she did on research. The result was something of a life in double time for both Colwins; full teaching schedules and late night hours at the electron microscope and in the darkroom were routine. Despite her preference for research, she enjoyed teaching and brought all her skills to it. For example, she sewed a model of the early chick embryo, with different-colored layers of cloth representing the three germ layers and drawstrings that could be pulled to simulate the foldings of the germ layers that produce the body and the extraembryonic membranes of the chick. Laura Colwin especially valued the contact with students of widely varied backgrounds; she found them interesting personally and discovered that preparing lectures that would be clear to all of them helped clarify her own thinking.

In addition to research and teaching, Colwin worked on college committees. She was a member of the Queens College Women's Club (originally, Faculty Wives' Club) and once served as its president. In the late 1960s, at the time of the student unrest, Colwin was active in efforts to resolve the differences between the students and the college and to prevent further violent acts.

The Colwins retired from Queens College (1973), were made Professors Emeriti, and became Adjunct Professors at the Rosenstiel School of Marine and Atmospheric Science, the University of Miami, near Key Biscayne, Florida. In 1975 they were voted Trustees Emeriti of the MBL. The contributions of Laura and Arthur Colwin to Queens College were recognized by the naming of the building in which they had worked the Laura H. and Arthur L. Colwin Building.

WORK

In graduate school she worked on two forms of a peritrich, *Urceolaria synaptae*, that lived in the digestive tract of sea cucumbers. She studied their differing morphology and behavior, describing their reproduction, including conjugation in the peritrich form that inhabited the posterior region of the gut.

In the two years (1938–40) between receiving her doctorate and marrying, she was an instructor and then assistant professor at Chatham College. Most of her time, except summers at the MBL in Woods Hole, was taken with teaching. When she and Arthur Colwin married, they decided that collaboration in research would be more enjoyable for them and more likely to be productive than if each pursued a separate course. Since Arthur Colwin was an embryologist by training, and Laura Colwin had long been drawn to embryology, they decided to go in that direction.

The Colwins' first research plan was to study the early development of the enteropneust *Saccoglossus kowalevskii*. Their goal was to discover information that would help clarify the evolutionary position of the Enteropneusta, commonly known as acorn worms, which in the past had been variously aligned with Echinodermata, placed in a subphylum of Chordata, or given a phylum of their own. Unfortunately, they had barely begun their planned research when World War II started. When they resumed their work after the war, it took them five or more years, consisting mostly of work during summers, to advance their initial plan.

As their work with *Saccoglossus* revealed the bases for body axes in the early embryos and the developmental potencies of the isolated blastomeres, they also became interested in its fertilization. In an abstract published in 1949 they described the fertilization reaction in *Saccoglossus*, including observation of a filament connecting the sperm and the egg. Such filaments had been seen by many since the time of H. Fol (1877), but their source and function had not yet been determined.

In the next several years they continued to investigate *Saccoglossus* development, but their interest in the unanswered questions of fertilization grew. By 1953 they had obtained data and written manuscripts for two papers, which were published in 1954. One dealt with changes in the egg cortex and its extraneous coats, and the other focused on the spermatozoon and the fertilization cone. In the latter, they presented a fuller account of the filament first discussed in their 1949 abstract. They conjectured that this filament was the result of a reaction of the acrosome of the sperm head in response to its approach to the egg. In 1952 Jean C. Dan had reported the breakdown upon exposure to eggwater of the acrosome of echinoderm spermatozoa with the release of a mass of substance and had proposed that this material played a role in penetrating the extracellular membranes (coats) of the egg. The Colwins further considered the possibility that when the *Saccoglossus* spermatozoon reached the outer egg coat, it reacted and formed a filament from material similar to that which Dan had described as a "labile mass of substance" and that this filament elicited the fertilization cone when it reached the plasma membrane of the egg.

While in Japan (1953–54) on fellowship to continue work with enteropneust embryology, the Colwins observed fertilization reactions among various echinoderms as they waited for mature specimens of acorn worms. Among the echinoderms they discovered that a species of sea cucumber had gametes that lent themselves to fertilization experiments. Studying them helped the Colwins sharpen their approach to the question of the filament. They concluded in 1955 that the filament, which they now, using Dan's term, referred to as the acrosomal filament, precedes the sperm into the egg cytoplasm and stimulates it to form the fertilization cone.

Upon their return to the United States and during summers at MBL, they extended their investigations of fertilization to include other echinoderms and certain polychaete worms. Recognizing that further definition of the sperm–egg interactions required fine structure studies, the Colwins collaborated with D. E. Philpott, an electron microscopist at MBL. In 1957 a National Institutes of Health grant allowed them to buy their own electron microscope. Their hours in the laboratory and darkroom became longer and more intense than ever.

In the late 1950s, the Colwins concentrated on fertilization in the polychaete *Hydroides hexagonus*, which resulted in three papers (1960) describing lytic activities of the sperm, reactions of egg coats, and the acrosome of the sperm during fertilization. In the summer of 1960 their years of work brought them to a discovery that spread like the elevation of a fertilization membrane around an egg in every direction over the globe of embryologists. From the painstaking work of making serial sections for electron microscopy of spermatozoa and eggs at many short intervals after insemination and evaluation of thousands of electron micrographs, they discovered the truth about fertilization. The sperm does not *enter* the egg as everyone from Fol on, including themselves, had believed;

the sperm and the egg *fuse* with each other at the point where they first meet, the tip of acrosomal *tubules*.

In September 1960, they presented their findings at a symposium on germ cells and development, organized by the International Institute of Embryology in Pallanza, Italy. At the same time, manuscripts of a series of three papers were submitted to the *Journal of Biophysical and Biochemical Cytology*, which accepted them immediately, although they were not published until June 1961. At the symposium and in these papers the Colwins described and illustrated the intricate morphology of the acrosome and the complicated sequence of changes it underwent as it approached the egg. The apex of the acrosome dehisced, lysing holes in the extraneous coats, and the acrosomal vesicle everted to form a projecting tuft of tubules. These tubules corresponded to what earlier had been called acrosomal filaments, a term that was no longer appropriate for membranous projections that were found to be hollow. The Colwins' electron micrographs showed that the tip of the acrosomal tubules fused with the egg plasma membrane and that the fused membranes parted, making the two cells confluent, forming a zygote with a cell membrane that was a mosaic of sperm and egg plasma membranes. They pointed out that the old idea of the sperm's penetrating the egg at fertilization was in error and proposed that *incorporation* be used instead of *penetration* to describe the process.

Next, they returned to *Saccoglossus*, where they again found that fertilization was a matter of membrane fusion, with the variation of one long acrosomal tubule instead of several short ones. During the next several years, they extended their researches to fertilization in several other invertebrates and, with I. Friedmann, in an alga, *Chlamydomonas reinhardi*. In all of these organisms they found fusion of gamete membranes producing the zygote. As others investigated the fine structure of fertilization in a variety of animals, including mammals, the Colwins' findings were confirmed again and again.

Since 1932 several workers had observed refertilization of blastomeres of early embryos from which the extraneous membranes had been removed. In the 1960s the Colwins investigated this phenomenon in *Saccoglossus* by electron microscopy. They found that some of the spermatozoa exposed to the denuded embryos became activated, dehiscing at the apex and forming an acrosomal tubule. The spermatozoa fused with the blastomere plasma membrane in the same way as in normal fertilization. This observation showed that contact with the extraneous coats of the unfertilized egg was not required for the spermatozoon's role in the normal fertilization process.

Their last publication, with R. G. Summers (1974), described fertilization in *Thyone briareus*, which was found to have a fibrous core to its acrosomal tubule. Again, they found that there was a mutual incorporation of the gametes by membrane fusion, not a penetration of the egg by the sperm. (Unfortunately, many authors still use the term ''penetration'' when writing about fertilization.)

Despite the heavy demands of teaching and research, Laura Colwin was a member of many learned societies, including the American Society for Developmental Biology, the American Society of Zoologists, the Electron Microscope Society of America, the New York Academy of Sciences, and Sigma Xi. She served as treasurer (1963–65) of the New York Society of Electron Microscopists and president (1970–72) of the Sigma Xi chapter at Queens College. She was very active as a corporation member of the MBL, especially as an elected trustee (1971–75).

BIBLIOGRAPHY

Works by Laura North Hunter Colwin

Scientific Works

Space does not permit the listing of the complete works of Laura North Hunter Colwin. Listed here are all works by Colwin except those cited in Colwin, Friedmann, and Colwin (1968) and in Colwin and Colwin ("Fertilization . . ." 1954). Included here are her dissertation and all references cited in the text.

"Diversion Conjugation and Reorganization in *Trichodine synaptae (Protozoa, Ciliata)* with Special Reference to the Nuclear Phenomena." Ph.D. diss., University of Pennsylvania, 1938 (as L. N. Hunter).

"Binary fission and conjugation in *Urceolaria synaptae.* Type II (*Protozoa, Ciliata*) with special reference to the nuclear phenomena." *Journal of Morphology* 75 (1944): 203–249.

    Based on doctoral dissertation.

"Note on the spawning of the holothurian, *Thyone briareus* (Lesueur)." *Biological Bulletin* 95 (1948): 296–306.

(with A. L. Colwin) "The fertilization reaction in the egg of *Saccoglossus (Dolichoglossus) kowalevskii.*" *Biological Bulletin* 97 (1949): 237.

(———) "Fertilization changes in the membranes and cortical granular layer of the egg of *Saccoglossus kowalevskii* (Enteropneusta)." *Journal of Morphology* 95 (1954): 1–46.

(———) "Sperm penetration and the fertilization cone in the egg of *Saccoglossus kowalevskii* (Enteropneusta)." *Journal of Morphology* 95 (1954): 351–372.

(———) "Sperm entry and the acrosome filament (*Holothuria atra* and *Asterias amurensis*)." *Journal of Morphology* 97 (1955): 543–568.

(———) "The acrosome filament and sperm entry in *Thyone briareus* (Holothuria) and *Asterias.*" *Biological Bulletin* 110 (1956): 243–257.

(———) "Morphology of fertilization: Acrosome filament formation and sperm entry." In *The Beginnings of Embryonic Development,* edited by A. Tyler et al., 135–168. Washington, DC: American Association for the Advancement of Science, 1957.

(——— and D. E. Philpott) "Electron microscope studies of early stages of sperm penetration in *Hydroides hexagonus* (Annelida) and *Saccoglossus kowalevskii* (Enteropneusta)." *Journal of Biophysical and Biochemical Cytology* 3 (1957): 489–502.

(with A. L. Colwin) "Egg membrane lytic activity of sperm extract and its significance in relation to sperm entry in *Hydroides hexagonus* (Annelida)." *Journal of Biophysical and Biochemical Cytology* 7 (1960): 321–328.

(———) "Fine structure studies of fertilization with special reference to the role of the acrosomal region of the spermatozoon during penetration of the egg (*Hydroides hexagonus*-Annelida)." In *Symposium on the Germ Cells and Development*, edited by A. Tyler et al., 220–222. Milan, Italy: Istituto Internazionale d'Embryologie e Fondazione A. Basalli, Istituto Lombardo, 1960.

(———) "Formation of sperm entry holes in the vitelline membrane of *Hydroides hexagonus* (Annelida) and evidence of their lytic origin." *Journal of Biophysical and Biochemical Cytology* 7 (1960): 315–320.

(———) "Changes in the spermatozoon during fertilization in *Hydroides hexagonus* (Annelida). I. Passage of the acrosomal region through the vitelline membrane." *Journal of Biophysical and Biochemical Cytology* 10 (1961): 231–254.

(———) "Changes in the spermatozoon during fertilization in *Hydroides hexagonus* (Annelida). II. Incorporation with the egg." *Journal of Biophysical and Biochemical Cytology* 10 (1961): 255–274.

(———) "Fine structure of the spermatozoon of *Hydroides hexagonus* (Annelida), with special reference to the acrosomal region." *Journal of Biophysical and Biochemical Cytology* 10 (1961): 211–230.

(———) "Cell–cell interaction in fertilization." In *Conference on Cellular Dynamics*, 353–369. New York: New York Academy of Sciences Interdisciplinary Communications Program, 1968.

(———) "Comparative studies of sperm–egg association and their bearing on some problems of fertilization." *Accademia Nazionale dei Lincei, Rome* 104 (1968): 11–24.

(with I. Friedmann and A. L. Colwin) "Fine structural aspects of fertilization in *Chlamydomonas reinhardi*." *Journal of Cell Science* 3 (1968): 115–128.

(with A. L. Colwin and R. G. Summers) "The acrosomal region and the beginning of fertilization in the holothurian *Thyone briareus*." In *Symposium on the Functional Anatomy of the Spermatozoon*, edited by B. A. Afzelius, 27–38. Oxford: Pergamon Press, 1974.

Other Works

"Bryn Mawr to me." In *Contributions to the Class of 1932 Reunion*, edited by C. T. Siepmann, 24–35. Bryn Mawr, PA: Bryn Mawr College, 1982.

Personal communications to author, 1994.

Other References

Dan, J. C. "Studies on the acrosome. I. Reaction to egg-water and other stimuli." *Biological Bulletin* 103 (1952): 54–66.

Fol, H. "Sur le commencement de l'henogenie chez divers animaux." *Archives de Zoologie Experimentale et Genetique* 6 (1877): 145–169.

GERTY THERESA RADNITZ CORI (1896–1957)

Rose K. Rose

BIOGRAPHY

Gerty Theresa Radnitz was born on August 15, 1896, in Prague, Austria-Hungary (now Czech Republic). She was the oldest daughter of Otto and Martha (Neustadt) Radnitz. Her father was a businessman who managed several sugar refineries. Her early education was obtained from tutors at home, and in 1906 she was enrolled in a school for girls, from which she graduated in 1912 at the age of 16.

Gerty Radnitz was interested in pursuing her studies at the university, probably in science or medicine. At the time, few women attended the university, but she was greatly encouraged by an uncle who was a physician and a professor of pediatrics. Unfortunately, her previous education was weak in several subjects, especially Latin, mathematics, and the sciences. In order to enter the university she had to pass the *matura*, a difficult, special, comprehensive examination. She studied for one year and passed the *matura* at the Tetschen Realgymnasium. This enabled her to enter the Medical School of the German University of Prague (Karl Ferdinand University) in 1914, from which she received the M.D. degree in 1920.

In an anatomy class at the university she met Karl Ferdinand Cori. They soon learned that they had a great deal in common: a fascination with science, a desire to pursue research, a love for the out-of-doors, and an interest in sports. They performed research together on the components of blood and published their first joint paper (1920). After obtaining their medical degrees, they both moved to Vienna, where they were married on August 5, 1920.

Karl Cori obtained an assistantship in the medical clinic at the University of Vienna, while Gerty Cori found a position as an assistant at the Karolinen Children's Hospital. This was the beginning of a pattern that would continue through most of their professional careers. Even though she was as good a student as Karl and was an equally good researcher, Karl Cori obtained better jobs and more recognition than his wife. He would find a post, and she would follow and be given a lesser position. They had a difficult time in getting equal treatment from the academic and scientific communities. Whenever possible they conducted their research together but never competed with each other or felt like rivals (Veglahn 1991, 59).

Karl Cori was unhappy about the lack of opportunities in postwar Europe. After two years in Vienna, he accepted a position as biochemist at the New York State Institute for the Study of Malignant Diseases (now Roswell Park Memorial Institute) in Buffalo. A few months later the Institute was persuaded to offer Gerty Cori a position as an assistant pathologist, and she came to the United States. In 1925 she was appointed assistant biochemist, and they began their joint research. At first they investigated the metabolism of tumors, and later their interest turned to carbohydrate metabolism, which occupied the rest of their lives.

In 1928 Karl and Gerty Cori became naturalized citizens of the United States. (He changed his first name to Carl.) They developed a great love for their adopted country. Later in her life Gerty Cori told an interviewer: "I believe that the benefits of two civilizations, followed by the freedom and opportunities of this country, have been essential to whatever contributions I have been able to make to science" (Opfell 1978, 192).

In 1931 Carl Cori was offered a position as professor of biochemistry and pharmacology at Washington University of St. Louis, Missouri. This time provisions were made so that Gerty Cori could continue to work with her husband. She was hired as a research assistant in pharmacology and was able to use the same laboratory. They continued their research on carbohydrate metabolism, and their laboratory attracted students from around the world. Their research collaborators included the future Nobel laureates Arthur Kornberg (1959), Severo Ochoa (1959), Luis Leloir (1970), Earl Sutherland, Jr. (1971), Christian deDuve (1974) and Edwin G. Krebs (1992). In addition, several women scientists collaborated with the Coris, including Jane Anne Russell* and Salome Gluecksohn Waelsch.*

An article in the *New York Post* described the cooperative research by the Coris as follows, "[I]t is hard to tell where the work of one leaves off and that of the other begins. They talk over their problems together, decide what is to be done, and then parcel the tasks out between them, checking and correlating with each other all the way" (Opfell 1978, 189).

The Coris had one son, Carl Thomas, who was born in August 1936, when Gerty Cori was 40 years old. He earned a Ph.D. degree in chemistry and became a research biochemist.

Carl Cori was promoted to professor of biochemistry and pharmacology, and in 1946 he became chairman of the Department of Biological Chemistry. In 1942 Gerty Cori was appointed associate professor of research in biochemistry and pharmacology, and in 1946 she was transferred to the Department of Biochemistry.

In 1947 the Coris shared the Nobel Prize in physiology or medicine with

Bernardo Houssay, an Argentinian doctor. The Nobel Prize was given to them in recognition of their elucidation of the catalytic conversion of glycogen. Gerty Cori was the first American woman to win a Nobel Prize in science. That year Gerty Cori was promoted to professor.

In the summer of 1947, while climbing Snow Mass Peak in Colorado, Gerty Cori noticed the first symptoms of the disease that was to take her life. She was suffering from myelofibrosis, a rare cancer of the bone marrow that usually proves fatal and requires many blood transfusions. In spite of her illness, she continued with her research. She died on October 26, 1957. Severo Ochoa described his former teacher's rich, open personality and her astonishing drive for action, blended with a universally kind and charming nature (Ochoa and Kalckar 1958).

Gerty Cori was elected to the National Academy of Sciences in 1948 and that same year received the Garvan Medal for women chemists from the American Chemical Society. Together with Carl Cori she received the Adler Prize, the Midwest Award, and the St. Louis Award, all from the American Chemical Society. They also received the Squibb Award of the American Society of Endocrinology and a Sugar Research Foundation Award. She was given the American Brotherhood Award by the National Conference of Christians and Jews and was named a Woman of Achievement by the Women's National Press Club. She received honorary degrees from Boston University (1948), Smith College (1949), Yale University (1951), Columbia University (1954), and the University of Rochester (1955). In 1950 President Harry S. Truman appointed her to the Board of Directors of the National Science Foundation, a post in which she served till her death.

At a memorial service for Gerty Cori, a recording was played of a radio program on which she had appeared and spoke about her work:

For a research worker the unforgotten moments of life are those rare ones, which come after years of plodding work, when the veil over nature's secret seems suddenly to lift and when what was dark and chaotic appears in a clear and beautiful light pattern. . . . I believe that in art and science are the glories of the human mind. I see no conflict between them. (Parascandola 1980, 167)

WORK

While still in Vienna, Gerty Cori became interested in helping children with congenital myxedema, a disease involving the reduced function of the thyroid. Her research, both clinical and experimental, dealt with studies of the influence of the thyroid on temperature regulation (1922). She continued this work in Buffalo and published the results in a paper describing the effect of the thyroid extracts on paramecia multiplication (1923).

With the discovery of insulin in 1921, scientists increased efforts to determine the role of sugar with regard to diabetes. Using the quantitative approach, Carl and Gerty Cori investigated *in vivo* the effect of insulin on various body chemicals, such as liver glycogen. They also analyzed venous and arterial blood for organic and inorganic phosphates, lactic acid, and lactates. They developed very accurate methods of analysis for carbohydrates and their metabolic products and investigated the rate of sugar absorption in the intestine.

In 1929 they proposed the Cori cycle to explain the interrelationship between glucose and glycogen, the form in which glucose is stored in the liver. Before their discoveries it was believed that the metabolic breakdown of glycogen involved its hydrolysis to glucose. The Coris suggested that blood glucose is changed to muscle glycogen, which then becomes blood lactic acid. The lactic acid can form liver glycogen, which becomes blood glucose, thus completing the cycle. The Coris searched for an active agent within insulin that caused the formation of liver glycogen. In 1949 they isolated the hyperglycemic-glycogenolytic factor.

In 1936 the Coris discovered and isolated a new phosphorylated intermediate of carbohydrate metabolism, glucose-1-phosphate (the Cori ester). In 1938 they described its enzymatic interconversion with glucose-6-phosphate, which was known to be formed by the phosphorylation of glucose in an enzyme-catalyzed reaction involving adenosine triphosphate (ATP). They then demonstrated that the formation of glucose-1-phosphate from glycogen is affected by a new enzyme, phosphorylase, which catalyzed the cleavage and synthesis of polysaccharides. Thus, the Coris showed the existence of an enzymatic mechanism for the phosphorolysis of the glycosidic bonds of a polysaccharide.

The Coris identified and isolated other enzymes involved in the formation and breakdown of the highly branched glycogen molecule. In 1939 they performed the first *in vitro* synthesis of glycogen from glucose-1-phosphate with phosphorylase and some glycogen as a primer. They produced both linear and branched-chain glycogen. This was an exciting achievement, because it was the first test-tube synthesis of a natural polymer from its precursor in a system that did not contain any cells. This also provided the proof of the structure of glycogen. Later the Coris also succeeded in the test-tube synthesis of glycogen from glucose.

Their research resulted in the isolation of other important enzymes: phosphoglucomutase, muscle aldolase, diphosphopyridine nucleotide pyrophosphatase, and certain hexokinases (enzymes that work in six-carbon sugars). Using various enzymes, they analyzed the structure of polysaccharides and elucidated the mechanism of metabolism of biochemical phosphates.

In 1953 Gerty Cori reported her results of the investigation of the nature of hereditary glycogen storage diseases in children. She recognized two types of disorders: one involving excessive amounts of normal glycogen and the other characterized by abnormally branched glycogen. She showed that both types of disorders result from the deficiency or changes in particular enzymes in the metabolic pathway. She is reported to be the first to prove that a disease could be caused by the lack of a specific enzyme. Her work demonstrated the importance of investigating enzymes to improve understanding of dysfunction of those metabolic processes in which these enzymes participate.

Gerty Cori was a member of the American Association for the Advancement of Science, the American Chemical Society, the American Philosophical Society, the American Society of Biological Chemists, the Harvey Society, and Sigma Xi, as well as an honorary member of Iota Sigma Pi.

BIBLIOGRAPHY

Works by Gerty Theresa Radnitz Cori

Scientific Works

Space does not permit the listing of complete works of Gerty Theresa Radnitz Cori. A bibliography of works by and about Cori can be found in Miller (1993). Included here are references cited in the text.
(with K. Cori) "Über den gehalt des menschlichen blutserums an komplement und nor-malambozeptor für hammelblutkorpchen." *Zeitschrift für Immunitatsforschung* 29 (1920): 445–462 (as G. Radnitz).
"Über den einfluss der schilddrüse auf die wärmeregulation." *Archiv für Experimentelle Pathologie und Pharmakologie* 95 (1922): 378–380.
"The influence of thyroid extracts and thyroxin on the rate of multiplication of para-mecia." *American Journal of Physiology* 65 (1923): 294–299.

Works about Gerty Theresa Radnitz Cori

Miller, Jane A. "Gerty Theresa Radnitz Cori (1896–1957)." In *Women in Chemistry and Physics*, edited by Louise S. Grinstein et al., 120–127. Westport, CT: Green-wood Press, 1993.
Ochoa, S., and H. M. Kalckar. "Gerty T. Cori, biochemist." *Science* 128 (1958): 16–17.
Opfell, Olga S. *The Lady Laureates*. New York: Scarecrow Press, 1978.
Parascandola, John. "Gerty Cori." In *Notable American Women*, edited by B. Sicherman and Carol Hurd Green, 165–166. Cambridge, MA: Belknap Press of Harvard University Press, 1980.
Veglahn, Nancy J. "Gerty Cori (1896–1957)." In *American Profiles: Women Scientists*, 57–65. New York: Facts on File, 1991.

LYDIA MARIA ADAMS DEWITT (1859–1928)

Mary R. S. Creese

BIOGRAPHY

Lydia Maria Adams was born on February 1, 1859, in Flint, Michigan. She was the second daughter among the three children of Oscar Adams, a Flint attorney, and Elizabeth (Walton) Adams. When Lydia Adams was five, her mother died, and her father remarried. His second wife was his first wife's sister, who brought up the children. After completing her early education in Flint public schools, Lydia Adams began to teach. In 1878, at the age of 19, she married a colleague, Alton D. DeWitt, also a native of Flint. She had two children: a daughter, Stella, and a son, Clyde, born in 1879 and 1880, respectively. Family commitments notwithstanding, she continued to teach and also took a two-year course at the Ypsilanti Normal College, from which she graduated in 1886 with high standing.

Over the next nine years she taught in the public school system, accompanying her husband as he moved to various school superintendent positions throughout Michigan, first in St. Louis, later in South Haven and Portland. In 1895, when her younger child reached the age of 15, Lydia Dewitt returned to a full program of study, enrolling at the University of Michigan, for a combined medical and science course. Her M.D. degree was awarded in 1898 and B.S. degree in 1899; she was then 40.

Their interests having diverged drastically, the DeWitts separated about this time. Lydia DeWitt stayed at the University of Michigan, joining the staff of George Dock, professor of medicine and pathology, first as a demonstrator in anatomy (1896–97) and then as assistant in histology until 1902, after which she was promoted to instructor. She held the latter position until 1910, with a year's break in 1906 for a study leave at the University of Berlin. Her early research during this period was carried out under the guidance of Carl Huber, junior professor (later, full professor) of anatomy and histology. Of special importance was her independent work in microscopic anatomy and neuroanatomy, which brought her a star in the 1906 edition of *American Men of Science* (Cattell 1906–). Though she was admitted to the Association of American Anatomists in 1902, DeWitt, along with several other women research workers at the University of Michigan, was excluded from both the institution's Faculty Research Club and the Junior [Faculty] Research Club (Rossiter 1982, 214). In 1902, therefore, very conscious of the need for a supportive organization, DeWitt started the Women's Research Club, which functioned successfully for decades at the University of Michigan, encouraging research by women. For several years she served as its president.

In 1910 she moved to St. Louis, Missouri, where she took up the positions of instructor of pathology at Washington University and assistant city pathologist and bacteriologist in the St. Louis Department of Health. Her move was doubtless related to the fact that her senior colleague at the University of Michigan, George Dock, became professor of medicine at Washington University the same year. Dock's strong interest in state and local public health work may also have influenced DeWitt and led to her taking her position with the city health department. Over the next two years she and her coworkers published notable reports on bacteriological procedures in public health practice, including work on the diphtheria organism and typhoid diagnosis. This resulted in her being invited by Harry Gideon Wells, professor of pathology at the University of Chicago, to join the university's Otho S. A. Sprague Memorial Institute, recently established for the experimental study of possible chemical treatments for tuberculosis. Accordingly, in 1912, she moved to the University of Chicago as assistant professor of pathology and member of the institute staff. Six years later, at the age of 59, she became associate professor, a considerable achievement for a woman at the time. A further distinction was her election to the presidency of the Chicago Pathological Society in 1924–25. In 1914, a little more than a decade after she had been blocked from joining its Research Club, the University of Michigan awarded her an honorary A.M. degree.

She retired in 1926 because of failing health and went to live with her daughter, Stella (Mrs. Ritcheson), in Winter, Texas, where she died on March 10, 1928, at the age of 69, of chronic high blood pressure and arteriosclerosis.

WORK

DeWitt's work separates into two distinct areas, her early studies in microscopic anatomy, carried out before 1910, and the later investigations in bacteriology, pathology, and chemotherapeutics, which occupied her for the rest of her career. Her considerable initiative, energy, and enthusiasm for exploring new intellectual horizons, which had taken her through her teacher training and her M.D. and B.S. courses while also bringing up a family, carried over fully into her research work. She was known as a thorough, persistent, and meticulous investigator and was highly respected by her colleagues both for her early anatomical studies and her work at the Sprague Institute. Although throughout her career she held university teaching positions as well as research appointments, it was as a research worker that she is remembered.

Her first work, involving joint studies with Huber on the structures of motor and sensory nerve endings, appeared in print as two short notes, even before she received her M.D. degree. These notes discussed the branching forms and patterns of the terminal fibers in both striated muscle and involuntary smooth muscle in a number of different species. Of particular interest was an examination of "muscle spindles," structures widespread in vertebrate muscle, in which special muscle fibers are connected to both a branched system of sensory nerves and motor nerve endings. The first paper DeWitt published by herself was a morphological study that used recently devised "reconstruction methods" to explore the shape and degree of branching of the pyloric glands in a variety of species. She went on to compare these glands with Brunner's glands of the duodenum.

Although she also published on improvements in histological procedures, two topics in microscopic anatomy particularly occupied her attention: the nerves involved in the transmission of impulse in the beating of the mammalian heart and structures in the pancreas. Her studies of the latter are of special interest (at least in retrospect) in that they constitute a sound, if largely forgotten, contribution to the long effort to understand the function of the pancreas and, ultimately, to alleviate the effects of diabetes. A connection between that disease and dysfunction of the pancreas had been recognized since well before the turn of the century, and a great deal of work by many investigators had gone into the preparation and study of pancreatic extracts and their potential use in treatment. However, no clinically useful material had been obtained. A major difficulty was the fact that most secretions of the pancreas are digestive enzymes, and extraction of the entire gland produced a mixture in which these enzymes overwhelmed the effect of a different secretion, the one that controls carbohydrate metabolism and whose absence results in diabetes. DeWitt's long paper in the *Journal of Experimental Medicine* in 1906 reported her isolation and extraction of the clusters of pancreatic cells known as the "islets of Langerhans" and her conclusion that these "islets" produced a substance that was a key factor in carbohydrate metabolism. She was probably the first person to test this "internal secretion" of the pancreas, although the facilities to which she had access at the University of Michigan allowed her to do so only in a very limited way, and she was unable to examine the influence of her extract on human diabetes. The subsequent isolation in 1921–22 of the hormone insulin from this internal secretion was carried out by a team of Canadian workers led by J.J.R. Macleod at the University of Toronto; their prompt demonstration that their extract was effective in treating diabetes brought the 1923 Nobel Prize in physiology or medicine to Macleod and Frederick Banting, one of the other key members of the team.

DeWitt's investigations of possible chemical treatment of tuberculosis, carried out at the Sprague Memorial Institute, were modeled on the classic work of Paul Ehrlich (Nobel laureate, 1908). Starting from the observation that dyes stain tissues selectively, Ehrlich had, over a number of years, searched for some that would stain parasites in preference to host tissue. Having found a dye that met this requirement, he went on to investigate other comparable substances in which a key grouping of two nitrogen atoms in the dye structure was replaced by two atoms of arsenic, a member of the same periodic group as nitrogen. His work led to the preparation in 1909 of the antisyphilitic drug patented in Germany as Salvarsan. Following Ehrlich's general principles, DeWitt and her coworkers began by searching for dyes that would penetrate tuberculous lesions. Later, in collaboration with chemist colleagues, they started a lengthy program of investigations in which appropriate dye compounds, modified by the incorporation of metal atoms, were systematically tested in animal studies as potential antituberculosis agents. Methylene blue, trypan red, and related compounds were the dyes of choice; these were linked with copper, gold, and mercury. The work was reported in a long series of papers, many under the general title "Studies in the biochemistry and chemotherapy of tuberculosis," published between 1913 and 1926. The monograph *The Chemistry of Tuberculosis* (1923), which DeWitt coauthored with H. G. Wells and Esmond R. Long, presented the findings in collected form. A number of her coworkers in these studies were women, including Hope Sherman and Gladys Leavell, in the earlier years, and Lauretta Bender, who later became well known for her work in child psychiatry.

The group's exhaustive investigations did not, in fact, lead to the discovery of an antituberculosis drug. Nevertheless, the meticulous testing procedures devised and followed provided a model for later work, which did result in the preparation of chemotherapeutic agents that were notably successful in the treatment of tuberculosis. Indeed, the overall approach of DeWitt and her coworkers to the project, involving multidisciplinary teamwork and the systematic synthesis and testing of a large number of compounds of the class under examination, is remarkably similar to that regularly followed at the present time in drug-design research.

DeWitt belonged to the Association of American Anatomists, the Chicago Pathological Society, the Michigan Medical Society, and the American Association of Pathologists and Bacteriologists. She was also an associate fellow of the American Medical Association.

BIBLIOGRAPHY

Works by Lydia Maria Adams DeWitt

Scientific Works

(with G. C. Huber) "Endings of sensory and motor nerves in the 'muscle spindles' of voluntary muscle with demonstration of preparations." *Science* 5 (1897): 908–909.

(———) "The innervation of motor tissues, with special reference to nerve-endings in the sensory muscle-spindles." *Reports of the British Association for the Advancement of Science* (1897): 810–811.

"Morphology of the pyloric glands as shown by reconstruction. Demonstration of models." *American Journal of Anatomy* 1 (1901–2): 514.

Atlas and Epitome of Human Histology and Microscopic Anatomy. Philadelphia and London: W. B. Saunders, 1903.
 Translation from the German of the work by Johannes Sobotta.

"Preliminary report of experimental work and observations on the areas of Langerhans in certain mammals." *American Journal of Anatomy* 4 (1904–5): viii.

"Morphology and physiology of areas of Langerhans in some vertebrates." *Journal of Experimental Medicine* 8 (1906): 193–239.

"Observations of the sino-ventricular connecting system of the mammalian heart." *Anatomical Record* 3 (1909): 475–497.

"The pathology of the sinoventricular system or bundle of His." *Physician and Surgeon* 32 (1910): 145–150.

"Some observations on phenol as a cleaning agent in histological work." *Journal of Medical Research* 23 (1910–11): 369–376.

"A case of generalized infection with a diphtheroid organism." *Journal of Infectious Diseases* 10 (1912): 36–42.

"Report of some experiments on the action of *Staphylococcus aureus* on the Klebs-Loeffler bacillus." *Journal of Infectious Diseases* 10 (1912): 23–35.

(with F. L. Evans) "Laboratory methods of diagnosis of typhoid fever." *Interstate Medical Journal* 19 (1912): 770–786.

"Preliminary report of experiments in the vital staining of tubercles. Studies in the biochemistry and chemotherapy of tuberculosis. IV." *Journal of Infectious Diseases* 12 (1913): 68–92.

"Report on some experimental work on the use of methylene blue and allied dyes in the treatment of tuberculosis. Studies in the biochemistry and chemotherapy of tuberculosis. VII." *Journal of Infectious Diseases* 13 (1913): 378–403.

"Vital staining of tubercles." *Transactions of the Chicago Pathological Society* 9 (1913): 22–24.

(with H. J. Corper and H. G. Wells) "The effect of copper on experimental tuberculosis lesions. Preliminary notes." *Journal of the American Medical Association* 60 (1913): 887–889.

"Therapeutic use of certain azo dyes in experimentally produced tuberculosis in guinea-pigs. Studies in the biochemistry and chemotherapy of tuberculosis. VIII." *Journal of Infectious Diseases* 14 (1914): 498–511.

(with H. Sherman) "Tuberculocidal action of certain chemical disinfectants. Studies in the biochemistry and chemotherapy of tuberculosis. IX." *Journal of Infectious Diseases* 15 (1914): 245–256.

(with H. G. Wells) "Studies in the chemotherapy of tuberculosis." *Zeitschrift für Chemotherapie und verwandte Gebiete* 2 (1914): 110–127.

"The present status of tuberculosis chemotherapy." *Journal of Laboratory and Clinical Medicine* 1 (1915–16): 677–684.

"The value of copper in the treatment of tuberculosis." *Transactions of the National Association for the Study and Prevention of Tuberculosis* 11 (1915): 237–241.

"Chemotherapy of tuberculosis." *Transactions of the Chicago Pathological Society* 10 (1916–17): 175–177.

"Some derivatives of methylene blue in tuberculosis chemotherapy." *Transactions of the National Association for the Study and Prevention of Tuberculosis* 12 (1916): 257–261.

(with H. Sherman) "The bactericidal and fungicidal action of copper salts. Studies on the biochemistry and chemotherapy of tuberculosis. XV." *Journal of Infectious Diseases* 18 (1916): 368–382.

"Gold therapy of tuberculosis." *American Review of Tuberculosis and Pulmonary Diseases* 1 (1917–18): 424–430.

"Mercury in the chemotherapy of experimental tuberculosis in guinea-pigs." *Transactions. National Tuberculosis Association* 14 (1918): 369.

"The use of gold salts in the treatment of experimental tuberculosis in guinea-pigs. Studies in the biochemistry and chemotherapy of tuberculosis. XVIII." *Journal of Infectious Diseases* 23 (1918): 426–437.

(with S. M. Cadwell and G. Leavell) "Distribution of gold in animal tissues. Studies in the biochemistry and chemotherapy of tuberculosis. XVII." *Journal of Pharmacology and Experimental Therapeutics* 11 (1918–19): 357–377.

"Inhibitory action on tubercle bacilli of some new mercury compounds." *Transactions of the Chicago Pathological Society* 11 (1919–22): 248.

(with J. H. Lewis) "The continuous injection method in the treatment of experimental tuberculosis." *American Review of Tuberculosis and Pulmonary Diseases* 3 (1919–20): 548–552; *Transactions. National Tuberculosis Association* 15 (1919): 277.

"Action of mercurochrome-220 and of mercurophen; a preliminary report of the effects on the human tuberculosis bacillus and on experimental tuberculosis in guinea-pigs." *Journal of the American Medical Association* 75 (1920): 1422.

"Weight curves of tuberculous guinea-pigs. Studies in the biochemistry and chemotherapy of tuberculosis. XX." *Journal of Infectious Diseases* 27 (1920): 503–512.

(with B. Suyenago and H. G. Wells) "The influence of creosote, guaiacol and related substances on the tubercle bacillus and on experimental tuberculosis. Studies in the biochemistry and chemotherapy of tuberculosis. XIX." *Journal of Infectious Diseases* 27 (1920): 115–135.

"Mercury compounds in the chemotherapy of experimental tuberculosis in guinea-pigs. I. Studies in the biochemistry and chemotherapy of tuberculosis. XXI." *Journal of Infectious Diseases* 28 (1921): 150–169.

"The inhibitory action of certain organic mercury compounds on the growth of human tubercle bacilli. Studies in the biochemistry and chemotherapy of tuberculosis. XXII." *Journal of Infectious Diseases* 30 (1922): 363–371.

(with L. Bender) "Blood changes during the progress of experimental tuberculosis in guinea-pigs." *Transactions. National Tuberculosis Association* 18 (1922): 444–449; *Transactions of the Chicago Pathological Society* 11 (1919–22): 256–258.

"The therapeutic and bactericidal value of organic mercurial compounds in experimental tuberculosis in guinea-pigs. Studies in the biochemistry and chemotherapy of tuberculosis. XXVIII." *American Review of Tuberculosis* 8 (1923–24): 234–244.

(with L. Bender) "Hematological studies on experimental tuberculosis in the guinea-pig. I. Blood morphology, Arneth counts and coagulation time in normal and untreated tuberculous guinea-pigs. Studies in the biochemistry and chemotherapy of tuberculosis. XXVII." *American Review of Tuberculosis* 8 (1923–24): 138–162.

(with H. G. Wells and E. R. Long) *The Chemistry of Tuberculosis; being a Compilation and Critical Review of Existing Knowledge on the Chemistry of the Tubercle Bacillus and Its Products.* Baltimore: Williams and Wilkins, 1923.

(with L. Bender) "Hematological studies in experimental tuberculosis of the guinea-pig. II. The effect of certain drugs on the blood picture in tuberculosis. Studies in the biochemistry and chemotherapy of tuberculosis. XXXI." *American Review of Tuberculosis* 9 (1924–25): 65–71.

(———) "Hematological studies in experimental tuberculosis of the guinea-pig. III. Protein concentration of the blood serum and erythrocyte volume in tuberculous guinea-pigs. Studies in the biochemistry and chemotherapy of tuberculosis. XXXII." *American Review of Tuberculosis* 9 (1924–25): 477–486.

(with H. G. Wells) "An attempt to establish tissue-specific strains of tubercle bacilli." *American Review of Tuberculosis* 13 (1926): 92.

Works about Lydia Maria Adams DeWitt

Cattell, J. McKeen (ed.). *A Biographical Dictionary of American Men of Science.* 1st, 2d, 3d eds. Garrison, NY: Science Press, 1906– .

Cox, P. B. "Pioneer women in medicine. Michigan. XVI." *Medical Woman's Journal* 56 (1949): 48–51, 64.

Rossiter, M. W. *Women Scientists in America. Struggles and Strategies to 1940.* Baltimore and London: Johns Hopkins University Press, 1982.

Other References

Bliss, M. *The Discovery of Insulin.* Chicago: University of Chicago Press, 1982.

NELTJE BLANCHAN DE GRAFF DOUBLEDAY (1865–1918)

Keir B. Sterling

BIOGRAPHY

Neltje (sometimes called "Nellie") Blanchan De Graff, the daughter of Liverius and Alice (Fair) De Graff, was born on October 23, 1865, in Chicago, Illinois. Her father, who was of Dutch descent, was the proprietor of a men's clothing store. Information concerning her early years is very limited. She was sent to two schools for girls in New York State: St. John's in New York City and the Misses Masters' School in Dobbs Ferry. On June 8, 1886, she married Frank Nelson Doubleday in Plainfield, New Jersey. They had three children, Felix Doty (b. 1887), Nelson, who later succeeded his father as head of the family publishing firm (b. 1889), and Dorothy (b. 1892).

Mrs. Doubleday, who used the pen name of Neltje Blanchan, was a sociable and vivacious woman with a great deal of fervor. The Doubledays maintained homes in both New York City and Oyster Bay, Long Island, where they entertained regularly. In addition to her writing career, she was involved in a number of philanthropic activities, among them the American Red Cross chapter in Nassau County, Long Island. Appointed with her husband as a commissioner for the National Red Cross, she embarked with him on a special assignment for that organization to the Philippines and to various parts of China early in December 1917. She died unexpectedly in Canton, China, on February 21, 1918, at the age of 52.

WORK

The partnership of Neltje and Frank Doubleday was both a personal and professional success. He had worked in the publishing field since boyhood, became an editor of several publications in his 20s, and in 1897 founded the publishing firm of Doubleday, McClure (later Doubleday, Page, and Company). Nicknamed "Effendi" by Rudyard Kipling, one of the authors published by the Doubleday firm, he was a dynamic and well-known figure in the publishing world. He was extremely supportive of his wife's writing career.

Neltje Doubleday developed a number of interests in the natural history field, principally in plant and bird study and American Indian lore. She came to popular notice just prior to the onset of the Progressive Era, when the early American conservation movement was beginning to attract broad public support. No doubt, the earlier successes of other women naturalists of the day, notably, Olive Thorne Miller (1831–1918), Mabel Osgood Wright (1859–1934) and Florence Merriam Bailey* (1863–1948), stimulated her work in natural history. Undoubtedly, she was also encouraged by her husband's willingness to publish her work.

The Piegan Indians, her first book, which appeared in 1889, was a popular account of a northern plains tribe. Her interest in Indian education and handicraft was shared with other women naturalists of her time. Neltje Doubleday then turned to writing about birds, which were probably her principal interest. *Bird Neighbors*, which appeared in 1897, was subtitled *An Introductory Acquaintance with One Hundred and Fifty Birds Commonly Found in the Gardens, Meadows, and Woods about Our Homes*. The naturalist John Burroughs, one of many authorities to whom she turned for scientific understanding of her subject, provided an introduction. Doubleday's objective was to provide "accurate and reliable" (Doubleday 1897, ix) information about her subject in clear and lively prose for a popular audience. *Bird Neighbors*, which remained in print for a number of years, sold more than a quarter of a million copies, and many of her other books enjoyed wide popularity, making her the most commercially successful woman nature writer of her generation. As with many other nature writers of the time, her work was anthropomorphic, and she was criticized then and later for being "shameless about endowing birds with human traits" (Kastner 1986, 167). The female black-throated blue warbler, for example, was described as being "not easy to distinguish . . . except as seen in company with her husband, whose name she has taken with him for better or for worse" (Doubleday 1897, 95). After discussing the habitats and seasons of birds in *Bird Neighbors*, Doubleday listed birds according to their size and then described them according to color, without regard to more conventional systems of classification. The somewhat uninspired colored illustrations were probably based on mounted specimens.

In *Birds That Hunt and Are Hunted* (1898), the author gave evidence of her considerable interest in conservation by suggesting that those who were knowledgeable about birds would have a great interest in encouraging their protection. Doubleday's next book, *Nature's Garden* (1900), organized information about wild flowers by color and also discussed the insects attracted to them. She employed poetry and folklore in trying to convey her feelings about each species. *How to Attract the Birds* (1902) was a slim volume in which the author discussed various birds and the plants that they favor. Here, perhaps, she is at her best in combining two of her greatest interests, birds and plants. Again, however, she sometimes wrote in rhapsodic terms about her subjects and in a manner with which some critics took issue.

While it is true, that manners improve steadily the higher birds ascend in the evolutionary scale; that hen-pecked husbands are treated with more consideration, overworked wives with greater respect and even tenderness until burdens become more evenly shared by both mates . . . ideal devotion is short lived, confined as it is to the nesting season. (Doubleday 1902, 85)

Birds That Every Child Should Know (1907) centered on the theme that "[i]nterest in bird life exercises the sympathies . . . it is nature sympathy, the growth of the heart, not nature study, the training of the brain, that does most for us" (Doubleday 1907, viii). This book would have made charming bedtime reading for intelligent youngsters, but the scientific content was modest.

The American Flower Garden (1909) was written with the interests of large-scale, propertied householders in mind. Produced with some care and illustrated with many photographs, it was a useful publication that sold well and continued in print into the 1930s. *Birds Worth Knowing*, the last book to be published during her life, appeared in 1917. Featuring paintings by Louis Agassiz Fuertes and other well-known illustrators, it consisted of selections from her earlier works. In her preface to this volume, she expressed great optimism that a rising tide of nationwide support for legislation protecting birds would lead to action being taken by all of the states. A similar compilation, entitled simply *Birds*, appeared posthumously in 1926. Most of her publications appeared in more than one edition under the various imprints produced by the Doubleday firm.

Neltje Doubleday wrote a number of articles on a wide range of subjects for such periodicals as *Country Life in America* and *Ladies' Home Journal*. She also authored a number of book reviews. Doubleday was friendly with the novelist Gene Stratton Porter, whom she encouraged in her natural history writing.

Neltje Doubleday was very supportive of her husband's work as a publisher, and he may have followed her advice in at least one instance when the firm was considering a work of fiction. In 1900, while she and her husband were abroad, his partner, Walter Hines Page, accepted Theodore Dreiser's stark first novel, *Sister Carrie*, for publication. After Doubleday and his wife had an opportunity to examine page proofs, he concluded that the book was inappropriate for their list and sought to break the contract for it. Unable to do this, with great reluctance he published a small run of the book, to generally negative reviews. Dreiser was convinced that Neltje Doubleday had been responsible for the fact that his novel was a commercial failure. There is no clear evidence to support this view, although the strong moral temper of much that Neltje Doubleday published may have prompted Dreiser to take this position. The public was not then ready for Dreiser's novel, but several years later, both the public and the critics were more receptive when *Sister Carrie* was brought out by another publisher.

Although her writings about birds were not considered as authoritative or consequential as those published by other women nature writers, notably Miller, Wright, and Bailey, Neltje Blanchan Doubleday enjoyed wide popularity and contributed a great deal to the national conservation movement.

BIBLIOGRAPHY

Works by Neltje Blanchan De Graff Doubleday

The Piegan Indians. New York: Doubleday, McClure, 1889.
Bird Neighbors. New York: Doubleday, McClure, 1897.
Birds That Hunt and Are Hunted. New York: Doubleday, McClure, 1898.
Nature's Garden. New York: Doubleday, Page, 1900.
"Village thrift." *Ladies' Home Journal* 18 (Nov. 1901): 30 (as N. Blanchan).
"Birds' Christmas dinner." *Ladies' Home Journal* 20 (Dec. 1902): 44 (as N. Blanchan).
"How birds care for their babies." *Ladies' Home Journal* 19 (June 1902): 3–4 (as N. Blanchan).
"How birds protect themselves." *Ladies' Home Journal* 19 (Mar. 1902): 7 (as N. Blanchan).
How to Attract the Birds. New York: Doubleday, Page, 1902.
"Why the birds come and go." *Ladies' Home Journal* 19 (Apr. 1902): 10 (as N. Blanchan).
Birds That Every Child Should Know. New York: Doubleday, Page, 1907.
"Lessons for Americans from the art of garden design in Italy." *Country Life* 11 (1907): 533.
"Formal garden in America." *Country Life* 14 (1908): 271–274.
"Garden any one may have." *Ladies' Home Journal* 25 (Apr. 1908): 44.
"Naturalistic garden." *Country Life* 14 (1908): 443–445.
The American Flower Garden. New York: Doubleday, Page, 1909.
"Perennials for a thought-out garden." *Country Life* 15 (1909): 475–480.
"Joy of gardening." *Country Life* 17 (1910): 541–544.
Birds Worth Knowing. New York: Doubleday, Page, 1917.
"Prussianizing the campaign against sparrows." *Country Life* 32 (1917): 82.
Birds. New York: Doubleday, Page, 1926.
Wild Flowers Worth Knowing. New York: Doubleday, Page, 1926. Revised ed., 1932 (as N. Blanchan).

Works about Neltje Blanchan De Graff Doubleday

Obituary. *New York Times* (Feb. 23, 1918).

Other References

Doubleday, Frank Nelson. *The Memoirs of a Publisher.* Garden City, NY: Doubleday, 1972.
Kastner, Joseph. *A World of Watchers.* New York: Knopf, 1986.
Lingeman, Richard. *Theodore Dreiser: At the Gates of the City, 1871–1907.* New York: Putnam, 1986.
Porter, Gene Stratton. *Homing with the Birds.* New York: Doubleday, Page, 1919.
Strom, Deborah. *Birdwatching with American Women.* New York: W. W. Norton, 1986.
Swanberg, W. A. *Dreiser.* New York: Scribner, 1965.
Welker, Robert H. *Birds and Men.* Cambridge, MA: Harvard University Press, 1955.

ALICE EASTWOOD (1859–1953)

Joel S. Schwartz

BIOGRAPHY

Alice Eastwood was born in Toronto, Canada, on January 19, 1859, the eldest child in a family of two daughters and one son. Her father, Colin Skinner Eastwood, a man with a limited education, had few resources. His father, Jim Eastwood, had emigrated from Yorkshire, England, to Toronto, where he helped found the Unitarian Church. Her mother, Eliza Jane (Gowdey) Eastwood, was born in Ballymacash, near Lisburn in Northern Ireland, and emigrated to Canada. Her father was steward of the Toronto Asylum for the Insane, the institution where Eliza Eastwood's cousin, Dr. Joseph Workman, an eminent neurologist, served as director. In a small cottage on the grounds of this forbidding place, Alice Eastwood was born.

Eastwood's first six years were quite pleasant in spite of these grim surroundings and her mother's chronic illness. At the age of six, showing remarkable composure under difficult circumstances, she promised her dying mother that she would always look after her younger brother and sister. Eastwood tried to keep this promise, even though she faced a great deal of hardship resulting from her family's nomadic and unsettled existence in the decade after her mother's death. Her extraordinary determination and self-reliance allowed her to cope with the many obstacles she would face during this very difficult period of her life.

Soon after her mother passed away, her father's grocery store in Toronto failed due to his lack of business acumen, and Eastwood and her sister, Kate, were sent to live in a convent in Oshawa, Ontario. Her father took her little brother, Sidney, to Denver, Colorado, in search of employment. For the next six years the two Eastwood girls were clothed, housed, fed, and educated in this austere setting. Instead of bemoaning her fate, Eastwood tried to make the best of her life, demonstrating at a tender age a capacity for turning a difficult situation to her advantage. A good example of this ability, which she exhibited her entire life, was when a Sister at the convent recognized Eastwood's musical talent and helped nurture this gift by giving her music lessons. Eastwood used this skill to earn some extra money later on by giving music lessons to a woman in a nearby village.

However, her real interest was in botany. Yearly visits to her uncle, Dr. William Eastwood, helped stimulate her interest in nature. He was an experimental horticulturalist, and the grounds of his property were filled with unusual hybrid plants. Her uncle noticed the delight with which she collected seeds from the plants in bloom, so he presented her with a book on botany as a gift. An old French priest, Father Pugh, the convent's gardener, also noticed young Eastwood's interest in botany. He made the Eastwood girls a make-believe playhouse amid the pleasant-smelling lilacs. Although both girls did not have any toys, Eastwood did not feel particularly deprived. While observing Father Pugh make grafts with the apple trees in the convent's orchard, Eastwood devised her own experiments by planting the seeds she had collected from her uncle's property. Under the tutelage of the kindly and knowledgeable priest, she developed into a dedicated young botanist.

When Eastwood was 14, her father gained steady and secure employment and was able to send for her. Unfortunately, she had to leave her sister behind, earning her train fare by caring for a very young child during the entire journey. Upon her arrival, she was quite upset by the conditions under which her father and brother were living. They were staying in a cheap hotel, certainly no place for a young girl to start her life in the still untamed West. Her father assured her that this would be a temporary arrangement, and he soon purchased property in order to build a store. The Eastwood family lived in the back of the store as they struggled for financial security.

Eastwood was generally pleased with her new life in this frontier town, which was emerging into a growing city. She became enchanted with the nearby Rocky Mountains and explored them extensively every summer. The Alpine vegetation was a revelation for her; she had never encountered such a variety of plant life. Also, she was able to attend school on a regular basis for the first time in her life, and here she soon excelled in all her subjects, displaying a particular gift for music.

Eastwood entered East Denver High School the following year (1875), and she quickly established herself as the most outstanding student in the school. She was able to maintain her standing in school in spite of having to work long hours in her father's store, in addition to attending to myriad household chores. When her father's store failed during her junior year, she resumed the hardscrabble existence that had been her fate for most of her life. She had to quit school and worked full-time in order to bring in enough money for her family. Adding to her distress, her father remarried, and initially she resented these changes in her life. However, her father traveled to Ontario to bring her sister Kate to Colorado, thereby reuniting the family. Additionally, her stepmother proved to be a warm and generous person, helping Eastwood with her studies at night so that she could keep up with her classmates while she was unable to attend school regularly. Eventually, her father became the janitor of the newly completed building for East High School, and the entire Eastwood family made its home in the school's basement quarters.

Eastwood was able to enter high school on a regular basis for her senior year. She earned her way by helping to stoke the furnace fires in the school before daybreak and by working in the dress department of a Denver department store after school. This latter experience enabled her to learn the art of dressmaking, and she utilized these valuable skills later on. She graduated as valedictorian in the class of 1879, and her valedictory address was published in the *Denver Tribune*. Her graduation gifts were two books, T. C. Porter and J. M. Coulter's *Synopsis of the Flora of Colorado* (1874) and Asa Gray's *Manual of the Botany* (1878), which helped form the foundation of her budding library in botany.

After her graduation Eastwood faced the responsibility of earning her own living. Although she was certainly qualified to pursue a formal education at the university, her family's precarious finances made this impossible. Also, in the latter part of the nineteenth century, talented and bright young women were not encouraged to continue their higher education. Eastwood did not give up her involvement with education altogether, as she became a teacher at her old school, teaching Latin, drawing, natural science, composition, and American literature over the next decade. With an annual starting salary of $475 a year, she was able to save enough for her summer explorations in the Rocky Mountains. Taking advantage of this opportunity, she collected many different Alpine species of plants, some of which had not been classified or cataloged. Her sewing skills proved useful, as she designed outfits that were both sturdy and comfortable enough to withstand the rigors of her arduous fieldwork.

Eastwood developed into a skillful field person, gaining a profound knowledge of the flora and fauna of the areas she annually explored, in addition to acquiring a good grasp of its geography and geology. She expected no one to defer to her because she was a woman. On her many trips, she proved that she could hold her own with any of the farmers, ranchers, and other settlers of the area, as she bounded over the rugged mountain terrain. The plants she collected eventually became the nucleus of Colorado University's herbarium at Boulder. Before she was 30, her reputation as a prominent local botanist was firmly established, and word about her abilities as a naturalist grew.

When the English naturalist and pioneer in evolutionary biology Alfred Russel Wallace visited Colorado during the Alpine flowering season in May 1887, Alice Eastwood, a 28-year-old botanist and schoolteacher, was selected to guide the 66-year-old world-renowned colleague of Charles Darwin and Thomas Henry Huxley on a botanical collecting trip in the nearby mountains. Wallace and Eastwood negotiated this difficult climb alone. Wallace was so impressed by his young companion's knowledge of the plant life they encountered and her skills in the field that he referred prominently to this incident in his autobiography, *My Life* (Wallace 1905, 180–184). When she traveled to England years later,

Eastwood visited Wallace and his family, and they both spoke about their trip up Gray's Peak.

With her salary giving her some financial security for the first time in her life, Eastwood invested in the booming real estate market in the fast-growing city of Denver. This nest egg afforded her some fiscal peace of mind she previously had not been able to enjoy. She also could now afford the train fare to distant places in order to conduct fieldwork in more remote regions and collect new species she had not previously encountered. She took advantage of low train fares that were the result of a railroad rate war and traveled east to visit her Uncle William and other family and friends from her childhood. She continued farther east and visited her stepmother's family in Newburyport, Massachusetts, as well as the Gray Herbarium in Cambridge. She actually was able to meet the 75-year-old revered botanist Asa Gray in a chance encounter in the garden. Trying valiantly to overcome her shyness and awe of such an eminent individual, she told him of her own discoveries in the mountains of Colorado.

In the winter of 1890, Eastwood journeyed to San Diego, California, to accompany a sick, elderly friend. Although the Denver school department was not happy with her extensive travels, they allowed her to take a leave of absence after a particularly hectic period of time in the school district. With Gray's *Synoptical Flora of North America* and *Botany by the Geological Survey of California* as her constant companions, she avidly collected much of San Diego's local flora. She became interested in the early French and British naturalists who explored the wildlife of the California coast in the eighteenth century, and she decided to retrace their steps. As she worked her way up the coast, she collected much of the plant life she encountered. She also visited local herbaria and nurseries, introducing herself to the local botanists. Other naturalists began to take note of this extraordinary young woman who had dedicated so much of her time and energy to the painstaking pursuit of unusual forms of vegetation.

Eastwood's final stop was at the California Academy of Sciences in San Francisco, in its old Market Street location. She was armed with an introduction from C. R. Orcutt, a naturalist whom she met in San Diego and for whom she wrote articles on the flora of the Colorado Rockies for the popular magazine he edited. She called on Katharine Brandegee, Curator of Botany at the Academy, and her husband, Townsend Stith Brandegee, a botanist who had established his reputation surveying the Colorado Rockies in the 1870s. He was the publisher of *Zoe*, a magazine of natural history, with his wife serving as editor. Eastwood hoped to contribute articles to this journal. Her meeting with the Brandegees went very well; they already knew about her remarkable collection of Colorado plants, and they were quite impressed by her expertise and enthusiasm. They assigned her to write an article for *Zoe*.

When Eastwood returned to Colorado, she did so with a new sense of purpose. Determined to devote more time to her investigations in botany, she sat down to write "Common Shrubs of Southwest Colorado" (1891). The Brandegees were impressed with her initial article, and Eastwood agreed to undertake further work for them. She found this experience so rewarding that it strengthened her interest in becoming a botanical writer. After receiving an extended leave from the Denver school system, she set out once more for California. While living in a sparsely furnished room, she wrote another article for *Zoe* and spent her spare time working at the Academy's herbarium.

Eastwood's association with the Brandegees proved to be a turning point in her life and career. They had long wanted to retire to San Diego, and in 1892, in order to follow through on their plans, they offered her the dual position of Curator of Botany at the Academy and editorship of *Zoe*. She was overwhelmed by this opportunity, but she was reluctant to leave her family and friends in Colorado. She had grown to love the Colorado landscape and had many friends there, including Alfred and Richard Wetherill, who had a ranch near some property she owned. These ranchers had discovered the ruins of the Anasazi people at Mesa Verde, and Eastwood accompanied them on several of their trips to this now world-famous archaeological site. She collected rare plants among the cliffs and ruins there and on the bluffs and used her skills to write "Notes on the cliff-dwellers" for *Zoe* (1893). Thus, by writing on a subject that was not strictly botanical, she demonstrated her versatility, utilizing her investigative skills to produce a first-rate study on the agricultural practices of the early native people of the American Southwest.

There was another reason for her reluctance to leave Colorado: reportedly, she had become romantically involved with a journalist from the East who had moved to the area in the hope of curing his tuberculosis. Eastwood and the young man planned to marry. Thus, a move to the damper San Francisco area was out of the question. Unfortunately, he was unable to survive even in the more salubrious climate of the Rockies. Although his death was a terrible blow for Eastwood, it did make her decision about the future a bit easier. Her plans to start a family and combine this with her botanical writing were no longer possible. Although little has been written about this difficult period in Eastwood's life, it is fair to speculate that it was crucial in influencing her choice to devote her life exclusively to botany. She accepted the Brandegees' offer and prepared to take up her posts as soon as her teaching and other responsibilities would allow.

Eastwood became Curator of Botany at the Academy in 1893, and she remained in this position for over a half a century. At the outset, she was faced with the very difficult task of reclassifying and relabeling the collection of plants in the Academy's possession. The Brandegees were rather unsystematic in organizing this material, so Eastwood had to sort out the many examples of California flora that had been collected over the years. She was able to put this material together in a more coherent fashion, because she understood the Brandegees' train of thought in creating such a chaotic system. While she relabeled the entire collection, she took full advantage of an unlimited travel railroad pass the Brandegees had provided and went on collecting trips in the region in her spare time. She thereby helped enrich the Academy's collection and began to familiarize herself with the area.

When the railroad did not reach her ultimate destination, she continued her journey by stagecoach, by horseback, and many times on foot. Her collecting methods and utilization of live plants gave her a distinct advantage over other botanists, even Asa Gray, the Brandegees, and others, in that they relied on dried plants in their classification. Eastwood worked with fresh material. She was fascinated by the plants of Mount Shasta, Mount Tamalpais in the San Francisco area, Del Norte County, the Santa Lucia Mountains, and the interior valleys of California, as well as the Alpine regions of Oregon, Nevada, Utah, and New Mexico. She traveled to Lower (Baja) California and Alaska, and the plant life she collected there was added to the Academy. Nothing missed her observant and well-trained eye. When she observed tiny plants forcing their way through the openings between the cobblestones in the streets of San Francisco, she developed a flora of such plants and included it in her paper ''Plant inhabitants of Nob Hill, San Francisco'' (1898).

Initially, her lodgings in Nob Hill, San Francisco, were rather sparse but functional. She had little need for more elaborate living quarters. During these years (1893–1906), she was constantly traveling to areas where few naturalists had been. She discovered new species of plants in the southern Sierras, the Tehipte Valley on the middle fork of the Kings River, the Trinity Alps, Mount St. Helena, and the Wasiaja River area. She made many friends among the local ranchers, settlers, lumbermen, and farmers, who often put her up for the night or gave her food as she roamed over this vast area. Her knowledge about the natural history of the far western states grew until she enjoyed a similar command of the area as she had in Colorado. Through her association with other naturalists and her growing publications, she became a well-known figure in her field.

In April 1906, Eastwood's life and career suffered a serious blow. The San Francisco earthquake shattered more than the city; it disrupted her life, and it nearly destroyed the Academy. She lost all her personal possessions except her lens, as she heroically tried to save as much as she could from the Academy's collection. The fires that resulted from the earthquake proved more destructive than the quake; the Academy building burned down and had to be completely rebuilt. Eastwood also had to rebuild her life. She had no home and no prospect for any meaningful work. The Academy was uncertain about rebuilding and about Eastwood's eventual role at the new Academy. She was given temporary space at the University of California to conduct research and store the material she collected. After arranging with various owners of unused and undamaged buildings to keep the few parts of the Academy's collections she had heroically saved, she began to travel during the next six years (1906–12) in an effort to restore what was lost. She supported herself during these troubled years with the income she received from property in Colorado.

Eastwood went to Cambridge, Massachusetts, in 1908 to study at the Gray

Herbarium. Through a botany club associate, Francis Howland, she was invited to live in nearby Concord with Sophia Ripley Thayer, widow of James Bradley Thayer, former Dean of the Harvard Law School. Sophia Thayer owned the "Manse," where Eastwood stayed, and was a second cousin of Ralph Waldo Emerson, whom Eastwood always admired. She spent some of her time speaking to various horticultural groups and associations about western vegetation. As a result, her reputation widened as she developed a new group of admirers among the more reserved easterners. After she went on a collecting trip to the Southeast and examined groups of plants with which she was not too familiar, she returned to the Gray Herbarium and served there as a staff assistant.

When Eastwood returned to California, she learned that the Academy's rebuilding efforts had not made much progress. What was more troubling to her was that her future role at the Academy was not clear, and there were no concrete plans to restore the herbarium. However, she was free to travel again, and as a consequence, she left for Europe. In England she studied at Kew Gardens and the Natural History Museum at South Kensington. She visited the eminent botanist Dr. Joseph Hooker at his home in Surrey shortly before his death. He enthusiastically showed her some plants from California that he had planted years before when he retired from the Royal Gardens at Kew. She renewed her friendship with Wallace at his home, Orchard House. Thus, she was able to make the acquaintance of two of Darwin's closest associates, in addition to her chance meeting years before with Darwin's chief advocate in America, Asa Gray.

After observing Linnaeus's own herbarium at the Linnaean Society in London, she went to Paris. There she studied the different varieties of the evening primrose in Jean Baptiste Lamarck's Herbarium. She went back to Britain to study at the Lindley Herbarium in Cambridge before returning to the United States. She again worked at the Gray Herbarium, and while she was there she received word from the Academy that they wanted her to return to California to rebuild the herbarium. She finished her studies in Cambridge, Massachusetts, in 1911 and stopped in New York at the New York Botanical Gardens, where she was warmly greeted by Dr. Nathaniel Britton, founder of this illustrious institution.

In May 1912, Eastwood returned to San Francisco and immediately set out to rebuild the Academy's herbarium, department of botany, and library. Working with a limited budget and in cramped, rented quarters, she made a good start on this difficult task. She was able to get the budget increased, and with the assistance of friends and other sympathetic parties, she exchanged material with other institutions (e.g., the Smithsonian, the Arnold Arboretum, Kew Gardens) so she could again build up the Academy's collection. She housed the material in various temporary quarters, often supplied by her circle of friends and admirers, before the Academy was able to be moved to its permanent quarters in Golden Gate Park, where it is still located today. In 1920 she purchased the entire Präger Herbarium from a private owner in Germany for an extremely low price. Her interest in acquiring this magnificent collection for the Academy was the result of her having observed it firsthand while she traveled abroad after the earthquake.

Eastwood served as Curator of Botany and Director of the Herbarium at the Academy for almost 40 years, until her retirement in 1949. She spent those years collecting, writing, teaching classes, and giving lectures to various horticultural societies in the Bay Area, in addition to her duties at the Academy. In 1950, after her retirement, she was selected to serve as honorary president of the Seventh International Botanical Congress in Stockholm, Sweden. At the age of 91, she made the long journey to Sweden to receive this recognition, and she was granted the additional privilege of being allowed to sit in Linnaeus's chair. This honor was awarded only to people of unusual achievement, and it marked the culmination of her full and successful career. Throughout her career, she was offered numerous honorary doctorates in several fields, but she turned all of them down. She did accept membership in the honor roll of the Native Daughters of the American West in spite of her Canadian birth. A grove of redwoods was named after her, and the herbarium she was so instrumental in developing is located in the Academy's Alice Eastwood Hall of Botany. On October 30, 1953, Eastwood died of cancer, after a life of great accomplishment and industry.

WORK

Eastwood's most important contribution in the field of botany was the work she conducted in identifying and classifying large groups of plants. Her investigations included developing taxonomic relationships among the different forms of plant life she examined. Her work ranged from compiling checklists of vascular plants in a particular geographical area, to a more thorough and complete taxonomic or biosystematic analysis of the vegetation in a large region, called a flora by botanists. In addition to Eastwood's observant eye and keen appreciation of detail, her understanding of the criteria for determining species was important in making decisions in classifying the organisms she studied.

Her disciple and successor at the Academy, John Thomas Howell, indicated that she was a rather daring taxonomist in that she was not afraid to express her opinion and make decisive judgments in taxonomy. This approach differed from that of more conservative botanists, such as Asa Gray and John Dalton Hooker. Unlike many of her more cautious colleagues, she was open to the ideas of other botanists. She did not summarily dismiss the proposals of others until she had determined conclusively that they were incorrect. Her indefatigable energy and superior powers of observation gave her an advantage over many others in the field.

Eastwood's major work in plant taxonomy centered on the Alpine plants of Colorado and central and northern California, although her work was by no means restricted to the plants of these regions. From the beginning of her career,

Eastwood's painstaking work was honored by her peers, who named genera, species, and varieties of plants she discovered after her. An early instance of this was her discovery on a collecting trip in California of a hitherto unknown shrub belonging to the sunflower family. Townsend Stith Brandegee named the genus of this previously unknown plant *Eastwoodia*, in honor of his promising successor. Her contributions of over 60 years to botanical research were focused on her identification and naming of many cultivated ornamentals, so when other botanists occasionally bestowed her name on a group of plants, this recognition was much deserved.

While settling in at the Academy after her move to California, Eastwood published numerous articles relating to her previous work in Colorado. She published updated botanical lists of the Rocky Mountain region and added important new information about the geographical distribution of plants in the Rockies. In addition to her research about the rare plants of this large region, she made a significant contribution to the field of botanical archaeology when she wrote about the cliff-dwelling people who inhabited southeastern Utah and southwestern Colorado from the twelfth to the fourteenth centuries (e.g., Mesa Verde). Her papers "General notes of a trip through southeastern Utah" (1893) and "Notes on the cliff-dwellers" (1893) are the result of these investigations. She determined that the native people who inhabited these unusual dwellings raised corn, squash, and beans for food, and their most important plant for textile production was *Yucca baccata*. She found that they likely used the fruit of yucca for food. She also discovered that the common rush, *Phragmites communis*, was used as matting for a second covering for the dead.

Eastwood's purely botanical research, conducted while in Colorado, was highlighted by her discovery of two new species of the genus *Aquilegia*, *A. ecalcarata* and *A. micrantha*, commonly known as columbines, the beautiful garden flowers of many colors and also the state flower of Colorado ("Two species . . ." 1895). She also reported on the reproductive habits of the genus *Nemophila*, a group of weak-stemmed herbs with hermaphrodite flowers belonging to the water leaf family (Hydrophyllaceae), which grow in shade and are found widely in western North America ("Observations . . ." 1895).

Her first significant work concerning plants indigenous to California was the study of the heteromorphic organs of *Sequoia sempervirens*, the well-known and popular redwoods ("On heteromorphic . . ." 1895). The paper dealt with the differences between the redwoods that thrive along the coastal fog belt and the sequoias, which are known for their wide girth and are found on the much drier slopes of the Sierra Nevada Mountains. Her field studies in the Mount Shasta area resulted in the publication of a flora, or taxonomic checklist, of the Alpine flowering plants in that region (1896). Subsequent fieldwork in the Santa Lucia Mountains led to her publication describing the different conifers and their distribution in this locality ("The coniferae . . ." 1897).

Eastwood's versatility was demonstrated by her work on the ferns of Yosemite and the nearby Sierra Nevada Mountains ("Ferns . . ." 1898). She furnished a complete checklist of these primitive vascular plants. She also examined and cataloged the plants she found near her home in San Francisco ("Plants..." 1898), and she published a checklist of the 64 plant species growing in the Nob Hill vicinity. She spent considerable time climbing nearby Mount Tamalpais and collecting plants from its slopes, which she described in her "Notes on the flora of Marin County, California" (1898). She often returned to Mount Tamalpais to study the manzanitas growing there. She was fascinated by this hardy shrub, a familiar ornamental plant found in chaparral communities, also called bear grape or mountain laurel. She reported the discovery of four new species of manzanitas and included it in her "Studies in the herbarium and the field," parts I and II (1897–98).

In addition to discovering new species, Eastwood was able to find organisms that were previously believed to be extinct, such as the sandwort, *Arenaria paludicola*, in a marsh near the Presidio, a fortress in San Francisco (1899). This low-growing perennial herb, which grows best in the sandy regions of the Mediterranean, particularly Spain, was believed extinct. Further work she conducted resulted in the discovery of herbs in regions far removed from their usual range or habitat, for example, *Dicentra chrysantha*, golden eardrops; *Trichostema lanceolatum*, a blue curl; and *Chrysopsis rudis*, a golden aster, which she found growing along the tracks of a narrow-gauge railroad in Alameda County ("Migrating Plants..." 1899). In a nearby area, she also found a sea blite, *Suaeda Californica* Watson; a heath, *Frankenia grandifolia*; and a perennial grass, *Spartina stricta* var. Eastwood developed a detailed checklist of plants that she or other botanists discovered ("New species..." 1900). The culmination of her work during her early years in California was her revision of the key and flora of the Rocky Mountain edition of *Bergen's Foundations of Botany* (1900), and she prepared a similar reworking of the Pacific Coast edition the following year (1901).

Eastwood continued to describe new species she had encountered on her numerous collecting trips in different regions. Among the new plants she discovered was *Trifolium tenerum*, a species of clover; *Monotropa Californica*, a species related to Indian pipe; and *Fritillaria Purdyi*, a species of fritillaries, an unusual plant with greenish-white flowers streaked with purple. She also discovered *Potentilla Hickmanii*, a very rare species of cinque foils belonging to a genus used for food and medicine by the Indians, and *Orthocarpus psittacinus*, a species of owl clovers. Eastwood collected some of these plants; others were sent to her at the Academy, and a few were in the Academy's collection and were long forgotten. She reported nine new species belonging to the genus *Ribes*, a group of wild currants, most of which were collected along the Pacific Coast in the 1890s and early 1900s by various botanists, including herself ("Some new species of Pacific..." 1902).

A significant contribution was her compilation of a flora of the Kings River, South Fork, an area adjacent to Sequoia National Park. The flora collection was the result of her trip to Kings River in 1899 and several other less intensive excursions. It was published by the Sierra Club (1902), and it remains an indispensable guide to the plants found along the trails and forests of this still remote and little-traveled region.

Eastwood continued her identification and classification of the plants she uncovered in the collection at the Academy. She conducted a similar analysis of the plant species she gathered on her collecting trips in California, as well as of the material she had collected in Colorado and Utah in the years before coming to California. She was perhaps enjoying the most productive period of her life. Her study of the genus *Garrya*, a group of shrubs or bushes in the arid hills of California, most frequently found in the foothills of Mount Tamalpais in the Bay Area, is an example of the exhaustive work she was able to accomplish before the earthquake. Her carefully constructed key to this genus included a number of new species that had previously been overlooked (1903). Her handbook of California trees was the result of further painstaking work accomplished in gathering data about the trees native to the state (1905).

Eastwood's last published work before the earthquake cited additional new species she had uncovered (1906). They included three previously overlooked species belonging to the genus *Silene* (the catchflies), *S. deflexa, S. lacustris*, and *S. pacifica*; a species of *Veronica, V. Copelandi*, an herb commonly known as speedwells; two species of *Erigeron, E. decumbens* and *E. Copelandi*, a group containing wild daisies and fleabanes. She also reported on *Chrysopsis gracilis*, a plant belonging to the golden asters; and *Senecio Millikeni*, a species from the large genus of flowering plants commonly called ragwort and groundsels, which are generally yellow-flowered.

After the earthquake in 1906, Eastwood devoted the next six years to study in Cambridge, Massachusetts, and at several gardens and herbaria in Great Britain, in addition to brief visits to Europe and New York. When she returned to the Academy in 1912, she spent a good deal of her time supervising the construction of new space for its Botany Department and restoration of the collection that had been lost. She also began to plan the building of a new herbarium. During the hiatus, her published work included a description of the destruction of the old Academy by the earthquake and the resultant fire and numerous short pieces reporting on the new species she had discovered. Much of her time was taken up with writing the annual reports of the Botany Department, which appeared in the Academy's *Proceedings*, once it began functioning normally again. At the Academy's new location in Golden Gate Park, with her position once again secure, she continued her descriptive work in taxonomy, much as she had done before.

After the daily drudgery of the reconstruction had diminished, Eastwood's published work became focused on the vegetation of the immediate area, particularly Mount Tamalpais. She returned to a favorite subject: the manzanitas. She also wrote about the rosaceous shrubs in this area. Many of these articles appeared in more popular publications like *California-Out-of-Doors*. She wanted to reach the general public and hoped this would stimulate greater interest in horticulture. She also actively promoted educational programs that stressed the importance of conserving the vegetation of the area and developing a greater sensitivity to these issues. Thus, her published work was an extension of the lectures, classes, and assistance she gave to the horticultural societies that began to flourish after the First World War.

Although Eastwood's administrative and other responsibilities increased, her scholarly work continued. Her work with *Ceanothus*, a group of woody deciduous evergreen shrubs (California lilacs), was particularly noteworthy. She reported the discovery of *C. cyaneus* in the San Diego area, *C. insularis*, and earlier she herself found *C. utahensis*, near Soldier's Summit in Utah, along the tracks of the Central Pacific Railroad (1927). She published a complete guide to a group of shrubs native to South America, the escallonias, which were in Golden Gate Park. In this work she included descriptions of new species, particularly new hybrids that had been formed naturally as well as those produced artificially, for example, *E. rockii* Eastwood, *E. franciscana* Eastwood, and *E. rubicalyx* Eastwood (1929).

In 1925 a biological survey of Lower (Baja) California and the Revillagigedo Islands off the Pacific coast of Mexico, was conducted under the auspices of the Academy. Although Eastwood did not accompany this expedition as its botanist, she was chosen to organize the data and write a flora of this region (1929). The flora consisted of a list of plants ranging from ferns and pines, to many different families of flowering plants such as grass, palm, rush, lily, nettle, mistletoe, buckwheat, oak, saltbrush, buttercup, and mustard.

In subsequent years Eastwood continued her painstaking research, describing and classifying thousands of plant species and developing their nomenclature. In addition to these efforts, she helped create a strong department of botany, editing and publishing *Zoe* and the *Leaflets of Western Botany* and leaving a magnificent legacy of scholarship and tireless investigation. She was a member of numerous horticultural organizations, serving as treasurer of the American Fuchsia Society and president of the California Botanical Club and the Tamalpais Conservation Club. She was the force in organizing the Save-the-Redwoods League and was the organization's treasurer. She was an honorary member of the Royal English Forestry Society and a fellow of the American Association for the Advancement of Science and the Royal Scottish Arboricultural Society. Her acquisition of the Präger Herbarium for a relatively inexpensive sum in 1920 provided the Academy with the nucleus of its herbarium and is another enduring monument to her lifelong dedication to science.

NOTE

I would like to thank Bernadette Callery, research librarian, and the staff at the library of the New York Botanical Gardens for their hospitality and assistance. I am grateful to Pennington Ahlstrand of the Archives and Special Collections Division of the library of the California Academy of Sciences for her help and useful suggestions. I would also like to thank Linda Oestry of the Missouri Botanical Gardens, in St. Louis, for her assistance.

BIBLIOGRAPHY

Works by Alice Eastwood

Scientific Works

Space does not permit the listing of the complete works of Alice Eastwood. Included here are all references cited in the text. Eliminated are those references included in MacFarland and Sexton (1943–49) and Howell and Noldeke (1968).

"The common shrubs of southwest Colorado." *Zoe* 2 (1891): 102–104.

"General notes of a trip through southeastern Utah." *Zoe* 3 (1893): 354–361.

"Notes on the cliff-dwellers." *Zoe* 3 (1893): 375–376.

"On heteromorphic organs of *Sequoia semipervirens.*" *Proceedings of the California Academy of Sciences* 2d s., 5 (1895): 170–176.

"Observations on the habits of *Nemophila.*" *Erythea* 3 (1895): 151–153.

"Two species of *Aquilegia* from the Upper Sonoran Zone of Colorado and Utah." *Proceedings of the California Academy of Science* 2d s., 4 (1895): 559–562.

"The alpine flora of Mt. Shasta." *Erythea* 4 (1896): 136–142.

"The coniferae of the Santa Lucia Mountains." *Erythea* 5 (1897): 71–74.

"Studies in the herbarium and the field, I." *Proceedings of the California Academy of Sciences*, Botany 3d s., 1 (1897): 71–88.

"Ferns of the Yosemite and the neighboring Sierras." *Erythea* 6 (1898): 14–15.

"Notes on the flora of Marin County, California." *Erythea* 6 (1898): 72–75.

"Plant inhabitants of Nob Hill, San Francisco." *Erythea* 6 (1898): 61–67.

"Studics in the herbarium and the field, II." *Proceedings of the California Academy of Sciences*, Botany 3d s., 1 (1898): 89–146.

"*Arenaria paludicola.*" *Erythea* 7 (1899): 149.

"Migratory plants in Alameda County." *Erythea* 7 (1899): 175–176.

"New species of California plants." *Zoe* 5 (1900): 80–90.

(with J. Y. Bergen) *Bergen's Foundations of Botany, Key and Flora, Rocky Mountain Edition.* Boston: Ginn, 1900.

(———) *Bergen's Foundations of Botany, Key and Flora, Pacific Coast Edition.* Boston: Ginn, 1901.

"A flora of the South Fork of Kings River from Millwood to the head waters of Bubbs Creek." *Sierra Club Bulletin Publication* 27 (1902): 1–96.

"Some new species of California plants." *Bulletin of the Torrey Botanical Club* 29 (1902): 75–82.

"Some new species of Pacific Coast *Ribes.*" *Proceedings of the California Academy of Sciences*, Botany 3d s., 2 (1902): 241–254.

"Notes on *Garrya* with descriptions of new species and key." *Botanical Gazette* 36 (1903): 456–463.

"A handbook of the trees of California." *California Academy of Sciences, Occasional Paper* 9 (1905): 1–86.

"New species of California plants." *Botanical Gazette* 41 (1906): 283–293.

"New species of *Ceanothus*." *Proceedings of the California Academy of Sciences* 4th s., 16 (1927): 361–363.

"The escallonias in Golden Gate Park, San Francisco, California, with descriptions of new species." *Proceedings of the California Academy of Sciences* 4th s., 18 (1929): 385–391.

"Studies in the flora of Lower California and adjacent islands." *Proceedings of the California Academy of Sciences* 4th s., 18 (1929): 393–484.

Other Works

Papers. California Academy of Sciences, San Francisco.

Works about Alice Eastwood

Bonta, Marcia. "Alice Eastwood." *American Horticulturalist* 62 (1983): 10–15.

Dakin, Susanna Bryant. *The Perennial Adventure: A Tribute to Alice Eastwood*. San Francisco: California Academy of Sciences, 1954.

Fletcher, Maurine S. "Portrait for a western album." *American West* 17 (1980): 30.

Howell, John Thomas. "Alice Eastwood, 1859–1953." *Sierra Club Bulletin* 39 (1954): 78–80.

————. "I remember when I think . . ." *Leaflets of Western Botany* 7 (1953): 153–176.

————, and Anita M. Noldeke (comps.). *Leaflets of Western Botany. Index to Volumes 1–10*, 114-116. Berkeley, CA: Gillick, 1968.

MacFarland, Frank M., and Veronica J. Sexton (comps.). "Bibliography of the writings of Alice Eastwood." *Proceedings of the California Academy of Sciences* 4th s., 25 (1943–1949): xv-xxiv.

Munz, Philip A. "A century of achievement." *Leaflets of Western Botany* 7 (1953): 69–78.

Reiter, Victor, Jr. "Horticulture and the California Academy of Sciences." *Leaflets of Western Botany* 7 (1953): 79–84.

Wallace, Alfred Russel. *My Life*, vol. 2. London: Chapman and Hall, 1905.

Wilson, Carol Green. *Alice Eastwood's Wonderland: The Adventures of a Botanist*. San Francisco: California Academy of Sciences, 1955.

————. "The Eastwood era at the California Academy of Sciences." *Leaflets of Western Botany* 7 (1953): 58–64.

————. "A partial gazetteer and chronology of Alice Eastwood's botanical explorations." *Leaflets of Western Botany* 7 (1953): 65–68.

Other References

Geological Survey of California. *Botany*. 2 vols. Cambridge, MA: Welch, Bigelow University Press, 1876–80.

Gray, Asa. *Manual of the Botany of the Northern United States*. 5th ed. New York: Ivison, Blakeman, Taylor, 1878.

————. *Synoptical Flora of North America*. New York: American Book, 1895–97.

Porter, Thomas C., and John M. Coulter. *Synopsis of the Flora of Colorado of the U.S.: Geological and Geographical Survey of the Territories*. Misc. pub. #4. Washington, DC: Government Printing Office, 1874.

SOPHIA HENNION ECKERSON
(ca. 1867–1954)

Bonnie Konopak

BIOGRAPHY

Sophia Hennion Eckerson was born in Old Tappan, New Jersey. Her parents, Albert Bogert and Ann (Hennion) Eckerson, were of Dutch and French descent. Her younger brothers became established in the fields of medicine and art.

Eckerson attended Smith College and received her B.A. degree (1905) and her M.A. (1907), specializing in the field of botany. During her stay at Smith College, she served as a fellow (1905–6), as a demonstrator (1906–8), and as an assistant (1908–9). While at Smith College she was influenced by her teacher, plant physiologist, William F. Ganong.

Later, she attended the University of Chicago and received her Ph.D. degree (1911) under the direction of Dr. William Crocker, again specializing in the field of botany. After graduating, she remained at the University as an assistant plant physiologist (1911–15) and as an instructor (1916–20).

Eckerson's experiences at the University of Chicago reflect the lives of other women scientists at that time. The University provided a rare opportunity to women in offering higher education programs and degrees in the sciences nearly equal to those for men. However, having chosen a profession such as the field of botany, women faced discrimination in terms of careers within the university setting.

In her account of women scientists at the University of Chicago, Hall (1979) focused on ten women botanists who had graduated in the seven years before, and in the seven years after, World War I and whose biographies are included in *American Men of Science*. Hall particularly singled out Eckerson, who remained at the University as "the most successful member of the group . . . [but] slow in advancing." That is, although Eckerson had a strong academic background and was an excellent scholar, she was not able to achieve promotion to a professorial rank within the university faculty.

During and after her stay at the University of Chicago, Eckerson held other appointments that helped broaden and advance her career. While at the University, she held one-term appointments as a microchemist with Washington State University (1914) and with the Bureau of Plant Industry, U.S. Department of Agriculture (USDA), Washington, D.C. (1919).

After leaving the University, she served as a microchemist with the USDA in its Cereals Division (1921–22). In addition, she held a faculty position at the University of Wisconsin (1921–23). Her last position was at the Boyce Thompson Institute for Plant Research (Yonkers, New York), beginning when the Institute was organized in 1924. Among ten scientists hired from the University of Chicago, she served as a plant microchemist until her retirement July 1, 1940. She was chair of the Department of Microchemistry at the Institute.

Similar to other career women scientists of her day, Eckerson never married but chose to devote her life to her career, relocating as necessary for advancement. Toward the end of her life she lived quietly in Pleasant Valley, Connecticut, enjoying her hobbies of reading, handiwork, and gardening. Following a week's illness, she died on July 19, 1954.

WORK

Eckerson's work generally focused on microchemical examinations of plants, including such specializations as germination, mineral nutrition, reduction of nitrates, nitrate reducase, and cellulose membranes. Her work was primarily published in the *Botanical Gazette* and *Contributions from Boyce Thompson Institute*.

In her dissertation research at the University of Chicago, Eckerson (1913) focused on germination, specifically "after-ripening," or the change in the embryo that allows germination to take place. Previous studies had indicated delayed germination was due to characteristics of the seed embryo itself. To examine this process, she conducted microchemical research on metabolic changes during after-ripening and found various substances in the experimental embryo at different periods. She concluded that there was a correlation among acidity of the embryo, its water-absorbing capacity, production of enzymes, and germinating power.

Eckerson (1914) then focused on another microchemical process in plants: the thermotropism of roots. Previous research had indicated different findings regarding the curvature of roots in response to heat. In her studies, she also found that curvatures varied according to temperature and plant species. In addition, she noted that permeability of the root cells to dissolved substances also varied with the same factors. She concluded that heat was not the direct stimulus for curvature but rather affected permeability.

In a series of investigations at Boyce Thompson Institute, Eckerson studied protein synthesis in plants, specifically nitrate reduction (1924, 1932). Because the process had not been sufficiently clarified in earlier research, she first hoped to examine nitrate reduction step by step on similar plants to obtain a generalized view. She conducted microchemical tests over periods of time using plants in which nitrate was supplied through the soil. She found that nitrates were reduced to nitrites and ammonia under different conditions.

In the later study Eckerson continued to address conditions affecting nitrate reduction in plants. She examined "reducase" activity, or the nitrate-reducing ability of plants in terms of light and mineral nutrition. She found wide variations in the amount and distribution of reducase under different conditions, such as degrees of light exposure and amount of nutrition.

Another series of investigations concerned microchemical examinations of tomato mosaic (1926, 1927). These were begun with the cooperation of Dr. Erik Johannes Kraus at the University of Wisconsin and continued with her coauthor, Dr. Henry Reist Kraybill, at Boyce Thompson Institute. Eckerson began by examining organisms seen in mosaic plants, especially the tomato plant. She found similar organisms developing in all plants, with slight changes in form and considerable differences in size, but the same general appearance and behavior.

In the later study with Kraybill, Eckerson inoculated tomato plants with juice from mosaic plants to study the effects. Two types of juice were used: one filtered without treatment and one with colloidal substances partially removed before filtration. She found that when the juice was filtered without treatment, the infectious principle that produced the mottling disease was retained on the filter. However, when the juice was allowed to settle or was centrifuged to remove some colloidal material, the infectious principle passed through the filter. These two conditions produced different results on the tomato plant tests: the former resulted in mottling symptoms, while the latter produced fern-leaf symptoms.

The last series of studies were conducted with Wanda K. Farr, a cotton technologist with the USDA ("Formation of . . ." 1934; "Separation of . . ." 1934). First they examined cellulose particles in the cytoplasm of young cotton and found that single strands of these pectic-coated particles form a fibril of the mature fiber wall. These fibrils could be separated through mechanical treatment into microscopic particles of uniform size identical with those present in the cytoplasm of the young fiber. The role these fibrils played in building up the fiber wall presented a new concept of the method of formation of cellulose walls in plant cells.

In their later research, Farr and Eckerson initially examined a microchemical method of removing noncellulose substances from cellulose particles in plant cell walls by means of mild pectic solvents. While this method worked to yield a relatively small number of particles, it did not work on a macrochemical scale to obtain larger quantities. The resulting material was not entirely freed from the noncellulose substances. In their next study, they attempted to use treatments with hydrochloric acid (HCl), which proved successful. Through microchemical, optical, and X-ray diffraction techniques, the researchers showed that the essential cellulose nature of the particles was unaltered.

In addition to doing research, Eckerson was considered an excellent teacher.

Among the courses that she taught on a regular basis at the University of Chicago were plant physics, plant chemistry, growth and movement, plant microchemistry, and research in physiology. According to Pfeiffer (1954), she influenced young scientists in undergraduate and graduate studies, particularly through her ability to teach the special methods she had developed for following metabolic processes in plants. Further, although she never published a draft of a book intended for class use, she circulated mimeographed copies of her *Outlines of Plant Microchemistry*. Her methods of teaching, as well as the content, have been included in many other textbooks on this topic.

During her long academic career, Eckerson held memberships in Phi Beta Kappa and Sigma Xi. In addition, she was elected vice-chair (1934) and chair (1935) of the Physiological Section of the Botanical Society of America, a rare situation for a woman at that time. Further, she was starred in the 1938 edition of *American Men of Science*, an indication of her national scientific standing.

BIBLIOGRAPHY

Works by Sophia Hennion Eckerson

Scientific Works

"Chlorophyll solutions and spectra." *Botanical Gazette* 40 (1905): 302–305.
"The number and size of stomata." *Botanical Gazette* 46 (1908): 221–224.
"Root pressures and exudation." *Botanical Gazette* 45 (1908): 50–54.
"The demonstration of starch formation in leaves." *Botanical Gazette* 48 (1909): 224–228.
"A physiological and chemical study of after-ripening." Ph.D. diss., University of Chicago, 1911.
"A physiological and chemical study of after-ripening." *Botanical Gazette* 55 (1913): 286–299.
"Thermotropism of roots." *Botanical Gazette* 58 (1914): 254–263.
"Microchemical studies in the progressive development of the wheat plant." *Washington Agricultural Experiment Station Bulletin* 139 (1917): 3–20.
(with J. E. Dickson and K. P. Link) "The nature of resistance to seedling blight of cereals." *Proceedings of the National Academy of Sciences* 9 (1923): 434–439.
"Protein synthesis by plants. I. Nitrate reduction." *Botanical Gazette* 77 (1924): 377–390.
"An organism of tomato mosaic." *Botanical Gazette* 81 (1926): 204–209.
(with H. R. Kraybill) "Tomato mosaic. Filtration and inoculation experiments." *American Journal of Botany* 14 (1927): 487–495.
"Influence of phosphorus deficiency on metabolism of the tomato (*Lycopersicon esculentum* Mill)." *Contributions from Boyce Thompson Institute* 3 (1931): 197–217.
"Seasonal distribution of reducase in the various organs of an apple tree." *Contributions from Boyce Thompson Institute* 3 (1931): 405–412.
"Conditions affecting nitrate reduction by plants." *Contributions from Boyce Thompson Institute* 4 (1932): 119–130.

(with W. K. Farr) "Formation of cellulose membranes by microscopic particles of uniform size in linear arrangement." *Contributions from Boyce Thompson Institute* 6 (1934): 189–203.

(———) "Separation of cellulose particles in membranes of cotton fibers by treatment with hydrochloric acid." *Contributions from Boyce Thompson Institute* 6 (1934): 309–313.

Works about Sophia Hennion Eckerson

American Men of Science, A Biographical Directory. 6th ed., edited by J. Cattell. Lancaster, PA: Science Press, 1938.

Hall, D. L. "Academics, bluestockings, and biologists: Women at the University of Chicago, 1892–1932." In *Expanding the Role of Women in the Sciences*, edited by A. M. Briscoe and S. M. Pfafflin, 300–320. New York: New York Academy of Sciences, 1979.

McCallan, S.E.A. *A Personalized History of Boyce Thompson Institute.* New York: Boyce Thompson Institute for Plant Research, 1975.

Pfeiffer, N. E. "Sophia H. Eckerson, plant microchemist." *Science* 120 (1954): 820–821.

Other References

Ganong, William Francis. *The Teaching Botanist.* 2d ed. New York: Macmillan, 1910.

U.S. Census—Index, 1880. Bergen County, NJ. Index card E262.
    Sophia Eckerson is listed as 13 years old.

GERTRUDE BELLE ELION (1918–   )

I. Edward Alcamo

BIOGRAPHY

Gertrude Elion was born on January 23, 1918, in a middle-class neighborhood of Manhattan, New York. Her father, Robert Elion, was a graduate of New York University Dental School. He had immigrated from Lithuania as a small boy. Her mother, Bertha (Cohen) Elion, was also an immigrant, having come from Russia at the age of 14. Elion has a brother who is six years her junior. Although her mother lacked a college education, she was a prodigious reader, and she encouraged Elion to read avidly. Shortly after Elion began school, the family moved to the Bronx. Because Elion had learned to read so well, she advanced rapidly in school, skipping several grades.

After grammar school Elion attended Walton High School, at that time an all-girls' school with an excellent reputation. (Coincidentally, Rosalyn Yalow,* another Nobel Prize-winning researcher, also attended Walton High School.) After high school, Elion attended Hunter College. She graduated in 1937 with an A.B. degree in chemistry, summa cum laude.

Elion spent the next summer job hunting, and she soon became aware of the limited opportunities for women in science. She decided to go to secretarial school to learn a profession. After six weeks in the school, she received a job offer from New York Hospital School of Nursing to become a laboratory assistant and instructor in biochemistry for nurses. Several months later she obtained a position as a chemist at a small pharmaceutical company, the Denver Chemical Company, for a salary of $12 per week. She later secured a position teaching chemistry and physics in a New York City high school and enrolled at New York University for a master's degree in biochemistry, which she received in 1941.

With the advent of World War II, doors for women gradually opened in chemistry laboratories. The men were away at war, and employers hired women to take their places. Elion worked as a quality-control chemist in a food technology laboratory, testing pickle acidity, measuring mayonnaise color, and determining sugar concentrations in preserves. After a year and a half she received a research position at Johnson and Johnson Laboratories in New Brunswick, New Jersey. While at that laboratory, she became familiar with the therapeutic activities of sulfur drugs called sulfonamides, and she developed an interest in therapeutic compounds. Six months later, her job was terminated, and she began job hunting once again.

Elion received a job interview with George H. Hitchings at Burroughs Wellcome Laboratories in Tuckahoe, New York, just north of New York City. The location allowed her to continue living in the Bronx. She could also think about an advanced graduate degree at a New York university.

During her interview with Hitchings, Elion first became interested in nucleic acids. Hitchings was attempting to stop cell division by synthesizing chemical substances that interfere with nucleic acid metabolism. He was looking for a chemist, and Elion fit his requirements for the job. On June 15, 1944, Elion accepted a position at Wellcome Laboratories at a yearly salary of $2,600.

Over the course of the next four decades, Elion became renowned for her work in nucleic acid biochemistry. She became an expert on treatments for leukemia, on methods for suppressing the immune system during transplants, and on therapies for gout, parasitic diseases, and viral diseases. During the 1950s, Elion decided to attempt her doctoral studies, while continuing to work at Burroughs Wellcome. She attended Brooklyn Polytechnic Institute, one of the few universities offering graduate classes in chemistry in the evening. After two years it became apparent that she would have to give up her position to continue her graduate work, so she decided to remain with Wellcome and forgo her graduate education.

By the late 1960s Elion had investigated myriad problems in nucleic acid metabolism and had achieved a prominent reputation in biochemical research. She served as senior research chemist (1944–63) and assistant to the director of the Division of Chemotherapy (1963–67). In 1967 Hitchings was promoted to vice president in charge of research at Burroughs Wellcome, and the Biochemistry Department was divided into two subdivisions, one being the Department of Experimental Therapy. Elion became the head of that department. A year later, in 1968, Burroughs Wellcome moved from Tuckahoe to new research facilities in North Carolina, close to Duke University, the University of North Carolina, and North Carolina State University. This area became known as Research Triangle Park. Elion eventually assumed teaching positions at Duke and the University of North Carolina, while continuing her research.

Elion received the Garvan Medal of the American Chemical Society (1968) and the Judd Award of the Sloan-Kettering Institute (1983). The next year, she won the Cain Award of the American Association for Cancer Research. In 1988 Elion was awarded the Nobel Prize in Physiology or Medicine, which she shared with Hitchings and Sir James Black of Kings College Medical School in London.

Only a few times in the history of the Nobel Prize has the award been granted to researchers who developed drugs or worked for drug companies. The Nobel committee named the three scientists for their "discovery of important principles of drug treatment" and for developing an intelligent method of designing new compounds based on an understanding of basic biochemical processes.

In 1983 Elion retired as department head but continued as a scientist emeritus. She remained very active in her retirement. She was in demand by the press and media as well as universities, committees, and boards. Elion was elected to the National Academy of Sciences, the National Inventors' Hall of Fame, and the National Women's Hall of Fame. She received the National Medal of Science and numerous honorary degrees, including those from Brown University, the University of Michigan, George Washington University, Hunter College of the City University of New York, Philadelphia College of Pharmacy and Science, and Rensselaer Polytechnic Institute. In 1989 an Elion scholarship was endowed at Hunter College for two women to pursue two years of graduate research and study in chemistry and biochemistry.

Elion continues to spend time in her offices both in Wellcome Laboratories and at Duke University. She is readily accessible to graduate and medical school students, and she freely gives of her time to ensure that young people interested in science retain and develop that interest.

WORK

In 1944, when Elion came to Burroughs Wellcome, very little was known about nucleic acids or their structure. It was fairly certain that deoxyribonucleic acid (DNA) is the main component of the cell nucleus, but the significance of DNA was poorly understood. Scientists knew that DNA was composed of four different building blocks called nucleotides, but many believed that the four nucleotides were repeated over and over again. The idea of a genetic message in DNA had not yet been accepted. Ironically, in 1944, Oswald Avery and his group at Rockefeller Institute in New York City reported that DNA from one strain of bacteria could transform another strain of the same bacterium and give it new genetic characteristics.

Elion's initial assignments at Burroughs Wellcome were to synthesize certain nucleic acids. Hitchings wished to test one nucleic acid for antithyroid activity, and he intended to use another to antagonize a known nucleic acid as a means of inhibiting its activity. Elion became excited at the prospect of creating new biochemical compounds. Her first paper, in 1946, dealt with the ability of certain organic compounds to absorb ultraviolet radiation.

One of the compounds that drew Elion's attention was a nucleic acid derivative called 2,6-diaminopurine (2,6-DAP). This compound prevents the uptake of a nucleic acid component by mimicking its structure, and Elion hypothesized that 2,6-DAP could inhibit the synthesis of nucleic acids in cancer cells because cancer cells are actively producing DNA. Tests at Sloan-Kettering Institute for Cancer Research confirmed her suspicion: 2,6-DAP had good pharmacologic activity against mouse leukemia. The drug also appeared to be effective in humans, and research on 2,6-DAP intensified. However, the drug had several unfavorable side effects, and it could not be used in broad-scale tests.

Interest in DNA biochemistry increased substantially in 1953 with the revelation of DNA's structure by James D. Watson and Francis H. C. Crick. Also, many enzymes for nucleic acid synthesis were described and isolated during the 1950s, and the process of protein synthesis was further elucidated. Researchers studying nucleic acid metabolism suddenly found themselves in the forefront of science, and Elion was a member of that group. During that time she continued to synthesize a variety of new nucleic acid derivatives, one of which was 6-mercaptopurine (6-MP).

Elion's new compound, 6-MP, proved very useful in treating sarcomas in mice. When it was combined with other drugs and used in children suffering from leukemia, it increased the mean survival time from three months to one year. In 1953, 6-MP was approved by the federal Food and Drug Administration (FDA) as a viable treatment for acute childhood leukemia, and it continues in use as a remission-inducing agent for maintenance therapy. Thioguanine, another related drug synthesized by Elion, has also been used in the treatment of leukemia in adults.

Her successes encouraged Elion to synthesize a variety of derivatives of 6-MP and to use the newly developed radioactive techniques to explore 6-MP activity in cells. This methodology contrasted with the processes employed by other researchers, who were inclined to synthesize new drugs and test them in animals without understanding how they functioned. The reasoning and thought processes employed by Elion's research group were later highlighted by the Nobel Prize committee.

In 1958 researchers at Tufts University in Boston requested samples of 6-MP and two derivatives, azathioprine and thiamiprine, to test them as antibody-suppressing agents in rabbits. Their rationale was that if 6-MP and its derivatives could inhibit the activity of leukemia cells, they might also inhibit antibody-forming cells, since the two types of cells are structurally and functionally related. The antibody response is a substantial problem in transplantation procedures, because a recipient's antibodies vigorously reject transplant tissue from a distantly matched donor. It is, therefore, imperative to restrict the antibody response.

The tests with azathioprine yielded highly significant results in rabbits. A young British surgeon, Roy Calne, read the results in a journal article and requested a sample to try in dogs that were receiving kidney transplants. Calne was working with the noted transplant surgeon Joseph Murray. The tests in dogs were successful, and the production of rejecting antibodies was dramatically curtailed. (Joseph Murray's group is renowned for successfully performing the first kidney transplants in humans in 1954, and Murray became a Noble laureate in 1992.) In 1962 azathioprine was used for the first time in humans. For the next decade azathioprine (commercially available as Imuran) remained the major antirejection drug for use in kidney transplant recipients.

During studies with 6-MP, Elion's research group noted that the compound was broken down in the body to yield a substance identified as thiouric acid. This breakdown to thiouric acid was catalyzed by an enzyme, xanthine oxidase. Elion wondered what would happen if xanthine oxidase activity was inhibited. She located a drug called allopurinol, which would inhibit the enzyme without having any cytotoxic side effects. Inhibiting the enzyme prevented the breakdown of 6-MP as predicted, but it also yielded two very interesting results. First, it increased the immunosuppressive activity of 6-MP severalfold by permitting the drug to remain in the tissues longer; second, it prevented the buildup of thiouric acid, also as predicted.

This second observation has an important application to the treatment of a physiologic disorder called gout. Patients with gout experience an accumulation of uric acid crystals in the joints, and there is substantial pain when the joint is used. Working with researchers at Duke University, Elion determined that allopurinol could reduce the amount of uric acid considerably and provide relief. In 1966 allopurinol (Zyloprim) was approved by the FDA for treating gout.

During the 1970s researchers at St. Louis University, J. Joseph Marr and Randolph L. Berens, found that allopurinol could also be used against pathogenic species of the protozoa *Leishmania*. *Leishmania* species cause a sandfly-transmitted disease called leishmaniasis (1977). Leishmaniasis occurs in tropical regions of the world, and it can be a severe skin disease characterized by painful skin ulcers. It can also be a visceral disease in which the protozoan infects multiple internal organs of the abdomen. The report from St. Louis intrigued Elion, and her research group set about to study the nature of allopurinol's activity. They discovered that in the protozoan, allopurinol is converted to a certain toxic compound (''Allopurinol . . .'' 1979). However, a similar biochemistry does not occur in animals or humans. The metabolic differences were of interest to zoologists and biochemists. The work on allopurinol continues to this day, and it has been found that allopurinol can also be toxic to a number of other protozoal parasites.

In 1968 researchers reported the therapeutic value of a compound called adenine arabinoside when used against certain DNA viruses. Elion read the report with interest, because 20 years before, she had worked with a compound that had antiviral activity. The compound was 2,6-DAP. It occurred to Elion that perhaps the arabinoside derivative of 2,6-DAP might be active against DNA viruses in much the same way as adenine arabinoside. She synthesized the DAP-arabinoside, and, in 1969, she sent a sample to a specialized laboratory for antiviral testing. Several weeks later the answers came: DAP-arabinoside was very inhibitory against several DNA viruses, especially the virus of herpes simplex.

The studies on DAP-arabinoside began in earnest, and the researchers soon discovered that other derivatives were equally potent. One derivative called acyclovir (acycloguanosine) was outstanding. It became a valued antiviral agent in contemporary medicine. Elion started studying acyclovir in 1979 ("Acyclovir . . ." 1979). The drug is extremely valuable because it prevents the replication of herpesviruses without toxic side effects on the body, and it is effective for treating genital herpes. In 1982 the FDA approved the use of acyclovir in ointment and intravenous formulations and several years later in oral form. It is commercially available as Zovirax. Today the drug is used to lessen the severity of herpes simplex attacks and to extend the time between attacks. In patients with chickenpox the drug reduces the symptoms, and in patients with herpes zoster (shingles) the drug relieves some of the severe pain associated with this disease.

The discovery that acyclovir has antiviral activity was significant, but it was also important to know how acyclovir inhibits viral replication. During the 1970s Elion and her collaborators discovered that acyclovir inhibits viral replication by interfering with an enzyme called thymidine kinase. In the replication of herpesviruses, thymidine kinase synthesizes new strands of DNA for new viruses. Acyclovir bears a strong structural resemblance to one of the nucleotides used in the synthesis of DNA, but where the normal nucleotide has a chemical group for attaching to the next nucleotide, acyclovir has no such chemical group. Because of the structural resemblance, thymidine kinase incorporates acyclovir into the DNA instead of the normal nucleotide. No additional nucleotides can be added, because acyclovir lacks the chemical group for attachment. In this biochemistry, acyclovir is a terminator of the DNA synthesis. Without DNA, no new viruses can be synthesized.

After her retirement in 1983, Elion continued to lend her expertise to the development of azidothymidine (AZT). AZT is one of the few licensed drugs used against human immunodeficiency virus (HIV). Developed by Wellcome researchers, AZT works against HIV in much the same way as acyclovir works against herpesviruses. The FDA approved AZT in 1987.

Elion has served as president of the American Association for Cancer Research and on the National Cancer Advisory Board. She served on committees for the investigation of parasitic diseases for the World Health Organization. She is a member of numerous professional societies, including the American Chemical Society and the New York Academy of Sciences (Goodman 1993). The Nobel Prize committee took note of the fact that Elion's work was fundamental in developing chemotherapy. It noted that she concentrated on the physiology and biochemistry of cells, and in doing so, she came to understand the essential metabolic pathways and how to interfere with them. This approach, the Nobel committee indicated, was the more rational approach to the discovery of new compounds and their therapeutic effects.

BIBLIOGRAPHY

Works by Gertrude Belle Elion

Space does not permit a listing of the complete works of Gertrude Belle Elion. A bibliography of scientific articles by Elion and a listing of works about her are included in Goodman (1993). Listed here are all works cited in the text.

Scientific Works

(with W. S. Ide and G. H. Hitchings) "The ultraviolet absorption spectra of thiouracils." *Journal of the American Chemical Society* 68 (1946): 2137–2140.

(with P. deMiranda, R. J. Whitley, et al.) "Acyclovir kinetics after intravenous infusion." *Clinical Pharmacology and Therapeutics* 26 (6) (1979): 718–728.

(with Donald J. Nelson, Stephen W. LaFon, et al.) "Allopurinol ribonucleoside as an antileishmanial agent." *Journal of Biological Chemistry* 254 (22) (1979): 11544–11549.

"The quest for a cure." *Annual Review of Pharmacology and Toxicology* 33 (1993): 1–23.

Works about Gertrude Belle Elion

Goodman, Miles. "Gertrude Belle Elion (1918–)." In *Women in Chemistry and Physics: A Biobibliographic Sourcebook*, edited by L. Grinstein et al., 169–179. Westport, CT: Greenwood Press, 1993.

Other References

Avery, Oswald, C. L. Macleod, et al. "Studies on the chemical nature of the substance inducing transformation of Pneumococcal types." *Journal of Experimental Medicine* 79 (1944): 137–158.

Marr, J. Joseph, and Randolph L. Berens. "Antileishmanial effect of allopurinol. II. Relationship of adenine metabolism in *Leishmania* species to the action of allopurinol." *Journal of Infectious Diseases* 136 (6) (1977): 724–732.

Watson, J. D., and F.H.C. Crick. "Molecular structure of nucleic acids: A structure for deoxyribose nucleic acid." *Nature* 171 (1953): 737–738.

KATHERINE ESAU (1898–   )

Ray F. Evert and Susan E. Eichhorn

BIOGRAPHY

Katherine Esau, the youngest of four children, was born on April 3, 1898, to a Mennonite family in Yekaterinoslav, now called Dnepropetrovsk, in the Ukraine. The city was named originally after Katherine the Great, who promoted agriculture in the steppes of the Ukraine by inviting settlers from Germany, among them the Mennonites.

Esau's great-grandfather, Aron Esau, was born in Prussia in 1783. In 1804 he immigrated to the Ukraine on foot from his home in the Danzig area. Her grandfather, Jacob Esau, born in 1814, lived in the town of Halbstadt in the colony of Gnadenfeld. Jacob Esau married Catharine Neufeld from Halbstadt. Esau's father, John, was born there in 1859. In 1889 John Esau married Margarethe Toews, who was born in Yekaterinoslav in 1870. Margarethe Esau's great-grandfather, Heinrich Heese, was Lutheran and lived in Prussia. He, too, immigrated to the Ukraine, where he taught school and became a Mennonite.

Some Russians complained that the Mennonite youth avoided Russian schools and had too little communication with the Russian people. In 1869, during a visit by the future czar, Alexander III, to a Mennonite agricultural exhibit honoring him, the Mennonites promised to send two of their young men to a Russian school, proving their patriotism. A message was sent to all Mennonite colonies to find two boys; a stipend would be paid to send the youths to a gymnasium. All the colonies feared that the young men would become entirely Russian, and, therefore, nobody wanted to send their sons. Jacob Esau, wanting to give his sons a good education but lacking the funds to do so, immediately announced the availability of his two oldest sons, John and Jacob. The Mennonite community was reluctant to send two boys from the same family. The call was twice repeated, but nobody else responded; thus, John and Jacob were chosen as the first Mennonites to attend a Russian school. Both were admitted to a classical gymnasium in Yekaterinoslav.

John Esau continued his studies at Technical College in Riga and completed them in 1884, with the title of engineer technologist (mechanical engineer). Jacob Esau continued with medical training as an eye physician in Kiev. Both Esaus felt it their duty to return to Halbstadt after completing their educations, but the colony had no work for them. John Esau worked in several cities before settling in Yekaterinoslav, where he was married. His two oldest children, Nicolai, born in 1890, and Marie, born in 1892, died in childhood. Paul was born in 1894. John Esau became a very successful businessman in Yekaterinoslav and built a home there for his family. As a homeowner and citizen he participated in the city elections, becoming a city councillor and then mayor.

In 1905, at the time of the first Russian revolution, there were various threats and uprisings against the government. The citizens were becoming antagonistic to all things German. Because of all his good works for the city, however, John Esau was not threatened. In the new elections a more politically experienced man was elected mayor. John Esau again became a businessman. He was working on a waterworks system in Yalta when World War I started. He returned to Yekaterinoslav and agreed to take over the management of the Red Cross.

The Germans succeeded in occupying the Ukraine, a situation most welcomed because it saved them from occupation by the Bolsheviks and invasions by unorganized bands massacring people and destroying property. John Esau was again elected mayor; the city government was in great disarray.

Occupied Yekaterinoslav had the appearance of peace and order, but the political situation was very uncertain. The commanding German general advised John Esau to leave Russia, promising that a car on each German train would be reserved for civilians. John Esau began preparations to leave. As the German troops were preparing to leave, order vanished again. Heavily armed workers came to John Esau and demanded that he step down from the mayoral position immediately. They insisted that he was an agent of the rich, not a representative of the people. John Esau now liquidated the rest of his possessions and prepared to take his family to Berlin. The evening before departure, Russian gangs forced the German troops to give up their weapons. Along with the weapons, the baggage, including the food and clothing packed by the Esaus, was taken away. On December 21, 1918, the day after the departure of the Esau family, posters appeared in Yekaterinoslav proclaiming that the new "managers" were looking for John Esau, evidently to take care of this "representative of the old regime." Moreover, the posters proclaimed that the Esaus belonged to the "counterrevolutionary bourgeoisie" and were "enemies of the country." The Esaus arrived in Berlin on January 5, 1919. What normally was a two-day trip took two weeks because the train was frequently detained by revolutionary governments in the cities through which the train passed. Bribes were paid each time to allow the train to continue.

While in Germany, Paul Esau studied in Charlottenburg, and Katherine Esau in Berlin. Their father was very active in the Mennonite community to help other displaced Mennonites. In October 1922, Katherine Esau and her parents emigrated from Germany to the United States. Her brother followed after he completed his university studies in Germany. The family destination was Reedley, California, a largely Mennonite town near Fresno. They arrived there on November 16, 1922.

Katherine Esau had learned to read and write at home before beginning school in 1905. As a child she had initially attended a Mennonite primary school for four years. A Russian teacher taught the Russian language and such subjects as arithmetic, geography, history, and natural science. The minister from the church taught high German and Bible stories.

When she was 11 years old, Esau had entered the gymnasium, from which she graduated in 1916. That fall she continued her education at the Golitsin Women's College of Agriculture in Moscow, starting with the natural sciences, physics, chemistry, and geology.

Esau says that she chose agriculture over botany because she thought agriculture dealt with plants in a more interesting way than botany. Her view of botany was that the field dealt largely with the naming of plants, an impression gained from one of her relatives attending the University of Saint Petersburg.

The Russian revolution of 1917 interrupted her schooling after the first year. As travel became impossible, she remained in Yekaterinoslav, waiting for further developments. In the meantime, she studied English, took piano lessons, attended a gardening school, and collected plants that she was supposed to present at school in the second year.

In Berlin, Esau registered in the Berlin Landwirtschaftliche Hochschule (Agricultural College of Berlin). Fortunately, she had gathered her school documents (including one stating that she had earned the gold medal from the gymnasium) and carried them with her to Germany. During the second year of study Esau spent two semesters in Hohenheim, near Stuttgart, where she enrolled in various agricultural courses. After two more semesters in Berlin and a final examination, she received the title Landwirtschaftlehrerin. With some additional studies she passed a Zusatzprüfung in plant breeding given by the then-famous geneticist Erwin Baur. From Berlin, Esau went to a large estate in northern Germany that housed a model seed-breeding station for wheat. There she joined the workers in the fields and barns doing various chores.

When she met with some of her German professors for the last time, two of them seemed concerned about her future. One of them offered her an assistantship in teaching. Baur thought that she should return to Russia, where she would be able to contribute a great deal to Russian agriculture. Little did he understand the Russian revolution and what was in store for that country. By then the Esaus had decided to settle in the United States, and preparations were under way to do so.

In California Esau's father talked about buying a farm to apply her agricultural training, but she persuaded him that it would be wiser for her to find a job in a seed company. In the meantime, she performed housework in Fresno. While there, she met the Siemens family (her brother later married the daughter, Esther). Mr. Siemens introduced Esau to a Mr. Sloan from Idaho, who was starting the Sloan Seed Company in Oxnard, California. With her training in plant breeding, Esau was just the kind of person Sloan was looking for, and she was hired.

The beginning was difficult. Esau had a great many responsibilities. She had to hire a farmer with teams of horses and equipment to prepare the soil for planting. She had to hire Mexicans to plant selected seed by hand; this required Spanish lessons so that she could communicate with the Mexicans. The Oxnard episode lasted only one year. Sloan declared bankruptcy and gave up the company.

Shortly after, Esau was hired by the Spreckels Sugar Company in Spreckels, California, near Salinas. Her main task was to develop a sugar beet resistant to the curly top disease. The disease was already recognized as caused by a virus transmitted by the beet leafhopper. In 1919 Spreckels had succeeded in obtaining a resistant strain of sugar beet, but it had a poorly shaped root and a low sugar content. Esau's task was to improve the strain by hybridization. The work would be entirely her responsibility.

Despite crude working conditions, genuine progress was being made on the curly top project. Unexpectedly, two visitors from the University of California at Davis (U.C.–Davis), Dr. W. W. Robbins (chairman of the Botany Division) and Dr. H. A. Jones (chairman of the Truck Crops Division), came to see what was being done about the curly top problem at Spreckels. Esau showed them her work and inquired about the chances of doing graduate work at U.C.–Davis. Robbins immediately offered her an assistantship in his division. Neither he nor Spreckels had any objection to her working on a project based on the sugar beet. When she left Spreckels in the fall of 1927, a truckload of beets and beet seed followed her car. In the spring of 1928 Esau registered as a graduate student in the Botany Division of the College of Agriculture. Because U.C.–Davis had no graduate school, registration had to be done through the University of California at Berkeley (U.C.–Berkeley). In that same year, she and her parents became citizens of the United States.

Dr. T. H. Goodspeed, the *Nicotiana* (tobacco) cytologist in the Botany Department at U.C.–Berkeley, served as her adviser. A summer and one semester were spent taking courses at U.C.–Berkeley. Most of the courses were selected to strengthen her botanical background.

At U.C.–Davis, Esau was allowed to use the grounds belonging to the Truck Crops Division for raising sugar beets. She soon discovered, however, that little chance existed for her plants becoming naturally infected with the curly top disease, because the Davis area was not favorable for propagation of the beet leafhopper. She turned for help to a U.C.–Berkeley entomologist, Dr. H.H.P. Severin, who taught her how to raise infective leafhoppers. Now a new obstacle developed. The Truck Crops Division embarked upon a project of standardizing varieties of table beets. From then on, she was not allowed to liberate infective leafhoppers over the Truck Crops grounds. With a new plan in mind, she went to Robbins, explaining to him that the U.C.–Davis campus was not suitable for her original project. She proposed to replace it with a study on the effect of the curly top virus upon the plant. Her research area would now be plant anatomy, or, more specifically, pathological anatomy.

The committee appointed to supervise Esau's research did not include a plant anatomist, for there were no plant anatomists at either U.C.–Davis or U.C.–Berkeley. Esau passed her qualifying examination in September 1930 and her final examination a year later at U.C.–Berkeley. The Ph.D. degree was awarded in December 1931 and granted at the U.C.–Berkeley commencement of 1932. In that same year Esau was elected to Phi Beta Kappa.

Robbins immediately offered Esau the position of Instructor in the Botany Division and Junior Botanist in the Experiment Station of the College of Agriculture. She was especially pleased with the latter appointment because it would afford her time for research. Esau was not sure whether she would enjoy teaching. She was assigned to teach plant anatomy, systematic botany, morphology of crop plants, and microtechnique. Despite her initial reluctance to teach, she came to enjoy teaching, and her students responded accordingly. She was an excellent teacher, with a keen sense of humor; her classes were conducted in a relaxed atmosphere.

Esau served six years (the maximum number) in each rank (instructor, assistant, associate), becoming a full professor in 1949. Robbins did not believe in accelerated promotions. As an associate professor in 1946, she was selected to give the Faculty Research Lecture, a high honor at U.C.–Davis. A few years later, the new Director of the Campus observed that she was underpaid and raised her salary.

At U.C.–Davis, the Botany Division was housed in the Horticulture Building, where little space was available for offices. Early on (1930s), Esau shared a room (a teaching laboratory) with a newly appointed plant physiologist, Dr. Alden S. Crafts. It was a profitable arrangement, because both Crafts and she were becoming increasingly interested in phloem tissue (the food-conducting tissue). Crafts was interested in the phloem as a possible pathway for the transport of chemicals used for weed control. Esau wanted to study the close relationship between phloem tissue and the spread of virus-induced tissue degeneration. The relation between the virus and phloem provided the primary stimulus for her becoming so greatly involved in phloem research. This research continued, however, to be combined with efforts to find and to explore the so-called phloem-limited viruses. In the early 1960s, Esau turned to electron microscopy, which greatly enhanced the understanding of virus–plant host relations. Electron microscopy began to reveal the role of the unique features of the sieve element as the food-conducting cell in the phloem. These two aspects of phloem research came to dominate her interest in plant science.

When the Botany Division outgrew its quarters in the Horticulture Building, it was moved to other temporary housing in a building intended to serve as a garage. A small part of one room was walled off as a darkroom for photographic work. Because there was no air-conditioning, the darkroom was unbearably hot much of the time. As Esau's demands for photomicrography increased, she solved the problem by buying her own photomicrography and darkroom equipment and setting it up at home, just off the campus, where she lived with her mother and father. During the 1940s and 1950s all of her published photomicrographs, including those for the first edition of *Plant Anatomy*, were "home products."

Esau moved to the University of California at Santa Barbara (U.C.–Santa Barbara) in 1963, to continue her collaborative research on the phloem with Dr. Vernon I. Cheadle. She considers her 22 years at U.C.–Santa Barbara her most productive and satisfying years.

Numerous honors came her way. In addition to election to the National Academy of Sciences (1957), she received honorary degrees from Mills College, Oakland, California (1962), and the University of California (1966). The Botanical Society of America elected her president in 1951 and in 1956 awarded her the Certificate of Merit "for outstanding contributions in the advancement of botanical sciences." Some of her other honors include election to the American Academy of Arts and Sciences, the American Philosophical Society, and the Swedish Royal Academy of Science. In 1989 she was awarded the National Presidential Medal of Science.

WORK

While studying infected and noninfected sugar beets, Esau discovered that the initial degenerative effect of the disease upon the plant involved the first-formed sieve tubes and companion cells of the phloem. This indicated that the virus was invading the phloem tissue. Moreover, cytologic studies indicated that the virus was being transported in the sieve elements of the phloem. Esau's studies on the effect of the curly top virus in the tobacco plant supported the concept of the dependence of the virus on the phloem tissue for initiating the infection and spreading it throughout the plant. The concept of a "phloem-limited virus" was clearly established.

Esau's research on diseased plants was interspersed with developmental studies on healthy plants, including celery, tobacco, carrot, and pear. She investigated several general problems of developmental anatomy. She studied the characters used to distinguish between procambium and cambium and between primary and secondary vascular tissues, the origin of internal phloem, leaf-trace differentiation, and the concept of pericycle. Her anatomical research received immediate attention and brought invitations to write several comprehensive reviews in the *Botanical Review* (1938, 1939, 1943), with "follow-up" reviews (1948, 1950) about ten years later.

During the Second World War Esau became involved with a project intended to develop more productive rubber-yielding strains of guayule. Her task was to determine why certain polyploid strains of guayule failed to yield hybrids when crossed with other strains but instead produced a maternal type of progeny.

Apomixis (reproduction without sexual union) was found to be the cause of the problem.

Esau's attention then turned to the grapevine, both normal and infected with Pierce's disease. (At the time the Pierce's disease agent was thought to be a virus. Later, electron microscopy revealed the agent to be a rickettsia-like organism.) Light microscope studies continued on other plants known to be infected with viruses, including barley yellow dwarf in Gramineae, yellow leaf roll of peach in celery, and beet yellows in beet and *Tetragonia*.

In the 1950s and early 1960s, Esau's collaboration with Cheadle resulted in a series of contributions on the comparative structure of secondary phloem in dicotyledons. These studies were especially important in discussions of the evolutionary specialization of the phloem tissue in relation to function.

Esau was introduced to electron microscopy at U.C.–Davis in 1960, and her move to U.C.–Santa Barbara in 1963 led to the development of electron microscopy in the Biology Department there. As before in U.C.–Davis, the main topics of Esau's research at U.C.–Santa Barbara were the structure and development of the phloem, especially of the sieve element, and the appearance and fate of the virus in plant cells. One of these investigations, in collaboration with Dr. James Cronshaw on tobacco mosaic virus (1967), contributed to the characterization of the P-protein, a common component in the sieve elements of dicotyledons. This research was followed by studies on the development of the P-protein in squash and then in mimosa, cotton, and bean. Other aspects of sieve-element development were also examined, including development of sieve-plate pores, of the nucleus and endoplasmic reticulum, and of nuclear crystalloids, especially in the Boraginaceae.

Beet yellows disease virus was the first virus studied by Esau with the electron microscope. Perhaps the most rewarding of the virus studies were those undertaken with Dr. Lynn Hoefert (1960s) on beet western yellows disease, involving another phloem-dependent virus that causes yellowing of sugar beet leaves. Comparing leaves of different ages from the same plant, they found a sequence in the systemic spread of the virus, which begins with virions transported in sieve tubes. This is followed by virions passing through plasmodesmata (cytoplasmic connections between adjacent cells) into parenchymatous cells capable of viral synthesis and ends with the formation of new virus in the nucleus.

No less important than her research contributions are her books, beginning with *Plant Anatomy*, which was first published in 1953. The book immediately became a classic and the bible for structural botanists worldwide. In it Esau introduced a new and exciting developmental approach to the subject. In 1960 *Anatomy of Seed Plants* was published. In these two books, Esau standardized and unified terminology as well as usage of the entire vocabulary of plant anatomy.

In the 1960s Esau wrote four more books: *Plants, Viruses, and Insects*; *Vascular Differentiation in Plants*; *Viruses in Plant Hosts*; and *The Phloem*. In *The Phloem* Esau reviews the structure and development of the phloem, beginning with the earliest records of this tissue. An excellent example of Esau's contributions is the thoroughness of her literature reviews, which she made available to others in her many publications. No structural botanist has had such mastery of languages, including English, Russian, German, French, Spanish, and even Portuguese. Through her literature reviews and books, Esau introduced English-speaking structural botanists to the significant work carried out by early German, Russian, and French botanists. The value of this contribution cannot be overstated.

Esau was a superb teacher, in part because she genuinely liked students. With her friendly, relaxed attitude, keen sense of humor, and enthusiasm, her course in plant anatomy was exceptional. Although she served as major professor for only 15 doctoral students, numerous structural botanists, including many who have never met her but have studied her papers and books, consider themselves her students. Esau instilled in her students an appreciation of the precision and rigor that go into truly excellent studies of plant structure and development. In every aspect of her work, she set new standards of excellence.

Her career has no parallel. The citation accompanying the President's National Medal of Science reads:

In recognition of her distinguished service to the American community of plant biologists, and for the excellence of her pioneering research, both basic and applied, on plant structure and development, which has spanned more than six decades; for her superlative performance as an educator, in the classroom and through her books; for the encouragement and inspiration she has given to a legion of young, aspiring plant biologists; and for providing a special role model for women in science.

BIBLIOGRAPHY

Works by Katherine Esau

Scientific Works

Space does not permit the listing of the complete works by Katherine Esau. Listed here are all works by Esau except those cited in Esau (1938, 1939, 1943, 1948, 1950, 1961, *Plant* . . . 1965; *Vascular* . . . 1965, 1967, 1968, 1969, 1977) and Esau et al. (1957). Included here are her dissertation and all references cited in the text.

"Studies of the breeding of sugar beets for resistance to curly top." *Hilgardia* 4 (1930): 417–441.

"Sugar beets resistant to curly top. A report of results of an experiment pertaining to the breeding of beets for resistance to curly top disease." *Facts about Sugar* 25 (1930): 610–612.

"Planting season for sugar beets in Central California." *Bulletin, California Agricultural Experiment Station* 526 (1932).

"Pathologic changes in the anatomy of leaves of the sugar beet, *Beta vulgaris* L., affected by curly top." *Phytopathology* 23 (1933): 679–712.
    From Ph.D. dissertation.

"Bolting in sugar beets. A determination of its effect upon the weight and quality of the roots, based on studies in Central California." *Facts about Sugar* 29 (1934): 155–158.

(with C. W. Bennett) "Further studies on the relation of the curly top virus to plant tissues." *Journal of Agricultural Research* 53 (1936): 595–620.

"Some anatomical aspects of plant virus disease problems." *Botanical Review* 4 (1938): 548–579.

"Development and structure of the phloem tissue." *Botanical Review* 5 (1939): 373–432.

(with W. B. Hewitt) "Structure of end walls in differentiating vessels." *Hilgardia* 13 (1940): 229–244.

"Origin and development of primary vascular tissues in seed plants." *Botanical Review* 9 (1943): 125–206.

"Apomixis in guayule." *Proceedings of the National Academy of Sciences* 30 (1944): 352–355.

"Morphology of reproduction in guayule and certain other species of *Parthenium*." *Hilgardia* 17 (1946): 61–120.

"Some anatomical aspects of plant virus disease problems. II." *Botanical Review* 14 (1948): 413–449.

"Development and structure of the phloem tissue. II." *Botanical Review* 16 (1950): 67–114.

Plant Anatomy. New York: John Wiley and Sons, 1953.

(with V. I. Cheadle and E. M. Gifford, Jr.) "A staining combination for phloem and contiguous tissues." *Stain Technology* 28 (1953): 49–53.

"The nonconformist plant cell." *Idea and Experiment* 3 (1954): 13–15.

(with H. B. Currier and V. I. Cheadle) "Physiology of phloem." *Annual Review of Plant Physiology* 8 (1957): 349–374.

Anatomy of Seed Plants. New York: John Wiley and Sons, 1960.

Plants, Viruses, and Insects. Cambridge, MA: Harvard University Press, 1961.

(with R. Namba and E. A. Rasa) "Studies on penetration of sugar beet leaves by stylets of *Myzus persicae*." *Hilgardia* 30 (1961): 517–529.

(with E. A. Rasa) "Anatomic effects of curly top and aster yellows viruses on tomato." *Hilgardia* 30 (1961): 469–515.

(with L. M. Srivastava) "Relation of dwarfmistletoe (*Arceuthobium*) to the xylem tissue of conifers. I. Anatomy of parasite sinkers and their connection with host xylem." *American Journal of Botany* 48 (1961): 159–167.

(———) "Relation of dwarfmistletoe (*Arceuthobium*) to the xylem tissue of conifers. II. Effect of the parasite on the xylem anatomy of the host." *American Journal of Botany* 48 (1961): 209–215.

(with A.H.P. Engelbrecht) "Occurrence of inclusions of beet yellows viruses in chloroplasts." *Virology* 21 (1963): 43–47.

(with T. Bisalputra) "Polarized light study of phloem differentiation in embryo of *Chenopodium album*." *Botanical Gazette* 125 (1964): 1–7.

Plant Anatomy. 2d ed. New York: John Wiley and Sons, 1965.

Vascular Differentiation in Plants. New York: Holt, Rinehart and Winston, 1965.

(with R. H. Gill) "Observations on cytokinesis." *Planta* 67 (1965): 168–181.

(with V. I. Cheadle and R. H. Gill) "Cytology of differentiating tracheary elements. II.

Structures associated with cell surfaces." *American Journal of Botany* 53 (1966): 765–771.

"Anatomy of plant virus infections." *Annual Review of Phytopathology* 5 (1967): 45–76.

(with J. Cronshaw) "Tubular and fibrillar components of mature and differentiating sieve elements." *Journal of Cell Biology* 34 (1967): 801–815.

(———) "Tubular components in cells of healthy and tobacco mosaic-virus infected *Nicotiana.*" *Virology* 33 (1967): 26–35.

Viruses in Plant Hosts: Form, Distribution, and Pathologic Effects. Madison: The University of Wisconsin Press, 1968.

The Phloem. Handbuch der Pflanzenanatomie. Histologie, Band V, Teil 2. Berlin-Stuttgart: Gebrüder Borntraeger, 1969.

(with V. I. Cheadle) "Secondary growth in *Bougainvillea.*" *Annals of Botany* 33 (1969): 807–819.

(with R. H. Gill) "Structural relations between nucleus and cytoplasm during mitosis in *Nicotiana tabacum* mesophyll." *Canadian Journal of Botany* 47 (1969): 581–591.

(———) "Tobacco mosaic virus in dividing cells of *Nicotiana.*" *Virology* 38 (1969): 464–472.

"On the phloem of *Mimosa pudica* L." *Annals of Botany* 34 (1970): 505–515.

(with R. H. Gill) "Observations on the spiny vesicles and P-protein in *Nicotiana tabacum.*" *Protoplasma* 69 (1970): 373–388.

(———) "A spiny cell component in the sugar beet." *Journal of Ultrastructure Research* 31 (1970): 444–455.

"Development of P-protein in sieve elements of *Mimosa pudica.*" *Protoplasma* 73 (1971): 225–238.

"The sieve element and its immediate environment: Thoughts on research of the past fifty years." *The Journal of the Indian Botanical Society* 50A (1971): 115–129.

(with R. H. Gill) "Aggregation of endoplasmic reticulum and its relation to the nucleus in a differentiating sieve element." *Journal of Ultrastructure Research* 34 (1971): 144–158.

(with L. L. Hoefert) "Composition and fine structure of minor veins in *Tetragonia* leaf." *Protoplasma* 72 (1971): 237–253.

(———) "Cytology of beet yellows virus infection in *Tetragonia*. I. Parenchyma cells in infected leaf." *Protoplasma* 72 (1971): 255–273.

(———) "Cytology of beet yellows virus infection in *Tetragonia*. II. Vascular elements in infected leaf." *Protoplasma* 72 (1971): 459–476.

(———) "Cytology of beet yellows virus infection in *Tetragonia*. III. Conformations of virus in infected cells." *Protoplasma* 73 (1971): 51–65.

"Apparent temporary chloroplast fusions in leaf cells of *Mimosa pudica.*" *Zeitschrift für Pflanzenphysiologie* 67 (1972): 244–254.

"Changes in the nucleus and endoplasmic reticulum during differentiation of a sieve element in *Mimosa pudica* L." *Annals of Botany* 36 (1972): 703–710.

"Cytology of sieve elements in minor veins of sugar beet leaves." *New Phytologist* 71 (1972): 161–168.

(with R. H. Gill) "Nucleus and endoplasmic reticulum in differentiating protophloem of *Nicotiana tabacum.*" *Journal of Ultrastructure Research* 41 (1972): 160–175.

(with L. L. Hoefert) "Development of infection with beet western yellows virus in the sugarbeet." *Virology* 48 (1972): 724–738.

(———) "Ultrastructure of sugarbeet leaves infected with beet western yellows virus." *Journal of Ultrastructure Research* 40 (1972): 556–571.

"Comparative structure of companion cells and phloem parenchyma cells in *Mimosa pudica* L." *Annals of Botany* 37 (1973): 625–632.

(with R. H. Gill) "Correlations in differentiation of protophloem sieve elements of *Allium cepa* root." *Journal of Ultrastructure Research* 44 (1973): 310–328.

(with L. L. Hoefert) "Particles and associated inclusions in sugarbeet infected with the curly top virus." *Virology* 56 (1973): 454–464.

"Ultrastructure of secretory cells in the phloem of *Mimosa pudica* L." *Annals of Botany* 38 (1974): 159–164.

(with I. B. Morrow) "Spatial relation between xylem and phloem in the stem of *Hibiscus cannabinus* L. (Malvaceae)." *Botanical Journal of the Linnean Society* 68 (1974): 43–50.

"Crystalline inclusion in thylakoids of spinach chloroplasts." *Journal of Ultrastructure Research* 53 (1975): 235–243.

"Dilated endoplasmic reticulum cisternae in differentiating xylem of minor veins of *Mimosa pudica* L. leaf." *Annals of Botany* 39 (1975): 167–174.

"The phloem of *Nelumbo nucifera* Gaertn." *Annals of Botany* 39 (1975): 901–913.

(with I. Charvat) "An ultrastructural study of acid phosphatase localization in cells of *Phaseolus vulgaris* phloem by the use of the azo dye method." *Tissue and Cell* 7 (1975): 619–630.

(———) "An ultrastructural study of acid phosphatase localization in *Phaseolus vulgaris* xylem by the use of an azo-dye method." *Journal of Cell Science* 19 (1975): 543–561.

(with L. L. Hoefert) "Plastid inclusions in epidermal cells of *Beta*." *American Journal of Botany* 62 (1975): 36–40.

(with H. Kosakai) "Laticifers in *Nelumbo nucifera* Gaertn.: Distribution and structure." *Annals of Botany* 39 (1975): 713–719.

(———) "Leaf arrangement in *Nelumbo nucifera*: A re-examination of a unique phyllotaxy." *Phytomorphology* 25 (1975): 100–112.

"Hyperplastic phloem and its plastids in spinach infected with the curly top virus." *Annals of Botany* 40 (1976): 637–644.

(with A. C. Magyarosy and V. Breazeale) "Studies of the mycoplasma-like organism (MLO) in spinach leaf affected by the aster yellows disease." *Protoplasma* 90 (1976): 189–203.

Anatomy of Seed Plants. 2d ed. New York: John Wiley and Sons, 1977.

"Membranous modifications in sieve element plastids of spinach affected by the aster yellows disease." *Journal of Ultrastructure Research* 59 (1977): 87–100.

"Virus-like particles in nuclei of phloem cells in spinach leaves infected with the curly top virus." *Journal of Ultrastructure Research* 61 (1977): 78–88.

"Developmental features of the primary phloem in *Phaseolus vulgaris* L." *Annals of Botany* 42 (1978): 1–13.

"The protein inclusions in sieve elements of cotton (*Gossypium hirsutum* L.)." *Journal of Ultrastructure Research* 63 (1978): 224–235.

(with I. Charvat) "On vessel member differentiation in the bean (*Phaseolus vulgaris* L.)." *Annals of Botany* 42 (1978): 665–677.

(with L. L. Hoefert) "Hyperplastic phloem in sugarbeet leaves infected with the beet curly top virus." *American Journal of Botany* 65 (1978): 772–783.

"Beet yellow stunt virus in cells of *Sonchus oleraceus* L. and its relation to host mitochondria." *Virology* 98 (1979): 1–8.

"Phloem." In *Anatomy of the Dicotyledons*, 2d ed., vol. 1. *Systematic Anatomy of Leaf and Stem, with a Brief History of the Subject*, edited by C. R. Metcalfe and L. Chalk, 181–189. Oxford: Clarendon Press, 1979.

(with A. C. Magyarosy) "A crystalline inclusion in sieve element nuclei of *Amsinckia*. I. The inclusion in differentiating cells." *Journal of Cell Science* 38 (1979): 1–10.

(———) "A crystalline inclusion in sieve element nuclei of *Amsinckia*. II. The inclusion in maturing cells." *Journal of Cell Science* 38 (1979): 11–22.

(———) "Nuclear abnormalities and cytoplasmic inclusion in *Amsinckia* infected with the curly top virus." *Journal of Ultrastructure Research* 66 (1979): 11–21.

(with L. L. Hoefert) "Endoplasmic reticulum and its relation to microtubules in sieve elements of sugarbeet and spinach." *Journal of Ultrastructure Research* 7 (1980): 249–257.

(with A. C. Magyarosy) "A study of the source of virus in the exudate appearing on leaves of *Amsinckia* infected with the beet curly top virus." *Plant, Cell and Environment* 3 (1980): 425–433.

(with S. Shepardson and R. McCrum) "Ultrastructure of potato leaf phloem infected with potato leafroll virus." *Virology* 105 (1980): 379–392.

(with L. L. Hoefert) "Beet yellow stunt virus in the phloem of *Sonchus oleraceus* L." *Journal of Ultrastructure Research* 75 (1981): 326–338.

(with J. Thorsch) "Changes in the endoplasmic reticulum during differentiation of a sieve element in *Gossypium hirsutum*." *Journal of Ultrastructure Research* 74 (1981): 183–194.

(———) "Nuclear degeneration and the association of endoplasmic reticulum with the nuclear envelope and microtubules in maturing sieve elements of *Gossypium hirsutum*." *Journal of Ultrastructure Research* 74 (1981): 195–204.

(———) "Ultrastructural studies of protophloem sieve elements in *Gossypium hirsutum*." *Journal of Ultrastructure Research* 75 (1981): 339–351.

(———) "Microtubules in differentiating sieve elements of *Gossypium hirsutum*." *Journal of Ultrastructure Research* 78 (1982): 73–83.

(———) "Nuclear crystalloids in sieve elements of species of *Echium* (Boraginaceae)." *Journal of Cell Science* 54 (1982): 149–160.

(———) "Nuclear crystalloids in sieve elements Boraginaceae: A protein digestion study." *Journal of Cell Science* 64 (1983): 37–47.

(with V. I. Cheadle) "Anatomy of the secondary phloem in Winteraceae." *IAWA Bulletin* n.s., 5 (1984): 13–43.

(with J. Thorsch) "The sieve plate of *Echium* (Boraginaceae): Developmental aspects and response of P-protein to protein digestion." *Journal of Ultrastructure Research* 86 (1984): 31–45.

(———) "Sieve plate pores and plasmodesmata, the communication channels of the symplast: Ultrastructural aspects and developmental relations." *American Journal of Botany* 72 (1985): 1641–1653.

(———) "An ultrastructural study of the phloem of *Drimys* (Winteraceae)." *IAWA Bulletin* n.s., 6 (1985): 255–268.

(———) "Ultrastructural aspects of primary phloem. Sieve elements in Poinsettia (*Euphorbia pulcherrima*, Euphorbiaceae)." *IAWA Bulletin* n.s., 9 (1988): 363–373.
(with D. Fisher and J. Thorsch) "Inclusions in nuclei and plastids of Boraginaceae and their possible taxonomic significance." *Canadian Journal of Botany* 67 (1989): 3608–3617.
(with R. H. Gill) "Distribution of vacuoles and some other organelles in dividing cells." *Botanical Gazette* 152 (1991): 397–407.

Works about Katherine Esau

Evert, R. F. "Katherine Esau." *Plant Science Bulletin* 31(5) (1985): 33–37.

ALICE CATHERINE EVANS (1881–1975)

Diana M. Colon

BIOGRAPHY

Alice Catherine Evans was born on January 29, 1881, in Neath in the farm country of northern Pennsylvania. Neath was named for Neath, Wales, which was the home of her paternal grandparents until 1831, when they immigrated to northern Pennsylvania. Her parents were William Howell and Anne B. (Evans) Evans. She had an older brother.

Her education began in local schools. After her primary education in the district school, she studied in the Susquehanna Collegiate Institute in Towanda, Pennsylvania, where she was one in a class of seven. At the time, despite her dream of attending college, it looked as if her formal education had come to an end, since she lacked the means to pursue it. As a result, she began her career in grade school teaching, which was one of the few professions open to women at the time. She taught for four years; then she took advantage of a two-year, tuition-free nature study course at the Cornell University College of Agriculture. This course was designed especially for rural schoolteachers to foster the love of nature in country children (O'Hern 1973, 573).

Before completing this nature study course, Evans's interest in science was stimulated to the point that she wanted to continue the study of science. Fortunately for Evans, the Cornell College of Agriculture was then accepting out-of-state students tuition-free. With no tuition expenses and with the help of a scholarship during her junior and senior years, Evans succeeded in obtaining the B.S. degree in agriculture in 1909. Seniors were also required to major in some branch of applied science, so Evans chose bacteriology.

During her senior year, her professor of dairy bacteriology was asked to recommend a graduating student for a scholarship at the University of Wisconsin. Alice Evans was his choice, and she was accepted. This was the first time such a scholarship was awarded to a woman. She was awarded the M.S. degree in bacteriology by the University of Wisconsin in 1910. Her professor of chemistry of nutrition, E. V. McCollum, encouraged her to continue her studies for the

Ph.D. degree on a university fellowship in chemistry, but she decided against it. She felt that the physical and financial strain of her past seven years' studies disinclined her to continue her formal education at this time. She left the option open to complete the Ph.D. degree someday, but until she was ready, she continued with her career (O'Hern 1973, 574). She eventually pursued her graduate studies at George Washington University and at the University of Chicago but never obtained an earned doctorate.

Evans accepted the offer of her adviser, Professor E. G. Hastings, for a research position with the Dairy Division of the U.S. Department of Agriculture (USDA), starting on July 1, 1910. Since the Dairy Division Laboratories in Washington, D.C., were not ready for occupancy, she continued to work at the University of Wisconsin. Evans started out with the investigation into better methods of cheese making, an important industry in Wisconsin.

In 1913 she moved to Washington, D.C., and, contrary to her expectations, the environment at the Bureau of Animal Industry proved to be quite congenial. A number of women scientists were already employed in the Department of Agriculture, and Evans became the first woman to hold a permanent appointment in the Dairy Division of the Bureau of Animal Industry. In addition, she found that her immediate supervisors had a favorable attitude toward women in science.

At the time, freshly drawn milk was thought to be particularly nutritious if not contaminated, and one of Evans's projects was to search for sources of bacteria in dairy products. Evans was also assigned a project of her own: the study of bacteria that occur in milk freshly drawn from the cow's udder. Her research was important because it proved for practical purposes that the bacterium that causes contagious abortion in cows and the one that causes Malta fever in humans may be regarded as nearly identical and that either of these may cause a painful and dangerous disease in man known as undulant fever. She also concluded that milk should be pasteurized since humans contract this illness by handling infected animals or by drinking unpasteurized milk.

In 1918, World War I was under way, and Evans sought to work more closely with the war effort. The Director of the U.S. Public Health Service urged Evans to join his staff in the Hygienic Laboratory, the forerunner of the National Institutes of Health, as an assistant bacteriologist. This new job dealt with the study of meningitis. She joined a group of doctors who were striving to improve an antiserum used in the treatment of epidemic meningitis, one of the dreaded diseases of World War I. Despite this change of active research agendas, Evans continued to follow the literature concerning the bacteria found in milk that cause undulant fever.

The great pandemic in 1918 of Spanish influenza spread to Washington and was very severe due to wartime overcrowding. Congress passed a resolution stating that the Public Health Service was to combat this disease by aiding state and local Boards of Health. All medical officers of the Hygienic Laboratory were put onto this case. Evans was to investigate the cause of the epidemic, but she barely began when she became ill. When she finally recovered, the war was over, and the peak of the epidemic had passed, so she returned to the work on meningococci.

She actively resumed her original research at the Hygienic Laboratory, and in 1922 the first case of human undulant fever was diagnosed in Baltimore. A culture was sent to Evans for study, and from then on her discoveries were taken more seriously. Just as Evans began to experience the satisfaction of her discovery being recognized by American scientists, she fell victim to undulant fever in 1922, and she was seriously ill for seven years. She was treated in five different hospitals during the next 21 years; the last attack occurred in 1943. She once said, "It seems as if those bugs had a special animosity toward me, since I made that discovery" (Block 1943, 200).

In 1939 she turned her attention to hemolytic streptococci and made notable contributions to the knowledge of this group of microorganisms until she retired in 1945 as a senior bacteriologist.

Some of the honors that Alice Evans received included an honorary M.D. degree from the Woman's Medical College in 1934 (now Medical College of Pennsylvania) and honorary D.Sc. degrees from Wilson College (1936) and the University of Wisconsin (1948). Evans was also elected honorary member of the American Society for Microbiology in 1975, the year proclaimed by the United Nations as International Woman's Year.

Her interest in education and in youth was manifested in 1969 by her gift through the American Association of University Women (AAUW) of a scholarship fund for Federal City College in Washington, D.C. Alice Evans remained single throughout her life, and she died of a stroke September 5, 1975, in Goodwin House, a nursing home in Alexandria, Virginia.

WORK

While Evans was working in the Dairy Division of the USDA, she studied the bacterial flora of freshly drawn milk. She found the bacterial flora to be varied, with several species commonly present. Her attention focused on the causal organism of contagious abortion (Bang's disease) in cows, *Bacillus abortus*, as then classified. This bacterium had been regarded as harmless to humans. One day Evans was discussing the matter with her supervisor, who casually asked her if she had ever compared the Malta fever coccus with *B. abortus*. Consequently, her study of the relationship of *B. abortus* to other pathogenic bacteria included the organism that caused Malta fever, *Micrococcus melitensis*, as then named. In comparing these two organisms, she discovered that all strains of each type of organism behaved essentially the same in all culture tests then available for differentiating bacteria and in cross-agglutinations. Both microorganisms are now classified in the genus *Brucella*, and the human illness brucellosis has been nearly eliminated by the universal pasteurization of milk.

Evans then requested animal cultures of these two bacterial species. She proceeded to inoculate pregnant guinea pigs with *B. abortus* and *M. melitensis*. Each of the animals aborted and produced specific antibodies, and positive cultures were obtained from various organs. She then demonstrated that the two microorganisms were similar, but they could be differentiated by agglutinin-absorption tests.

"Sort of vaguely I knew what it meant," said Evans, speaking of her discovery. "I knew most American milk wasn't pasteurized. I'd just proven there was no practical difference between the abortion germ and the Malta fever microbe" (Block 1943, 199).

Evans reported her results at the 1917 meeting of the Society of American Bacteriology. According to Evans, her reported results were greeted with skepticism on the part of bacteriologists, dairymen, and physicians on the grounds that if the two organisms were the same, someone else would have discovered that fact already. In addition, Evans's lack of a doctorate and her government employment, as well as her female gender, undercut her credibility (O'Hern 1973, 575; Rossiter 1982, 229–230). Another question raised was: If the microbes were alike and would produce Malta fever in humans, why was there no such illness in the United States? It was later realized that Malta fever (brucellosis) was more prevalent than anyone had suspected. In its mild form, the disease resembles influenza, and severe cases are similar to typhoid fever, tuberculosis, malaria, or rheumatism; therefore, they were often misdiagnosed.

In 1922, when a case of human brucellosis was diagnosed at Johns Hopkins Hospital in Baltimore, a culture was sent to Evans for study. This first case of human brucellosis was studied by both Theobald Smith and Alice Evans. Smith was one of the most important scientists of the time and one of the most vocal scientists in the United States who opposed the idea that *B. abortus* in cow's milk might cause human disease. Indeed, Smith had been working on brucellosis of cattle for several years before Evans's first publication on the subject in 1918, and he published reports disagreeing with the premise that brucellae in milk might be hazardous to the health of those drinking it (O'Hern 1973, 576). Smith was so strong in his feelings that he refused to chair the Committee on Infectious Abortion because Alice Evans was to be a member.

About six months later in Rhodesia and in other parts of the world, human brucellosis was reported after ingestion of raw milk. This information, along with Evans's data, changed Smith's skepticism about Evans and her conclusions. In 1925 Smith became the chairman of the committee, with Evans as a member, and she remained a member until 1931.

Around 1939 Evans began investigation of immunity to streptococcal infection. In addition, her contributions to the knowledge and classification of the hemolytic streptococcal group are notable. Her investigations began when there were about 30 known types of streptococci. By 1945, when she retired, there were 46 types. Sulfa drugs and penicillin had been added to the armament against infection. She continued the study and found that, when unbroken bacterial cells are the antigens, the immune sera fall into groups (O'Hern 1973, 578).

Evans was active in professional societies as well. As a token of recognition for her research, she was the first woman elected to the presidency of the Society of American Bacteriologists (1928). From 1925 to 1931, she was a member of the Committee on Infectious Abortion of the National Research Council, and she was a delegate to the First and Second International Microbiological Congress in Paris (1930) and London (1936). She was also elected honorary president of the First Inter-American Congress on Brucellosis in Mexico City (1946) and remained a member of the organization for 11 years. Evans was also a fellow of the American Association for the Advancement of Science and a member of the American Academy of Microbiology, Washington Academy of Sciences, AAUW, Sigma Delta Epsilon, Sigma Xi, American Association of United Nations, and United World Federalists.

In 1947 an NBC radio program, "Cavalcade of America," presented the story of Evans's life and work.

BIBLIOGRAPHY

Works by Alice Catherine Evans

Scientific Works

Space does not permit the listing of the complete works of Alice Catherine Evans. Listed here are all works by Evans except those cited in Morgan and Corbel (1977). Also included are all references cited in the text.

(with E. G. Hastings and E. B. Hart) "Cheese. Bacteriology of cheddar cheese." Circular no. 150. Washington, DC: Public Health Service, 1912.

(with E. G. Hastings) "Milk. Comparison of acid test and rennet test for determining condition of milk for cheddar type of cheese." Circular no. 210. Washington, DC: Public Health Service, 1913.

(with I. A. Rogers and W. M. Clark) "Colon bacteria and streptococci and their significance in milk." *American Journal of Public Health* 6 (1916): 374–380.

"Further studies on *Bacterium abortus* and related bacteria. I. II." *Journal of Infectious Diseases* 22 (1918): 576–579, 580–593.

"Further studies on *Bacterium abortus* and related bacteria. III." *Journal of Infectious Diseases* 23 (1918): 354–372.

"Determination of bacteritropic content of antimeningococcic serum." *Public Health Reports* 34 (1919): 2375–2377.

"Lack of correlation among meningococci between serological grouping and lethal effects on rabbits." *Journal of Medical Research* 42 (1920): 33–48.

"Serological grouping of meningococcus strains isolated in New York City in 1921 and 1922." *Public Health Reports* 37 (1922): 1247–1250.

"Toxicity of acids for leucocytes, as indicated by tropin reaction." *Journal of Immunology* 7 (1922): 271–304.

"Life cycles in bacteria." *Journal of Bacteriology* 17 (1929): 63–77.

"Effect of hemolytic streptococci and their products on leucocytes." *Public Health Reports* 46 (1931): 2539–2557.

"New subspecies, radicans of *Alcaligenes faecalis.*" *Public Health Reports* 46 (1931): 1676–1680.

(with S. L. Cummins) "Intradermal tuberculin test in nontuberculosis adults." *British Medical Journal* 1 (1933): 815–817.

"Brucellosis (undulant fever): Public health problem." *Journal of the Medical Association of the State of Alabama* 8 (1938): 97–98.

"Studies on chronic brucellosis." *Archives Internationales des Brucelloses* 1 (1938): 174–179.

"Brucellosis (undulant fever)." Circular no. T27.2:B83/939. Washington, DC: Public Health Service, 1939.

"Comparison of *Streptococcus pyogenes* and *Streptococcus epidemicus.*" *Southern Medical Journal* 33 (1940): 318–321.

"Descripción de las técnicas de pruegas especificas para la brucelosis." *Boletín de la Oficina Sanitaria Panamericana* 19 (1940): 233–240.

"Potency of nascent streptococcus bacteriophage B." *Journal of Bacteriology* 39 (1940): 597–604.

(with E. M. Sockrider) "Another serologic type of streptococcic bacteriophage." *Journal of Bacteriology* 44 (1942): 211–214.

(with R. R. Trail et al.) "Mass miniature radiography in Royal Air Force, report on 250,027 consecutive examinations of R.A.F. and W.A.A.F. personnel." *British Journal of Tuberculosis* 38 (1944): 116–140.

Works about Alice Catherine Evans

"Alice Evans, 94, bacteriologist, dies." *Washington Post* (Sept. 8, 1975): B4.

"Alice Evans, who found cause of undulant fever in man, dies." *New York Times* (Sept. 7, 1975): 51.

Block, M. (ed.) *Current Biography Yearbook.* New York: H. W. Wilson, 1943.

De Kruif, P. H. "Before you drink a glass of milk: The story of a woman's discovery of a new disease." *Ladies' Home Journal* 46 (1929): 8–9, 162, 165–166, 168–169.

"Dr. Alice Evans dies at 94: Joined PHS in 1918: Among 1st women to pick medical research career." *United States, National Institutes of Health Record* (1975): 3.

Eddy, B. E. "Obituary: Alice Catherine Evans." *American Society of Microbiology News* 42 (1976): 166–168.

MacKaye, M. "Undulant fever." *Ladies' Home Journal* 61 (1944): 69–70.

Morgan, W.J.B., and M. J. Corbel. "Miss Alice C. Evans." *Annali Sclavo* 19 (1977): 3–11.
    Contains an extensive bibliography.

O'Hern, E.M. "Alice Evans and the brucellosis story." *Annali Sclavo* 19 (1977): 12–19.

———. "Alice Evans, pioneer microbiologist." *American Society of Microbiology News* 39 (1973): 573–578.

Other References

Barnard, E. F. "Women microbe hunters." *Independent Women* 15 (1936): 379, 396–397.

Rossiter, M. R. *Women Scientists in America: Struggles and Strategies to 1940.* Baltimore: Johns Hopkins University Press, 1982.

Smith, T. "Science opens its doors to women." *Literary Digest* 124 (1937): 17–18.

HONOR BRIDGET FELL (1900–1986)

Suzanne E. Moshier

BIOGRAPHY

Honor Bridget Fell was born near Filey at Fowthorpe, Yorkshire, England, on May 22, 1900. Her parents were Colonel William Edwin Fell and Alice (Pickersgill-Cunliffe) Fell (Burgess 1989). Honor Fell was the youngest of nine children; she had six sisters and two brothers. As a youngster she demonstrated a love of nature that foreshadowed her eventual decision to become a biologist. One story of her childhood recalls the time she brought one of her pet ferrets as an uninvited and unwelcomed guest to her sister's wedding.

Fell was a student at Wychwood School, Oxford; Madras College, St. Andrew's; and then Edinburgh University, Scotland, where she received her B.Sc. degree in zoology with first-class honors (1923) and also her Ph.D. degree (1924) and D.Sc. degree (1932). In 1955 she was awarded her M.A. degree from Cambridge University. While studying toward her Ph.D. degree, she worked with Dr. F.A.E. Crew, who was the director of the Animal Breeding Research Department. Her first published research dealt with the sexual development of the fowl (Fell and Crew 1922).

One of Fell's projects at Edinburgh University was an investigation of the development of bone and cartilage in fowl. She thought the methods of T.S.P. Strangeways would be of use in her work, and she visited the Cambridge Research Hospital to learn them from him. When she published her results, she acknowledged him for "much valuable advice and friendly criticism" (Fell, "The histogenesis . . ." 1925). He invited her to become his assistant and promised to arrange a grant for her. When no position was available for Fell at Edinburgh University, she accepted Strangeways's offer.

As the recipient of a grant from the Medical Research Council, Fell became research assistant to Strangeways in 1923 at the Cambridge Research Hospital. After Strangeways died in 1926, Dr. J. A. Andrews, and then Fell, became temporary directors of the laboratory. She was formally appointed director of the small, scantily funded, newly renamed Strangeways Research Laboratory at age 29. Fell and all of the staff wanted to keep the laboratory running because "we knew the effort and sacrifice that had gone into founding and maintaining it, and . . . at that time it was the only institution in the country entirely devoted to the study of cell biology, and finally it seemed to us so silly to close the Laboratory when we were perfectly capable of running it—such is the self-confidence of youth" (Fell 1981). Fell remembers as crucial the early support from the Medical Research Council, which elected to continue funding the small operation in 1927 despite Strangeways's death. She dedicated herself to the laboratory's survival and her own research and training activities there until she resigned the directorship in 1970. During her tenure the laboratory gained stature and became internationally known for its cell biology research.

When Fell arrived at Cambridge to work with Strangeways, he already had in progress experiments that employed embryo extract combined with plasma clots to provide nourishment to tiny pieces of tissue removed from both embryonic and mature specimens. The technique, called organ culture, was intriguing to Fell because it retained intact the naturally occurring sets of cells and, in the case of cartilage and bone, their extracellular matrix. The culture medium was for maintenance of function, rather than cellular multiplication. Consequently, organ culture allowed the study of interacting groups of cells and experimental tests of the effects of various biologically active chemicals, such as vitamins, hormones, and drugs. These features set it apart from the tissue culture technique, in which cells are kept constantly proliferating, and normal tissue architecture is abolished. Fell saw the potential of organ culture to address certain scientific problems in which she was interested. She used the technique from the beginning of her career and ultimately became the world's foremost authority on organ culture.

In one of her applications of organ culture, Fell was the first biologist to study skeletal tissue biochemically (Fell and Robison 1929). At that time only a handful of biologists were using biochemical approaches to cell biology. She is also credited as the first to "apply biological techniques to the study of pathological processes" (Lasnitzki 1986).

The quality of Fell's work was recognized quickly. In 1931 F. G. Hopkins, president of the Royal Society of London, announced in the Anniversary Address that £800 per year for five years had been awarded to Honor B. Fell "for the support of her valuable work on Tissue Culture" (Hopkins 1931).

In devising new experiments, Fell improved the organ culture technique, which she used extensively. She introduced the use of defined media plus serum in place of plasma clots and embryo extract and of small, stainless steel mesh grids to support the tiny specimens, which were sometimes only one or two cubic millimeters in size. By the end of her career, many scientists had learned the utility of organ culture, either directly from her or by reading her articles.

Fell's endless fascination and delight in scientific experimentation were well known to her associates and students. A. R. Poole and A. I. Caplan (1987) recall her as "naturally inquisitive and enormously energetic and imaginative." She dedicated herself to tireless investigation, patiently and painstakingly conducted, year after year, in a professional life that spanned some 60 years. Even though she was the director of Strangeways Research Laboratory, she "was always in and out of the laboratory, planning experiments, setting up cultures, nursing them, analyzing their histology, drawing conclusions and discussing the results with colleagues, and planning the 'next' experiment" (Poole 1989). Administrative duties never kept her out of the laboratory.

After World War II Fell shared her expertise with many scientists, British and foreign. Fell was an excellent choice as a mentor, for she was a consummate scientist who was outstanding in all phases of research. Poole (1989) remembers especially her "remarkable ability to see simply through a maze of experimental observations and details and ask fundamentally important questions from which answers could be obtained." Lasnitzki (1986) credits her with providing advice, guidance, and encouragement in addition to training. Over the years her generous, kindly ways established for her a scientific family, far-flung but still important to her. These scientists "trained with her, were inspired by her, and will forever profit from her wisdom and example" (Poole and Caplan 1987).

While the public record contains only a little information on Fell's personal life, many personal attributes in the conduct of her professional career are recorded. She was well known for "meticulous precision and care" in her work, maintenance of "strict asepsis," and "minute observation of morphological and biochemical changes" in her experimental cultures (Waymouth 1986). She remained an active bench scientist throughout her career, even though she was heavily involved in editing monographs and writing review articles, in addition to her many years of administrative duties. She told Poole (Poole and Caplan 1987) "that she would never agree to being an author on a paper unless she had personally made a substantial contribution to the practical work." The number of original research articles on which she is first author testify to that standard. She devoted much effort to transmitting her high standards, techniques, methodology, and scientific philosophy to her students and to visiting investigators. Fell was proud of having trained scientists who became international leaders known for their own excellence ("Obituary . . ." 1986).

After Fell retired from the directorship of the Strangeways Research Laboratory, she spent nine years working with R.R.A. Coombs in the Immunology Division of the Department of Pathology at Cambridge. They collaborated in the investigation of complement-mediated action in cartilage breakdown. When Fell's interests became less directly related to immunology, she accepted the offer of John Dingle (by then the director of Strangeways Research Laboratory) to return, and she spent the final years of her career there. She worked full-time, experimenting and training students, up until illness sent her to the hospital for the last two weeks of her life. Among her final concerns was her doctoral student, who was writing his dissertation (Poole and Caplan 1987).

Remembrances record that Honor Fell regarded scientific research as fun. Her account of the history of the Strangeways Research Laboratory concludes, "Few people can have enjoyed their working life more than I have, both in the Strangeways and during the happy interlude in Robin Coombs's laboratory, and at 80 I am still having fun in quite a big way" (Fell 1981).

Fell's last seven years of work, from 1979, when she returned to the Strangeways Laboratory, until her death on April 22, 1986, were among her happiest. No longer burdened with administration, she was doing what she liked best, scientific experimentation. "It was only a few weeks before her death that her colleagues heard the familiar voice saying: 'It's worked. Isn't it exciting, come and see the results!' " ("Obituary . . ." 1986). Poole and Caplan (1987) said, "Dame Honor Fell was a woman who flourished in what was then largely a male-dominated profession by the sheer force of her excellence as a scientist and as a person." Ilse Lasnitzki (1986) called her "one of the most outstanding biologists of this century" and concluded that "her death means the end of an era in biomedical science."

Fell received many honors in her long career. Her earliest award was the Trail Medal of the Linnaean Society (1948). Two of the most significant awards were election as a fellow of the Royal Society in 1952, the ninth woman granted this status (Burgess 1989), and being created a Dame of the British Empire in 1963. After this latter honor was bestowed, her colleagues called her "Dame Honor," thus signifying both their respect and their familiarity with her.

Fell was awarded eight honorary degrees, including LL.D., 1959, Edinburgh; D.Sc., 1964, University of Oxford; Sc.D., 1964, Harvard University; D.Sc., 1967, University of London; Sc.D., 1969, Cambridge University; and M.D., 1975, University of Leiden. She was a fellow of Girton College, Cambridge (1955), honorary fellow of Somerville College, Oxford (1964), and a fellow of King's College, London (1967). She was awarded the Prix Charles-Leopold Mayer of the Académie des Sciences de l'Institute de France. She was also made a foreign member of the Royal Netherlands Academy (1964) and of the Serbian Academy of Sciences and Arts (1975).

WORK

At Edinburgh University Fell's first research projects dealt with the development of the reproductive system of chickens. Among her earliest publications were studies of sex reversal in chickens (Fell "Histological . . ." 1923; "A histological . . ." 1923), the histology of hen's ovaries (Fell 1924), and the development of bone and cartilage in fowl (Fell, "The histogenesis . . ." 1925). Her work on cartilage and bone development continued in the early years at Strangeways Research Laboratory and related well to the Cambridge Research Hospital's original mission, which was to learn the causes of arthritis. Her work was envisioned as providing essential background knowledge that would lead ultimately to a means of treatment. Fell herself became much more caught up in basic science rather than its applications, both in these early years and for the remainder of her career.

The first project published, in conjunction with the Strangeways Research Laboratory, investigated the development of cartilage and bone of chick limb buds (Strangeways and Fell, I., 1926). She used the organ culture technique to determine whether mesenchyme from limb buds would organize and differentiate *in vitro*. She found that it did and thus discovered for the developing chick limb skeleton what was then called self-differentiation (now called determination). This refers to the fact that prior signals had already ''programmed'' the mesenchyme to differentiate in an appropriate physiologic environment. Her descriptions of bone histology are classic, and derivative summaries of her observations are still in use. Fell published other developmental studies, but development was seldom the focus of her work after these early investigations. Fell's groundbreaking work on the biochemical characterization of ossifying tissue was her first investigation into molecular mechanisms in cell biology. With Robert Robison, she discovered that alkaline phosphatase activity is pronounced in ossifying chick femora and absent in a cartilaginous structure that does not ossify (Fell and Robison 1929, 1930, 1934).

During World War II, Fell's research activities were redirected to war-related projects. In collaboration with James Frederick Danielli, she investigated the enzymes of healing wounds and burns. With E. M. Brieger she studied avian and bovine tubercle bacillus, apparently as a model of the human disease, which was a persistent problem in hospitals. She worked with C. B. Allsopp on the effects of various chemical warfare agents. Under the auspices of the Chemical Defence Research Department (1939–44) she investigated effects of the arsenical vesicant lewisite and of British anti-lewisite (BAL). They also studied the effects of mustard gas.

With the end of the war, Fell returned to her pursuit of the cell biology of the skeletal system. She undertook studies with Edward Mellanby to learn whether vitamin A had a direct effect on skeletal tissue, since it was well known that excess amounts caused skeletal deformities *in vivo*. Their experiments dramatically demonstrated the resorption of bone matrix in cultures treated with high levels of vitamin A. Following this work, Fell collaborated with Lewis Thomas (Fell and Thomas 1960, 1961) in the beginnings of an inquiry that occupied her through the mid-1960s. Their research led them to suggest that cartilage matrix breakdown in the presence of excess vitamin A was due to proteolytic activity. Confirmation of this idea resulted from additional experiments by Fell and collaborators, principally John Dingle and Jack Lucy. They discovered that vitamin A, retinol, acts on the membranes of lysosomes, causing the release of their degradative enzymes. The cathepsins in particular were found to digest the proteoglycans of cartilage.

A serendipitous finding (Hardy 1989) prompted Fell to study the effects of vitamin A on skin and ectoderm as well as skeletal tissue. While experiments on the effects of high concentrations of vitamin A on cartilage were being conducted, a bit of chick skin was accidentally left on a piece of cartilage. Fell alertly noticed that the skin was not the normal, keratinized epithelium that one would expect, but rather a mucosa, that is, a mucus-secreting epithelium. That observation led to extensive and fruitful studies on the effects of vitamin A on epithelia.

Knowledge of retinol's action on epithelia paved the way to a diverse array of research by many others. The path has led to present knowledge of the important biological actions not only of retinol but also of the spectrum of compounds known as retinoids, as retinol and its chemical relatives are called collectively. These actions are currently thought to include, besides regulation of the differentiation of certain epithelia, operating as a morphogen (a substance that provides positional information for pattern formation during development) in limb development and preventing cancer in some cells. In addition, retinoids are now used therapeutically for some skin diseases.

After working on retinol, Fell turned her attention to immunological sources of cartilage degradation. Organ cultures of fetal mouse bones showed resorption of both bone and cartilage in the presence of complement-sufficient antisera against mouse tissues (Fell and Weiss 1965). In 1970 Fell decided to continue these studies on adult pig cartilage, since pigs develop a form of arthritis similar to rheumatoid arthritis in humans. She found that when bone marrow was included with the specimens, the antiserum produced destruction of both proteoglycan and collagen. Suspecting that cartilage destruction was initiated by living cells, Fell, working with Ronald W. Jubb (1977), established that living synovial cells initiated the process by releasing an unknown substance that, in turn, promoted the release from chondrocytes of degradative compounds.

When this work was continued by her, Dingle, and his associates (1979), they identified the active molecules, which had at first been called catabolins, because they promoted the catalysis of cartilage. Later these were identified as two forms of interleukin 1 (Saklatvala, Sarsfield, and Townsend 1985). This cytokine is now well known for its pivotal role in the initiation of inflammation and its ability to activate lymphocytes.

Fell's final experiments extended the investigations into destruction of porcine cartilage. Her last two articles dealt with the role of cyclic adenosine monophosphate (cAMP) in cartilage destruction (Fell et al. 1986) and of microtubules, by inference from treatment with colchicine and related chemicals (Fell et al. 1989).

Fell was president of the International Society for Cell Biology, another indication of her renown and of the great respect for her work. When the society met in Paris, she delivered her presidential address, "Fashion in Cell Biology" (1960), and warned colleagues against being caught up by fashion in their research and, at the same time, indulged her low-key sense of humor by speaking on scientific fashion in the center of sartorial fashion. Fell was a Messel Research Fellow of the Royal Society (1931), a Foulerton Research Fellow of the Royal Society (1941–67), and a Royal Society Research Professor (1963–67).

BIBLIOGRAPHY

Works by Honor Bridget Fell

Scientific Works

Space does not permit a more expansive presentation of Fell's research or the listing of her complete works. Listed here are works that represent the scope of Fell's research undertakings. Forty-four published works not cited here and a lengthy discussion of her research are in Vaughn (1987).

(with F.A.E. Crew) "The nature of certain ovum-like bodies found in the seminiferous tubules." *Quarterly Journal of Microscopical Science* 66 (1922): 557–578.

"Histological studies on the gonads of the fowl. I. The histological basis of sex reversal." *British Journal of Experimental Biology* 1 (1923): 97–130.

"A histological study of the testis in cases of pseudo-intersexuality and cryptorchism with special reference to the interstitial cells." *Quarterly Journal of Experimental Physiology* 13 (1923): 145–158.

"Histological studies on the gonads of the fowl. II. The histogenesis of the so-called 'luteal' cells in the ovary." *British Journal of Experimental Biology* 1 (1924): 293–312.

"The histogenesis of cartilage and bone in the long bones of the embryonic fowl." *Journal of Morphology and Physiology* 40 (1925): 417–459.

"Histological studies on the gonads of the fowl. III. The relationship of the 'luteal' cells of the ovary of the fowl to the tissue occupying the atretic and discharged follicles, and the question of the homology of the latter tissue and mammalian corpus luteum." *Quarterly Journal of Microscopical Science* 69 (1925): 591–609.

(with T.S.P. Strangeways) "Experimental studies on the differentiation of embryonic tissues growing *in vivo* and *in vitro*. I. The development of the undifferentiated limb-bud (a) when subcutaneously grafted into the post-embryonic chick and (b) when cultivated *in vitro*." *Proceedings of the Royal Society of London* B 99 (1926): 340–366.

(————) "Experimental studies on the differentiation of embryonic tissues growing *in vivo* and *in vitro*. II. The development of the isolated early embryonic eye of the fowl when cultivated *in vitro*." *Proceedings of the Royal Society of London* B 100 (1926): 273–283.

(with J. A. Andrews) "Cytological study of cultures *in vitro* of Jensen's rat sarcoma." *British Journal of Experimental Pathology* 8 (1927): 413–428.

(with T.S.P. Strangeways) "Study of direct and indirect action of X-rays upon tissues of embryonic fowl." *Proceedings of the Royal Society of London* B 102 (1927): 9–29.

"Development *in vitro* of isolated otocyst of embryonic fowl." *Archiv fuer Experimentelle Zellforschung, Besonders Gewebezuechtung* 7 (1928): 69–81.

"Experiments on the differentiation *in vitro* of cartilage and bone." *Archiv fuer Experimentelle Zellforschung, Besonders Gewebezuechtung* 7 (1928): 390–412.

(with R. Robison) "The growth, development, and phosphatase activity of embryonic avian femora and limb-buds cultivated *in vitro.*" *Biochemical Journal* 23 (1929): 767–784.

(————) "The development and phosphatase activity *in vivo* and *in vitro* of the mandibular skeletal tissue of the embryonic fowl." *Biochemical Journal* 24 (1930): 1905–1921.

"Osteogenesis *in vitro*." *Archiv fuer Experimentelle Zellforschung, Besonders Gewebezuechtung* 11 (1931): 245–252.

(with R. Chambers) "Micro-operations on cells in tissue cultures." *Proceedings of the Royal Society of London* B 109 (1931): 380–403.

"Morphological and Experimental Studies on Skeletogenesis of the Fowl." D. Sc., diss., University of Edinburgh, 1932.

"The osteogenic capacity *in vitro* of periosteum and endosteum isolated from the limb-skeleton of fowl embryos and young chicks." *Journal of Anatomy* 66 (1932): 157–180.

(with R. G. Canti) "Experiments on the development *in vitro* of the avian knee-joint." *Proceedings of the Royal Society of London* B 116 (1934): 316–351.

(with R. Robison) "The development of the calcifying mechanism in avian cartilage and osteoid tissue." *Biochemical Journal* 28 (1934): 2243–2253.

"Tissue culture: Advantages and limitations as a research method." *British Journal of Radiology* 8 (1935): 27–31.

(with W. Landauer) "Experiments on skeletal growth and development *in vitro* in relation to the problem of avian phokomelia." *Proceedings of the Royal Society of London* B 118 (1935): 133–154.

"The origin and developmental mechanics of the avian sternum." *Philosophical Transactions of the Royal Society of London* B 229 (1939): 407–463.

(with J. F. Danielli) "Enzymes of healing wounds: Distribution of alkaline phosphomonoesterase in experimental wounds and burns in the rat." *British Journal of Experimental Pathology* 24 (1943): 196–203.

(with E. M. Brieger) "Warm-stage observations on initial development of avian tubercle bacillus cultivated in embryo extract." *Journal of Hygiene* 44 (1945): 158–169.

(with J. F. Danielli and E. Kodicek) "Enzymes of healing wounds: Effect of different degrees of vitamin C-deficiency on phosphatase activity in experimental wounds in the guinea-pig." *British Journal of Experimental Pathology* 26 (1945): 367–376.

(with C. B. Allsopp) "Effects of lewisite and of lewisite oxide on living cells *in vitro*." *British Journal of Experimental Pathology* 27 (1946): 305–309.

(——) "Therapeutic effect of British anti-lewisite (BAL) on tissue cultures grown in medium containing lewisite oxide." *British Journal of Experimental Pathology* 27 (1946): 310–315.

(with E. M. Brieger) "Warm-stage observations on development of pseudo-mycelia in cultures of avian tubercle bacilli grown in dilute embryo extract." *Journal of Hygiene* 44 (1946): 256–263.

(——) "Effect of phagocytosis on growth and survival of avian tubercle bacilli in embryonic chicken tissue cultivated *in vitro*." *Journal of Hygiene* 45 (1947): 359–370.

(with C. B. Allsopp) "Action of mustard gas (B,B'-dichlorodiethylsulphide) on living cells *in vitro*. Immediate cytological effects of mustard gas and its hydrolysis products. Effect on cell growth of adding small concentrations of mustard gas to the culture medium." *Cancer Research* 8 (1948): 145–161.

(——) "Effect of repeated application of minute quantities of mustard gas (B,B'-

dichlorodiethylsulphide) on the skin of mice." *Cancer Research* 8 (1948): 177–181.

(———) "Tissue culture experiments on biological action of methyl *bis* (B-chlorethyl) amine and its hydrolysis products." *Cancer Research* 9 (1949): 238–246.

(with A.F.W. Hughes) "Mitosis in the mouse: A study of living and fixed cells in tissue cultures." *Quarterly Journal of Microscopical Science* 90 (1949): 355–380.

(with C. B. Allsopp) "Toxicity of product of reaction of mustard gas with serum proteins." *British Journal of Experimental Pathology* 31 (1950): 258–262.

(with E. Mellanby) "Effects of hypervitaminosis A on foetal mouse bones cultivated *in vitro*." *British Medical Journal* 2 (1950): 535–539.

(with E. M. Brieger) "Comparative study of reaction *in vivo* and *in vitro* of rabbit tissues to infection with bovine tubercle bacilli: Observations on rabbit spleen infected *in vitro*." *Journal of Hygiene* 49 (1951): 181–188.

(——— and B. R. Smith) "Comparative study of reaction *in vivo* and *in vitro* of rabbit tissues to infection with bovine tubercle bacilli: Observations on cultures of spleen and lymph glands from infected rabbits." *Journal of Hygiene* 49 (1951): 189–200.

(with E. Mellanby) "The effect of hypervitaminosis A on embryonic limb bones cultivated *in vitro*." *Journal of Physiology* 116 (1952): 320–349.

"Recent advances in organ culture." *Scientific Progress* 162 (1953): 212–231.

(with E. Mellanby) "Metaplasia produced in cultures of chick ectoderm by high vitamin A." *Journal of Physiology* 119 (1953): 470–488.

"The effect of hormones and vitamin A on organ cultures." *Annals of the New York Academy of Sciences* 58 (1954): 1183–1187.

(with E. Mellanby and S. R. Pelc) "Influence of excess vitamin A on the sulphate metabolism of chick ectoderm grown *in vitro*." *British Medical Journal* 2 (1954): 611.

"The effect of hormones on differentiated tissues in culture." In *The Hypophyseal Growth Hormone, Nature and Actions*, edited by R. W. Smith et al., 138–148. New York: McGraw-Hill, 1955.

(with E. Mellanby) "The biological action of thyroxine on embryonic bones grown in tissue culture." *Journal of Physiology* 127 (1955): 427–447.

"Effect of excess vitamin A on organized tissues cultivated *in vitro*." *British Medical Bulletin* 12 (1956): 35–37.

"The physiology of skeletal tissue in culture." *Lectures on the Scientific Basis of Medicine* 6 (1956–57): 28–45.

(with E. Mellanby) "The effect of L-triiodothyronine on the growth and development of embryonic chick limb-bones in tissue culture." *Journal of Physiology* 133 (1956): 89–100.

(——— and S. R. Pelc) "Influence of excess vitamin A on the sulphate metabolism of bone rudiments grown *in vitro*." *Journal of Physiology* 134 (1956): 179–188.

"The effect of excess vitamin A on cultures of embryonic chicken skin explanted at different stages of differentiation." *Proceedings of the Royal Society of London* B 146 (1957): 242–256.

(with V. A. Galton and R. Pitt-Rivers) "The metabolism of some thyroid hormones by limb-bone rudiments cultivated *in vitro*." *Journal of Physiology* 144 (1958): 250–256.

"Fashion in cell biology." *Science* 132 (1960): 1625–1627.

(with S. R. Pelc) "The effect of excess vitamin A on the uptake of labelled compounds by embryonic skin in organ culture." *Experimental Cell Research* 19 (1960): 99–113.

(with L. Thomas) "Comparison of the effects of papain and vitamin A on cartilage. II. The effects on organ cultures of embryonic skeletal tissue." *Journal of Experimental Medicine* 111 (1960): 719–744.

(with J. T. Dingle and J. A. Lucy) "Studies on the mode of action of excess of vitamin A. 1. Effect of excess of vitamin A on the metabolism and composition of embryonic chick-limb cartilage grown in organ culture." *Biochemical Journal* 79 (1961): 497–500.

(with J. A. Lucy and J. T. Dingle) "Studies on the mode of action of excess of vitamin A. 2. A possible role of intracellular proteases in the degradation of cartilage matrix." *Biochemical Journal* 79 (1961): 500–508.

(with L. Thomas) "The influence of hydrocortisone on the action of excess vitamin A on limb bone rudiments in culture." *Journal of Experimental Medicine* 114 (1961): 343–362.

"Some effects of environment on epidermal differentiation." *British Journal of Dermatology* 74 (1962): 1–7.

"The influence of hydrocortisone on the metaplastic action of vitamin A on the epidermis of embryonic chicken skin in organ culture." *Journal of Embryology and Experimental Morphology* 10 (1962): 389–409.

(with J. T. Dingle and M. Webb) "Studies on the mode of action of excess of vitamin A. 4. The specificity of the effect on embryonic chick-limb cartilage in culture and on isolated rat-liver lysosomes." *Biochemical Journal* 83 (1962): 63–69.

(with G. Weissman) "The effect of hydrocortisone on the response of fetal rat skin in culture to ultraviolet irradiation." *Journal of Experimental Medicine* 116 (1962): 365–380.

(with J. T. Dingle) "Studies on the mode of action of excess of vitamin A. 6. Lysosomal protease and the degradation of cartilage matrix." *Biochemical Journal* 87 (1963): 403–408.

(with S. Fitton Jackson) "Epidermal fine structure in embryonic chicken skin during atypical differentiation induced by vitamin A in culture." *Developmental Biology* 7 (1963): 394–419.

"The role of organ cultures in the study of vitamins and hormones." *Vitamins and Hormones* 22 (1964): 81–127.

"The effect of vitamin A on the breakdown and synthesis of intercellular material in skeletal tissue in organ culture." *Proceedings of the Nutrition Society* 24 (1965): 166–170.

(with L. M. Rinaldini) "The effects of vitamins A and C on cells and tissues in culture." In *Cells and Tissues in Culture. Methods, Biology, and Physiology*, vol. 1, edited by E. N. Willmer, 659–699. London: Academic Press, 1965.

(with L. Weiss) "The effect of antiserum alone and with hydrocortisone on foetal mouse bones in culture." *Journal of Experimental Medicine* 121 (1965): 551–560.

"The role of lysosomes in pathology. Epilogue." *Proceedings of the Royal Society of Medicine* 59 (1966): 880.

(with R.R.A. Coombs, and J. T. Dingle) "The breakdown of embryonic (chick) cartilage and bone cultivated in the presence of complement-sufficient antiserum. I. Morphological changes, their reversibility and inhibition." *International Archives of Allergy and Applied Immunology* 30 (1966): 146–176.

(with J. T. Dingle and J. A. Lucy) "Synthesis of connective-tissue components. The effect of retinol and hydrocortisone on cultured limb-bone rudiments." *Biochemical Journal* 98 (1966): 173–181.

(with R. L. Spooner, R.R.A. Coombs, et al.) "Effect of hydrocortisone and epsilon-caproic acid on immune cytolysis *in vitro*." *International Archives of Allergy and Applied Immunology* 30 (1966): 231–237.

"Preface." In *La resorption osseuse et l'hormone parathyroidienne; approche des mecanismes biochimiques et cytologiques de l'osteoclasie et de l'osteolyse,* by Gilbert Vaes. Bruxelles: A. de Visscher, 1967.

(with J. T. Dingle and R.R.A. Coombs) "The breakdown of embryonic cartilage and bone cultivated in the presence of complement-sufficient antiserum. II. Biochemical changes and the role of the lysosomal system." *International Archives of Allergy and Applied Immunology* 31 (1967): 283–303.

"The effect of environment on skeletal tissue in culture." *Embryologia* 10 (1969): 181–205.

"Role of biological membranes in some skeletal reactions." *Annals of the Rheumatic Diseases* 28 (1969): 213–227.

"Vitamin A and enzymes, membranes, differentiation and reproduction." *American Journal of Clinical Nutrition* 22 (1969): 975–977.

(with T. C. Appleton, S. R. Pelc, et al.) "Endocytosis of sugars in embryonic skeletal tissues in organ culture. III. Radioautographic distribution of [^{14}C] sucrose." *Journal of Cell Science* 4 (1969): 133–137.

(with R.R.A. Coombs) "Lysosomes in tissue damage mediated by allergic reactions." In *Frontiers of Biology,* vol. 14B. *Lysosomes in Biology and Pathology,* pt. 2, edited by J. T. Dingle and H. B. Fell, 3–18. Amsterdam: North-Holland, 1969.

(with J. T. Dingle) "Endocytosis of sugars in embryonic skeletal tissues in organ culture. I. General introduction and histological effects." *Journal of Cell Science* 4 (1969): 89–103.

(———, eds.) *Frontiers of Biology,* vols. 14A, 14B. *Lysosomes in Biology and Pathology,* pts. 1 and 2. Amsterdam: North-Holland, 1969.

(with J. T. Dingle and A. M. Glauert) "Endocytosis of sugars in embryonic skeletal tissues in organ culture. IV. Lysosomal and other biochemical effects." *Journal of Cell Science* 4 (1969): 139–154.

(with A. M. Glauert and J. T. Dingle) "Endocytosis of sugars in embryonic skeletal tissues in organ culture. II. Effect of sucrose on cellular fine structure." *Journal of Cell Science* 4 (1969): 105–131.

(with P. J. Lachmann, R.R.A. Coombs, et al.) "The breakdown of embryonic (chick) cartilage and bone cultivated in the presence of complement-sufficient antiserum. III. Immunological analysis." *International Archives of Allergy and Applied Immunology* 36 (1969): 469–485.

"The chain of discovery." *Advancement in Science* 27 (1970): 129–137.

"The direct action of vitamin A on skeletal tissue *in vitro*." In *The Fat-Soluble Vitamins,* edited by H. F. de Luca and J. W. Suttie, 187–202. Madison: University of Wisconsin Press, 1970.

(with M. B. Hille, A. J. Barratt, et al.) "Microassay for cathepsin D shows an unexpected effect of cycloheximide on limb-bone rudiments in organ culture." *Experimental Cell Research* 61 (1970): 470–472.

"Tissue culture and its contribution to biology and medicine." *Journal of Experimental Biology* 57 (1972): 1–13.

(with J. T. Dingle and D. S. Goodman) "The effect of retinol and of retinol-binding protein on embryonic skeletal tissue in organ culture." *Journal of Cell Science* 11 (1972): 393–402.

"Commentary on 'Epidermal chalone: Cell cycle specificity of two epidermal growth inhibitors' by K. Elgjo." *National Cancer Institute Monographs* 38 (1973): 77–78.

(with M.E.J. Barratt) "The role of soft connective tissue in the breakdown of pig articular cartilage cultivated in the presence of complement-sufficient antiserum to pig erythrocytes. I. Histological changes." *International Archives of Allergy and Applied Immunology* 44 (1973): 441–468.

(with A. R. Poole and M.E.J. Barratt) "The role of soft connective tissue in the breakdown of pig articular cartilage cultivated in the presence of complement-sufficient antiserum to pig erythrocytes. II. Distribution of immunoglobulin G (IgG)." *International Archives of Allergy and Applied Immunology* 44 (1973): 469–488.

"The effect of complement-sufficient antiserum against pig erythrocytes on pig articular tissues in organ culture." In *Future Trends in Inflammation II*, edited by J. P. Giroud et al., 86–90. Basel: Birkhauser, 1975. Reprinted, *Agents and Actions* 6 (1976): 86–90.

(with J. T. Dingle, eds.) "A discussion of the pericellular environment and its regulation in vertebrate tissues." *Philosophical Transactions of the Royal Society of London* B 271 (1975): 233–410.

(with J. T. Dingle, P. Horsfield, et al.) "Breakdown of proteoglycan and collagen induced in pig articular cartilage in organ culture." *Annals of the Rheumatic Diseases* 34 (1975): 303–311.

""The development of organ culture." In *Symposium of the British Society for Cell Biology. 1. Organ Culture in Biomedical Research: Festschrift for Dame Honor Fell*, edited by M. Balls and M. Monnickendam, 1–13. London: Cambridge University Press, 1976.

(with M. E. Barratt, H. Welland, et al.) "The capacity of pig articular cartilage in organ culture to regenerate after breakdown induced by complement-sufficient antiserum to pig erythrocytes." *Calcified Tissue Research* 20 (1976): 3–21.

(with A. M. Glauert, M. E. Barratt, et al.) "The pig synovium. I. The intact synovium *in vivo* and in organ culture." *Journal of Anatomy* 122 (1976): 663–680.

"The evolution of the International Federation for Cell Biology." *Cell Biology International Reports* 1 (1977): 9–11.

(with M. E. Barratt, R.R.A. Coombs, et al.) "The pig synovium. II. Some properties of isolated intimal cells." *Journal of Anatomy* 123 (1977): 47–66.

(with R. W. Jubb) "The effect of synovial tissue on the breakdown of articular cartilage in organ culture." *Arthritis and Rheumatism* 20 (1977): 1359–1371.

"Synoviocytes." *Journal of Clinical Pathology* 31 (Suppl. 12) (1978): 14–24.

(with J. T. Dingle, J. Saklatvala, et al.) "A cartilage catabolic factor from synovium." *Biochemical Journal* 184 (1979): 177–180.

(with R. W. Jubb) "The breakdown of collagen by chondrocytes." *Journal of Pathology* 130 (1980): 159–167.

(———) "The effect of synovial tissue on the synthesis of proteoglycan by the articular cartilage of young pigs." *Arthritis and Rheumatism* 23 (1980): 545–555.

"The Strangeways Research Laboratory and cellular interactions." In *Cellular Interactions*, edited by J. T. Dingle and J. L. Gordon, 1–14. Amsterdam/New York: Elsevier/North Holland, 1981.

(with J. A. Tyler and C. E. Lawrence) "The effect of cortisone on porcine articular tissues in organ culture." *Journal of Pathology* 137 (1982): 335–351.

(with J. J. Reynolds, C. E. Lawrence, et al.) "The promotion and inhibition of collagen-breakdown in organ cultures of pig synovium. The requirement for serum components and the involvement of cyclic adenosine 3',5'-monophosphate (cAMP)." *Collagen and Related Research* 6 (1986): 51–75.

(with C. E. Lawrence, M. R. Bagga, et al.) "The degradation of collagen in pig synovium *in vitro* and the effect of colchicine." *Matrix* 9 (1989): 116–126.

Works about Honor Bridget Fell

Burgess, Patricia (ed.). *The Annual Obituary 1986*, 250–251. Chicago and London: St. James Press, 1989.

"Fell, Honor Bridget." *Lancet* (8490) (May 17, 1986): 1166.

Lasnitzki, Ilse. "Dame Honor Fell F.R.S. (1900–1986)." *Nature* 322 (1986): 214.

Leighton, J. "Function in organized tissue in culture: Introduction." *In Vitro Cellular and Developmental Biology* 25 (1989): 449.

"Obituary, Dame Honor Fell, advances in cell biology." *Times* (London) 26 (Apr. 1986): 19.

Poole, A. R. "Honor Bridgett [*sic*] Fell, Ph.D., D.Sc., F.R.S., D.B.E., 1900–1986: The scientist and her contributions." *In Vitro Cellular and Developmental Biology* 25 (1989): 450–453.

———, and A. I. Caplan. "An appreciation: Dame Honor B. Fell, F.R.S. (1900–1986)." *Developmental Biology* 122 (1987): 297–299.

Vaughn, Janet. "Honor Bridget Fell." *Biographical Memoirs of Fellows of the Royal Society* 33 (1987): 235–259.

Waymouth, C. "Honor Bridgett [*sic*] Fell, D.B.E., F.R.S., Ph.D., D.Sc. (1900–1986)." *In Vitro Cellular and Developmental Biology* 22 (1986): 499.

Other References

Balls, M., and M. Monnickendam (eds.). *Symposium of the British Society for Cell Biology. 1. Organ Culture in Biomedical Research: Festschrift for Dame Honor Fell*. London: Cambridge University Press, 1976.

Hardy, M. H. "The use of retinoids as probes for analyzing morphogenesis of glands from epithelial tissues." *In Vitro Cellular and Developmental Biology* 25 (1989): 454–459.

Hopkins, F. G. "Anniversary address." *Proceedings of the Royal Society of London* B 109 (1931): 412.

Saklatvala, J., S. J. Sarsfield, and Y. Townsend. "Pig interleukin. I. Purification of two immunologically different leukocyte proteins that cause cartilage resorption, lymphocyte activation, and fever." *Journal of Experimental Medicine* 162 (1985): 1208–1222.

DIAN FOSSEY (1932–1985)

Soraya Ghayourmanesh-Svoronos

BIOGRAPHY

Dian Fossey was born on January 16, 1932, in San Francisco, the daughter of George III and Kitty (Kidd) Fossey. Her father was a big, affable insurance agent who loved his little daughter but hated the impoverished life his job gave him. He eventually turned to heavy drinking, which led to divorce in 1938. Fossey remained with her mother, who, one year later, married Richard Price, an ambitious building contractor.

After the divorce her father tried to keep in touch by sending pictures of himself in his navy uniform during World War II, but even his name was deleted from the daily conversations in the Price house. Nevertheless, Fossey always thought of her father and yearned to see him. In 1959 she received a letter from him, informing her that two years before, he had married Kathryn (Kay) and was happily settled. Fossey met with him only three times more, first in Chicago (1965), then in New York, and later in San Francisco. She maintained a sporadic correspondence until the summer of 1968, when she received a telegram from her stepmother announcing that her father had committed suicide. His death was deeply grieved by Fossey, who at that time was alone in the Rwandan mountains.

Growing up at the Price home was difficult. Fossey was never adopted by her stepfather, yet she dutifully called him "Daddy." He was a stern traditionalist who, until her tenth birthday, had her eat her dinner with the housekeeper in the kitchen. She developed a love for animals but was never given one. She generally resented both her mother and stepfather and felt starved for affection. She mentioned several times that she felt safe in the Price home only if she carried a pistol or a can of mace for protection. However, Kitty Price always maintained that both she and her husband loved Fossey dearly.

After graduating from high school, she enrolled in a pre-veterinary medicine program at the University of California at Davis, contrary to her stepfather's request that she seek a career in business. She supported herself by work such as clerking weekends in the White House Department store, doing clerical or laboratory work, and even being a machine operator in a factory. Having academic difficulty with chemistry and physics courses, Fossey transferred to San Jose State College (now San Jose State University) to study occupational therapy. She received her bachelor's degree in 1954. She also became a first-class equestrienne. After her postcollege clinical training, she moved to Kentucky and worked as the director of the occupational therapy department at the Kosair Crippled Children's Hospital in Louisville.

Her love for Africa was inspired by a book about mountain gorillas by American zoologist George Schaller. In 1963 she took a three-year, $8,000 bank loan to finance a seven-week safari trip to Africa. At Olduvai Gorge, Tanzania, she made her first stop in order to meet Dr. Louis S. B. Leakey and his wife, Mary, who were involved with the search for hominid fossils. On one of the excavation trips she fell and broke her ankle. Two weeks later she left the Leakey camp, hobbling with the aid of a walking stick, for Zaire (then known as the Democratic Republic of Congo), hoping to meet the great apes. Her first encounter with a mountain gorilla had a tremendous impact on her. At the end of her African trip, she went back to Kentucky and resumed her work with handicapped children. Three years later (1966) Leakey came to Louisville and tried to convince her to study the gorillas in the wild as part of a long-term expedition. Fossey indicated to Leakey that she had no experience or training in studying primates, but he stated that he wanted someone with a fresh approach. She readily agreed to undertake the task, and, to prove her decisiveness, she had her appendix removed as a prophylactic measure. Much later Fossey found out that his suggestion to remove the appendix had been a facetious attempt to test her enthusiasm (Mowat 1987, 22–23).

After paying a short visit to Jane Goodall in Tanzania to learn her methods of data collection, she set up her first campsite and workstation at Kabara, Zaire. She managed to approach and study the gorillas in the remote mountain areas for about seven months, until political unrest took over the country. On July 10, 1967, she was escorted by armed guards from her campsite and was kept in custody for two weeks, during which time she saw all the white male detainees tortured and killed. She herself was subjected to repeated raping. Eventually she convinced her guards to accompany her to Uganda but managed to escape when they got drunk during the trip. Fossey found refuge in the Travellers' Rest Hotel, operated by Walter Baumgartel, whom she had befriended during her first trip to Africa. A message was eventually relayed to her that she would be shot on sight should she be spotted again in Zaire.

On September 24, 1967, Fossey established the Karisoke Research Center in Rwanda, with diplomatic help from Leakey and Leighton Wilkie. In 1970 she left Africa temporarily for Great Britain. She started her doctoral studies at Cambridge University, earning a Ph.D. degree in zoology six years later without receiving a master's degree. Fossey continued her work in Rwanda until 1980, when she went back to the United States to recuperate from physical exhaustion.

She accepted a visiting associate professorship at Cornell University, while still acting as the project coordinator at the Karisoke Research Center.

While at Cornell, she heard bad news concerning increased poaching and the rapid deterioration of the research center. By 1983 the price of a gorilla skull had risen to $500, and the Belgian government had refused to ratify the 1975 international treaty making poaching illegal. Deciding that she would be the person to improve the situation in Rwanda, Fossey returned to Karisoke in June 1983. The acting director, Richard Barnes, was removed, and he vehemently protested his expulsion to the National Geographic Society. As a result of this event, Fossey's support was terminated.

Smoking three packs of cigarettes a day took its toll on Fossey, and she often needed oxygen to combat her chronic emphysema. She also suffered from continuous insomnia and recurring pneumonia, and she developed a drinking problem. Her incessant, passionate antipoaching activities created opposition and many enemies. In December 1985, to her surprise, she was given a two-year, rather than a two-month, visa extension by the Rwandan government.

Shortly thereafter, on December 27, 1985, Fossey was found hacked to death by a machete in her camp at the Viruga Mountains. Four days later she was buried in the gorilla cemetery she had built near her camp. Engraved on her tombstone is her African nickname, Nyiramachabelli, which means ''the woman who lives alone in the mountain.''

On August 21, 1986, the Rwandan government announced that Wayne Richard McGuire, an American wildlife researcher who was there to collect data for his Ph.D. thesis, was being sought as a primary suspect, together with Rwelekana, her tracker. Soon after his arrest, Rwelekana allegedly hung himself in his cell. McGuire, fearing the worst and proclaiming his innocence, fled the country. His attorney maintained that the poachers, who would have wanted Fossey dead, and the U.S. government may have been duped by Rwandans into making McGuire a scapegoat. On December 4, 1986, the U.S. Embassy announced that McGuire would be tried in absentia by the government of Rwanda. McGuire was tried and found guilty after a forty-minute trial. The motive ascribed to him was the stealing of Fossey's research papers. The government of Rwanda never asked for McGuire's extradition from the United States. On September 22, 1989, Diane Doran was named to direct the Karisoke Research Center in Rwanda. The center was financed by the Fossey-Digit Fund.

Fossey was at least six feet tall, with thick black hair, dark eyebrows, and a tentative smile. Her unusual voice was low and husky but turned to shrieks when she felt that her gorillas were in danger. Her physical appearance was described as being close to Jackie Kennedy's, but she liked to wear casual clothes. Although both her parents were Protestants, she converted to Catholicism during her stay in Louisville. Later she denounced her Catholicism and declared her agnosticism. Friends attributed this to a disappointing affair she had had with a Catholic priest.

Fossey never married and had no offspring. In the mid-1960s she almost married Alexie Forrester, whom she met in Africa during her first trip. However, she was too involved, first with her occupational therapy job and then with her mission, to save the mountain gorillas. Forrester eventually proposed to her but was rebuffed when he asked her to choose between himself and the gorillas. Later, she had a long affair with Robert Campbell, a married photographer sent to her camp by the National Geographic Society. He photographed the mountain gorillas during the period 1969–72. She had two abortions in Rwanda, and the second one almost killed her.

WORK

Fossey's loneliness and lack of affection as a child turned her interest to animals, especially horses. Her early goal was to become a veterinary student, but she later switched to helping physically disadvantaged children. Louis Leakey's input was instrumental in her decision to go to Africa, where she attained the independence she sought. She had to overcome tremendous natural obstacles: acrophobia on 45-degree slopes, torrential rains, foot-deep mud, personal discomfort, and wild animals. African regimes at the time were unstable, and the increasing antiwhite animosity of many ethnic leaders in the 1960s contributed to the insecurity. She witnessed several of her close friends die after being tortured. More than once she almost lost her own life. Poachers, witchcraft, slaughters, and revolutions were part of the everyday routine. But above all, she had to overcome loneliness in the mountains.

The efforts of the American naturalist Carl Ethan Akeley led to the first preservation center in the Belgian Congo during the 1920s. The natives and poachers had constantly abused the habitat in order to establish a greater degree of farmland for the rapidly growing population. At the time Central Africa had only three parks: Zaire's Parc National des Virungas, Rwanda's Parc National des Volcans, and Uganda's Kigezi Gorilla Sanctuary. Leakey's conclusion was that such a lack of habitat preservation would lead to the extinction of the mountain gorilla.

After establishing the Karisoke Research Center in 1967, Fossey studied more than 50 gorillas, which she described as rather peaceful, charging at humans only when threatened. After several years of isolation, with only occasional help from native trackers and African tribesmen, Fossey felt that she needed research assistance. Many student volunteers came, but few of them could survive the physical exhaustion and the long-term isolation, as well as Fossey's highly demanding and militaristic style. Nevertheless, a few offered help in developing a parasitology project. In 1981 Fossey completed a census of the mountain gorilla population. These results intensified her drive to fight poachers and continue her conservation efforts.

Fossey became the world's leading authority on the mountain gorillas, after devoting 13 years (1967–80) to their study in the mountains of Zaire, Uganda, and Rwanda. During this period she amassed a tremendous wealth of information, which she published for the first time in 1983 under the title *Gorillas in the Mist*.

Through patience, inspired by her love for the gorillas, Fossey was capable of locating them and familiarizing them with her presence. The data she collected dramatically enlarged the contemporary knowledge of the gorillas' habits, their communication, and social structure. She described her first encounter with the gorillas in 1963 as follows:

Sound preceded sight. Odor preceded sound in the form of an overwhelming musky-barnyard, humanlike scent. The air was suddenly rent by a high series of screams followed by the rhythmic rondo of sharp *pok-pok* chestbeats from a great silverback male. ... Immediately I was struck by the physical magnificence of the huge jet-black bodies blended against the green palette wash of the thick forest foliage (Fossey 1983, 3).

Fossey believed that gorillas are fine, altruistic, and regal animals whose family structure is very strong. She also felt that great apes were threatened by Rwanda's human population, which was growing at the high rate of about 3.4% a year.

Fossey studied four different groups of gorillas that were living near her camp. She concluded that each of the families was a cohesive unit dominated by a silverback, which was a sexually mature male at least 15 years old whose leadership was supported by a blackback, a sexually immature male about eight years old. The silverback's nucleus consisted of four females and their offspring, which remained with him permanently. Fossey also found that they are vegetarians that rested about 40% of the time and traveled a distance of about 400 meters daily. Finally, she identified about 15 types of sounds, including chuckling during playtime, grunting when scolding rambunctious juveniles, and growling to warn about sensed danger.

Fossey's greatest contribution was her continuous anti-poaching struggle. In 1978 Digit, one of her favorite gorillas, was slaughtered, and Fossey decided to publicize the poaching technique and its future consequences. CBS Evening News announced the death, as well as the establishment of the "Digit Fund" to support anti-poacher patrols and equipment (Fossey 1983, 209). Despite the public awakening, two more gorillas were killed during six months. Rumors started spreading that Fossey suffered a nervous breakdown, but she quickly refuted these stories.

Fossey conducted an incessant, courageous battle with the poachers, which often led to violent confrontations. In one case two infant gorillas, Coco and Pucker, were captured for a European zoo, at the cost of 18 other gorilla lives. Fossey rescued them and nursed them back to health, only to see them removed by Rwandan officials and then shipped out to Europe. Both gorillas died in captivity soon thereafter. Friends recalled one incident in which she kidnapped the child of a poacher and offered to return her in exchange for a baby gorilla taken by poachers. Fossey was murdered 18 months after this incident. The gorillas' future is much more secure than in the days before Fossey started her campaign, despite the expectations that the Rwandan population would double by the year 2000.

BIBLIOGRAPHY

Works by Dian Fossey

Scientific Works

Space does not permit the listing of the complete works of Dian Fossey. Those are cited in Hausfater and Kennedy (1986). Included here are her dissertation and all references cited in the text.

"The Behaviour of the Mountain Gorilla." Ph.D. diss., Darwin College, Cambridge University, 1976.
Gorillas in the Mist. Boston: Houghton Mifflin, 1983.

Works about Dian Fossey

Borst, B. A. "Rwandan justice: A show of pomp and circumstantial evidence." *Discover* 8 (1987): 45.
Carr, R. "Lonely struggle: An old friend remembers Dian Fossey." *International Wildlife* 18 (1988): 12–13.
"Dian Fossey." In *Contemporary Newsmakers: 1986 Cumulation*, edited by Peter M. Gareffa, 122. Detroit: Gale Research, Book Tower, 1986.
Hausfater, G., and K.A.R. Kennedy. "Dian Fossey." *American Anthropologist* 88 (1986): 953–956.
Hayes, H.T.P. *The Dark Romance of Dian Fossey.* New York: Simon and Schuster, 1990.
"Inside a vanishing kingdom." *Newsweek* (Aug. 29, 1983): 61.
Le Bihan, Jill. "Gorilla girls and chimpanzee mothers: Sexual and cultural identity in the primatologist's field." *Journal of Commonwealth Literature* 27 (1992): 139–148.
Montgomery, S. *Walking with the Great Apes: Jane Goodall, Dian Fossey and Biruté Galdikas.* Boston: Houghton Mifflin, 1991.
Morell, V. "Dian Fossey: Field science and death in Africa." *Science* 86 (7) (1986): 17–19.
———. "Primatology: Called 'trimates,' three bold women shaped their field." *Science* 260 (1993): 420–425.
Mowat, F. *Woman in the Mists: The Story of Dian Fossey and the Mountain Gorillas of Africa.* New York: Warner Books, 1987. Also published as *Virunga, the Passion of Dian Fossey.* Canada: Warner Books, 1987.
Shoumatoff, A. *African Madness.* New York: A. A. Knopf, 1988.
"Three who have chosen life in the wild." *New York Times* (May 1, 1981): B4.

Veit, P.G. "A life among gorillas." *Natural History* 92 (Aug. 1983): 66–67.
"Zoologist is slain in Central Africa." *New York Times* (Dec. 29, 1985): 15.

Other References

Schaller, G. B. *The Mountain Gorilla: Ecology and Behavior*. Chicago: University of Chicago Press, 1963.

ROSALIND ELSIE FRANKLIN (1920–1958)

Maureen M. Julian

BIOGRAPHY

Rosalind Franklin's great-great grandfather, Abraham Franklin, a successful merchant banker and son of the rabbi of Breslau, settled in London in 1763. She was born in London on July 25, 1920, the second child and first daughter of well-to-do Jewish parents, Ellis and Muriel (Waley) Franklin. David was born first and then, after Rosalind, came Colin, Roland, and Jenifer. The Franklins were happily married. They not only were committed to the importance of domestic life but, as orthodox Jews, felt obligated to engage in philanthropic activities. After Ellis Franklin returned from World War I, he devoted most of his energy to voluntary work at the Working Men's College, an institution that brought together working-class and university men. Rosalind Franklin attended the fashionable St. Paul's Girls' School, where she received an excellent background in physics and chemistry. In the 1930s her father devoted his volunteer efforts to the Center at Woburn House, which aided Jews who had escaped from Nazi Germany. Rosalind Franklin was sympathetic with this cause and did volunteer work there while she was at school.

Shortly after Franklin entered Newnham College of Cambridge University in 1938, much of the science faculty's time was absorbed in war research. This situation left the undergraduates on their own, more so than usually, which suited her. After obtaining her undergraduate degree in 1941, she remained at Cambridge for her graduate work. Through her volunteer work, Rosalind Franklin met metallurgist Adrienne Weill, a wartime refugee at Cambridge who gave Franklin French lessons. At the age of 22, Franklin gave up her fellowship in order to take a position as a physical chemist at the British Coal Utilisation Research Association (CURA). CURA was directed by Dr. D. H. Bangham, who brought together a team of brilliant young university graduates, fresh with the latest scientific knowledge. The organization greatly expanded during the war because fuel was so important to the defense of Britain. Three years later, in 1945, Franklin received her Ph.D. degree from Cambridge. The title of her dissertation was "The Physical Chemistry of Solid Organic Colloids with Special Relation to Coal and Related Materials."

In 1946 Rosalind Franklin wrote to Adrienne Weill, who by then had returned to Paris: "If ever you hear of anybody anxious for the services of a physical chemist who knows very little about physical chemistry, but quite a lot about the holes in coal, please let me know" (Sayre 1975, 69). Weill suggested Marcel Mathieu at the Sorbonne, who in the 1920s had worked with William Henry Bragg (Nobel laureate, 1915) at the Royal Institution in London. At Mathieu's invitation Franklin came to Paris in February 1947 as a *chercheur* (researcher) in the Laboratoire Central des Services Chimiques de l'État. She worked closely with crystallographer Jacques Méring and began using X-ray diffraction techniques to study the graphitizing and nongraphitizing carbons. Franklin particularly enjoyed the informal relationships that existed among the laboratory workers who ate in local bistros and spent much of their free time together.

Meanwhile, the crystallographic research frontier lay in the biological substances. With mixed feelings, Franklin left France in 1951 and accepted Professor John Randall's offer of a three-year Turner-Newall Research Fellowship at King's College, London. The newly formed Biophysics Department was beginning to work on deoxyribonucleic acid (DNA), which was generally thought to be associated with heredity. Franklin chose DNA for the subject of her work. DNA, like poorly crystallized graphite, gave poor diffraction patterns and demanded the special X-ray techniques Franklin had developed in France.

Life at King's College was not easy for Franklin. She found it difficult to work with her colleague, Maurice Wilkins. This problem was compounded by the "men's club" atmosphere of King's College, which excluded her from almost anything but formal laboratory contacts with her fellow scientists. The contrast was great compared with the informal atmosphere she had enjoyed in France. She remained at King's College from January 1951 to the spring of 1953. During this time Maurice Wilkins secretly showed James Watson, an American studying at Cambridge, some of Franklin's pivotal X-ray photographs of DNA. This dishonest action is one of the low points in scientific research.

Professional conditions between Franklin and Wilkins at King's deteriorated so badly that in the spring of 1953 she took her fellowship, but not the DNA problem, to Birkbeck College, London. John D. Bernal had invited her to study the structure of plant viruses, especially the tobacco mosaic virus. Again, these materials are of borderline crystallinity. At Birkbeck she had a number of successful collaborations, especially with Aaron Klug, who won the 1982 Nobel Prize in Chemistry.

In the autumn of 1956, Franklin learned that she had cancer. She refused to let up on her work and, even near the end, began a personally dangerous project on the structure of polio virus. After her cancer operation, she took a happy and longed-for vacation in Switzerland. However, her reprieve was short-lived. Later, in too much pain to be alone, she moved in with her brother and his family. Finally, on April 16, 1958, at the age of 37, Franklin died at the Royal

Marsden Hospital. Much of Franklin's life was sad, not only in its cruel short-ness but in the difficult position in which she found herself at King's College. Her contributions were many and brilliant. Rosalind Franklin had an uncanny ability to work with poorly crystallized materials. Such samples are difficult to prepare, their X-ray diffraction patterns are tricky to record, and their analysis is usually next to impossible. Franklin began her professional career as a phys-ical chemist studying coal and graphite. She published a series of papers on carbon and coal, many of them recognized as classics today. Then she turned her attention to DNA. Her two years of research on DNA produced publications associated with work that later won the Nobel Prize, but not for herself. Finally, during her last five years, she wrote numerous papers on viruses. Her last three papers were published posthumously.

In December 1962 James Watson, Francis Crick, and Maurice Wilkins re-ceived the Nobel Prize in Physiology or Medicine for their discovery of the double helix structure of the DNA molecule. These men based their work, in considerable part, on experimental data obtained surreptitiously from Franklin's files. At the time of the award, she had been dead for four years, and her crucial X-ray photographs and calculations were not acknowledged.

Watson, Crick, and Wilkins's three Nobel Prize lectures contain a total of 98 references, and not one of Franklin's papers is specifically mentioned. Wilkins makes the only textual reference to her in this casual remark: "Rosalind Franklin . . . made some very valuable contributions to the X-ray analysis." Only Wilkins included her in his acknowledgments (Wilkins 1975).

Watson published *The Double Helix* (1980), in which he told his version of the DNA story. In this book the incident of the viewing of Franklin's B-form X-ray picture was publicly revealed for the first time. The date of the original printing was 1968. In an interview with Anne Sayre on June 15, 1970, Wilkins discussed the pirate viewing of this picture: "Perhaps I should have asked Ros-alind's permission and I didn't. Things were very difficult. Some people have said that I was entirely wrong to do this without her permission, without con-sulting her at least, and perhaps I was" (Sayre 1975).

WORK

In 1941 Rosalind Franklin began her research career with a fellowship awarded by Newnham College, Cambridge University. She used it to study gas-phase chromatography of organic materials, including coal, with Professor Ron-ald G. W. Norrish, who became a Chemistry Nobel Prize laureate (1967). In 1942 Franklin resigned her fellowship to work on coal at CURA. She began by measuring the thermal expansion, density, and chemical composition, as well as the structure of coals and carbonized coals. Later, in France, she learned and applied X-ray crystallographic techniques to coals.

An important result was her recognition of similar characteristics in seemingly unlike materials, which led her to classify coal and some organic materials into graphitizable and nongraphitizable carbons. Under controlled pyrolysis, the non-graphitizable carbons, including low-rank coals (materials rich in oxygen or poor in hydrogen and polyvinylidine), have low density and large, fine-structure porosity and are very hard. On the other hand, the graphitizable carbons, including coking coals, substances rich in hydrogen and polyvinyl chloride, form soft, compact, and high-density carbons. She discovered that the unusual behavior of nongraphitizing carbons comes from a system of cross-linking between carbon crystallites, which in this fine-pored structure prevented the formation of graphite. At high temperatures the pores shrink, so that even helium atoms are unable to enter.

In 1951 Franklin returned to England and utilized her invaluable experience with poorly crystallized materials. In only eight months she set up an Ehrenberg fine-focus X-ray tube, with a Phillips microcamera, for taking high-resolution photographs of single fibers of DNA. To further facilitate data collection, she designed a microcamera specifically to photograph crystals inclined to the X-ray beam at several angles. Franklin then took an unrivaled series of X-ray diffraction photographs of DNA. Under carefully controlled humidity conditions, she discovered the hydrated B-form of DNA, with its diffuse pattern clearly indicating a helix. She defined the conditions under which the two distinct reversible phases, A and B, existed. Previous DNA photographs were a mixture of the two phases. This careful experimental work and lucid analysis cleared up the confusion of earlier workers, who had assumed that this A + B mixture of phases was a single phase.

Franklin also showed that the sugar phosphate backbone was on the outside of the DNA molecule and that hydrogen bonding played an important role. She then concentrated her efforts on performing a Patterson analysis of the A-form of DNA, which showed a more detailed crystalline pattern. The mathematics involved is complex and subtle, so it was distinctly possible that the A-form might assume a very different shape (like rods or sheets inclined to the fiber axis) from the helix shown in the hydrated B-form (Klug 1968). Although her notebooks from this time indicate that she knew the B-form was helical and contained ten nucleotides per turn, she was not sure of the number of intertwined chains. The correct analysis of the A-form would yield more precise information. Because of the reversible phase change between A and B, the molecular model would have to be easily adaptable to crystallize in either form.

Meanwhile, at Cambridge, Crick and Watson were building their now-famous models of DNA. Their helical DNA model building needed an experimental foundation. Wilkins secretly showed Watson the unique picture of the B-form that had been taken by Franklin. Also at this time, Professor Max Perutz gave Crick and Watson a privileged copy of Franklin's Medical Research Council report, containing calculations from the spacings of the critical reflections on the B-form photograph. Watson and Crick then fitted the DNA molecule into the space defined by the helical repeat distance and helical diameter. Franklin's picture formed the major experimental link that culminated in Watson and

Crick's solution of the structure of DNA. Their contribution was the splendid assembling of the double helix in which they matched the base pairs, adenine to thymine and guanine to cytosine.

By agreement, three papers on DNA appeared in the famous April 25, 1953, issue of *Nature*: the first by Watson and Crick; the second by Wilkins and his coworkers, A. R. Stokes and H. R. Wilson; and the third by Franklin and her graduate student, Raymond Gosling. Watson and Crick's paper listed the helical pitch, the diameter of the helix, and the number of turns between repeats along the helix but did not say where the data originated. Franklin rejoiced at Watson and Crick's elucidation of the DNA structure and immediately, without suspicion, confirmed its agreement with her own experimental results. She never dreamed that their double helix model was based on her experimental evidence. Three months later she and Gosling published a confirmation of the helix in the A-form (with Gosling, "The Structure . . . I, II" 1953). Of course, science is built on many workers' research, but it is heinous that Franklin was never given credit for her crucial experimental information, either during her lifetime or at the time the Nobel Prize was awarded in 1962.

In 1953 Franklin turned to another biological problem, tobacco mosaic virus (TMV). Professor Bernal, the director of the laboratory at Birkbeck College and an X-ray crystallography student of Professor W. Henry Bragg, had begun studies on TMV long before, in 1941. He was eager for Franklin to apply her ability to work with materials of borderline crystallinity to the TMV problem. She immediately took a series of superb X-ray diffraction photographs, which confirmed its helical structure and showed, surprisingly, that ribonucleic acid (RNA) was not in the central core but was embedded in the protein. Furthermore, she showed that the virus was a hollow structure rather than a solid particle. TMV was the first virus to be subjected to this type of analysis.

Franklin learned her crystallography in France and applied the techniques first to coal, then to DNA, and finally to plant viruses. Her work represents an attack on substances that were on the borderline of crystallinity.

BIBLIOGRAPHY

Works by Rosalind Elsie Franklin

Scientific Works

Space does not permit the listing of the complete works of Rosalind Elsie Franklin. Listed here are all works by Franklin except those cited in Miksic (1993). Included here are her dissertation and all references cited in the text.

"The Physical Chemistry of Solid Organic Colloids with Special Relation to Coal and Related Materials." Doctoral thesis, Cambridge University, 1945.

(with R. G. Gosling) "Evidence for a 2–chain helix in crystalline structure of sodium deoxyribonucleate." *Nature* 172 (1953): 156–157.

(————) "The structure of sodium thymonucleate fibres. I. The influence of water content." *Acta Crystallographica* 6 (1953): 673–677.

(————) "The structure of sodium thymonucleate fibres. II. The cylindrically symmetrical Patterson function." *Acta Crystallographica* 6 (1953): 678–685.

(————) "The structure of sodium thymonucleate fibres. III." *Acta Crystallographica* 8 (1955): 151–156.

(with A. Klug) "Order-disorder transitions in structures containing helical molecules." *Discussions of the Faraday Society* 25 (1958): 104–110.

(with A. Klug, J. T. Finch, et al.) "On the structure of some ribonucleoprotein particles." *Discussions of the Faraday Society* 25 (1958): 197–198.

(with A. Klug and S.P.F. Humphreys-Owen) "The crystal structure of tipula iridescent virus as determined by Bragg reflection of visible light." *Biochimica et Biophysica Acta* 32 (1959): 203–219.

Works about Rosalind Elsie Franklin

Bernal, J. D. "Dr. Rosalind E. Franklin." *Nature* 182 (1958): 154.

Judson, H. F. *The Eighth Day of Creation: The Makers of the Revolution in Biology.* New York: Simon and Schuster, 1979.

Julian, Maureen M. "Four women in crystallography." In *Women's Contribution to Chemistry: An Historical Perspective and Women at the Forefront,* edited by V. I. Birss et al., 1–29. Sherbrooke, Quebec: Canadian Society for Chemistry, 1993.

————. "Profiles in chemistry: Rosalind Franklin, from coal to DNA to plant viruses." *Journal of Chemical Education* 60 (1983): 660–662.

————. "Women in crystallography." In *Women of Science: Righting the Record,* edited by G. Kass-Simon and P. Farnes, 335–383. Bloomington: Indiana University Press, 1993.

Klug, A. "Rosalind Franklin and the discovery of the structure of DNA." *Nature* 219 (1968): 808–810, 843–844.

Miksic, M. C. "Rosalind Elsie Franklin (1920–1958)." In *Women in Chemistry and Physics: A Biobibliographic Sourcebook,* edited by L.S. Grinstein et al., 190–200. Westport, CT: Greenwood Press, 1993.

Sayre, Anne. *Rosalind Franklin and DNA.* New York: W. W. Norton, 1975.

Watson, J. D. *The Double Helix.* New York: W. W. Norton, 1980.

Other References

Watson, J. D., and F.H.C. Crick. "A structure for deoxyribose nucleic acid." *Nature* 171 (1953): 737–738.

Wilkins, M. *Nobel Lectures in Molecular Biology 1933–1975.* New York: Elsevier, 1975.

————, A. R. Stokes, and H. R. Wilson. "Molecular structure of deoxypentose nucleic acids." *Nature* 171 (1953): 738–740.

FRANCES CARNES FLINT HAMERSTROM (1907–)

Barbara Mandula

> When I was a small child I longed one day to become so famous that I did not have to hide how odd I was—how unlike other people. Few people really held my attention. It was birds and mammals, reptiles and insects that filled my dreams and eternally whetted my curiosity.
>
> Other, more normal loves intruded, often to my delight, and so it was that I have led a double life from childhood, when grownups forbade wild pets and tried to squelch my companionship with creepy crawly creatures, to the present, when I still cannot explain my passion for the wild and free.
> —Hamerstrom, *My Double Life* 1994.

BIOGRAPHY

Frances Flint Hamerstrom was raised in one of America's wealthiest families. A bright, inquisitive, daring child, she knew from an early age that she wanted to study wild animals. Her life was a series of conscious choices. Instead of becoming the international hostess she was trained for, she became a wildlife field biologist, a licensed falconer, and an international expert on raptors (hawks, owls, and eagles). Hamerstrom and her husband were largely responsible for saving the prairie chicken from extirpation in Wisconsin in the early 1950s, when they recognized that loss of nest brood habitat threatened the bird's survival. The Hamerstroms worked as a scientific team for 59 years, publishing many joint papers. When she was more than 60 years old, Hamerstrom wrote the first of a dozen popular books on natural history. She has received several prestigious book awards for her special ability to share her intimate knowledge of the natural world.

Born on December 17, 1907, in the Boston suburb of Needham, Frances Carnes Flint was the oldest of four children and the only daughter. She developed a close relationship with Bertram, who was one and one-half years younger than herself. Vasmer was seven years younger; Putnam, ten years.

Both parents came from wealthy Boston families of English, Welsh, Scottish, and Irish descent who could trace their lineage to the Revolutionary War. They tended toward Republican views. Frances Flint's mother was an heiress, Helen (Chase) Flint. She was educated in a private school in Farmington, Connecticut, and was brought up Episcopalian (Personal communication 1993). Mrs. Flint was interested in art and literature and gardening but could also climb trees and identify unfamiliar species in Asa Gray's *Manual of Botany* (*My Double Life* 1994, 113).

Hamerstrom remembers her father as a brilliant, frustrated man. Both Laurence Flint and his daughter were atheists. Mr. Flint was an international criminologist who loved detective work and assisted the United States during World War I. Freed from the necessity of earning a living, he became a master photographer and an authority on the French Revolution. But his wife did not respect any of his accomplishments, doting instead on the lifestyle of *her* father, who went to an office every day (Personal communication 1993).

Although she was raised in a privileged household, Frances Flint was an unhappy child. Because her father's avocation required frequent travel, she spent her early years in various countries in Europe. By the time she returned to the United States in 1914, she spoke French, German, and English and was surprised that American children spoke only one language. Hamerstrom recalls that her governess, Fraulein A.E.A. Taggesell, brought up the children very strictly in the large house in Milton, a Boston suburb: "It might have been fun except that I had personal tutors. I had to do everything correctly—riding, tennis, tatting, knitting, skiing, music, speaking French and German and English. They were always after you. I wouldn't be a child again for anything" (Personal communication 1993).

Because her parents disapproved of her interest in wildlife, young Frances Flint developed a double life (*My Double Life* 1994). In one life, she was trained for the role of international hostess, even learning correct carriage by walking around balancing a book on her head. In her hidden life she took advantage of the acres of woods and stables surrounding her parents' home to pursue her infinite curiosity about birds and other animals. She kept her animals and guns in a concealed area behind the stables. To keep prying adults away, she planted poison ivy across the path; she was immune.

As a child she was determined to do what she wanted without adult interference. She hated her weekly dancing lessons because the boys did not want to dance with such a tall girl. After she discovered that she could appear flushed and feverish from eating certain berries, she regularly became ill on the day of dancing class. When the berries were gone, she avoided dancing class by scraping a bramble across her face to make deep scratches (*My Double Life* 1994, 60). Despite her dislike of dancing school, she was recognized as a spectacular ballroom dancer once the boys caught up to her in height.

She also recalls that when she was about eight years old, she wanted to cut open a dead blue jay to see if it had a heart, lungs, and liver, as people did. But a neighbor child insisted on a little funeral, complete with hymns and a cross.

Frances Flint reluctantly agreed, but later that evening her insatiable curiosity drove her to slip back to the burial site, unearth the dead bird, and dissect it.

Several people nurtured Flint's early interests in wildlife. A gardener taught her about plants and gardening and showed her how to mount bird skins. Charles Johnson, an entomologist at the Boston Museum of Natural History, also took an interest in the curious child. He helped her to identify the insects she collected and showed her how to mount them. (She ingeniously maneuvered her governess into taking her to the museum regularly. By using a lead pencil to keep her gums irritated, Flint frequently needed to visit the dentist, whose office was near the museum.)

When Flint was about 12 years old, she received her first falcon, a kestrel. Determined to nurture and train the bird, she went to the Milton Public Library to read everything available on keeping and training falcons. A librarian was sure the child would be unable to read a book printed in Old English script. But the book was not a serious challenge to Flint, who could read Old German print and could read and write Old German script. She quickly decoded the text. Using an English dictionary, she looked up such falconry terms as austringer, creance, and rangle (*My Double Life* 1994, 68). This kestrel marked the first of dozens of birds of prey that she would nurse and fly.

Flint's father was a fervent hunter but never recognized that his daughter, rather than his sons, shared this passion. Flint secretly learned about guns and hunting by listening to her father instruct her brothers. She became a gun owner by pawning a fur stole that her mother had discarded.

In addition to lessons from her governess, Flint attended Milton Academy, an expensive private finishing school. She was the only student in her class who did not graduate. She concentrated on wildlife instead of the standard curriculum. Admitted to Smith College in 1925, she left in her sophomore year because she was "too interested in the two b's—birds and boys." She did very well in the subjects that interested her but failed the subjects she didn't care about (Paulson 1994).

Her father was furious. He told her she was not trained for any kind of real work. But his daughter rose to the challenge. She entered the most expensive dress shop in Boston and said she wanted to model their clothes. When asked about her experience, she replied, "I have been accustomed to wearing beautiful clothes all my life." Offered a job, she was soon earning more than her father (*My Double Life* 1994, 107).

Frances Flint met her husband, Frederick Nathan Hamerstrom, Jr., while he was an undergraduate at Dartmouth. He was a descendant of the Vikings and a nephew of the famed criminal lawyer Clarence Darrow. Frances Hamerstrom recalls: "I wanted to work with wild animals and to marry a tall dark man—I did both. I saw Frederick at a football game smoking a pipe and knew I would die if I couldn't live with him. He didn't propose until the third date, commenting that '[w]e Norwegians are slow, we like to think things over' " (Personal communication 1993).

The wedding took place in June 1931, against the wishes of both sets of parents. The Hamerstroms subsequently completely merged their professional and personal lives, forming one of the closest and longest lasting research teams known to science.

Frederick and his wife shared a passionate love of hunting, wildlife, and dancing. When Frederick was offered his first real job, his wife was astonished and pleased when he insisted on a guaranteed vacation during hunting season. The pair almost never missed a hunting season, not even when they were expecting their first child during that time of year. The baby waited until hunting season was over (*My Double Life* 1994, 176).

The Hamerstroms had two children, Alan (b. 1940) and Elva (b. 1943). A good hunter but with a strong interest in the sea, Alan is now a yacht broker. Elva (Hamerstrom) Paulson inherited her mother's interest in animals and became a professional wildlife artist. She sometimes illustrates her mother's books. She is married to a soil scientist.

When their children were young, the Hamerstroms started a children's exchange program with several European biologists. "Twelve seems the perfect age for children to live in a foreign country. Our arrangement was simple: one year we'd have one child at home and the next year we might have three or even four (one or two of our own and the others European)" (*My Double Life* 1994, 286).

Soon after their marriage, Frederick and Frances Hamerstrom spent a year at the New Jersey Game Conservation Institute, hoping to find a direction for their consuming interest in wildlife. They went on to study wildlife management with the pioneers in the field. One of these was Paul Errington at Iowa State College, where Frances Hamerstrom worked double time to finish her undergraduate degree while doing research, and Frederick worked on his Master's thesis; they received their respective degrees in 1935.

They then were accepted as students by famed ecologist Aldo Leopold at the University of Wisconsin, who assigned them both to work on prairie chickens. In her spare time Frances Hamerstrom studied the flocks of chickadees outside their old farmhouse, inventing colored leg bands so that she could identify individual birds. Not satisfied, she glued colored feathers onto wintering chickadees to more easily identify the birds as individuals. Many of the colored feathers had been gleaned from a badly stuffed multicolored parrot skin.

The time was ripe for behavior studies on wild animals. Frances Hamerstrom, Ralph King, and Marguarite Baumgartner pioneered in devising the necessary techniques in the 1930s, techniques still used today. Hamerstrom observed chickadee interactions for two winters, discovering that flocks had a pecking order. When it became obvious that the prairie chicken work could not be split into two theses, she mentioned her work on chickadees to Leopold. He responded, " '*Color-marked chickadees*, Fran? . . . I had no idea you were interested in chickadees' " (*Birding with a Purpose* 1984, 8). It was one of the first studies in the world using individual color-marked birds. It earned her a master's degree in 1940; a year later Frederick received his doctorate for their joint prairie chicken research.

Frances Hamerstrom received an honorary doctorate in 1961 from Carroll College, Waukesha, Wisconsin. She and her husband used to give public lectures on prairie chickens to the visitors who were going to count the birds the following morning. One evening, a group specifically requested that she give the lecture and afterward asked her detailed questions. The questioners turned out to be the Regents of Carroll College, testing Hamerstrom before awarding her an honorary doctoral degree. She passed with flying colors.

Most women today would be unwilling to emulate Hamerstrom's career path: getting paid when she could; not worrying when others received credit for her work; concerned only with having the opportunity and freedom to do her research. But it suited her. She wanted to give something to the world.

During the 1940s she worked with her husband and carried out her own research while he was curator of the Edwin S. George Reserve, University of Michigan. From 1949 to 1972 she was employed 60% of the time as a game biologist for the Wisconsin Department of Natural Resources, where her husband was employed full-time. Since 1982 she has been an adjunct professor at the University of Wisconsin, Stevens Point.

Hamerstrom has received many awards for her varied contributions. She received the Wildlife Society Award twice: in 1940 for her first major published paper ("The great horned owl . . .") and in 1957 for *A Guide to Prairie Chicken Management*. Other awards include the Josselyn Van Tyne Award of the American Ornithologists' Union (1960); honorary doctorate, Carroll College (1961); Chapman Award of the American Museum of Natural History (1964); Distinguished Service to Conservation Award of the National Wildlife Federation (1970); United Peregrine Society Conservation Award (1980); Edwards Prize of the Wilson Ornithological Society (1985); and Notable Author, Wisconsin Library Association (1992). She is a fellow of the Wisconsin Academy of Sciences, Arts and Letters. In 1991 she was one of two authors honored by Governor Tommy Thompson of Wisconsin. In 1994 she was inducted into the Wisconsin Hall of Fame.

Until the 1960s wildlife management was dominated by males. In addition to Frances Hamerstrom, only three women were recognized as wildlife field biologists in the United States: Elizabeth Schwartz, Margaret Altman, and Elizabeth (Baird) Locey. "We were quite a novelty," Hamerstrom quipped (Personal communication 1993).

Hamerstrom's success is due, in part, to her physical strength and endurance, as well as to her tenacity and determination. She learned to hoist 100-pound sacks of grain to her shoulder and carry them to animal pens when she and Frederick were students at the New Jersey Game Conservation Institute. (She was also the only woman admitted among 40 men, after she assured the board members she would not flirt with the boys because she would be married when school started.) When her children were young, she took them with her to the field, "carrying them one at a time, with a dozen traps, from tussock to tussock when trapping in a tamarack swamp" (Personal communication 1993). She did more than her share of difficult home chores, such as carrying in water or chopping wood. Whenever she beat Frederick to an unpleasant job, he would quietly say, "You should have let me do that."

Frances Hamerstrom benefited greatly by working with her husband. She says:

It would have been extremely difficult if I'd had to do it alone. People were afraid to hire a woman biologist because they feared they'd be criticized. I was lucky I could afford to work without a salary, or with a partial salary. For example, during the 1930s in Wisconsin, I didn't have a salary but I had permission to do my research and to have a crew. I felt lucky to be doing what I wanted. The result is that my husband has a series of publications on central Wisconsin game animals where I had done half the work. But he was on a government salary and they said I couldn't be an author, although he could acknowledge my help. Later we worked on prairie chickens for the Wisconsin Conservation Department. I was paid for studying prairie chickens 60% time. That left 40% in case the children had measles, but since they were healthy, I spent that 40% doing raptor research—a very good deal, but I always worried that some bureaucrat would complain about it. (Personal communication 1993)

Frances Hamerstrom tried to fit in with the men, taking long steps, wearing olive-drab clothing, and keeping her voice low-pitched. The men accepted her more readily than did the women; the wives of the biologists were especially envious because she spent time with her husband in the field while they were home with the babies. "I had babies and I just kept right on working," she explains. "Neither Errington nor Leopold held it against me that I was a woman. And my husband never held it against me" (Personal communication 1993).

WORK

Hamerstrom has taught the skills of field research to more than 100 apprentices (gabboons) from the United States and abroad. Many of the gabboons came under her influence early in their studies, and some have become today's leaders in raptor research and other fields of wildlife biology. Her gabboons became accustomed to her creativity and unpredictability, as she asked questions nobody had thought of before and then developed the experimental techniques to answer the questions. Her style of doing science always depends on careful observation and description and on explaining function based on data; she leaves to others the development of ecological theories and hypotheses (Schmutz, Personal communication 1994). Although she was never interested in formal teaching, Hamerstrom has found other ways of sharing her knowledge. Unlike some scientists, she does not make a special effort to publish only in prestigious technical journals but also writes accurately in popular magazines.

The color-marking techniques Hamerstrom invented for studying individual chickadees were described in the previous section. Later, she and her apprentices developed portable traps and the use of live bait (pigeons, mice, and a pet rat) for catching raptors efficiently while the bait remained unharmed. She discovered that women's nylon stockings were perfect for temporarily storing owls until they could be weighed, measured, banded, and released. Hawks did better in rigid tubes or cans. These trapping and holding techniques were crucial for banding the raptors and thereby being able to study individual wild raptors for the first time.

The Hamerstroms recognized that long-term, consistent observations were the only way to study the needs and natural variation of wildlife populations. They began their prairie chicken studies in the 1930s with Leopold and continued these studies for more than 30 years. They learned that, like a number of other wild species, prairie chickens undergo 10-year population cycles. Nonetheless, early in the 1950s they realized that saving habitat was crucial to making sure the prairie chicken did not become extirpated in Wisconsin. Knowing the characteristics of good prairie chicken habitat and publicizing the problem, the Hamerstroms helped convince the appropriate people to buy or donate the land needed to assure survival of the prairie chicken population (Hayes, "Saving . . ." 1990; "Chickens . . ." 1990).

In 1960 Frances Hamerstrom began studying harrier populations on 50,000 acres of Buena Vista Marsh near their home in Wisconsin. She summarized her 25 years of observations in a popular book published by the Smithsonian Institution in Washington, D.C. (*Harrier . . .* 1986). She discovered that harrier populations are controlled naturally by the abundance of voles (field mice). When there are many mice, young male harriers mate in their first season, and older males may have three or four mates.

She recounts also that she was marking birds and monitoring breeding success when she noticed a precipitous decline in the population and in nests in 1965. She suspected that harriers were responding differently than peregrines to DDT (dichlorodiphenyltrichloroethane); while peregrines were laying soft-shelled eggs that broke easily, harriers were instead behaving listlessly and not nesting at all. She developed a two-minute surgical procedure to obtain muscle biopsies for DDT analyses. Without funds to pay for the analyses, she arranged to barter her famous home-baked apple pies for the data. The results provided increasing evidence of the widespread effect of DDT on raptors and helped get the pesticide banned in the United States.

Since 1968 Hamerstrom has also studied kestrel nesting at Buena Vista Marsh. With old woodpecker holes and old buildings becoming increasingly scarce for nest sites, she put out 50 nest boxes for kestrels. By monitoring the populations, she learned that paucity of nest sites, rather than food, is the prime limiting factor for kestrels in central Wisconsin. After the Hamerstroms retired from the Wisconsin Department of Natural Resources in 1972, they regularly spent some of the winter studying Harris hawks in Texas (they banded more than 1,200) and ospreys in Mexico.

In her long-term studies, Hamerstrom correlates activity and mating success

with weather and other environmental factors. "Any time there's an environmental emergency, only long-term studies let you know if it's just a local fluke
or something much more serious. We need a catalog of these studies, which
will be especially important for understanding the effects of climate change,"
she explains (Personal communication 1993). Field biologists quickly notice if
something changes in a population they know well; Hamerstrom did not need
statistical analyses to tell her that DDT was harming harrier populations.

Hamerstrom agrees with Leopold about the need to forge links between wildlife and the general public. She, too, has written popular award-winning books,
including several children's books. She deplores the lack of accurate natural
history books that could stimulate children's interest in the natural world. Her
writing style is clear and direct, filled with humorous anecdotes of her activities
and fascinating facts about birds she has known. *My Double Life* (1994) is
autobiographical. The early part describes for the first time her life before her
marriage, while the second part is a reprint of her earlier book, *Strictly for the
Chickens*.

She often uses public appearances to educate the audience about raptors and
other wildlife. Hamerstrom has appeared on television shows, including "Late
Night with David Letterman," where she plucked a pigeon despite Letterman's
jibes. She recalls being on "What's My Line" with a giant female golden eagle
that made so much noise that the blindfolded panel quickly guessed the guest
had an eagle (Personal communication 1993).

Frances Hamerstrom is one of the last wildlife field biologists whose career
began when wildlife management meant killing everything that humans thought
they did not need. This was a time when bounties were offered for dead "var-
mints," such as coyotes, wolves, and raptors, because they were seen primarily
as livestock predators. "Wildlife management was just developing as a field in
the 1930s," she explained. "The public had no idea yet about conservation and
wildlife research. I thought it was my duty to kill raptors, and I was a very good
shot. Paul Errington helped me see the importance of raptors. I moved from the
shotgun to learning about the live animals" (Personal communication 1993).

Like many other wildlife biologists, Hamerstrom recognizes that wildlife
hunting and wildlife conservation have been closely linked historically, although
now they are usually seen as opposing forces. She believes that photographers
are potentially a greater danger to wildlife than hunters are. "There's closed
season on hunting. But there's no closed season on photographers. Unless
they're really scrupulous, photographers can break up nests and cause no end
of damage," she says. "They strive for the rarities, which increases the harm
they can do" (Personal communication 1994).

Hamerstrom's varied research career was possible, in part, because she and
her husband chose to live frugally, coping happily with old farmhouses and not
worrying about the trappings of civilization they had left behind. Having already
known a privileged lifestyle, they did not need to achieve it. Given the scientific
significance of the Hamerstroms' raptor research, it is surprising to realize that

almost all of it was conducted outside their paid positions and without outside funding (Mayr 1992).

Hamerstrom has been an active member of the worldwide ornithological community. From 1945 to 1948 she and her husband organized and ran a relief committee for European ornithologists, providing help to many who were on the verge of death by starvation. ''For Fran and Hammi it was simply something 'one has to do' '' (Mayr 1992). Hamerstrom's other professional activities and memberships include *Bird Banding* editorial board, for which she has reviewed dozens of German articles; Wisconsin Society for Ornithology, cochair, Research Committee, 1962–79; Citizens' Natural Resources Association, councilor, 1964–76; Wisconsin Society of Ornithology, director, 1965–89; Society of *Tympanuchus cupido pinnatus* (Prairie Chicken Society), board member; Raptor Research Foundation; British Falconers Association; North American Falconers' Association; American Ornithologists' Union; Wilson Ornithological Society; Ecological Society of America; Nature Conservancy; National Wildlife Federation; and Wildlife Society.

Today Hamerstrom is still finding creative ways to explore the natural world. While continuing her writing and her Wisconsin and southern raptor studies, she is spending time in remote parts of the Amazon and Africa. ''Frederick never liked to visit the hot, wet places on Earth. After his death [in 1990], I figured it was my chance to see parts of the world that I hadn't been to yet. I consult my travel agent to get started—after that I travel alone with one guide'' (Personal communication 1993). Hamerstrom plans to write a book describing how indigenous peoples of the rain forest have lived and hunted for centuries without destroying their habitat.

Hamerstrom eloquently expresses her concern about unchecked population growth: ''If we are to preserve this beautiful world of ours, with its creatures great and small and their wondrous homes, we must have fewer people on earth, we must have fewer children, or the beauty of the wild will be gone'' (*My Double Life* 1994, 315–316).

NOTE

Special thanks are due Frances Hamerstrom, who took the time to meet with me in the fall of 1993 on her way to Texas and Brazil. Elva Hamerstrom Paulson generously provided valuable personal insights I could not have gained any other way. Casey Martin helped me understand Hamerstrom's worldview. I am grateful also to David Blockstein, Richard Clark, Mike Kochert, Joe Schmutz, and Stan Temple.

BIBLIOGRAPHY

Works by Frances Carnes Flint Hamerstrom

Scientific Works

Space does not permit the listing of the complete works of Frances Flint Hamerstrom. Listed here are all works by Hamerstrom except those cited in Hamerstrom (1986). Included here are all references cited in the text.

"What eats what?" *Bird Lore* 41 (1) (1939): 31–33.

(with O. E. Mattson) "Food of central Wisconsin horned owls." *American Midland Naturalist* 22 (3) (1939): 700–702.

"Dominance in Winter Flocks of Chickadees." Master's thesis, University of Wisconsin, Madison, 1940.

(with P. L. Errington and F. N. Hamerstrom, Jr.) "The great horned owl and its prey in north-central United States." *Iowa Agriculture and Home Economics Experiment Station Research Bulletin* 277 (1940): 758–850.

"Dominance in winter flocks of chickadees." *Wilson Bulletin* 54 (1) (1942): 32–42.

(with J. Van Tyne and F. N. Hamerstrom, Jr.) "Breeding bird census of jack pine barrens." *Audubon Magazine* (Suppl.) 44 (1942): 28, 30.

(———) "Breeding bird census of jack pine barrens." *Audubon Magazine* 45, Sec. 2 (1943): 23–24.

(with F. N. Hamerstrom, Jr.) "Daily and seasonal movements of Wisconsin prairie chickens." *Auk* 66 (4) (1949): 313–337.

"Hahenjagd in North Dakota." *Der Deutsche Jäger* 14 (1950).

"Range habits and food requirements of prairie chicken." *Wisconsin Conservation Bulletin* 15 (11) (1950): 9–11.

(with F. N. Hamerstrom, Jr.) "Food of young raptors on the Edwin S. George Reserve." *Wilson Bulletin* 63 (1) (1951): 16–25.

(———) "Grouse of the brushlands." *Wisconsin Conservation Bulletin* 16 (10) (1951): 7–9.

(———) "Mobility of the sharp-tailed grouse in relation to its ecology and distribution." *American Midland Naturalist* 46 (1) (1951): 174–226.

"Of wilderness and the sharp-tailed grouse." *Passenger Pigeon* 14 (3) (1952): 97–99.

(with F. N. Hamerstrom) "What's a wilderness for?" *Wisconsin Conservation Bulletin* 17 (3) (1952): 19–20.

(——— and O. E. Mattson) *Sharptails into the Shadows*. Madison: Wisconsin Conservation Department, 1952.

"The ecology of resort country." *The Wilderness News* 1 (3) (1953): 5.

(with F. N. Hamerstrom) "Myiasis of the ears of hawks." *Falconry News and Notes* 1 (3) (1954): 4–8 .

(———) "Population density and behavior in Wisconsin prairie chickens (*Tympanuchus cupido pinnatus*)." *Proceedings of the International Ornithological Congress* 11 (1955): 459–466.

(with S. C. Kendeigh and T. C. Kramer) "Variations in egg characteristics of the house wren." *Auk* 73 (1) (1956): 42–65.

"The influence of a hawk's appetite on mobbing." *Condor* 59 (3) (1957): 192–194.

"Transportation of a shrike pair." *Auk* 74 (1957): 502.

(with F. N. Hamerstrom and O. E. Mattson) *A guide to prairie chicken management.*
  Technical Wildlife Bulletin #15. Madison: Wisconsin Conservation Department,
  1957.
"Mending Jupiter's leg." *Falconer* 3 (6) (1960): 191–193.
(with F. N. Hamerstrom) "Comparability of some social displays of grouse." *Proceed-
  ings of the International Ornithological Congress* 12 (1960): 274–293.
(———) "Operation snowy owl." *Passenger Pigeon* 22 (3) (1960): 126–128.
(———) "Buena Vista Marsh." In *Wisconsin's Favorite Bird Haunts*, edited by S. D.
  Robbins, Jr., 44–45. Madison: Wisconsin Society of Ornithology, 1961.
(———) "Status and problems of North American grouse." *Wilson Bulletin* 73 (3)
  (1961): 284–294.
(——— and D. D. Berger) "Nesting of short-eared owls in Wisconsin." *Passenger
  Pigeon* 23 (2) (1961): 46–48.
"Winter visitors from the far North." *Audubon Magazine* 64 (1) (1962): 12–15.
(with D. D. Berger) "Protective goshawk trapping." *Falconer* 4 (2) (1962): 55–57.
"The use of great horned owls in catching marsh hawks." *Proceedings of the Interna-
  tional Ornithological Congress* 13 (1963): 866–869.
(with F. N. Hamerstrom) "Grouse." In *A New Dictionary of Birds*, edited by A. L.
  Thomson, 343–345. London: Thomas Nelson and Sons, 1963.
(———) "Range of the red-bellied woodpecker in Wisconsin: 1963." *Passenger Pigeon*
  25 (4) (1963): 131–136.
(———) "The symposium in review." *Journal of Wildlife Management* 27 (4) (1963):
  868–887.
(———) "Wisconsin owl survey." *Passenger Pigeon* 25 (2) (1963): 50, 52–55.
"Legislation versus N.A.F.A. control." *Hawk Chalk* 4 (2) (1965): 11–13.
"Status of falconry in Anglo-America in 1964." *Hawk Chalk* 4 (1) (1965): 12–14.
"A white-tailed kite in Wisconsin." *Passenger Pigeon* 27 (1) (1965): 3–8.
(with D. D. Berger and F. N. Hamerstrom) "The effect of mammals on prairie chickens
  on booming grounds." *Journal of Wildlife Management* 29 (3) (1965): 536–542.
(with H.H.D. Heiberg, Jr., D. Hunter, et al.) "Report of the Legislative Committee."
  Hawk Chalk 4 (1) (1965): 11–12.
"Hawking in the Highlands." *Journal of North American Falconers' Association* 5
  (1966): 17–20.
"The legislative outlook." *Hawk Chalk* 5 (2) (1966): 12–13.
(with T. Ennenga, H.H.D. Heiberg, Jr., et al.) "Report of the Legislative Committee."
  Hawk Chalk 5 (1) (1966): 16–19.
(with F. N. Hamerstrom) "Stove in the popples." *Wisconsin Conservation Bulletin* 31
  (5) (1966): 3–5.
"Brief report on a harrier population study in central Wisconsin." *Raptor Research News*
  1 (4) (1967): 60.
"Raptor selling." *Hawk Chalk* 6 (1) (1967): 22.
"Report from the Legislative Committee." *Hawk Chalk* 6 (2) (1967): 13–15.
"On the use of fault bars in aging birds of prey." *Inland Bird Banding Association News*
  39 (2) (1967): 35–41.
(with D. W. Anderson) "The recent status of Wisconsin cormorants." *Passenger Pigeon*
  29 (1) (1967): 3–15.
(with R. K. Anderson and F. N. Hamerstrom) "Hen decoys aid in trapping cock prairie

chickens with bownets and noose carpets." *Journal of Wildlife Management* 31
   (4) (1967): 829–832.
(with F. N. Hamerstrom) "Trapping and banding." *Wisconsin Conservation Bulletin* 32
   (1) (1967): 14–15.
"Beautification." *Conservation and Natural Resources Association (CNRA) Report* 14
   (1968): 1–3.
(with J. H. Enderson) "Comments on sources of peregrines and prairie falcons for falconry." *Hawk Chalk* 7 (2) (1968): 29–31.
(with J. D. Weaver) "Ageing and sexing rough-[legged] hawks in Wisconsin and Illinois." *Ontario Bird Banding* 4 (4) (1968): 133–138.
"Bumblefoot." *Raptor Research News* 3 (2) (1969): 24.
"The dance of the cranes." *Purple Martin Capital News* 4 (7) (1969): 12–13.
"A golden eagle and the rearing of redtails." *Natural History* 78 (5) (1969): 62–69.
"A hawk in Louisiana." *The Falconer* 5 (3) (1969): 152–153.
"I knew the heath hen . . . and other treasures of the Great Plain, too." *Purple Martin
   Capital News* 4 (8) (1969): 12–13.
"A peripatetic robin nest." *Passenger Pigeon* 31 (4) (1969): 159–160.
"Singing the dogs." *American Field* (Nov. 1, 1969): 483.
"Stripping roadsides . . . creates more problems than it solves." *Purple Martin Capital
   News* 4 (9) (1969): 18–19.
(with F. N. Hamerstrom) "Es war einmal ein Balzplatz." *Wild und Hund* 72 (11) (1969):
   266–267.
"Ambrose." In *Alive in the Wild*, edited by V. H. Cahalane, 9–13. Englewood Cliffs,
   NJ: Prentice-Hall, 1970.
"Dandelion greens." *Wisconsin Conservation Bulletin* 35 (5) (1970): 28–29.
"Dead on my kitchen floor." *Purple Martin Capital News* 5 (11) (1970): 14.
An Eagle to the Sky. Ames: Iowa State University Press, 1970. Republished, Trumans-
   burg, NY: Crossing Press, 1978.
"First aid for birds." *Purple Martin Capital News* 5 (8) (1970): 10.
"Some hints to reduce nest predation." In *Memorandum to All Banders (MTAB)-l4*, 10–
   11. Laurel, MD: U.S. Fish and Wildlife Service, Patuxent Wildlife Refuge, 1970.
   Reprinted by W. S. Clark in *Raptor Banding Information Exchange* 1 (1974): 2–3.
"Ein Steinadler und die Aufzucht von Rotschwanzbussarden." *Das Tier* (1970).
"Think with a good nose near a nest." *Raptor Research News* 4 (3) (1970): 79–80.
"Three poisons." *Passenger Pigeon* 32 (1) (1970): 38.
"Time to cook deer meat." *Wisconsin Conservation Bulletin* 35 (6) (1970): 26.
(with F. N. Hamerstrom) "Not on the far side of the earth." *Wisconsin Academy Review*
   16 (4) (1970): 33–34.
(———) "Scar sac." In *Alive in the Wild*, edited by V. H. Cahalane, 3–7. Englewood
   Cliffs, NJ: Prentice-Hall, 1970.
Birds of Prey of Wisconsin. Madison, WI: Department of Natural Resources, 1972.
"Eat mushrooms of the autumn." *Purple Martin Capital News* 7 (10) (1972): 8.
"In the land of the Seri Indians." *Purple Martin Capital News* 7 (8) (1972): 10–11.
"An osprey banding trip." *Purple Martin Capital News* 7 (9) (1972): 16–17.
"Out of the mews." *The Falconer* 6 (1) (1972): 63.
"Triggers for egg-laying." *Raptor Research* 6 (Suppl. C) (1972): C41–42.
(with F. N. Hamerstrom) "A male hawk's potential in nest building, incubation and
   rearing young." *Raptor Research* 6 (4) (1972): 144–149.

"Some plants for the table." *Purple Martin Capital News* 8 (5) (1973): 5.
"Time to cook grouse." *Wisconsin Sportsman* 2 (5) (1973): 29.
"Why not give them birds?" *Purple Martin Capital News* 8 (5) (1973).
(with F. N. Hamerstrom) *The Prairie Chicken in Wisconsin: Highlights of a 22-Year Study of Counts, Behavior, Movements, Turnover and Habitat.* Madison, WI: Department of Natural Resources Technical Bulletin 64, 1973.
(——— and J. Hart) "Nest boxes: An effective management tool for kestrels." *Journal of Wildlife Management* 37 (3) (1973): 400–403.
(with K. Janick) "Diurnal sleep rhythm of a young barred owl." *Auk* 90 (4) (1973): 899–900.
"Aldo Leopold's recipe for woodcock." *Wisconsin Conservation Bulletin* 39 (5) (1974): 13.
"Hacking to the glove." *Journal of the North American Falconers' Association* 12–13 (1974): 112.
"Scrapple." *Wisconsin Conservation Bulletin* 39 (6) (1974): 11.
Walk When the Moon Is Full. Trumansburg, NY: Crossing Press, 1974. Translation. Tokyo: Kaisee sha, 1978.
"Weeds for eating." *Wisconsin Conservation Bulletin* 39 (2) (1974): 6–7.
"Dead trees are part of our world." *Wisconsin Conservation Bulletin* 40 (4) (1975): 16–17.
"The eggs of fishes." *Wisconsin Conservation Bulletin* 40 (6) (1975): 14–15.
"How to stuff a partridge." *Wisconsin Conservation Bulletin* 40 (5) (1975): 15.
"Porcupine." *Wisconsin Conservation Bulletin* 40 (1) (1975): 13.
"The use of color-code marking in hawk studies." *Proceedings of the American Hawk Migration Conference 1974* (1975): 103–106.
(with T. Ray, C. M. White, et al.) "Conservation Committee report on status of eagles." *Wilson Bulletin* 87 (1) (1975): 140–143.
(with W. Scharf) "A morphological comparison of two harrier populations." *Raptor Research* 9 (1/2) (1975): 27–32.
(with F. N. Hamerstrom) "Buena Vista Marsh." In *Wisconsin's Favorite Bird Haunts*, edited by D. D. Tessen, 140–142. Madison: Wisconsin Society for Ornithology, 1976.
(———) "Hand-rearing hawks and owls—some comparisons." *The Living Bird* 14 (1976): 257–261.
Adventures of the Stone Man. Trumansburg, NY: Crossing Press, 1977.
"Introducing captive-reared raptors into the wild." *World Conference on Birds of Prey, 1975.* Vienna. Report of Proceedings, edited by R. D. Chancellor, 348–353. London: International Council for Bird Preservation, 1977.
"This raptor keeps a low profile." *National Wildlife* 15 (6) (1977): 20–25.
"The tale of grey." In *Ways of Wildlife*, edited by Eleanor Horwitz, 17, 19–22. New York: Citation Press, 1977.
(with F. N. Hamerstrom) "On the art of Harris' hawk trapping." *Inland Bird Banding News* 49 (1977): 4–8.
(———) "External sex characters of Harris' hawks in winter." *Raptor Research* 12 (1/2) (1978): 1–14.
"But what can I do?" *Defenders* 54 (5) (1979): 268–269.
"On choosing the bird to fly." *The Falconer* 7 (3) (1979): 147–148.
Strictly for the Chickens. Ames: Iowa State University Press, 1980.

(with K. L. Bildstein) "Age and sex differences in the size of northern harriers." *Journal of Field Ornithology* 51 (4) (1980): 356–360.

(with M. Kopeny) "Harrier nest-site vegetation." *Raptor Research* 15 (3) (1981): 86–88.

"Harrier." In *Handbook of Census Methods*, edited by D. E. Davis. Boca Raton, FL: CRC Press, 1982.

Birding with a Purpose: Of Raptors, Gabboons, and Other Creatures. Ames: Iowa State University Press, 1984.

Eagles, Hawks, Falcons and Owls of America: A Coloring Album. New York: Robert Rinehart, 1984.

"Only a whistle away." *Wildfowl Carving and Collecting* 1 (Fall 1985): 8–26.

"Ruth Louise Hine: Team player extraordinary." *Wisconsin Natural Resources* 9 (6) (1985): 4–6.

(with F. N. Hamerstrom and C. J. Burke) "Effect of voles on mating systems in a central Wisconsin population of harriers." *Wilson Bulletin* 97 (3) (1985): 332–346.

Harrier, Hawk of the Marshes: The Hawk That Is Ruled by a Mouse. Washington, DC: Smithsonian Institution Press, 1986.

(with F. N. Hamerstrom) "The dance of the prairie chicken." *Wildfowl Carving and Collecting* 2 (Summer 1986).

"The evolution of an artist." *Stevens Point Wisconsin Journal* (May 27, 1987).

"Flop of the month." *Wisconsin Woman* (Aug. 1987).

"Leopold's influence." *Wisconsin Academy Review* 34 (1) (Dec. 1987): 27–29.

Is She Coming Too? Memoirs of a Lady Hunter. Ames: Iowa State University Press, 1988.

"Touchstone." In *Aldo Leopold, Mentor*, edited by E. McCabe, 28–31. Madison: Department of Wildlife Ecology, University of Wisconsin-Madison, 1988.

Wild Food Cookbook. Ames: Iowa State University Press, 1988.

(with K. Wuertz-Schaeffer, trans.) *Bird Trapping and Bird Banding*, by H. Bub. Ithaca, NY: Cornell University Press, 1991.

My Double Life—Memoirs of a Naturalist. Madison: University of Wisconsin Press, 1994.

Wild Food Cookbook from the Fields and Forests of the Lake States Region. Amherst, WI: Palmer, 1994.

Other Works

"Poems of a New England town." *The Parchment* 6 (1) (1934): 29–30.

"Lullaby 1969." *Beloit Poetry Journal* 20 (3) (1970): 5.

"The need for revision?" In *The Rites of Writing*, edited by D. J. Dieterich, 53–55. Stevens Point: Writing Lab, University of Wisconsin-Stevens Point, 1982.

Personal communications to the author, 1993, 1994.

Works about Frances Carnes Flint Hamerstrom

Hayes, P.J. "Chickens and city slickers." *Milwaukee Journal* Magazine Sec. (July 15, 1990): 19–25.

———. "Saving a feathery lover." *Milwaukee Journal* Magazine Sec. (July 8, 1990): 5–11, 22.

Janz, W. "Biologist dies in spot befitting him." *Milwaukee Sentinel*, pt. 1 (Mar. 29, 1990): 5, 10.

Kincaid, D. "Biologists' mutual career is love story." *Milwaukee Sentinel*, pt. 4 (Dec. 25, 1980): 1, 6.

Mayr, E. "Preface." *Journal of Raptor Research* 26 (1992): 106–107.

Paulson, Elba Hamerstrom. Personal communication, 1994.

Schmutz, J.K. Personal communication, 1994.

Other References

Gray, Asa. *Manual of Botany*. 8th ed. Portland, OR: Dioscorides Press, 1987.

ALICE HAMILTON (1869–1970)

Robin M. Haller

BIOGRAPHY

Alice Hamilton was born February 27, 1869, to Montgomery and Gertrude (Pond) Hamilton during a visit to Gertrude's mother in New York City. At six weeks she went home to Fort Wayne, Indiana, and grew up on the family compound built by her paternal grandfather, Allen Hamilton. He had settled in eastern Indiana in 1823 when it was still a frontier and helped to develop the city of Fort Wayne. His wife, Emerine J. (Holman) Hamilton, was from a politically prominent Indiana family and was known to her granddaughter as "an ardent advocate of the temperance movement [and] of woman's suffrage" (*Exploring the Dangerous Trades* 1943, 24). Friends and houseguests of Emerine Hamilton included Susan B. Anthony and Frances Willard.

Alice Hamilton's father joined the Union army in 1862. Ill health forced him to resign, however, and he traveled to Europe, where he met Gertrude Pond, daughter of Loyal Sylvester Pond, a sugar importer from Vermont, and Harriet Sarah (Taylor) Pond. They married in 1866 and settled in Fort Wayne, where Montgomery joined the family's wholesale grocery business. Four daughters arrived in short order: Edith in 1867; Alice in 1869; Margaret in 1871; and Norah in 1873. The only son, Arthur, known throughout his life as Quint, followed in 1886. Within the family compound were eleven cousins, eight of whom were very close in age. Hamilton and her cousins Agnes and Allen, known as the "three As," were close friends and maintained a lifelong correspondence. The Hamiltons were socially prominent in Fort Wayne, especially within the Presbyterian Church.

The Hamiltons were deeply interested in their children's education and, disliking the public schools, tutored them at home. Hamilton later commented that home teaching left gaps in her education because her parents emphasized their own interests: language, literature, poetry, and history, with an emphasis on the Bible. Although her early schooling contained little science, Hamilton was trained in the scientific approach by her father.

She was very close to her mother, describing her as an early and lasting influence. Gertrude Hamilton, an unconventional woman, had spent much of her life in Europe and was free of Victorian prudery. She spoke openly on topics considered unfit for gentlewomen to discuss, like pregnancy and childbirth, and on public issues such as police brutality, lynching, and child labor. Gertrude Hamilton also impressed upon her daughters the right of women to individual ambition, even above the claims of family.

In 1886 following several aunts and cousins and her older sister, Edith, Hamilton left Indiana for Miss Porter's School in Farmington, Connecticut. Her younger sisters and cousin Agnes soon followed. At Miss Porter's, she and Edith both decided that they wanted "a larger life" than marriage and children alone could bring, one that could come only through a profession (Sergeant 1926, 766). Edith Hamilton chose teaching and later became a noted classical scholar and writer and founder of a girls' school in Baltimore. Alice Hamilton chose medicine because, she said, "as a doctor I could go anywhere . . . to far-off lands or to city slums—and be quite sure that I could be of use anywhere. I should meet all sorts and conditions of men, I should not be tied down to a school or a college as a teacher is, or have to work under a superior, as a nurse must do" (*Exploring the Dangerous Trades* 1943, 38).

Hamilton entered the Fort Wayne College of Medicine in 1890 in a class including five other women and studied anatomy for a year. She then transferred to the University of Michigan in Ann Arbor for a regular course of study. In 1891 Michigan had one of the few medical schools in the country with a graded curriculum emphasizing laboratory and clinical instruction. Here Hamilton, one of 14 women in a class of 47, attended classes in physiology, biochemistry, bacteriology, and pharmacology, following the German approach of laboratory rather than didactic teaching. Hamilton graduated from Michigan in 1893 with a medical degree.

While in medical school Hamilton decided that she would never marry, realizing that the opposing claims of family and career would suffer if she tried to do both. Although young women in her generation had few models on which to base decisions like this, Hamilton may have been influenced by her mother. Of the four Hamilton sisters, none married, and all chose careers. Margaret became a science teacher in Edith's academy, and Norah an artist. Arthur, the only one of Hamilton's siblings to marry, became a professor of foreign languages at the University of Illinois in Urbana.

After medical school, Hamilton decided to do research in bacteriology and pathology rather than practice, perhaps after watching a patient die in childbirth. After one of her teachers at Michigan suggested that she study in a hospital, Hamilton spent two months at the Hospital for Women and Children in Minneapolis and then nine months near Boston at the New England Hospital for Women and Children. She wanted more work in bacteriology, however, and with her sister Edith traveled to Leipzig and Munich, Germany. While Edith studied the classics, Alice Hamilton did laboratory work and attended lectures.

She then returned to the University of Michigan in February 1895 as a resident graduate and assistant to Dr. F. G. Novy, her former professor of bacteriology.

Hamilton then decided to apply to the Ph.D. program at Johns Hopkins University. Although Johns Hopkins did not award doctorates to women (with one exception in 1893), she was allowed to work in pathological anatomy with Simon Flexner, who was later director of laboratories at the Rockefeller Institute. Hamilton produced her first scientific papers while at Johns Hopkins, on tubercular ulcers and on neuroglioma of the brain. Her work at Johns Hopkins brought her into contact with some of the leading medical men of the day: Flexner, who taught her how to do research; William H. Welch, professor of pathology; and William Osler, whose clinics she attended.

In 1897 Hamilton accepted a teaching position in pathology at Northwestern University's separate medical school for women. When she arrived in Chicago, Hamilton moved into Hull House, the famous settlement house on Chicago's South Side established by Jane Addams. Hamilton had heard Addams speak on the settlement house movement and decided she would like to be part of it. Thus began a 38-year-long association that influenced Hamilton's work for the rest of her life. Her association with Hull House led to her interest in industrial and urban diseases. Hamilton lived at Hull House for 22 years, and even after going to teach at Harvard in 1919, she spent part of the year there until 1935, when Jane Addams died.

At Hull House, Hamilton met Florence Kelley, a social worker who investigated industrial conditions in the neighborhood. Hamilton adopted Kelley's study habits for her own and used them later in her own investigations of the poisonous trades, along with the persuasive advocacy techniques of Julia Lathrop, another Hull House worker. The political and social influence of these women shaped Hamilton's approach to the conditions she found in the factories and plants she later investigated.

Hamilton's work at Hull House led Illinois Governor Charles S. Deneen to appoint her to a commission to study occupational diseases in Illinois. Her role was to survey industries that used lead. Her work on the Illinois commission brought her to the attention of Charles O'Neill, Commissioner of the Bureau of Labor in the Department of Commerce, who asked her to expand her survey of the lead industry to the entire country. Hamilton then began an eight-year career as special investigator of dangerous trades for the Department of Labor (established in 1912 as a cabinet-level department, separate from Commerce). She studied first the lead industry, then other poisonous trades, reporting her work in Bureau of Labor Statistics bulletins, as well as in technical and popular journals.

In 1910 her alma mater, the University of Michigan, awarded her an honorary master's degree. In 1911–12, Hamilton was president of the Chicago Pathological Society, and in 1914 she became vice president of the hygiene section of the American Public Health Association.

In 1915 Hamilton accompanied Jane Addams and 47 other members of the

Women's Peace Party to the International Congress of Women at The Hague, in which women from 12 belligerent and neutral nations met in an attempt to influence the course of the war. Although Hamilton was not an official delegate, she wrote of the conditions in the various war capitals of Europe. The Congress wanted neutral nations to offer continuous mediation to end the war. Addams presented their resolutions to President Woodrow Wilson and tried to persuade him to lead a conference of neutral nations. Wilson refused, but several of the conference's suggestions appeared later in his Fourteen Points. Addams received the Nobel Peace Prize for her efforts. Hamilton's experiences in Europe in 1915 crystallized her convictions about war and militarism.

Pacifism did not, however, interfere with her work. Hamilton never openly took an antiwar stand, because she felt it might jeopardize her investigations of conditions in industry. In 1917 Hamilton began a survey for the Department of Labor of munitions factories and other war industries. Several protests were lodged with the Labor Department because an avowed pacifist was entering munitions plants, but Hamilton's supervisor at Labor, Dr. Royal Meeker, Commissioner of Labor Statistics, supported her.

In 1919, after an unsuccessful effort to find a competent man, the Harvard University Medical School appointed Hamilton assistant professor of industrial medicine, the first woman professor in a school that did not even accept women students. This caused much stir among the faculty and administration, but she was clearly the best qualified candidate for the position. Nevertheless, her appointment had three conditions: that she never enter the Harvard Club, that she not ask for football tickets, and that she not walk in the faculty procession or sit on the platform at commencement. Hamilton made a condition of her own: she would teach at Harvard for one semester each year and spend the rest of the year at Hull House. After finishing two reports for the Labor Department, she severed her ties with the federal government and moved to Cambridge. She remained at Harvard until mandatory retirement age, in 1935, and retired, still an assistant professor.

At Harvard she continued her work in the poisonous trades, especially on carbon monoxide and mercury. She also wrote two textbooks: *Industrial Poisons in the United States* (1925) and *Industrial Toxicology* (1934). Her role now shifted from investigator to consultant and adviser, but she continued her field studies. She also lectured at Bryn Mawr's Graduate School of Social Economics and Social Research and at other universities on industrial poisons. In 1924 she traveled to Europe and Russia to study industrial medicine. She also traveled to Germany again and remarked on the ravages of war and the hostility of the vanquished Germans.

In the 1920s and 1930s Hamilton continued to write in scientific and popular journals on topics that concerned her: the effects of war, especially on children; the Sacco-Vanzetti case; the equal rights amendment; and birth control, a subject that had interested her since her early days at Hull House. In 1925 she served on a council of the American Birth Control League. Because of her writings,

Hamilton was considered a radical, even for Harvard, but she had the support of Dr. David Edsall, Dean of the Medical School, and of Harvard's tradition of academic freedom. In the 1920s she served on the Health Committee of the League of Nations, studying malaria in postwar Europe and diseases in the former German colonies in Africa, especially tuberculosis and sleeping sickness. She spoke on the need for an international medical organization to deal with the effects of war on civilian populations and public health problems, like contagious diseases. She also wrote articles about infringement of civil liberties, objecting to the government's treatment of aliens and dissenters. In 1930 Hamilton became a member of President Herbert Hoover's Research Committee on Social Trends, with specific responsibility for surveying the special needs of women in industry.

In the early 1900s, Hamilton had begun spending summers in rural Connecticut. In 1916, after the family compound in Fort Wayne was sold, she bought a house in Hadlyme, Connecticut, for summer use at first and also as a place where the Hamilton cousins and their offspring could gather. After retiring from Harvard, Hamilton spent the rest of her life in Hadlyme with her sisters Margaret and Norah and good friend Clara Landsberg. Frequent visitors included Edith, Agnes, Allen, and the younger generation of Hamiltons.

Hamilton turned down the opportunity to supervise Hull House after Jane Addams's death, although she did agree to become a trustee. In 1935 she returned to the Department of Labor, now headed by Frances Perkins, whom she admired. She served as medical consultant to the Division of Labor Standards, conducting surveys, offering advice, attending conferences, and testifying at hearings. She also continued her work in the dangerous trades, concentrating on volatile solvents, especially in the viscose rayon industry.

In 1938 she made another trip to Germany and saw the effects of Nazism on the country. Her writings reflected her fear of Hitler's growing power. After this trip, in contrast to her previous pacifist stance, she decided that America's isolation in the pre–World War II years had been a mistake. As Hitler continued to march against the countries of Western Europe, Hamilton organized efforts to send food and medical supplies to those nations. During World War II, she worked for the American Red Cross and wrote her autobiography, *Exploring the Dangerous Trades* (1943), which was illustrated by her sister Norah.

Hamilton, even in retirement, still gave talks and wrote papers. She lectured yearly at Bryn Mawr College until 1943 as well as at the Woman's Medical College of Pennsylvania and at Tufts University. In 1944 she became president of the National Consumer's League. After Norah's death in 1945, she remained with Margaret and Clara in Hadlyme, and various nieces and nephews moved in and out. She maintained long friendships through correspondence, including one with Justice Felix Frankfurter. After the war she continued as a consultant to the Department of Labor on standards of hygiene in industry. Although at the age of 78 she gave up the presidency of the National Consumer's League (1947), feeling they needed someone younger, she continued to write on issues that concerned her. In 1949 a revised edition of her 1934 textbook appeared, with Harriet L. Hardy as coauthor.

Hamilton enjoyed good health into her 80s, although she complained about having to use a hearing aid. The 1940s and 1950s brought recognition for her many contributions to the health of industrial workers: the Lasker Award (1947), an honorary doctorate from the University of Michigan (1948), the Knudsen Award of the Industrial Medical Association (1953), and the Elizabeth Blackwell Citation of the New York Infirmary (1954). In 1956 she was named New England Medical Woman of the Year. In honor of her 90th birthday, Harriet Hardy and other friends established the Alice Hamilton Fund for Occupational Medicine at the Harvard School of Public Health to sponsor special lectures.

The 1960s brought ill health and the loss of close friends and family. In 1961 Hamilton suffered several small strokes. Her sister Edith died in 1963 at the age of 95. Hamilton continued writing, though, and in 1963, she protested the war in Vietnam in a letter to President John F. Kennedy and complained to Senator Abraham Ribicoff about the use of poisonous gases there. By 1966, however, she was almost totally bedridden, attended by her sister Margaret. Both her brother Arthur and her good friend Clara Landsberg had died, leaving the two sisters the last of their generation. The centenary of her birth in February 1969 brought a notice in the *New York Times* and a telegram from President Richard M. Nixon. Margaret Hamilton died in July 1969. Alice Hamilton survived her sister by 15 months and died on September 22, 1970, at the age of 101, in Hadlyme. Three months later, President Nixon signed the Occupational Safety and Health Act, 60 years after Alice Hamilton had begun her lifelong crusade.

WORK

Alice Hamilton's work is impossible to separate from her life, because each grew out of the other, from the moment she walked into Hull House in 1897. Her experiences there formed her values, her standards, and the goals she set for herself thereafter. During her first years at Hull House, while she taught at Northwestern University, and later when she became a bacteriologist at Memorial Institute for Infectious Diseases in 1902, Hamilton researched typhoid, scarlet fever, and tuberculosis, diseases associated with workers or those who lived in cramped urban environments. Although she considered herself a medical researcher, not a practitioner, she opened a well-baby clinic at Hull House and soon found that the older brothers and sisters needed her services as well.

Her first research topic concerned a typhoid fever epidemic in 1903 in the area around Hull House. Feeling that a local condition was the cause, Hamilton collected flies in the area and found typhoid bacillus on them. She presented her findings to city authorities, to the Chicago Medical Society, and in the *Journal of the American Medical Association.* Local conditions led to research in other areas. The children in the Hull House clinic presented subjects for her studies on scarlet fever, gonorrheal vulvovaginitis, and pseudodiphtheria bacilli.

She joined the effort in Chicago to eliminate tuberculosis and was one of the first to link it to occupation, citing as contributory factors unsanitary conditions and excessive fatigue. Other public health efforts by Hamilton at Hull House linked excessive childbearing to infant mortality. This led to her advocacy of birth control at a time when dispensing information on contraception was illegal. She also sponsored training standards for midwives, hoping to regulate their activities.

About this time, in 1907, Hamilton read Sir Thomas Oliver's book *Dangerous Trades* (1902) on the efforts in Europe to control conditions leading to industrial diseases. She could find no similar work on conditions in the United States and thought it unlikely that American factory owners would voluntarily take precautions that had been forced upon their foreign competitors. Reading this book changed the direction of Hamilton's interests from bacteriology to industrial diseases and from the laboratory to the field. Shortly thereafter she was appointed by Governor Deneen to survey the lead industry in Illinois. Since the United States, unlike the countries of Western Europe, had no workers' compensation laws or insurance, there were no medical records to draw upon in researching industrial diseases, and Hamilton had to find other sources of information.

Following advice from Hull House social workers, she developed a "shoe leather" technique of researching hospital records, interviewing doctors and local apothecaries, and meeting in the homes of workers with their wives and families. Rather than accept the assertions of factory managers, she visited workplaces herself to see conditions firsthand. The sight of a diminutive woman of gentle breeding climbing ladders, inspecting smelters, and walking along catwalks to peer over the edges of bubbling vats must have seemed strange to the workers who saw her (Sergeant 1926, 763). Hamilton soon discovered that in many industries the use of lead or production of lead dust was not immediately obvious (for instance, in the making of freight-car seals, coffin trim, or cigar wrappings, all of which contained lead). From the first, she was convinced that lead poisoning resulted from inhaled dust, not dirty hands, and no matter how many times workers washed their hands, they would suffer plumbism if they continually breathed lead-laden air. Accumulated lead dust in the lungs caused severe poisoning; symptoms included wrist-drop (a form of palsy), pallor, weight loss, digestive upset, and premature senility. Hamilton, remarking that while workers ate three times a day, they breathed 14 to 16 times a minute, recommended controlling dust by ventilating systems and issuing respirators and protective clothing to workers (*Industrial Toxicology* 1934, xii).

Hamilton's reports on the dangers of inhaling lead dust led to the passage in Illinois of an occupational disease law in 1911, providing compensation for industrial diseases caused by poisonous fumes. Slowly, other states passed similar laws. Changes in the factories soon followed, as it became too expensive for employers to ignore dangerous conditions. Reforms included using machines rather than men to open and close furnaces; reducing the amount of dust; using less lead in enameling processes; reducing the amount of sandblasting; sealing the sandblasting chambers or making them dustproof; providing respirators; and giving breaks in the eight-hour day to provide respite from lead-laden dust.

She broadened the study from Illinois to the entire country when she began working for the Department of Labor. In July 1911 her report on the white lead industry documented 358 cases of poisoning in less than a year and a half. Her shoe-leather investigations of the lead industries took her from pottery, tile, and enamelware, to the painters' trade, smelting and refining, and the making of storage batteries.

In Hamilton's survey of smelting, she discovered that the fumes were as dangerous as lead dust, especially for those workers who cleaned out the flue systems in the smelters. Smelting was done on an open hearth, and worker protection was scant. Wetting down the area was the only way to control the dust and fumes. Hamilton produced statistics on the rate of tuberculosis in lead smelter areas, linking it and other respiratory ailments to lead dust and fumes. She continued her studies of lead poisoning by investigating the painting trades. Although painters themselves recognized that certain procedures were dangerous, such as mixing dry white lead with oil, sandpapering, or chipping old paint (all dusty jobs), no studies had ever proved their fears. Hamilton showed that lead poisoning threatened those workers who had to rub down one coat of paint before the next could be applied.

During World War I, Hamilton was asked by the Bureau of Labor Statistics (BLS) to study poisons in munitions industries. Other than the risk of explosion, especially when nitroglycerine or mercury fulminate was involved, production of munitions was not considered dangerous, as the toxic effects of trinitrotoluol, picric acid, and nitrous fumes were not yet known. In 1917 Hamilton published a report listing nitrous fumes as the leading cause of death by poisoning in war industries, with trinitrotoluene (TNT) poisoning second. Again, Hamilton followed her shoe-leather techniques. Secrecy surrounding munitions production meant she had to discover for herself the locations of plants, visiting those she heard of and picking up gossip about others in saloons, union halls, or other places where workers gathered. Often great clouds of yellow and orange fumes marked the location of picric acid factories. When Hamilton found men stained with yellow picric acid, she literally followed them home and interviewed them.

Hamilton studied the production of explosives: shells, mines, picric acid, dinitrobenzol, trinitrotoluol, smokeless powder, guncotton, and fulminate of mercury, which was used as the detonator to start an explosion. These nitrated compounds, made by the action of nitric acid on other substances, were highly toxic. Breathing of fumes or absorption through the skin caused damage to the central nervous system and to organs. Nitric acid, a caustic, emitted fumes when it encountered organic material. These fumes were powerful enough to blow lids from containers and cause explosions. Hamilton witnessed several incidents when fumes sent workers fleeing from buildings.

TNT poisoning occurred after absorption through the skin. While British man-

ufacturers had long recognized this and provided bathing and housekeeping facilities to ensure cleanliness, Hamilton found that American manufacturers did
not. TNT victims suffered skin eruptions and liver damage, and death resulted
from toxic jaundice or aplastic anemia. After the United States entered the war
in 1917, Hamilton was asked by the National Research Council of the National
Academy of Sciences for ways to prevent TNT poisoning. She recommended
frequent medical testing of employees and housekeeping methods to reduce or
eliminate dust and contamination through the skin.

Hamilton felt that World War I had one meritorious effect: industrial hygiene
became respectable. The Public Health Service began to show interest in industrial hygiene, and medical journals published articles on the dangers of industrial
poisons and processes. The medical profession also began to realize that company doctors should be advocates of the workers they treated rather than defenders of management policy, something Hamilton had preached for years.

While teaching in the industrial hygiene department at Harvard, which moved
to the university's new School of Public Health in 1922, Hamilton continued
investigating the dangerous trades with a study of the effects of carbon monoxide poisoning on workers. Not strictly a poison, but an asphyxiant, carbon
monoxide affected more workers in industry than any other toxic substance
except lead. Carbon monoxide acted as a strangulant, displacing oxygen in the
red blood cells and leading to oxygen starvation. Hamilton found evidences of
gassing in blast-furnace steel mills and in gas works, as well as in the printing
industry. Carbon monoxide was especially prevalent in coal mines, where it was
a by-product of dynamite blasting.

Hamilton's study of mercury as an industrial poison began in 1923. Prevailing
methods of dealing with the toxic effects of mercury were to shorten the work-
day in order to limit exposure. Physical symptoms of mercury poisoning, found
in miners, hatters, and makers of thermometers and dry batteries, were saliva-
tion, swelling and pain in the gums and lips, jerking of limbs, tremors, and a
condition called "erythrism" because it caused the victim to blush. Mental
symptoms of chronic mercury poisoning were long known to makers of felt
hats, giving rise to the expression "mad as a hatter." Mercury fumes were
slowly emitted during the tanning process of felt-hat manufacture. Quicksilver
miners were exposed to mercury as the dust got on their hands, and those who
smoked faced added danger as heat from a cigarette could volatilize the mercury.
Laboratory technicians who worked with quicksilver were exposed to its dangerous fumes if the quicksilver spilled. Often the only treatment for mercury
poisoning was to leave the work site and to literally "sweat out" the condition.
Hamilton recommended ventilating procedures and reduced exposure to fumes.

During her tenure at Harvard, Hamilton assumed a new role as industry consultant. At the request of the General Electric Company, she surveyed health
conditions at their plants from 1923 to 1933. This allowed her to focus on less
obvious conditions of employment, such as fatigue, improper lighting or posi-
tioning, noise, vibration, and the effect of long hours without break. She con-

tinued to prod both employers and governmental agencies to reform working conditions by reducing dust and fumes and instituting safety features, such as respirators, frequent medical examination of workers, and substituting safer chemicals for the more toxic ones (Sicherman 1984, 240–242).

In 1924 she began a study of petroleum and coal-tar solvents in industrial use, particularly benzol. These solvents, when inhaled, affected the fatty substances in the brain and acted as narcotics. Benzol also attacked the bone marrow, causing anemia and lowering the body's resistance to disease. Benzol poisoning was often not detected until it became irreversible. Symptoms included bleeding from the gums or the nose, a visible sign of benzol's effect on the clotting ability of the blood. Hamilton stressed the need to guard against the chronic reaction (as opposed to the acute), arguing that slow exposure over many months or years to a poison could be far more dangerous than an acute reaction to short exposure.

In 1925 Hamilton published her first book, *Industrial Poisons in the United States*, intended as a reference book for industrial physicians. It stressed the physician's role in preventing industrial poisoning. Her second book, *Industrial Toxicology*, in 1934, also outlined the duties of the company physician and listed Hamilton's remedies for the prevention of industrial poisoning, with strong emphasis on the regulation of fumes and dust.

Hamilton's last major study of the dangerous trades (1937–38) was on carbon disulphide poisoning in the viscose rayon industry. Carbon disulphide affected the central nervous system, leading to a manic-depressive psychosis, paralysis of the legs, and loss of vision. Again, she recommended ventilating procedures and routine testing of the air to detect both carbon disulphide and hydrogen sulphide, a by-product of the viscose rayon process that caused eye inflammations.

Late in her career, Hamilton acted as consultant and adviser to various business, government, and consumer groups interested in industrial hygiene. In 1928, along with her old friend Florence Kelley from Hull House, she had initiated a campaign on behalf of the National Consumers' League to study the use of radium in luminous paint on watches and clocks. She urged colleagues at Harvard and elsewhere to publish their findings on industrial poisons. Hamilton felt that public exposure of the dangers workers faced would be enough to change attitudes and alter behavior (Sicherman 1984, 312–314). She continued her efforts to get laws passed, guaranteeing compensation for occupational diseases, particularly in Pennsylvania, a major industrial state, which until 1938 had no workers' compensation laws. In 1938 Hamilton represented the Division of Labor Standards on a committee to set standards for allowable concentrations of toxic fumes and dust. Even in retirement, she persisted in her efforts to provide men and women with safe working environments.

BIBLIOGRAPHY

Works by Alice Hamilton

Scientific Works

"Multiple tubercular ulcers of the stomach, with a report of three cases." *Johns Hopkins Hospital Bulletin* 8 (1897): 75–79.

(with H. M. Thomas) "The clinical course and pathological histology of a case of neuroglioma of the brain." *Journal of Experimental Medicine* 2 (1897): 635–646.

"Peculiar form of fibrosarcoma of the brain." *Journal of Experimental Medicine* 4 (1899): 597–608.

"On the presence of new elastic fibres in tumors." *Journal of Experimental Medicine* 5 (1900): 131–138.

"The pathology of a case of polioencephalomyelitis." *Journal of Medical Research* 3 (1902): 11–30.

"The fly as a carrier of typhoid: An inquiry into the part played by the common house fly in the recent epidemic of typhoid fever in Chicago." *Journal of the American Medical Association* 40 (1903): 576–583.

"An inquiry into the causes of the recent epidemic of typhoid fever in Chicago." *Charities and the Commons* 8 (1903): 3–7.

"The common housefly as a carrier of typhoid." *Journal of the American Medical Association* 42 (1904): 1034.

"The question of virulence among the so-called pseudodiphtheria bacilli." *Journal of Infectious Diseases* 1 (1904): 690–713.

"Surgical scarlatina." *American Journal of Medical Science* 128 (1904): 111–129.

"The toxic action of scarlatinal and pneumonic sera on paramoecia." *Journal of Infectious Diseases* 1 (1904): 211–228.

"Dissemination of streptococci through invisible sputum: In relation to scarlet fever and sepsis." *Journal of the American Medical Association* 44 (1905): 1108–1111.

"Milk and scarlatina." *American Journal of Medical Science* 130 (1905): 879–890.

"The industrial viewpoint: Occupational conditions of tuberculosis." *Charities and the Commons* 16 (1906): 205–207.

"The role of the house fly and other insects in the spread of infectious diseases." *Illinois Medical Journal* 9 (1906): 583–587.

(with J. M. Horton) "Further studies on the virulent pseudodiphtheria bacillus." *Journal of Infectious Diseases* 3 (1906): 128–147.

"The Hull-House war on cocaine." *Charities and the Commons* 17 (1907): 1034–1035.

"A new weapon against cocaine." *Charities and the Commons* 17 (1907): 1045.

"The opsonic index and vaccine therapy of pseudodiphtheric otitis." *Journal of Infectious Diseases* 4 (1907): 313–325.

"Pseudodiphtheria bacilli as the cause of suppurative otitis, especially the postscarlatinal." *Journal of Infectious Diseases* 4 (1907): 326–332.

"The social settlements and public health: Pathology and bacteriology." *Charities and the Commons* 17 (1907): 1037–1040.

"Gonorrheal vulvo-vaginitis in children; with special reference to an epidemic occurring in scarlet-fever wards." *Journal of Infectious Diseases* 5 (1908): 133–157.

"Industrial diseases: With special reference to the trades in which women are employed." *Charities and the Commons* 20 (1908): 655–658.

"On the occurrence of thermostable and simple bactericidal and opsonic substances." *Journal of Infectious Diseases* 5 (1908): 570–584.

"Pathology and bacteriology." *Charities and the Commons* 21 (1908): 186–189.

(with Jane Addams) "The 'piece-work' system as a factor in the tuberculosis of wage-workers." *Transactions of the Sixth International Congress on Tuberculosis Sept. 28 to Oct. 5, 1908* 3 (1908): 139–140.

(with Jean M. Cooke) "Inoculation treatment of gonorrheal vulvovaginitis in children." *Journal of Infectious Diseases* 5 (1908): 159–174.

(with Rudolph W. Holmes, et al.) "The midwives of Chicago." *Journal of the American Medical Association* 50 (1908): 1346–1350.

"Experiments in antityphoid inoculation." *Transactions of the Chicago Pathological Society* 8 (1909–11): 151–158.

"Industrial lead poisoning." *Transactions of the Chicago Pathological Society* 8 (1909–12): 293–311.

"Medical charity from the point of view of the patient." *Illinois Medical Journal* 16 (1909): 333–336.

"The sociological aspects of medical charities." *Chicago Medical Recorder* (1909): 509–512.

"Chronic overwork." *Illinois Medical Journal* 17 (1910): 739–743.

"Excessive child-bearing as a factor in infant mortality." *Bulletin of the American Academy of Medicine* 11 (1910): 181–187.

"The opsonic index of bacillus-carriers." *Journal of the American Medical Association* 54 (1910): 704–705.

"The value of opsonin determinations in the discovery of typhoid carriers." *Journal of Infectious Diseases* 7 (1910): 393–410.

"Venereal diseases in institutions for women and girls." *Proceedings of the National Conference of Charities* 37 (1910): 53–56.

"Lead poisoning in Illinois." *Journal of the American Medical Association* 56 (1911): 1240–1244.

"Occupational diseases." *Human Engineering* 1 (1911): 142–149.

"Occupational diseases." *Proceedings, National Conference of Charities and Correction* (1911): 197–206.

"Report of Dr. Alice Hamilton on investigations of the lead troubles in Illinois, from the hygienic standpoint." In *Report of Commission on Occupational Diseases to His Excellency Governor Charles S. Deneen*. Chicago: Warner Printing, 1911.

The White-Lead Industry in the United States, with an Appendix on the Lead-Oxide Industry. Bulletin of the U.S. Bureau of Labor Statistics, no. 95. Washington, DC: Government Printing Office, 1911.

"Fatigue: Smoke: Motherhood." *Survey* 29 (1912): 152–154.

"Flies and privy vaults." *Transactions of the American Association for the Study and Prevention of Infant Mortality* 2 (1912): 157–162.

"Industrial lead poisoning in the light of recent studies." *Journal of the American Medical Association* 59 (1912): 777–782.

"Industrial plumbism." *Journal of the American Medical Association* 59 (1912): 1316.

Lead Poisoning in Potteries, Tile Works, and Porcelain Enameled Sanitary Ware Fac-

tories. Bulletin of the U.S. Bureau of Labor Statistics, no. 104. Washington, DC: Government Printing Office, 1912.

"Plumbism in the industries of the Middle West." *Monthly Bulletin of the Ohio State Board of Health* 2 (1912): 10–16.

"Protection from lead poisoning." *American Labor Legislation Review* 2 (1912): 537.

"Friedmann cure." *Survey* 30 (1913): 123–124.

"Heredity and responsibility." *Survey* 29 (1913): 865–866.

Hygiene of the Painters' Trade. Bulletin of the U.S. Bureau of Labor Statistics, no. 120. Washington, DC: Government Printing Office, 1913.

"Lead poisoning in the United States." *Transactions of the XV International Congress of Hygiene and Demography, Washington, 1912* 3 (1913): 809–820.

"Leadless glaze: Pottery and tile workers." *Survey* 31 (1913): 22–26.

"Rehabilitation of boiled milk." *Survey* 31 (1913): 303.

"Tuberculosis and the hookworm in the cotton industry." *Survey* 30 (1913): 734–735.

"Unpaid medical service." *Survey* 30 (1913): 727–728.

"Dammerschlaf." *Survey* 33 (1914): 158–159.

"The economic importance of lead poisoning." *Bulletin of the American Academy of Medicine* 15 (1914): 299–304.

Lead Poisoning in the Manufacture of Storage Batteries. Bulletin of the U.S. Bureau of Labor Statistics, no. 165. Washington, DC: Government Printing Office, 1914.

Lead Poisoning in the Smelting and Refining of Lead. Bulletin of the U.S. Bureau of Labor Statistics, no. 141. Washington, DC: Government Printing Office, 1914.

"Lead poisoning in the United States." *American Journal of Public Health* 4 (1914): 477–480.

"Radium treatment of cancer." *Survey* 31 (1914): 533–534.

"War surgery of yesterday." *Woman's Medical Journal* 24 (1914): 253.

"Bollinger case." *Survey* 35 (1915): 265–266.

Industrial Poisons Used in the Rubber Industry. Bulletin of the U.S. Bureau of Labor Statistics, no. 179. Washington, D.C: Government Printing Office, 1915.

"Occupational disease clinic of New York City Health Department." *Monthly Review of the U.S. Bureau of Labor Statistics* 1 (1915): 7–19.

"What we know about cancer." *Survey* 35 (1915): 188–189.

"As one woman sees the issues." *New Republic* 8 (1916): 239–241.

"Attitude of social workers toward the war." *Survey* 36 (1916): 307–308.

"Health and labor." *Survey* 37 (1916): 135–137.

"Is science for or against human welfare?" *Survey* 35 (1916): 560–561.

"Race suicide." *Survey* 35 (1916): 407–408.

"Wartime economy and hours of labor." *Survey* 36 (1916): 638–639.

(with R. V. Luce) "Industrial anilin poisoning in the United States." *Journal of the American Medical Association* 66 (1916): 1441–1445.

(with Gertrude Seymour) "The new public health." *Survey* 37 (1916): 166–169, 456–459; 38 (1917): 59–62.

"Dope poisoning, a new industrial hazard in the making of airplane wings." *Survey* 39 (1917): 168–169.

"Dope poisoning in the manufacture of airplane wings." *Monthly Review of the U.S. Bureau of Labor Statistics* 5 (1917): 37–64.

"Industrial poisoning in aircraft manufacture." *Journal of the American Medical Association* 69 (1917): 2037–2039.

"Industrial poisons encountered in the manufacture of explosives." *Journal of the American Medical Association* 68 (1917): 1445–1451.

"Industrial poisons used in the making of explosives." *Monthly Review of the U.S. Bureau of Labor Statistics* 4 (1917): 177–198.

Industrial Poisons Used or Produced in the Manufacture of Explosives. Bulletin of the U.S. Bureau of Labor Statistics, no. 219. Washington, DC: Government Printing Office, 1917.

"Prostitutes and tuberculosis." *Survey* 37 (1917): 516–517.

"Toxic jaundice in munition workers: A review." *Monthly Review of the U.S. Bureau of Labor Statistics* 5 (1917): 263–274.

"Trinitrotoluene poisoning." *Medicine and Surgery* 1 (1917): 761–766.

(with Charles H. Verrill) *Hygiene of the Printing Trades.* Bulletin of the U.S. Bureau of Labor Statistics, no. 209. Washington, DC: Government Printing Office, 1917.

"Causation and prevention of trinitrotoluene (TNT) poisoning." *Monthly Review of the U.S. Bureau of Labor Statistics* 6 (1918): 237–250.

"Dinitrophenol poisoning in munition works in France." *Monthly Labor Review* 3 (1918): 242–250.

"Dope poisoning in the making of airplanes." *Monthly Review of the U.S. Bureau of Labor Statistics* 6 (1918): 37–64.

Effect of the Air Hammer on the Hands of Stone Cutters. Bulletin of the U.S. Bureau of Labor Statistics, no. 236. Washington, DC: Government Printing Office, 1918.

"The fight against industrial diseases: The opportunities and duties of the industrial physician." *Pennsylvania Medical Journal* 21 (1918): 378–381.

"Industrial poisoning in aircraft workers." *International Association of Medical Museums Bulletin* 7 (1918): 97–102.

"Industrial poisoning in American dye manufacture." *Proceedings of the Institute of Medicine* 2 (1918–19): 183–190.

"Industrial poisoning in munitions workers." *International Association of Medical Museums Bulletin* 7 (1918): 86–97.

"Prophylaxis of industrial poisoning in the munitions industry." *American Journal of Public Health* 8 (1918): 125–130.

(with C. E. Nixon) "Optic atrophy and multiple peripheral neuritis developed in the manufacture of explosives (binitrotoluene)." *Journal of the American Medical Association* 70 (1918): 2004–2006.

"Dangers other than accidents in the manufacture of explosives." *Journal of Industrial and Engineering Chemistry* 8 (1919): 1064–1067.

"Health hazards in the manufacture of dyestuffs." *Pennsylvania Medical Journal* 22 (1919): 655–661.

"Hygienic control of anilin dye industry in Europe." *Monthly Labor Review* 9 (1919): 1–21.

"Industrial poisoning by compounds of the aromatic series." *Journal of Industrial Hygiene* 1 (1919): 200–212.

"Industrial poisoning in American anilin dye manufacture." *Monthly Review of the U.S. Bureau of Labor Statistics* 8 (1919): 199–215.

"Inorganic poisons, other than lead, in American industries." *Journal of Industrial Hygiene* 1 (1919): 89–102.

"Lead poisoning in American industry." *Journal of Industrial Hygiene* 1 (1919): 8–21.

"Medical and surgical lessons of the war: War industrial diseases." *Medical Record* 95 (1919): 1053–1059.

"New scientific standards for protection of workers." *Proceedings of the Academy of Political Science* 8 (1919): 157–162.

"Occupational diseases in Illinois." In *Report of the Health Insurance Commission of the State of Illinois.* Springfield: Illinois State Journal, 1919.

"Occupational diseases in Pennsylvania." *Monthly Labor Review* 9 (1919): 170–180.

"Practical points in the prevention of TNT poisoning." *Monthly Labor Review* 8 (1919): 248–272.

"Prevention of TNT poisoning." *American Journal of Public Health* 9 (1919): 394.

Women in the Lead Industries. Bulletin of the U.S. Bureau of Labor Statistics, no. 253. Washington, DC: Government Printing Office, 1919.

(with G. R. Minot) "Ether poisoning in the manufacture of smokeless powder." *Journal of Industrial Hygiene* 2 (1920–21): 41–49.

Carbon-Monoxide Poisoning. Bulletin of the U.S. Bureau of Labor Statistics, no. 291. Washington, DC: Government Printing Office, 1921.

"A discussion of the etiology of so-called aniline tumors of the bladder." *Journal of Industrial Hygiene* 3 (1921): 16–28.

Industrial Poisoning in Making Coal-Tar Dyes and Dye Intermediates. Bulletin of the U.S. Bureau of Labor Statistics, no. 280. Washington, DC: Government Printing Office, 1921.

"Trinitrotoluene as an industrial poison." *Journal of Industrial Hygiene* 3 (1921): 102–116.

"Women in the lead industries." *Medical Woman's Journal* 28 (1921): 1, 23, 57.

"The growing menace of benzene (benzol) poisoning in American industry." *Journal of the American Medical Association* 78 (1922): 627–630.

"Hazards in American potteries." *New Republic* 31 (1922): 187.

"Industrial diseases of fur cutters and hatters." *Journal of Industrial Hygiene* 4 (1922): 219–234.

"The industrial hygiene of fur cutting and felt hat manufacture." *Journal of Industrial Hygiene* 4 (1922): 137–153.

"The scope of the problem of industrial hygiene." *Public Health Record* 37 (1922): 2604–2608.

"Sunbaths for rickets." *Survey* 49 (1922): 261.

"The danger of industrial dusts in the light of recent studies." *Medical Woman's Journal* 30 (1923): 99–103.

"Progress report of committee on benzol poisoning." *Proceedings of the National Safety Council* (1923): 220–221.

"Lead poisoning in the United States." "Petroleum and its derivatives." "Coal tar benzene (benzol) poisoning." "Phosphorus poisoning: Sources, symptoms and treatment." In *Industrial Health,* edited by George M. Kober and Emery R. Hayhurst, 440–457, 495–499, 500–503, 531–534. Philadelphia: P. Blakiston's Son, 1924.

"Mercurialism in quicksilver production in California." *Journal of Industrial Hygiene* 5 (1924): 399–407.

"The prevalence and distribution of industrial lead poisoning." *Journal of the American Medical Association* 83 (1924): 583–588.

"Protection for women workers." *Forum* 72 (1924): 152–160.

"Protection for working women." *The Woman Citizen* 8 (1924): 16–17.

"Concerning motor car gasoline." *The Woman Citizen* 10 (1925): 14.

"Doctor's word on war." *The Woman Citizen* 9 (1925): 15.

Industrial Poisons in the United States. New York: Macmillan, 1925.

"What price safety? Tetra-ethyl lead reveals a flaw in our defenses." *Survey* 54 (1925): 333–334.

(with Rebecca E. Hilles) "Industrial hygiene in Moscow." *Journal of Industrial Hygiene* 7 (1925–26): 47–61.

(with Paul Reznikoff and Grace M. Burnham) "Tetra-ethyl lead." *Journal of the American Medical Association* 84 (1925): 1481–1486.

"The prevalence of lead poisoning in the United States." In *Lead Poisoning*, edited by Joseph C. Aub et al., 232–240. Baltimore: Williams and Wilkins, 1926.

Women Workers and Industrial Poisons. Woman's Bureau Bulletin, no. 57. Washington, DC: Government Printing Office, 1926.

"Fifteen years of industrial toxicology." In *Contributions to Medical Science*, edited by Willard J. Stone, 23–38. Ann Arbor: G. Wahr, 1927.

"Recent advances in industrial hygiene in Russia." *Rehabilitation Review* 1 (1927).

"Recent advances in industrial toxicology in the United States." In *De Lamar Lectures.* Johns Hopkins University School of Hygiene, 19–42. Baltimore: Williams and Wilkins, 1927.

"Storage battery industry." *Journal of Industrial Hygiene* 9 (1927): 346–369.

"Eight-hour day for women in industry." *Mid-Pacific Magazine* 36 (1928): 333–337.

"The lessening menace of benzol poisoning in American industry." *Journal of Industrial Hygiene* 10 (1928): 227–233.

"Cost of medical care." *New Republic* 59 (1929): 154.

"Enameled sanitary ware manufacture." *Journal of Industrial Hygiene* 11 (1929): 139–153.

Industrial Poisons in the United States. 2d ed. New York: Macmillan, 1929.

"Nineteen years in the poisonous trades." *Harper's* 159 (1929): 580–591.

"Volatile solvents used in industry." *American Journal of Public Health* 19 (1929): 523–526.

"Vasomotor disturbance in fingers of stonecutters." *Archiv für Gewerbepathologie und Gewerbehygiene* 1 (1930): 348–358.

"Benzene (benzol) poisoning." *Archives of Pathology* 11 (1931): 434–454, 601–637.

"Benzene poisoning in industry." *Medical Woman's Journal* 38 (1931): 221–224.

"Methanol poisoning." *New Republic* 62 (1931): 277.

"American and foreign labor legislation: A comparison." *Social Forces* 11 (1932): 113–119.

"Below the surface." *Survey Graphic* 22 (1933): 449–454.

"Formation of phosgene in thermal decomposition of carbon tetrachloride." *Industrial and Engineering Chemistry* 25 (1933): 539–541.

"Industrial hygiene." *American Journal of Public Health* 23 (1933): 332–334.

"Industrial poisons." *New England Journal of Medicine* 209 (1933): 279–281.

Industrial Toxicology. New York: Harper, 1934.

"Industrial poisons." *American Federationist* 43 (July 1936): 707–713.

"Medico-legal aspects of industrial poisoning." *Bulletin of the New York Academy of Medicine* 12 (1936): 637–649.

Recent Changes in the Painters' Trade. Bulletin of the U.S. Bureau of Labor Statistics, no. 7. Washington, DC: Government Printing Office, 1936.

"Some new and unfamiliar industrial poisons." *New England Journal of Medicine* 215 (1936): 425–432.

"Healthy, wealthy—if wise—industry." *American Scholar* 7 (1938): 12–23.

"Some new developments in the field of volatile solvents." *Proceedings of the Occupational Diseases Symposium at Northwestern University (1937)* (1938): 54–59.

"Occupational diseases and industrial surgery." *Industrial Medicine* 7 (1938): 46.

Survey of Carbon Disulphide and Hydrogen Sulphide Hazards in the Viscose Rayon Industry. Occupational Disease Prevention Division Bulletin, no. 46. Harrisburg: Pennsylvania Department of Labor and Industry, 1938.

Occupational Poisoning in the Viscose Rayon Industry. Bulletin of the U.S. Department of Labor Division of Labor Standards, no. 34. Washington, DC: Government Printing Office, 1940.

(with M. Bowditch et al.) "Code for safe concentrations of certain common toxic substances used in industry." *Journal of Industrial Hygiene and Toxicology* 22 (1940): 251.

"Pains of industry." *New Republic* 107 (1942): 291–292.

"Death in the factories." *Nation* 157 (1943): 66–68.

Exploring the Dangerous Trades: The Autobiography of Alice Hamilton. Boston: Little, Brown, 1943.

"International control of disease after the war." *Connecticut Medical Journal* 7 (1943): 383–387.

"Toxicity of chlorinated hydrocarbons." *Yale Journal of Biology and Medicine* 15 (1943): 787–801.

"New problems in the field of the industrial toxicologist." *California and Western Medicine* 61 (1944): 55–60.

"Occupational poisoning in the viscose rayon industry." *Bulletin of Hygiene* 19 (1944): 212–213.

"Science in the Soviet Union." *Science and Society* 8 (1944): 69–73.

"Diagnosis of industrial poisoning." *California and Western Medicine* 62 (1945): 110–112.

"Looking at industrial nursing." *Public Health Nursing* 38 (1946): 63–65.

"To the health of the worker." *Nation's Business* 35 (1947): 56.

"Pioneering in industrial medicine." *Journal of the American Medical Women's Association* 2 (1947): 292–293.

"Forty years in the poisonous trades." *American Industrial Hygiene Association Quarterly* 9 (1948): 5–17.

"Industry is health conscious." *Medical Woman's Journal* 55 (1948): 33–35.

"From a pioneer in the poisonous trades." In *Public Health in the World Today*, edited by James Stevens Simmons and Irene M. Kinsey. Cambridge, MA: Harvard University Press, 1949.

(with Harriet L. Hardy) *Industrial Toxicology.* 2d ed. New York: P. B. Hoeber, 1949.

"Developments in the field of industrial medicine." *Journal of the American Medical Women's Association* 6 (1951): 313–314.

Hamilton and Hardy's Industrial Toxicology. 4th ed. Revised by Asher J. Finkel. Boston: J. Wright, 1983.

Exploring the Dangerous Trades: The Autobiography of Alice Hamilton. Boston: Northeastern University Press, 1985.

Other Works

"What one stockholder did." *Survey* 28 (1912): 387–389.
"The boy who does not like to read." *Home Progressive* 4 (1914): 720–723.
"At the war capitals." *Survey* 34 (1915): 417–422.
(with Jane Addams and Emily Balch) *Women at The Hague: The International Congress of Women and Its Results, by Three Delegates to the Congress from the United States,* 22–54. New York: Macmillan, 1915.
"Angels of victory." *New Republic* 19 (1919): 244–245.
"On a German railway train." *New Republic* 20 (1919): 232–233.
"Visit to Germany." *British Journal of Childhood Diseases* 16 (1919): 129–132.
(with Jane Addams) "After the lean years: Impressions of food conditions in Germany." *Survey* 42 (1919): 793–797.
"Medieval industry in the twentieth century." *Survey* 51 (1924): 456–464.
"Colonel House and Jane Addams." *New Republic* 47 (1926): 9–11.
"Witchcraft in West Polk Street." *American Mercury* 10 (1927): 71–75.
"State pensions or charity?" *Atlantic Monthly* 145 (1930): 683–687.
"What about the lawyers?" *Harper's* 163 (1931): 548–549.
"Forward with Hitler." *Living Age* 344 (1933): 484.
"Hitler speaks: His book reveals the man." *Atlantic Monthly* 152 (1933): 399–408.
"An inquiry into the Nazi mind." *New York Times* (Aug. 6, 1933): 1–2.
"Sound and fury in Germany." *Survey Graphic* 22 (1933): 549–554.
"The youth who are Hitler's strength." *New York Times* 6 (Oct. 8, 1933): 3, 16.
"Plight of the German intellectuals." *Harper's* 168 (1934): 159–169.
"Woman's place in Germany." *Survey Graphic* 23 (1934): 26–29.
"Because war breeds war." In *Why Wars Must Cease,* edited by Rose Young. New York: Macmillan, 1935.
"Letters of a woman citizen." *Survey Graphic* 27 (1938): 282–283.
"Feed the hungry." *Nation* 151 (1940): 596–597.
"A mid-American tragedy." *Survey Graphic* 29 (1940): 434–437.
"Should we outlaw anti-Semitism?" *New Masses* 54 (1945): 16.
"Why I am against the equal rights amendment." *Ladies' Home Journal* 62 (1945): 23.
"Holiday with strings." *Ladies' Home Journal* 64 (1947): 34–35.
"The encroachment of the state upon the conscience." *Christianity and Crisis* 11 (l951): 61–62.
"Message bearer." *Saturday Review* 36 (1953): 42–43.
"Words lost, strayed, or stolen." *Atlantic Monthly* 194 (1954): 55–56.
"English is a queer language." *Atlantic Monthly* 203 (1959): 51–52.
"Woman of ninety looks at her world." *Atlantic Monthly* 208 (1961): 51–55.
"Edith and Alice Hamilton: Students in Germany." *Atlantic Monthly* 215 (1965): 129–132.

Works about Alice Hamilton

Addams, Jane. *My Friend, Julia Lathrop.* New York: Macmillan, 1935.
"Alice Hamilton: Crusader, pioneer, physician, scientist, teacher, author, humanitarian, gentlewoman." *Industrial Medicine* 4 (1935): 421–427.

Drinker, Philip. "Alice Hamilton." *Journal of Occupational Medicine* 14 (1972): 114.

Evans, Elizabeth Glendower. "People I have known: Alice Hamilton, M.D., pioneer in a new kind of human service." *The Progressive* 2 (Nov. 29, 1930): 2; (Dec. 20, 1930): 3.

Felton, Jean Spencer. "Alice Hamilton, M.D.—a century of devotion to humanity." *Journal of Occupational Medicine* 14 (1972): 106–110.

Goldwater, Leonard J. "Alices and hatters." *Journal of Occupational Medicine* 14 (1972): 112–113.

Grant, Madeleine P. *Alice Hamilton: Pioneer Doctor in Industrial Medicine.* New York: Abelard-Schuman. 1967.
  For younger readers.

Hardy, Harriet. "Alice Hamilton." *Journal of Occupational Medicine* 14 (1972): 97.

Johnstone, Rutherford T. "One of America's greatest physicians." *Journal of Occupational Medicine* 14 (1972): 105.

Mayers, May R. "Alice Hamilton, M.D." *American Industrial Hygiene Association Journal* 19 (1958): 449–452.

McCord, Carey P. "Alice Hamilton." *Journal of Occupational Medicine* 11 (1969): 2, 69.

Sergeant, Elizabeth Shepley. "Alice Hamilton, M.D., crusader for health in industry." *Harper's* 152 (1926): 763–770.

Sicherman, Barbara. *Alice Hamilton: A Life in Letters.* Cambridge, MA: Harvard University Press, 1984.

Slaight, Wilma Ruth. "Alice Hamilton: First Lady of Industrial Medicine." Ph.D. diss., Case Western Reserve University, 1974.

Young, Angela Nugent. "Interpreting the Dangerous Trades: Workers' Health in America and the Career of Alice Hamilton, 1910–1935." Ph.D. diss., Brown University, 1982.

Other References

Beecher, Henry K., and Mark D. Altschule. *Medicine at Harvard: The First Three Hundred Years.* Hanover, NH: University Press of New England, 1977.

Curran, Jean Alonzo. *Founders of the Harvard School of Public Health with Biographical Notes, 1901–1946.* New York: Josiah Macy, Jr. Foundation, 1970.

Oliver, Thomas. *Dangerous Trades.* New York: E. P. Dutton, 1902.

ETHEL NICHOLSON BROWNE HARVEY (1885–1965)

Paula Ford

BIOGRAPHY

Ethel Nicholson Browne was born in Baltimore, Maryland, on December 14, 1885. Her father, Bennet Bernard Browne, was a physician who practiced obstetrics and gynecology. He was a professor of gynecology at the Woman's Medical College of Baltimore. Her mother was Jennie (Nicholson) Browne. She had two brothers and two sisters. Both of her sisters became physicians, and she recognized one sister in her 1956 book (*American Arbacia*), saying, "My sister, Dr. Mary N. Browne, a classical scholar, has been most helpful with the section on history" (Harvey 1956).

Browne attended the Bryn Mawr School in Baltimore, from which she graduated in 1902. She received her A.B. degree from Goucher College in 1906 (known then as the Women's College of Baltimore). That summer she took a course at the Marine Biological Laboratory (MBL) at Woods Hole, Massachusetts, and in the autumn entered Columbia University as a graduate student in zoology. She studied with Edmund Beecher Wilson and Thomas Hunt Morgan (Nobel laureate, 1933), receiving her A.M. degree in 1907 and her Ph.D. degree in 1913. Her doctoral dissertation was "A Study of the Male Germ Cells in *Notonecta*," published in the *Journal of Experimental Zoology* (Harvey 1913).

While in graduate school Browne was awarded several fellowships. The first was from Goucher (1906–7), and a second was from the Society for the Promotion of University Education for Women (1911–12). She also received the Sarah Berliner Fellowship (1914–15), which was intended to support study in Germany. Because of World War I, she did not go to Germany but used the fellowship at the Hopkins Marine Station of the University of California.

Browne was an instructor of science and mathematics at the Bennett School for Girls in Millbrook, New York (1908–11). She worked as a laboratory assistant in biology at Princeton University (1912–13). From 1913 to 1914 she was an instructor in biology at Dana Hall, Wellesley, Massachusetts. Later

Browne worked as a laboratory assistant in histology at Cornell Medical College in New York (1915–16). She became an instructor in biology at Washington Square College, New York University (1928–31). For most of her professional life, she worked as an unsalaried, independent investigator in the Biology Department of Princeton University (1931–62).

On March 12, 1916, Browne married Edmund Newton Harvey, a professor of biology at Princeton University, who was best known for his work on bioluminescence. Their honeymoon trip to Japan included a stay at the Misaki Marine Biological Station, where they conducted research. The Harveys later had two sons: Edmund Newton Harvey, Jr. (b. 1916), who became a physical chemist, and Richard Bennet Harvey (b. 1922), who became a physician. While the children were young, she conducted unpaid, part-time research, relying on the help of maids and governesses.

She was associated with the MBL at Woods Hole for over 50 years, being elected to membership in 1909 while still a graduate student. In 1944 she became the third woman to deliver a Friday Evening Lecture at the laboratory. She was the second woman elected a trustee of Woods Hole, a position she held until 1958, when she became a trustee emeritus. She won prizes in diving contests at Woods Hole as a young woman and was known there for her skill at tennis and the liveliness of the Harveys' table conversation at the "mess" (Butler 1966).

Harvey also conducted research at marine laboratories in Bermuda, North Carolina, California, and Italy. From 1925 to 1926 she had the use of the American Women's Table in Italy, established by Ida Henrietta Hyde* and other women scientists. In 1933, 1934, and 1937, she worked at the Jacques Loeb Memorial Table of Rockefeller Institute, located at the Naples Zoological Station in Naples, Italy.

Edmund Newton Harvey died at Penzance Point, Massachusetts, on July 22, 1959, of a heart attack. A memorial to her husband points out that they "shared laboratories at Princeton and Woods Hole over the years, except for the period when it was necessary for her to devote full time to raising their two sons, Ned and Dick" (Chase 1960).

Honors awarded to Ethel Browne Harvey included election as a fellow of the American Association for the Advancement of Science, the New York Academy of Sciences, and l'Institut International d'Embryologie in Utrecht. In 1956 Goucher College awarded her an honorary Doctor of Science degree. Ethel Browne Harvey died in Falmouth, Massachusetts, on September 2, 1965, of peritonitis caused by acute appendicitis.

WORK

Some of Ethel Browne Harvey's important work is not well known. Her 1909 study of hydra transplantation, published while she was a graduate student, was the first to demonstrate that a transplant could induce a secondary axis of polarity (Browne 1909). She transplanted a piece of *Hydra viridissima* mouth onto another individual of the same species, inducing the development of a new hydranth on the recipient. To determine whether the transplanted tissue had induced the formation of the new hydranth, Browne developed an experiment using pigmented and nonpigmented hydra. She showed that the pigmented host tissue was organized by the nonpigmented transplant.

Later, similar work with amphibian embryos, done by Hans Spemann and Hilde Mangold,* received much more recognition (1924). In fact, Hans Spemann won the Nobel Prize in 1935 for his discovery of the phenomenon he called "organization." Spemann almost certainly knew of Browne's work shortly after she published her results, although he never acknowledged her influence (Lenhoff 1991).

In the 1920s Harvey began to study sea urchins. Her first paper on this subject dealt with the effects of a lack of oxygen on development of sea urchin eggs. She continued to do research on the sea urchin for the next three decades.

Harvey was best known for her research on the embryology of sea urchin eggs and for her pioneering techniques (with Alfred Lee Loomis and Edmund Newton Harvey) in using a centrifuge to study sea urchin eggs (Lenhoff 1983).

She centrifuged the eggs, separating them into nucleated and enucleated fractions. She found that "[a]t least one nucleus seems to be necessary for differentiation and complete development, but the early stages of development involving cell multiplication can take place without any nucleus" (Harvey 1956). The popular press presented Harvey's work as the "creation of life without parents," but she saw the significance of her research on parthenogenesis (the stimulated development of an unfertilized egg) in determining which characteristics of living matter were cytoplasmic and which were genetic (Haraway 1980).

Her book, *The American Arbacia and Other Sea Urchins* (1956), includes extensive information about her experiments in sea urchin embryology as well as the natural history of the Echinoidea.

Harvey was a member of the American Society of Zoologists, American Society of Naturalists, American Genetic Association, Society of Genetic Physiologists, International Society for Cell Biology, Societa Italiana de Biologia Sperimentale (honorary), Bermuda Biological Station, Biological Photographic Association, Phi Beta Kappa, and Sigma Xi.

BIBLIOGRAPHY

Works by Ethel Nicholson Browne Harvey

Scientific Works

Listed here are all works by Harvey except those cited in Harvey (1956). Space does not permit the listing of the complete works of Ethel Nicholson Browne Harvey. Included here are her dissertation and all works cited in the text.

"The production of new hydranths in hydra by the insertion of small grafts." *Journal of Experimental Zoology* 7 (1909): 1–37 (as E. Browne).

(with T. H. Morgan and F. Payne) "A method to test the hypothesis of selective fertilization." *Biological Bulletin* 18 (1910): 76–78 (as E. Browne).

"A Study of the Male Germ Cells in *Notonecta.*" Ph.D. diss., Columbia University, 1913 (as E. Browne).

"A study of the male germ cells in *Notonecta.*" *Journal of Experimental Zoology* 14 (1913): 61–121 (as E. Browne).

"A review of the chromosome numbers in the Metazoa, Part I." *Journal of Morphology* 28 (1916–17): 1–63.

"Mitotic division of binucleate cells." *Biological Bulletin* 37 (1919): 96–101.

"Fertilization." In *Encyclopaedia Britannica*, 14th ed., vol. 9, 188–191. Chicago: Encyclopaedia Britannica, 1946.

(with G. I. Lavin) "The eggs and half-eggs of *Arbacia punctulata* and the plutei as photographed by ultraviolet, visible and infrared light." *Experimental Cell Research* 2 (1951): 393–397.

(———) "Nuclei of *Arbacia* and *Chaetopterus* eggs as photographed by infrared light." *Experimental Cell Research* 2 (1951): 398–402.

The American Arbacia and Other Sea Urchins. Princeton: Princeton University Press, 1956.
    Includes an extensive bibliography.

"Cleavage with nucleus intact in sea urchin eggs." *Biological Bulletin* 119 (1960): 87–89.

"Proflavin and its influence on cleavage and development." *Biological Bulletin* 123 (1962): 132.

Works about Ethel Nicholson Browne Harvey

Butler, E. G. "Ethel Browne Harvey." *Biological Bulletin* 133 (1966): 9–11.

Haraway, Donna J. "Ethel Browne Harvey." In *Notable American Women: The Modern Period*, edited by Barbara Sicherman et al., 319–321. Cambridge, MA: Belknap Press of Harvard University Press, 1980.

Lenhoff, H. M. "William Farnsworth Loomis." In *Hydra: Research Methods*, edited by H. M. Lenhoff, xiii. New York: Plenum Press, 1983.
    Includes information on the work of Alfred Lee Loomis.

———. "Ethel Browne, Hans Spemann, and the discovery of the organizer phenomenon." *Biological Bulletin* 181 (1991): 72–80.

Obituary. *New York Times* (Sept. 3, 1965): 27.

Other References

Chase, A. M. "Edmund Newton Harvey." *Biological Bulletin* 119 (1960): 9–10.

Lentz, T. L. *The Cell Biology of Hydra.* New York: John Wiley and Sons, 1966.

Lillie, Frank R. *The Woods Hole Marine Biological Laboratory.* Chicago: University of Chicago Press, 1944.

Spemann, H., and H. Mangold. "Über Induktion von Embryonalanlagen durch Implantation artfremder Organisatoren." *Archiv für Mikroskopische Anatomie und Entwicklungsmechanik* 100 (1924): 599–638.

HILDEGARD OF BINGEN (1098–1179)

Katalin Harkányi

BIOGRAPHY

Hildegard of Bingen, one of the great abbesses of the Middle Ages, was a very talented woman with a complex personality. She lived in an era of religious, political, and intellectual tension. When she was born, the German people were involved in the Crusades, witnessed a constant shift of power between the Church and the State, and lived in a time of social tensions created by civil wars. Despite all the turbulence in the German duchies and throughout Europe, interest in the natural world, antiquity, and church reform began to thrive. Hildegard was a woman with poor health, diverse talents, and a strong ego. Despite the obstacles, she became one of the leading intellectuals of her time. She was the first great woman mystic writer, creator of the first morality play, poet, composer of music to hymns she wrote, natural historian, and healer. She served God in everything she did and dedicated all her work to Him. With her prophecies she influenced politics and ecclesiastic history. She foretold the fall of the Holy Roman Empire and, as an advocate of church reform, she anticipated the Reformation.

She was born in the summer of 1098 in Bermersheim, central Germany, a small community situated northwest of Mainz, not far from Frankfurt am Main. Little is written about her family; what is known has been gathered from her letters, autobiography, and fragmentary writings about her. Hildegard's family was distantly related to Emperor Frederick Barbarossa. Her father, Hildebertus of Bermersheim, was a member of the medieval free aristocracy, which status bestowed upon him the right to own land. Hildebertus was also a knight in the court of Meginhard, Count of Sponheim (Spanheim). Bermersheim was their family seat, and the church, which they probably owned, still survives in the town (Flanagan 1989, 25).

Hildegard was the tenth child of Hildebertus and Mechthilde. Her brothers were Drutwin, Hugo—a precentor of the Mainz Cathedral, and Roricus—who became the canon in Tholey, Saarland; her four sisters were Irmengard, Odilia,

Jutta and Clementina—who was ordained a nun and resided at the Rupertsberg cloister (Flanagan 1989, 215). Two of her siblings are not discussed in any of the sources concerning Hildegard.

A sickly child, Hildegard was ill throughout life. On the basis of her writings, it is supposed that she suffered from migraines. She had her first vision at the age of three and one again at five. Early in her youth she realized that having visions was rare; thus, she discussed them only with her benefactor and aunt, Jutta. She suppressed her visions and began sharing them with others only on the direct order from God.

At the age of eight Hildegard was entrusted to the care and guidance of her aunt Jutta, the beautiful, recluse daughter of Count Stephan of Sponheim. Their small, three-person (including a servant) community was attached to the Benedictine monastery at Disibodenberg, established by Archbishop Ruthard (ca. 1105). The place was built where the Nahe and Glan Rivers meet, and by the year 1000 the first monastery was already founded at Disibodenberg. It was named after a Celtic monk and hermit who lived there in the seventh century. Hildegard was enclosed for life on All Saints Day, November 1, 1106. At the time, being enclosed was equal to being buried alive. In the ceremony she was given the last rites of the dead, and the entrance to her cell was sealed.

Jutta was also responsible for her pupil's education. She instructed her charge according to the Benedictine rules and taught her Latin, the Scriptures, devotions, and music. Although later in life Hildegard called herself a *homo simplex*, and her works are dispersed with German words, Hildegard's writings attest to a great knowledge and extensive learning.

About 1113 Hildegard decided to take her vows and become a Benedictine nun. Otto, Bishop of Bamberg, performed the ordination ceremonies. As Jutta's reputation grew, more and more young women joined their community, and the restrictions on their lives lessened. Soon the retreat became a convent, with Jutta as its prioress. When Jutta died in 1136, Hildegard, only 38 years old, was elected abbess at Disibodenberg. In the intervening years she devoted herself to nursing and doctoring; she was the infirmarian at the convent caring for the nuns and the neighborhood sick. Her cures and medical skills became well known (Labarge 1986, 171).

Five years later, in 1141, through an especially brilliant vision, she received the divine command to write down her visions. Just as on previous occasions, she resisted, became ill, and recovered only when she began to record her visions. Her hesitation was rooted in her strong critical views of charlatans. Pope Eugenius III, aware of the written account and on the recommendation of Bernard of Claivaux, read the first part of her *Scivias* (a contraction of *Sci vias Domini* or Know the ways of the Lord) before the Synod of Trier (1147–49). At the same time, he sent a papal commission to Disibodenberg to study the authenticity of Hildegard's visions. When he became certain of her genuineness, the Pope gave the Church's approval in a letter and encouraged Hildegard to continue her writing.

Hildegard's fame spread from the German lands to Flanders, France, England, and Greece, and her spiritual advice was sought from faraway places. Because of her visions, Hildegard was taken seriously by all and was respected, and people listened to what she said. At that time, her correspondence with a great number of people began; it extended for three decades and ended only with her death. The surviving manuscript of her letters is known as the Riesenkodex (Wiesbaden). Since she corresponded with many people of humble origin as well as high officials of the Church and State, nearly 300 of her letters have been preserved. Some clergy sought her counsel in solving problems of their congregations; others asked for personal help, medical cures, or resolutions of a political nature. She received letters from the patriarch of Jerusalem, Frederick Barbarossa and his father Conrad III, Queen Bertha of Greece, St. Bernard, professors from the University of Paris, Henry II (King of England), Eleanor of Aquitaine, and Thomas Becket.

In 1151 she completed *Scivias*, the book instructing people as to how they should live. Hildegard soon began working on an extensive scientific and medical encyclopedia called, since the sixteenth century, *Physica*. It is mentioned in the literature by several other titles: *Liber simplicis medicinae* or *Subtilitates diversarum naturarum creaturarum* or *Book of Simple Medicine* or *Nine Books on the Subtleties of Different Kinds of Creatures*. Her complementary work, *Causae et Curae*, is also known as *Liber compositae medicinae* or *Book of Compound Medicine* or *Causes and Cures*.

As Hildegard's reputation grew, the number of residents at the Disibodenberg convent increased. Conditions and accommodations became crowded and inadequate; therefore, Hildegard and 18 of her nuns decided to establish a new cloister at Rupertsberg, across the river Nahe from the important medieval town of Bingen. The buildings were erected on the site of a no-longer-existing monastery and the ruined castle of the Dukes of Bingen. Kuno, the abbott at Disibodenberg, opposed the plan at first; if the nuns moved, the monastery would lose materially and spiritually. Finally, the abbott reluctantly gave permission in 1147, after Hildegard became ill, and the support of the Archbishop of Mainz for the move was secured. Construction began in 1148, and Hildegard and the nuns moved in 1150. The new convent had workrooms with water piped in, and the refuse was removed by water. These remarkable and unique features were installed at Hildegard's request (Butler 1962, 582).

As an abbess she had equal political power and jurisdiction over her domain as did a feudal lord over his. Hildegard became a strong and forceful administrator of the new convent. Despite the initial hardships and the dissatisfaction of her nuns (who lived under much less comfortable conditions at the beginning), she remained creative and wrote her scientific works and hymns with their accompanying music. She also created *lingua ignota*, an unknown language of 900 words, and *litterae ignotae*, an alternative alphabet.

In 1151, Richardis (Rikkarda) of Stade, a favorite assistant and friend, was called to become the abbess of the convent at Bassum and left Rupertsberg.

Hildegard strongly opposed Richardis's move and expressed her disfavor to Richardis's mother and brother Hartig (Archbishop of Bremen), to Henry (Archbishop of Mainz), and even to Pope Eugenius III. Her opposition was in vain, and she reluctantly stopped her protests after receiving direct orders from her superiors. The loss of her friend and Richardis's death a year later deeply affected Hildegard.

During 1158 to 1163 she wrote her second book of visions, *Liber vitae meritorium* (The Book of Life's Merits), which concerned itself with life's vices and virtues. During these years she had a number of visitors and went on four journeys, a rare activity for a nun, especially a twelfth-century woman. She traveled on horseback and on boat, which were difficult modes of transportation for an often ill and elderly woman. On these journeys, in addition to preaching, she practiced medicine and assisted a number of people seeking her help. Her apostolic tours earned her the name "Sibyl of the Rhine."

About 1165 Hildegard established a daughter house at Eibingen near Rüdesheim, across the Rhine from Rupertsberg. The convent stood at the site of the Augustinian monastery. It still exists today and is known as the Abbey of St. Hildegard.

Her third and final book of visions, *Liber divinorum operum* (The Book of Divine Works), was written between 1163 and 1173. This is the most mature of her major visionary works. It presents her theology and theories of the human mind and discusses the organization of the universe. While working on this, she also wrote the biographies of Saint Rupert and Saint Disibod, as well as her commentaries on the "Benedictine Rule."

In 1173 Volmar died; he had been her friend, confessor, scribe, and secretary. He was replaced by Gottfried, who began to write her biography, entitled *Vita Sanctae Hildegardis* (Life of St. Hildegard). Gottfried died in 1176 before finishing the biography; almost a decade later Theoderich finished this *Vita*. Guibert of Gembloux became her last secretary from 1177 to her death. He also wrote a biography of Hildegard, which, however, was never finished. In his work this scribe had stylistic freedom. Interestingly enough, Hildegard did not allow her previous collaborators to change her words or the contents of her writings; she accepted only grammatical corrections.

Disagreement with the clergy of Mainz marred the last year of Hildegard's life. She and her convent were placed under an interdict for refusing to cast out the body of a man from the monastery's cemetery after he had reconciled with the Church at his death. The canons disputed the reconciliation, and the interdict was lifted only in the spring of 1179, after Hildegard accused the clergy of siding with the devil. Later the same year, on September 17, after a short illness, Hildegard died. She was buried before the altar in the monastery church at Rupertsberg. When the church "was destroyed by the Swedes in 1632, her relics were removed to Eibingen" (Baring-Gould 1882, 288). To this day they lie in Rüdesheim/Eibingen: the heart and tongue in the parish church on Eibingerstrasse.

Three separate canonization processes were conducted on her behalf under Gregory IX (1237), Innocent IV (1243), and John XXII (1317) but were never completed. Notwithstanding, in church and secular literature she is referred to as Saint Hildegard and was even accorded a saint's day on September 17.

WORK

Hildegard's works, products of a Renaissance mind, are diverse and unique for a twelfth-century woman, and the range and extent of her writings are extraordinary. Although she was a mystic and lived a religious life, she was interested in concrete facts and tried to understand scientific and medical principles. Her healing affected the body and the soul, and she prescribed remedies for both.

She cared about social disease and injustice, was disturbed by the social conditions of her times, and believed that social evil needed to be obliterated and that morality should be improved (Eckenstein 1963, 257). She was against divorce and believed that marriage should be based on love. She did not see, like the Galenists (followers of the ancient Greek physician Galen), males and females as being parallel, nor as opposites, as the Aristotelians did (Cadden 1984, 173). Hildegard believed that men and women complemented each other. She advocated moderation in everything and condemned any deviation from the natural: transvestitism and perversion.

Her astronomical knowledge was remarkable and full of foresight. She knew that the sun was the center of the universe (centuries before Copernicus) and that the moon controlled the tides (Crook 1995, 21). Hildegard maintained that when it is cold in the Northern Hemisphere, it is warm in the Southern and wrote that the stars shine with unequal brightness because they are different in size. She also declared that the winds blew from east to west, thus keeping all other forces of the planet in balance and order ("S. Hildegardis, abbatissae" 1855, vol. 197: 791–795).

Hildegard had rudimentary knowledge of Latin grammar and had to dictate her visions and other works to her secretaries. Especially in her scientific works, she used many German words. Because she mixed her languages, many scholars, such as Charles Singer, doubted her authorship, even though she dated each of her works. Others thought it was only natural for Hildegard to use her own vernacular when writing about wildlife, herbs, flowers, and diseases (Thorndike 1923–58, vol. 2:128). Modern scholars, such as Schrader, Liebschütz, Cannon and others, however, have authenticated her works.

Her biological knowledge was rooted in the Galenic tradition, but her theories were quite different from the classic Hippocratic and Galenic ones. She maintained that a healthy person has both a healthy body and mind, and, contrary to many of her contemporaries, she believed that reason resides in both the heart and the brain. Hildegard was convinced that those illnesses that came from God could be treated only by God's grace but those that came from a person's own faults and errors needed to be treated by medicine. She had a scientific interest in human sexuality, wrote frankly about sexual behavior, and described human passion in a lyrical style.

The nature of her observations was rare for her age, and in this, she was an exception to other twelfth-century medical authors. She stated that sexuality influenced the health of men and women and that both parents influence the character and appearance of their children. Hildegard made unique contributions to gynecology: she described four female stereotypes, had unusual theories of conception (was interested in the time of conception and not in the child's birth date), discussed menstrual problems and events at parturition, and advised against pregnancy before the age of 20 and after 50.

Hildegard's remedies were herbal in nature, and her cures employed plants, nutrition, and prayer. Her medications also incorporated the animal world and minerals. She recommended pulverized salmon bones for bad teeth and warned against the use of lead and brass because of their toxic properties. However, she realized the medicinal value of copper and iron. Hildegard prescribed licorice to cure eye and voice problems; the lily to help leprosy; and the yarrow for healing external and internal wounds, fever, and in combination with other herbs for nosebleed, menstrual problems, and sleeplessness (Cannon 1993, 146). Her treatment of diabetes omitted sweets and nuts from the afflicted person's diet. She approved surgery only for bloodletting and lancing of boils and infections. In using narcotics, like henbane and hemlock, she was conservative, because large quantities may cause great disturbance in the body and possible death. Usually, she prescribed medicine in small doses. "[S]he came near to certain later discoveries, such as the circulation of the blood" (Butler 1962, 583), causes of contagion and autointoxication, nerve action beginning in the brain, and chemistry of the blood (Marks and Beatty 1972, 49).

Hildegard's true scientific works were *Physica* and *Causae et Curae*. Their foundation came from Benedictine tradition, folklore, and her personal observations. She was well versed in ancient and medieval medicine. But according to Gottfried Hertzka, Hildegard was not influenced by the school of Salerno or the school of Chartres, whose teachings spread to Paris and Oxford, or the school at Toledo, which disseminated Arabic medical recipes (Cannon 1993, 149). This assumption is probable, because awareness of current events in the Middle Ages was difficult. There were few scientists, and communication among them was hindered by distance and the problems of transportation, language, and politics.

Physica was the title given to *Liber simplicis medicus* by Johannes Schott of Salzburg, a sixteenth-century printer. It was well known and used as a text at the Montpelier medical school (Alic 1986, 66), and even Paracelsus, the sixteenth-century physician, might have been familiar with this work (Pagel 1982, 211). *Physica* is an encyclopedia of natural history and medicine; it contains nine books describing herbs, trees, animals, fishes, birds, reptiles, gems, metals, and elements. Botany is especially well treated: almost a thousand plants and animals are named in German (Sarton 1927–48, vol. 2:304). Because of the medieval usage of German words, this work is also an important source for philologists. The encyclopedia's fourth book, *De Lapidibus*, consists of 26 short chapters covering precious stones. Twenty-five stones were used to cure ailments of both physical and psychological nature. Hildegard indicated that sapphires should be used for sore eyes, gout, and to increase intellectual powers and that emeralds eased the pain in the heart.

Physica discussed food values and medicinal uses of water, air, and earth, listed properties of plants (curative and toxic), and gave the medical uses of each plant. Hildegard explained that illness (indigestion, fever, cough, delusion, and leprosy) was caused by an imbalance between *flegma* (burned matter) and *livor* (poisonous blood). She advocated a balance among the elements (air, water, earth, excluding fire) and the qualities (hot, cold, moist, and dry) for the health of human beings; she even described plants according to their humoral qualities: hot or cold. For example, the herb tansy is hot and a little damp; therefore it should be used for curing coughs.

Causae et Curae covers pathology and therapeutics. These writings discuss treatments based on natural resources along with her cosmology, which implies that the human body is a copy of the universe; therefore it is the microcosm of the macrocosm. The book also covers medical theory, definition of diseases, and diagnostics. Included are Hildegard's recipes for curing maladies and her explanations why the treatments work.

Causae et Curae is divided into five books. The contents of the first one are cosmological medicine and cosmography. Book two discusses Hildegard's adaptation of Hippocratic humoral theory. In the twelfth century physiology and cosmology were linked together through the four elements (fire, air, water, and earth), the four qualities (hot, cold, moist, and dry), and the four humors (choleric, sanguine, phlegmatic, and melancholic). But Hildegard's humors are the wet, the dry, the foamy, and the warm. The dominant ones she calls *flegmata*, and the subordinate ones *livores*. The humors can interact with the emotional and physical characteristics of a person and may influence sexuality, reproduction, and fertility. She classified human beings also by the humors; accordingly, there are four types of people.

In book three Hildegard discussed diseases, such as insanity and the common cold; procreation and birth; bloodletting, fevers, and asthma. Headaches, migraines, bone tumors, sterility, and ulcers are also treated here. Book four comments on epilepsy, leprosy (which she believed was a combination of skin diseases, ulcers, and skin rashes), fevers, and some veterinary medicine concerning sheep, horses, and goats. Book five speaks of life and death, urinalysis, and drinking water. Hildegard emphasized the necessity of boiling drinking water, which may be unclean, especially if it comes from rivers or swamps. She stressed the importance of hygiene, diet, exercise, rest, and moderation.

Hildegard of Bingen is the earliest known German medical writer and the most distinguished natural historian of the twelfth century. Her medical books are unique in her literary corpus, because she never claimed that they were the product of her visions or that they were divinely inspired. However, just like her religious writings, the scientific ones reflected her religious philosophy: namely, man was God's creation, and everything on Earth is here to serve man. Her writing was factual, realistic, and without judgment of the human being. Hildegard's scientific works have interested people for more than eight centuries. Her influence was considerable in her time and continues to this day.

BIBLIOGRAPHY

Works by Hildegard of Bingen

Scientific Works

Space does not permit the listing of Hildegard of Bingen's complete works. Her scientific writings and the most important publications about her are included here. For additional sources consult individual entries in this bibliography.

Physica Elementorum . . . Strassburg: J. Schott, 1533. Reprint. *Liber Beatae Hildegardis subtilitatum diversarum naturarum . . .* Edited by F. A. Reuss. Paris, 1856. Reprint. Wiesbaden, 1859.

Experimentarius medicinae . . . Edited by G. Kraut. Strassburg: J. Schott, 1544.

Physica. In *Patrologia Latina,* vol.197. Edited by Jacques P. Migne. Paris: Petit-Montrogue, 1855. Reprint. Paris: Garnier, 1882.

"S. Hildegardis, abbatissae opera omnia." In *Patrologia Latina,* vol.197, edited by Jacques P. Migne. Paris: Petit-Montrogue, 1855. Reprint. Turnhout, Belgium: Brepols, 1976.

Die Physica der heiligen Hildegard. Translated by J. Berendes. Vienna: Pharmazeutische Post, 1896–97.

Die naturwissenschaftlichen Schriften der Hildegard von Bingen. Berlin: R. Gätner, H. Heyfelder, 1901.

Causae et Curae. Edited by Paul Kasier. Leipzig: Teubner, 1903.

Der Äbtissin Hildegard von Bingen Uhrsachen und Behandlung von Krankenheiten (Causae et Curae). Translated by Hugo Schultz. München: Verlag der Ärztlichen Rundschau Otto Gmelin, 1933. 5th ed. Ulm: Haug, 1955.

Welt und Mensch: Das Buch "De operatione Dei," aus dem Genter Kodex [Liber divinorum operum]. Translated and edited by Maura Böckeler. Salzburg: Müller, 1954.

Heilkunde: Das Buch von dem Grund und Wesen und der Heilung der Krankheiten [Causae et Curae]. Translated by Heinrich Schipperges. Salzburg: Müller, 1957. 4th ed. Salzburg: Müller, 1981.

Naturkunde: Das Buch von dem inneren Wesen der verschiedenen Naturen in der Schöpfung (Physica). Translated and edited by Peter Riethe. Salzburg: Müller, 1959. 3d ed. Salzburg: Müller, 1980.

Das Buch von den Steinen. Translated and edited by Peter Riethe. Salzburg: Müller, 1979. 2d ed. Salzburg: Müller, 1986.

Le Livre des Subtilites des Creatures Divines: Physique. Translated by Pierre Monat. Grenoble: Jerome Millon, 1988–89.

Das Buch von den Fischen. Translated and edited by Peter Riethe. Salzburg: Müller, 1991.

Other Works

"Liber Scivias." In *Liber trium virozum et trium spiritualium virginum,* edited by J. Faber Stapulenis. Paris: Etienne, 1513.

"Epistolarum liber." In *Magna Bibliotheca . . .* vol.15, edited by Margarino de la Bigne. Parisiis: I. Billaine, S. Piget, F. Leonard, 1654.

Liber divinorum operum. In *Patrologia Latina,* vol.197, edited by Jacques P. Migne. Paris: Petit-Montrogue, 1855.

"Riesenkodex, Marburg, Westdeutsche Bibliothek, Berlin, Pr St B Cod. lat. (MS B)." In *Patrologialatina,* vol.197, edited by Jacques P. Migne. Paris: Petit-Montrogue, 1855.

Liber vitae meritorium. In *Sanctae Hildegardis Opera,* edited by J. B. Pitra. Rome: Monte Cassino, 1882.

Wisse die Wege—Scivias—nach dem Originaltext des illuminierten Rupertsberger Kodex ins Deutsche. Translated and edited by Maura Böckeler. Berlin: Sankt Augustinus Verlag, 1928; 8th ed. Salzburg: Müller, 1987.

Briefwechsel von Hildegard von Bingen. Edited by Adelgundis Führkötter. Salzburg: Müller, 1965; Salzburg: Müller, 1980.

Sanctae Hildegardis Opera . . . In *Analecta Sacra,* vol. 8, edited by J. B. Pitra. Rome: Monte Cassino, 1882. Reprint. Farnborough, England: Gregg Press, 1966–67.

"Epistolae variorum ad S. Hildegardem cum ejusdem ad eos responsis." In *Veterum Scriptorum et Monumentorum, Dogmaticorum, Moralium,* 9 vols., edited by Edmundi Martene and Ursini Durand. New York: B. Franklin, 1968 (first published 1724).

Scivias. In *Corpus christianorum continuatio medievalis,* vol.43, edited by Adelgundis Führkötter and Angela Carlevaris. Turnhout, Belgium: Brepols, 1978.

Hildegard of Bingen's Book of Divine Works, with Letters and Songs. Translated by Robert Cunningham, edited by Matthew Fox. Santa Fe, NM: Bear, 1987.

Scivias. Translated by Columbia Hart and Jane Bishop. New York: Paulist Press, 1990.

Epistolarium/Hildegardis Bingensis. Edited by L. van Acker. In *Corpus christianorum continuatio medievalis,* vol. 91. Turnhout, Belgium: Brepols, 1991.

Schriften der Hildegard von Bingen. Translated and edited by Johannes Bühler. Hildesheim; New York: Olms, 1991 (originally published Leipzig: Insel, 1922).

The Letters of Hildegard of Bingen. Translated by Joseph L. Baird and Rodd K. Ehrman. New York: Oxford University Press, 1994– .

Works about Hildegard of Bingen

Alic, Margaret. "The Sibyl of the Rhine." *Hypatia's Heritage,* 62–76. London: Women's Press, 1986.

Allen, Prudence. "Hildegard of Bingen's philosophy of sex identity." *Thought* 64 (Sept. 1989): 231–242.

Annales Sancti Disibodi, Monumenta Germanie Historica (MGH) SS. XVII. A chronicle describing contemporary events of Hildegard's time.

Baring-Gould, Sabine. *The Lives of the Saints*. 3d ed. 15 vols. London: J. Hodges, 1872–82.

Brede, Maria. "Die Klöster der hl. Hildegard Rupertsberg und Eibingen." In *Hildegard von Bingen, 1179–1979*, edited by Anton Ph. Brück, 77–94. Mainz: Selbstverlag der Gesellschaft für Mittelrheinische Kirchengeschichte, 1979.

Butler, Alban. *Lives of the Saints*. Edited, revised, and supplemented by Herbert Thurston and Donald Attwater. 4 vols. New York: Kenedy, 1962.

Cadden, Joan. "It takes all kinds: Sexuality and gender differences in Hildegard of Bingen's 'Book of Compound Medicine.' " *Traditio* 40 (1984): 149–174.

Cannon, Sue S. "The Medicine of Hildegard of Bingen." Ph.D. diss., University of California at Los Angeles, 1993.

Chevalier, Ulysse. *Repertoire des Sources Historiques du Moyen Age*. 2d ed. 2 vols. Paris: A. Picard, 1905–7.
   66 studies about Hildegard.

Crook, Marion. *My Body: Women Speak about Their Health Care*. New York: Plenum Press, 1995.

Dronke, Peter. *Women Writers of the Middle Ages: A Critical Study of Texts from Perpetua (+203) to Marguerite Porete (+1310)*. Cambridge, England: Cambridge University Press, 1984.

Eckenstein, Lina. *Woman under Monasticism*. Cambridge, England: Cambridge University Press, 1896; New York: Russell and Russell, 1963.

Engbring, Gertrude M. "Saint Hildegard, twelfth-century physician." *Bulletin of the History of Medicine* 8 (1940): 770–784.

Fischer, Hermann. *Die heilige Hildegard von Bingen: Die erste deutsche Naturforscherin und Ärtzin; ihr Leben und Werk*. Münchener Beiträge zur Geschichte und Literatur der Naturwissenschaften und Medizin, 7/8. München: Verlag der Münchener Drucke, 1927.
   Critical study of the works of Hildegard of Bingen. Includes a bibliography of secondary sources.

Flanegan, Sabina. *Hildegard of Bingen, 1098–1179. A Visionary Life*. London and New York: Routledge, 1989.

Gembloux, Guibert. "Vita Sanctae Hildegardis." In *Sanctae Hildegardis Opera*, edited by J. B. Pitra. Rome: Monte Cassino, 1882.

Gottfried and Theoderich. *Vita sanctae Hildegardis*. In *Patrologia Latina*, vol.197, edited by Jacques P. Migne. Paris: Garnier, 1855. German translation by Adelgundis Führkötter. Düsseldorf: Patmos, 1968. 3d ed. Salzburg: Müller, 1980; Latin text with German editorial matters by Monika Klaes. Turnhout, Belgium: Brepols, 1993; English translation by James McGrath. Collegeville, MN: Liturgical Press, 1995.

Hildegard von Bingen 1179–1979. Festschrift zum 800. Todestag der Heiligen. Edited by Anton Ph. Brück. Quellen und Abhandlungen zur mittelrheinischen Kirchengeschichte, Bd. 33. Mainz: Selbstverlag der Gesellschaft für Mittelrheinische Kirchengeschichte, 1979.

Kraft, Kent T. "The Eye Sees More Than the Heart Knows: The Visionary Cosmology of Hildegard of Bingen." Ph.D. diss., University of Wisconsin, 1977.

Labalme, Patricia H. (ed). *Beyond Their Sex: Learned Women of the European Past*. New York: New York University Press, 1980.

Labarge, Margaret W. *Women in Medieval Life: A Small Sound of the Trumpet.* London: Hamish Hamilton, 1986.

Liebschütz, Hans. *Das allegorische Weltbild der heiligen Hildegard von Bingen.* Studien der Bibliothek Warburg, 16. Leipzig: B.G. Teubner, 1930. Reprint. Darmstadt: Wissenschaftliche Buchgesellschaft, 1964.

> Discusses the authenticity of Hildegard's writings. Traces her sources of cosmology to Persian and Gnostic ideas.

Marks, Geoffrey, and William K. Beatty. *Women in White.* New York: Scribner's Sons, 1972.

May, Johannes. *Die heilige Hildegard von Bingen aus dem Orden des heiligen Benedikt (1098–1179).* Kempten: J. Kösel, 1911.

Meyer, E.H.F. *Geschichte der Botanik,* III. Königsberg: Gebruder Borntruger, 1856. Reprint. Amsterdam: Asher, 1965.

> Covers herbs listed in *Physica.*

Newman, Barbara. *Sister of Wisdom: St. Hildegard's Theology of the Feminine.* Berkeley: University of California Press, 1987.

Pagel, Walter. *Paracelsus: An Introduction to Philosophical Medicine in the Era of the Renaissance.* 2d rev. ed. Basel: Karger, 1982.

Petroff, Elisabeth A. *Medieval Women's Visionary Literature.* Oxford: Oxford University Press, 1986.

Portman, Marie L. *Der Darstellung der Frau in der Geschichtschreibung der früheren Mittelalters.* Basel: Helbing and Lichtenhahn, 1958.

Sarton, George. *Introduction to the History of Science.* 3 vols. Baltimore: William and Wilkins for the Carnegie Institution of Washington, 1927–48.

Schrader, Marianna, and Adelgundis Führkötter. *Die Echtheit des Schrifttums der heiligen Hildegard von Bingen.* Beihefte zum Archiv für Kulturgeschichte, Hft. 6. Cologne: Böhlau Verlag, 1956.

> Studies early sources to determine the authenticity of Hildegard's works.

Singer, Charles. *Studies in the History and Method of Science.* Oxford: Clarendon Press, 1917–21. 2d ed. London: W. Dawson, 1955. Reprint. New York: Arno Press, 1975.

Thorndike, Lynn. *A History of Magic and Experimental Science during the First Thirteen Centuries of Our Era.* 6 vols. London: Macmillan, 1923. Reprint. 8 vols. New York: Columbia University Press, 1923–58.

> The life of Hildegard, influences from earlier times, relationship of science and religion in her writings. Well documented.

Ulrich, Ingeborg. *Hildegard of Bingen; Mystic Healer.* Collegeville, MN: Liturgical Press, 1993.

Wilson, Katharina (ed.). *Medieval Women Writers.* Athens: University of Georgia Press, 1984.

IDA HENRIETTA HYDE (1857–1945)

Mary R. S. Creese

BIOGRAPHY

Ida Henrietta Hyde was born on September 8, 1857, in Davenport, Iowa. She was one of the four children of German immigrants from Würtemberg, Meyer H. and Babette (Lowenthal) Heidenheimer, who changed their name to Hyde soon after arriving in the United States. When the children were still young, the father, a merchant, abandoned his family, leaving the mother to support them as best she could by taking in mending and cleaning. They moved to Chicago, where Mrs. Hyde gradually developed a fairly prosperous small business. She sent her children to public schools and gave them what was essentially a middle-class upbringing, although only her one son, Ben, was expected to go to university. In 1871, however, the Great Fire of Chicago destroyed their home and wiped out the business, leaving them virtually destitute; and so from the age of 14 Ida Hyde had to work for her living. Indeed, she assumed the major role in maintaining the family and supporting her brother, while he completed his schooling and took a degree at the University of Illinois. Starting off as a milliner's apprentice in a clothing factory, she rose to a position as factory buyer and saleslady. This experience was valuable to her later, since she was always able to make her own clothes.

About this time she discovered, in a packing box stored at the factory where she worked, an English translation of Alexander von Humboldt's great work *Ansichten der Natur* (Views of Nature 1849). It became her lunch-hour reading and started her lifelong fascination with biology. It also stimulated her to complete her secondary education, and in 1875–76 she attended night school classes at the Chicago Athenaeum, housed in a building near her place of work. Five years later, in 1881, while attending her brother's graduation at the University of Illinois, she met several young women who were doing academic work. She decided then to try to do so herself, despite very strong opposition from her mother and brother. After a summer of concentrated study on her own, she passed the examinations for entrance to the College Preparatory School and later was admitted to first-year classes at the University of Illinois. Her escape to academe was short-lived, however; in 1882 her brother became ill, and she had to go home to nurse him. She remained in Chicago for the next six years, earning money as a teacher of second- and third-grade classes in the public school system. One of her special projects was the introduction of nature study into the elementary school curriculum, and she did notable work in helping to establish a school science program in Chicago.

In 1888, at the age of 31, she finally returned to formal academic studies, enrolling at Cornell University, where she proceeded to complete the four-year A.B. degree course in three years. Although she then considered studying medicine, she instead accepted a biology scholarship offered her by Bryn Mawr College and began graduate studies with Thomas Hunt Morgan (Nobel laureate, 1933) and Jacques Loeb. Much of her research was done at the Woods Hole Marine Biological Laboratory, an institution with which she was to maintain close ties from then on.

Her study of the anatomy and embryology of three species of scyphozoans, completed in 1893, settled a long-standing controversy between two European zoologists, Alexander Goette and Carl Claus, on the development of medusae (Hyde, "Entwicklungsgeschichte . . ." 1894), and resulted in Goette's inviting her to continue her work in his laboratory at the University of Strassburg. A European fellowship for 1893–94, provided by the Association of Collegiate Alumnae, the forerunner of the American Association of University Women (AAUW), made it possible for her to accept the invitation. She made full use of the excellent facilities at Strassburg and did well in her research. Goette offered to accept her results as a doctoral dissertation, but he failed to persuade the rest of the Strassburg faculty to agree to this, and so Hyde transferred to the University of Heidelberg. That institution was somewhat more liberal in its attitude toward women students, having accepted several as auditors since 1891. Thus, in 1896, after two years of full-time study as an auditor and after considerable difficulty in obtaining the instruction she wanted in physiology (the professors of zoology and chemistry were much more accommodating), Hyde was awarded her degree multa cum laude (a designation slightly below summa cum laude). She was the third woman to receive a doctorate from the University of Heidelberg, following two Germans, Käthe Winschied (1894, modern languages) and Marie Gernet (1895, mathematics).

Hyde then spent six weeks at the Davis Table at the Naples Zoological Station in Italy and from there proceeded to the University of Bern, where she carried out research on muscle physiology under the guidance of Hugo Kroenecker. She returned to the United States in the early summer of 1897; Henry P. Bowditch, former dean of the Harvard Medical School, whom she had met in Bern, helped her obtain an Irwin Research Fellowship for work at Harvard. She became the first woman to carry out research at that institution's medical school. While there she also took some courses, particularly in bacteriology, and taught part-time as an instructor in histology and anatomy in two Cambridge preparatory schools.

In 1898 she applied for a permanent position at the University of Kansas in Lawrence. At the time it was a small state institution, in the process of establishing a medical school and in need of a physiologist. Having excellent academic credentials and outstanding recommendations, she obtained an initial appointment as assistant professor of zoology. The following year she became associate professor of physiology and in 1905 was promoted to full professor and head of the Physiology Department. She was one of the first women, possibly *the* first, to become a full professor and head of a science department in an American state university.

Hyde never married, but she maintained close relationships with her nieces and nephews. In general philosophical outlook she would seem to have been something of an agnostic and was out of sympathy with "organized religion." However, she did occasionally attend meetings of the Society of Ethical Culture in Lawrence. After she retired in 1920 at the age of 63, she did a considerable amount of traveling. In 1922–23, using Heidelberg as her base, she took trips to Switzerland, Austria, and various German cities and even went as far as Egypt and India. While in Heidelberg she also discussed research possibilities with faculty at the university but does not appear to have actually carried out any laboratory work. A few years later she settled in California, first in San Diego and then in Berkeley. She died of a cerebral hemorrhage on August 22, 1945, shortly before her 88th birthday.

WORK

Hyde's research covered a wide range of topics, and she was original and inventive. Her first notable work was her undergraduate thesis research on mammalian heart structure, carried out at Sage College, Cornell University. Working on a number of different species, she discovered the coronary valve, a structure distinct from the valve of the coronary sinus and located over the orifice of the middle cardiac vein (Hyde 1891). Her graduate studies were in two areas, respiratory patterns and the mechanism that controlled respiration in *Limulus* (the horseshoe crab), and the development of scyphozoans (jellyfishes). Among the observations she made on *Limulus* was the fact that signals were transmitted to the effector organs both by nerves and by the blood. At the Naples Zoological Station she spent a short time looking into methods for collecting octopus secretions. However, during her first postdoctoral year, when she was at Harvard, she returned to her earlier subject of mammalian heart structure. The topic she investigated was the contraction of the ventricle in association with increased blood volume during forced influx of fluid into the heart. She was the first to point out that the coronary blood vessels are compressed by this muscular action, and the flow of blood through them diminished as a consequence. This produces an increase in heart rate and in the force and degree of ventricular contraction.

The tendency of the ventricle to eject whatever volume is put into it had been demonstrated earlier by Otto Frank (1885), but Hyde showed the importance of heart tissue itself in that response. These observations, considered some of her most important findings, proved relevant much later in studies to determine the cause of sudden death in athletes whose coronary vessels happen to be, congenitally, deeply embedded in the cardiac muscle.

Her work on blood circulation continued after she moved to the University of Kansas and included further studies on the response of the cardiovascular system to stress. She also studied the functioning of the nervous system and, in addition, put considerable effort into investigations of respiratory processes, one of the areas to which she had been introduced during her year at Bryn Mawr. Much of the work was done during summers, which she regularly spent at the Woods Hole Laboratory. Her findings on the functioning of the respiratory center of the skate appeared in 1904, a year when she spent a summer at the University of Liverpool. Several other papers discussing nerve distribution in a number of animals came out about the same time.

Among Hyde's most original studies were her pioneering investigations by micromethods of processes in single cells. The work began from a chance observation that specimens kept in seawater undergoing concentration by evaporation showed abnormal cell development—in particular, multinuclear cells were formed in embryos. Hyde conjectured that the presence of electrolytes in high concentrations might prevent cytokinesis without affecting karyokinesis. From there she went on to study minute electrical charges at different stages in the development of eggs, work that led her to the conclusion that ionic imbalances in concentrated salt solutions caused viscosity changes that, in turn, affected the division of the cell. These findings foreshadowed current controversies about the effect of electrical potential on development.

The instrument Hyde used in her cellular-level work, the micropipette stimulating electrode, to the development of which she contributed, is generally regarded as one of the key tools that have facilitated advances in neurophysiology. Unfortunately, much of the early work by her and others that involved its use was overlooked, and the instrument was essentially reinvented in the 1940s.

Hyde also investigated the relationship of diet to health, the physiological effects of commonly used substances such as caffeine, alcohol, and nicotine, and the effects of sensory input. Her studies in the latter area included an examination of the effects of music on electrocardiograms and blood pressure. One of the first investigations of its kind, it might well be looked on as marking the start of work in music therapy. Hyde went as far as to suggest that music might come to be selectively prescribed as a useful adjunct to psychotherapy. Her 1924 paper on the subject, published after her retirement, was awarded a prize by the American Psychological Association.

Hyde's relatively heavy commitment to research notwithstanding, much of her energies while she was at the University of Kansas went into teaching and administration as she built the Department of Physiology and organized the premedical physiology curriculum. Over the course of several years she took summer classes in surgery and clinical medicine at the Rush Medical School in Chicago and at the University of Kansas Medical School. She completed most of the requirements for an M.D. degree, the lack of which she felt put her at a disadvantage when dealing with her medical school colleagues. Indeed, she claimed that she had completed the requirements but that the degree was withheld because she did part of the final year work in absentia (Maltby 1929, 14–15).

She applied her medical training to her work. Mindful of her public service obligations as a faculty member of a state university, she worked on a number of public health projects in Kansas schools and at Haskell Indian College in Lawrence, Kansas. With the cooperation of local doctors, she established a program of medical inspection of schoolchildren for communicable diseases, particularly tuberculosis and spinal meningitis. However, this program aroused the opposition of influential citizens opposed to "compulsory medicine." As a result, the university Chancellor, fearful that the controversy might affect the institution's state funding, withdrew his backing. Hyde's lectures in Kansas high schools and at the university on social hygiene (sex education) were pioneering, not to mention courageous. She felt strongly that young people of both sexes would benefit greatly from correct information on the subject and spoke out openly about such matters as the sexual transmission of diseases. In 1918 the governor of Kansas appointed her to the chair of the Woman's Committee on Health and Sanitation of the State National Defense Committee. Her interest in public health also led her to push for a number of basic necessities at the University of Kansas, such as adequate toilet facilities for women students.

She wrote two textbooks, *Outlines of Experimental Physiology* (1905) and *Laboratory Outlines of Physiology* (1914). Although the university's Board of Regents sometimes covered the costs of printing such faculty publications at the time, it declined to do so for her.

Throughout her career she set herself high professional standards and demanded much from students and assistants. She taught many women students, several of whom went on to further work in science. Her teaching extended to summer work, and from 1897 until about 1901 she served as instructor at preparatory schools in Cambridge, Massachusetts. She also put much effort into securing educational opportunities for women in biology at the national level, particularly by taking steps to have research tables endowed for women students at the Woods Hole and Naples Laboratories. A life member of the Naples Table Association, she served as secretary from 1897 to 1900. In her will she left $2,000 to fund a University of Kansas scholarship for women in biology and $25,000 to endow the Ida H. Hyde Woman's International Fellowship of the AAUW.

During her years at Kansas, however, Hyde had continual trouble dealing with administrators, and there was frequent friction between her and her colleagues in the schools of medicine and pharmacy. Her rather blunt and dogmatic style, evident in her correspondence with the university's Chancellor, Frank Strong, may well have been something of a disadvantage in the world of campus politics. A constant irritation to her was the fact that her salary was about 15% lower than that of many other department heads. She lodged a number of complaints with the administration on the matter, but these never had any effect. From 1915 onward her situation became less and less satisfactory. That year the university decided to merge Hyde's department in the College of Liberal Arts and Sciences with the Medical School's Physiology Department, the combined unit to be administered by a committee. Not surprisingly, she had strong objections to giving up the control of the department she had built. Worse was to follow. Within a year a rumor reached her ears that the university was looking for a new head for its Physiology Department. She resigned from her administrative duties in 1917 and the next year worked only part-time. She was then 61 and doubtless not yet ready to retire. She took a year's leave of absence (1919–20), spending some time at the Scripp's Institution in La Jolla, California, and then returned to Kansas for one semester before going on another leave of absence, which turned into permanent retirement.

Hyde was elected to the American Physiological Society in 1902 as its first woman member, and she was the only one until 1913. She also belonged to Sigma Xi, the American Biological Society, the American Medical Association, and the American Geographical Society. Despite her somewhat uneasy career in academe and her difficulties during her last years at Kansas, her achievements, both as a woman student in Germany in the 1890s and later as a department builder and research worker, are undeniably impressive.

NOTE

I am especially grateful to Barry Bunch, John Nugent, and the staff of the University of Kansas Archives office for making available their records on Ida Hyde. I also thank Gail Tucker for telling me about Hyde's activities in Europe in 1922–23.

BIBLIOGRAPHY

Works by Ida Henrietta Hyde

Scientific Works

"Notes on the hearts of certain mammals." *American Naturalist* 25 (1891): 861–863.
"Entwicklungsgeschichte einiger Scyphomedusen." *Zeitschrift für wissenschaftliche Zoologie* 58 (1894): 531–564.
"The nervous mechanism of the respiratory movements in *Limulus polyphemus.*" *Journal of Morphology* 9 (1894): 431–446.
(with J. R. Ewald) "Zur Physiologie des Labyrinths. IV. Mitteilung die Beziehungen des Grosshirns zum Tonuslabyrinth." *Archiv für die gesammte Physiologie des Menschen und der Tiere* (Pflüger) 60 (1895): 492–508.

"Entwicklungsgeschichte einiger Scyphomedusen." Inaugural diss., Heidelberg, 1896. Leipzig: n.p., n.d.

"Beobachtungen ueber die Secretion der sogenannten Speichel druesen von *Octopus macropus.*" *Zeitschrift für Biologie* 35 (1897): 459–477.

"Collateral circulation in the cat after ligation of the post cava." *Kansas University Quarterly* 9 (1900): 167–171.

"The effect of distention of the ventricle on the flow of blood through the walls of the heart." *American Journal of Physiology* 1 (1900): 215–225.

"The nervous system of *Gonionema murbachii.*" *Biological Bulletin* 4 (1902): 40–45.

"The nerve distribution in the eye of *Pecten irradians.*" In *Mark Anniversary Volume*, article 24, 471–482. New York: Holt, 1903.
    Invited paper.

"Differences in electrical potential in developing eggs." *American Journal of Physiology* 12 (1904): 241–275.

"Localization of the respiratory center in the skate." *American Journal of Physiology* 10 (1904): 236–258.

Outlines of Experimental Physiology. Lawrence, KS: n.p., 1905.

"Recent scientific contributions to social welfare. Modern aspects of physiology." *Chautauquan* 41 (1905): 244–250.

"A reflex respiratory center." *American Journal of Physiology* 16 (1906): 368–377.

"The educational importance of physiology." *Interstate Schoolman* (1907): 18–20.

"The effect of salt solutions on the respiration, heart beat and blood pressure in the skate." *American Journal of Physiology* 23 (1908): 201–213.

(with R. Spray and I. Howat) "The influence of alcohol upon the reflex action of some cutaneous sense organs in the frog." *Kansas University Science Bulletin* 7 (1913): 229–238.

Laboratory Outlines of Physiology. Lawrence, KS: Department of Journalism Press, University of Kansas, 1914.

"The development of a tunicate without nerves." *Kansas University Science Bulletin* 9 (1915): 175–184.

(with C. Spreier) "The influence of light upon reproduction in *Vorticella.*" *Kansas University Science Bulletin* 9 (1915): 398–399.

(with C. B. Root and H. Curl) "A comparison of the effects of breakfast, of no breakfast and of caffeine on work in an athlete and a non-athlete." *American Journal of Physiology* 43 (1917): 371–394.

(with W. Scalopino) "The influence of music upon electrocardiograms and blood pressure." *American Journal of Physiology* 46 (1918): 35–38.

"A micro-electrode and unicellular stimulation." *Biological Bulletin* 40 (1921): 130–133.

"Effects of music upon electrocardiograms and blood pressure." *Journal of Experimental Psychology* 7 (1924): 213–224.

Other Works

"Before women were human beings." *Journal of the American Association of University Women* 31 (1938): 226–236.
    Reminiscences of personal experiences as a student in Germany.

Works about Ida Henrietta Hyde

"Ida H. Hyde, pioneer." *Journal of the American Association of University Women* (Fall 1945): 42.

Johnson, E. E. "Ida Henrietta Hyde: Early experiments." *Physiologist* 24 (1981): 10–11.

Maltby, M. E. (comp.). *History of the Fellowships Awarded by the American Association of University Women, 1888–1929, with Vitas of the Fellows*, 14–15. Washington, DC: American Association of University Women, 1929.

Tucker, G. S. "Ida Henrietta Hyde: The first woman member of the society." *Physiologist* 24 (1981): 1–10.

———. "Reflections on the life of Ida Henrietta Hyde, 1857–1945; the woman scientist in the twentieth century." *Creative Woman Quarterly* 1 (1978): 5–8.

Other References

Albisetti, J. C. *Schooling German Girls and Women: Secondary and Higher Education in the Nineteenth Century*. Princeton: Princeton University Press, 1988.

Frank, O. "Zur Dynamik des Herzmuskels." *Zeitschrift für Biologie* 24 (1885): 374–437.

Humboldt, F.W.H. Alexander von. *Ansichten der Natur, mit Wissenschaftlichen Erläuterungen*. Tübingen: J. G. Cotta, 1808. Translation, *Views of Nature*. London: Longmans, 1849.

Hyde, I. H. *Diaries*. AAUW Archives Library, Washington DC.

Sloan, J. B. "The founding of the Naples Table Association for promoting scientific research by women." *Signs* 4 (1978): 208–216.

University of Kansas Archives (Ida Hyde file; Chancellor's Office files, correspondence, faculty, 1903–20).

LIBBIE HENRIETTA HYMAN (1888–1969)

Soraya Ghayourmanesh-Svoronos

BIOGRAPHY

Libbie Henrietta Hyman was born on December 6, 1888, in Des Moines, Iowa, the daughter of Joseph and Sabina (Neumann) Hyman, both Jewish immigrants. Her father was a tailor who was born in Konin, Poland, and had left his country at age 14 for England and then for the United States, where his opportunities for success were greater than back home. His daughter remembers that he was able to speak English without an accent. Her mother was born in Stettin, Germany, one of eight children whose father died at a young age. She migrated to the United States and moved to Iowa, where her older brother was living. After several years of a difficult life, she decided to move in with a family named Posner. Eventually, she met and married Joseph Hyman (1884), who was 20 years her senior. She believed it was a mother's duty to rule her family and that daughters should stay at home and do the housework. As a result Hyman, the only daughter, had a harder life than her brothers Samuel, Arthur, and David.

Her father's business was not successful, especially when the family moved to Sioux Falls, South Dakota. In spite of her mother's domineering attitude, Hyman was a happy child. She enjoyed walking in the woods and studying trees, flowers, and weeds, fascinated by their richness and variety. She learned all their scientific names from a high school botany book that her brothers had acquired. In her teens she started a collection of butterflies and moths, which she arranged in a frame. She also enjoyed reading the great collection of books her father owned.

Hyman was the valedictorian and the youngest member of her graduating class (1905) in Fort Dodge High School. She had her father's love for learning, but it never occurred to her that she could go to college, even though her scholastic record was outstanding. She knew, however, that her high school offered advanced classes and that often students would come back after their graduation for a year's post–high school work. She, therefore, went back to school and took all the science courses that were available at the time. However, she was still too young to be eligible for teaching in an Iowa school. She accepted a job at minimum wage, pasting colored labels on boxes at Mother's Rolled Oats factory.

One day, in the summer of 1906, while on the way home from work, she met Mary Crawford, her English and German high school teacher, who asked about her current life. Crawford was surprised that Hyman did not think of going to college. Her teacher indicated that she could arrange for a scholarship to the University of Chicago but that Hyman would have to arrange for room and board. Hyman obtained the scholarship and moved to Chicago, where she supported herself by working as a cashier in the Women's Commons, the big dining room serving women students. The scholarship was renewable on a yearly basis, and Hyman managed to earn her bachelor's degree and, eventually, her Ph.D. degree without paying a penny in tuition fees.

In 1907 Hyman's father died, and her mother and brothers moved to Chicago. Hyman did not return to Fort Dodge until 50 years later. She lived at home until her mother's death in 1929, pursuing her career, despite heavy opposition by her mother and bachelor brothers.

At the university Hyman tried to become a botanist, but she perceived an anti-Semitic atmosphere after making enemies with a laboratory assistant. She then switched to chemistry. Still unimpressed with this lifeless science, she decided to try the zoology department. In her senior year, a course with Dr. Charles Manning Child convinced her to remain in zoology. (Child had earned his Ph.D. degree in Germany, after switching from graduate studies in philosophy.) He was the first to indicate that any organism was more than the sum of all its parts. Child was impressed with Hyman's intelligence and progress and encouraged her to pursue further studies in zoology.

Hyman graduated in 1910 with honors in zoology and decided to work for her Ph.D. degree in the same field. She was the first doctoral student to work with Child. Many other scientists at the time were discouraged because of his supposed radical and unsound ideas. She obtained a laboratory assistant fellowship from the University that took care of her expenses. Hyman suggested her own doctoral thesis topic and earned her Ph.D. degree in five years. She decided to stay in Child's laboratory, first as his assistant and then as his research assistant in elementary zoology and comparative vertebrate anatomy. She retained this position for 14 years (1917–31), publishing two texts that were outstanding successes.

Child retired in 1931, and Hyman realized that she had a choice to make. Teaching was not appealing to her, so she decided to write books on the invertebrates, the so-called "backboneless animals." She knew that there were no advanced treatises on invertebrates written in English. Hyman resigned from the University of Chicago, although the university was willing to give her time and space to do her work. She preferred to start in fresh surroundings. This enabled her to be away from her two older bachelor brothers, who expected her to take care of them.

After resigning from the University of Chicago, she traveled to Europe, where she stayed for 15 months; part of the time she visited the Naples Zoological Station in Italy. Upon her return to the United States, she rented an apartment in New York City near the American Museum of Natural History in order to use its library. In 1937 the Museum offered her a small laboratory and made her a titular member of its staff (a nonpaid staff member). She was later given the rank of research associate. At the Museum she worked on her series of volumes dealing with invertebrates, until the time of her death.

Hyman was fluent in several languages and had an excellent photographic memory, which helped her coordinate a great deal of information. She never had a secretary, an assistant, an illustrator, or a technician. During her stay in New York she published six volumes. The last book was delayed by her failing health, which confined her to a wheelchair.

In 1941 Hyman settled near the Croton Reservoir in a five-room house set on an acre of ground with many trees and flowers. The love for botany had never left her. Every day she used to walk the distance between her house and the station, except in the worst weather. She cooked for herself and even tended her own furnace. However, it took her four hours daily to commute from her home to the Museum. Eventually, this forced her to sell her house (1952) and return to New York City, where she lived in a hotel apartment.

To many people Hyman appeared as a proud, fair, and honest person, but not vain or abrasive. Although her appearance was unprepossessing, she was well respected by all who knew her. She carried on extensive correspondence with scientists from around the world. Despite the fact that she lived and worked alone, she was not a recluse. Many of her colleagues believed that she smoked cigars; in actuality, she never smoked or drank. When asked to comment on her work, Hyman once modestly said, "The treatise on the invertebrates has brought me much fame and many honors, but has given the zoological public an exaggerated idea of my scientific abilities" ("Dr. Libbie . . ." 1969). More than any other scientist, Hyman was the person whom professional zoologists from all over the world wished to meet when they visited the Museum. She carefully chose the few people she respected and considered friends. They knew her as a warm, gentle, and generous person who was liberal in her charitable contributions. Hyman lived to enjoy the appreciation of the scientific community. She received honorary Sc.D. degrees from the University of Chicago (1941), Goucher College (1958), and Coe College (1959), and an LL.D. degree from Uppsala College (1963). She was the first woman to be awarded the Daniel Giraud Elliot Medal of the National Academy of Sciences for Zoology and Paleontology (1955). She was also awarded the Gold Medal of the Linnaean Society of London (1960). On April 9, 1969, at the centennial celebration of the American Museum of Natural History, she was awarded its Gold Medal for Distinguished Achievements in Science.

In 1961 Hyman's health was permanently impaired by surgery, and for the rest of her life she was never well again. She had Parkinson's disease for more than ten years. She died on August 3, 1969, at her home. In fact, she was in a wheelchair when she accepted her award from the American Museum of Natural History a few months before her death. Her brother, Arthur, of San Diego, was the only sibling alive at the time.

WORK

In her junior year at the University of Chicago, Hyman switched to zoology, where she met the second teacher who had an impact on her career. This was Mary Blount, a Ph.D. candidate in zoology and a laboratory assistant teaching elementary zoology. Blount gave her the encouragement she needed. In her senior year, upon Blount's recommendation, Hyman took a course on invertebrates with Dr. Child. This was the turning point of her career.

Earning her doctorate with Child was a truly great experience, because he was one of the top zoologists at the time. Her 15 years in his laboratory resulted in a series of scientific studies that were instrumental in establishing the field of invertebrate zoology. Initially, Hyman knew little about invertebrates but soon became the foremost taxonomist of nonparasitic worms.

Most of Hyman's early articles were contributions to Child's projects. Studies of regeneration and the metabolism of adjacent tissues were performed. Of great assistance in her research was her knowledge of chemistry. For instance, she determined that water-dwelling worms have a respiratory system that involves oxygen. She placed worms in a water tank and at defined periods of time measured the oxygen content of water. Thus, she determined the quantity of oxygen used in their respiratory processes. The worms were then cut into segments, and different parts were placed in several water tanks. She determined that the respiratory rate of the front segment of the worm is different from the rate of its rear segment and that (as with humans) respiratory rates decline with age. She also showed that worms can generate electric currents that can be measured with a delicate instrument.

While working on worms, Hyman wrote two textbooks, as suggested by the University of Chicago Press. The first, *A Laboratory Manual for Elementary Zoology* (1919), was received enthusiastically. The second, *A Laboratory Manual for Comparative Vertebrate Anatomy* (1922), was later expanded as *Comparative Vertebrate Anatomy* (1942). Both editions had a great impact on the academic field. Although she was more involved with the invertebrates, she still provided a text that was quickly adopted by the majority of premedical schools. Her style of writing, coupled with rich, well-planned information, was greatly admired by most vertebrate zoology instructors. Many scientists consulted with Hyman and sent her specimens for identification. Her judgment and criticisms were always well respected. She often took it upon herself to explain why a term was unacceptable. She was the one to prove that the hemichordates are not part of the chordates, as generally believed, by outlining their differences.

The income from the royalties made her financially independent. In 1931 she resigned from her position at the University of Chicago and never earned a salary thereafter. Her interests in the protozoans, sponges, coelenterates (jellyfish), and turbellarians (flatworms) made her decide to publish a major monographic series in English. She insisted that she never liked vertebrate anatomy, and since 1942 she abandoned all contact with the subject.

Coming to New York as a nonpaid member of the American Museum of Natural History staff was a great experience for the enthusiastic Hyman. She started the monumental task of writing an account of the morphology, physiology, embryology, and biology of the invertebrates. Initially, she thought it was going to be only two volumes but quickly realized that it would be much longer. Her first volume (published in 1940) sold for a mere seven dollars and was followed by five more volumes, until her project was interrupted by her death. Because of high publishing costs, she decided to do all the illustrations herself, including many fine pen drawings. To make her drawings from living or prepared material, Hyman spent several summers at the Marine Biological Laboratory in Woods Hole, Massachusetts, and at other marine stations. The preface of the first volume states that "it is obviously impossible for any one person to have a comprehensive first-hand knowledge of the entire range of the invertebrates, and consequently a work of this kind is essentially a compilation from the literature." Nevertheless, her work involved the analysis, evaluation, and uniform integration of information necessary for this undertaking. Volumes 2 and 3 were published in 1951, volume 4 in 1955, volume 5 in 1959, and volume 6 in 1967. Between 1914 and 1966 she also published approximately 150 papers on invertebrates.

Hyman was a member of the National Academy of Sciences (elected in 1961), the American Society of Zoologists (vice president, 1953), the American Microscopical Society, the American Society of Naturalists, the American Society of Limnology and Oceanography, the Marine Biological Association, the Society of Protozoologists, Phi Beta Kappa, and Sigma Xi. During her career she was on the editorial boards of several publications, the editor of *Systematic Zoology* (1959–63), and the president of the Society of Systematic Zoology (1959).

Hyman was quoted as saying:

I don't like vertebrates. It's hard to explain but I just can't get excited about them, never could. I like invertebrates. I don't mean worms particularly, although a worm can be almost anything, including the larva of a beautiful butterfly. But I do like the soft delicate ones, the jellyfishes and corals and the beautiful microscopic organisms. (Winsor 1980)

After her death, the publisher of *The Invertebrates* proceeded with four more volumes, which were completed by a group of scientists headed by Dr. Joel H. Hedgepeth of the Marine Science Laboratory at Oregon State University. The preface to her last volume ends with the words, "I now retire from the field, satisfied that I have accomplished my original purpose—to stimulate the study of invertebrates."

NOTE

The author would like to express her appreciation for the valuable information provided by Joel Sweimler, Special Collections, the American Museum of Natural History.

BIBLIOGRAPHY

Works by Libbie Henrietta Hyman

Scientific Works

Space does not permit the listing of the complete works of Libbie Henrietta Hyman. Listed here are all works by Hyman except those cited in Emerson (1974). Included here are her dissertation and all references mentioned in the text.

"An Analysis of the Process of Regeneration in Certain Microdrilous Oligochaetes." Ph.D. diss., University of Chicago, 1915. Also in *Journal of Experimental Zoology* 20 (1916): 99–163.

A Laboratory Manual for Elementary Zoology. Chicago: University of Chicago Press, 1919; 2d ed., 1929.

A Laboratory Manual for Comparative Vertebrate Anatomy. Chicago: University of Chicago Press, 1922.

The Invertebrates: Protozoa through Ctenophora, vol. 1. New York: McGraw-Hill, 1940.

Comparative Vertebrate Anatomy. Chicago: University of Chicago Press, 1942.

The Invertebrates: Platyhelminthes and Rhynchocoela, vol. 2. New York: McGraw-Hill, 1951.

The Invertebrates: Acanthocephala, Aschelminthes, and Entoprocta, vol. 3. New York: McGraw-Hill, 1951.

The Invertebrates: Echinodermata, vol. 4. New York: McGraw-Hill, 1955.

The Invertebrates: Smaller Coelomate Groups, vol. 5. New York: McGraw-Hill, 1959.

The Invertebrates: Mollusca I, vol. 6. New York: McGraw-Hill, 1967.

(with G. E. Hutchinson) "Libbie Henrietta Hyman." *Biographical Memoirs of the National Academy of Sciences* 60 (1991): 102–114.

Works about Libbie Henrietta Hyman

"Dr. Libbie Hyman, zoologist, dead." *New York Times* (Aug. 5, 1969): 37.

Emerson, William K. "Bibliography of Libbie H. Hyman." In *Biology of Turbellaria,* edited by N. W. Riser and M. P. Morse, xv–xxv. New York: McGraw-Hill, 1974. The entire book is a Libbie H. Hyman memorial symposium volume.

"Geneticist tells of fall-out harm." *New York Times* (Apr. 26, 1955): 17.

Marcus, J. R. *The American Jewish Woman: A Documentary History.* New York: KTAV Publishing House, 1981.

"Museum worries about man on its centennial." *New York Times* (Apr. 10, 1969): 52.

"Obituaries. Dr. Libbie Henrietta Hyman." *Nature* 225 (Jan. 24, 1970): 393–394.

Stunkard, H. W. "In memoriam: Libbie Henrietta Hyman." *Journal of Biological Psychology* 12 (Oct. 1970): 1–23.

Winsor, M. P. "Hyman, Libbie Henrietta." In *Notable American Women: The Modern Period*, edited by Barbara Sicherman et al., 365– 367. Cambridge, MA: Belknap Press of Harvard University Press, 1980.

EMMY KLIENEBERGER-NOBEL (1892–1985)

Gary E. Rice

BIOGRAPHY

Emmy Klieneberger was born on February 25, 1892, in Frankfurt, Germany, the fourth and youngest child of (Abraham) Adolf and Sophie (Hamberger) Klieneberger. Her father's parents were from Austria (Bohemia), while her mother's parents came from Hanau am Main. Her siblings were much older: Carl was 16, Otto was 13, and Anna was 11 when Emmy was born.

The father came from a family of modest means. He joined the Austrian army at 18 and attained the rank of sergeant in the cavalry before leaving the service at age 38 to begin a career as a wine merchant. He eventually established his own successful wine cellar in Frankfurt, thereby providing his family with a comfortable life.

The mother, herself youngest of five children, was married at age 26. As a mother she had wanted only sons, because educational opportunities available to males were better than those for females. However, when two girls followed her two boys, she made certain that money was set aside for the girls' education, in amount equal to that set aside for her sons. The mother was the dominant force in the home; she was not domineering but rather kind, intelligent, and energetic.

While Emmy Klieneberger's parental grandfather was known as a rabbi, both her parents considered themselves "freethinkers" who favored assimilation. To this end, she herself was baptized at age seven, though she never practiced formal religion of any kind. Her childhood was virtually free of anti-Semitism.

Emmy Klieneberger was an average student, a bit of a dreamer who enjoyed sports and sewing, as well as drawing pictures of flowers and leaves found in the family garden. She was also an avid reader of works on the natural sciences. Teacher-training college followed her early education, and in 1911, at the age of 19, she was certified to teach. Lacking self-confidence and being somewhat shy, she deferred teaching at this point, opting instead for university classes. First she studied in the Unterprima of the Schillerschule, where she improved her skills in mathematics and Latin. Two years later she studied in Göttingen, which offered attractive lectures in mathematics and the sciences.

The advent of World War I and a three-month bout with pleurisy sent Klieneberger home to Frankfurt in 1914, where she attended classes at the new Goethe University. Botany and zoology were now her main interests. The next year she was assigned a problem that led to her dissertation (in German): "The Dimensions and Continuation of Cell Nuclei with Particular Reference to Classification." Two happy years of laboratory work followed, and in January of 1917, she earned her doctorate in botany, with mathematics and zoology as areas of special emphasis.

Because the term was only half over, she spent the next few months working in the zoology laboratory at the university, studying single-celled marine life. She then returned to Göttingen for the summer, where she attended more lectures in mathematics. By the autumn of 1918, she went back to Frankfurt and passed the examination to be a teacher in the higher schools. This necessitated a year as a teaching "probationer," observing teachers at the Schiller school and practice-teaching.

Her first and only teaching job was at the Nolden School for Girls in Dresden (1919). There she taught physics, chemistry, and biology to the upper-level girls and arithmetic to the middle schoolers. She found the work pleasant, but she missed the laboratory. Therefore, at the end of three years, she resigned and took a position as a bacteriologist at the Hygiene Institute in Frankfurt. Here she worked for the next 11 years. Although she knew little about bacteriology when she began, by 1930 she had become a member of the German Society for Hygiene and Bacteriology and was lecturing scientists on the subject as a member of the Institute's medical faculty.

The anti-Semitism that accompanied Hitler's rise to power devastated her family and caused her to lose her position. Both of her brothers and her sister later suffered because of the anti-Semitic atmosphere. Two of them committed suicide. The other brother eventually emigrated with her help to South America, where he spent his final years in lonely isolation.

In 1933 Klieneberger emigrated to England, securing a position at the Lister Institute in London. There she stayed for 29 years, the remainder of her professional life. Klieneberger was awarded a Ph.D. degree (1942) by the University of London, prior to receiving her D.Sc. degree and being appointed a full member of the Lister Institute staff. One year later she met Professor Edmund Nobel, a pediatrician from Vienna. He had become known for his success in controlling the incidence of endemic goiter among Viennese and Belgian children, as well as for his work on behalf of children's health issues through the League of Nations. At the time of their meeting Nobel was living with, and caring for, his aged mother, who died in 1943. The two married early in 1944, when Nobel was 60, and Klieneberger was 52. Two years later Nobel was dead. An obituary appeared in the *Lancet* on March 16, 1946.

In the spring of 1947 Klieneberger-Nobel began a year's work at the Hygiene

Institute in Zurich. By this time her reputation as a leader in the field of bacteriology was established; subsequent years were spent solidifying and maintaining that reputation.

She retired in October 1962 and spent her last years writing, learning Spanish, growing flowers, and receiving accolades. On her 75th birthday she was made an honorary member of the Robert Koch Institute, and soon after that she became a corresponding member of the German Society for Bacteriology and Hygiene. She was among the first honorary members of the International Organization for Mycoplasmology, and an award lectureship was established by that organization in her name. She also was awarded the Robert Koch Medal for her achievements in the field of microbiology. At her retirement dinner, Sir Ashley Miles, Director of the Lister Institute, spoke of Klieneberger-Nobel's "courage, honesty, and charm." She died in 1985, at the age of 93.

WORK

Klieneberger-Nobel was one of the fortunate scientists to enter her profession at a time when molecular biology was expanding greatly. She is acknowledged by her peers as a pioneer in mycoplasma research. Her success in this area can be attributed to her unique qualities of imagination and intellectual genius. Most notable of Klieneberger-Nobel's discoveries were the L-forms of mycoplasmas and bacteria, which she named after the Lister Institute. While other researchers were also examining these bacterial variants, Klieneberger-Nobel succeeded in profiling their cellular and colonial morphology.

Her first notable work was published in 1934, when she wrote a paper entitled "The colonial development of the organisms of pleuropneumonia and agalactia . . ." It was the first published paper that identified differences within mycoplasmas (pleuropneumonia-like organisms, or PPLOs, as they were known then). She realized that there were probably other organisms that were similar but different from the other bacterial forms known at that time. Continuing her work at the Lister Institute under the direction of John C. G. Ledingham, she began to explore systematically the similarities among these PPLOs, despite discouragement from Ledingham. Klieneberger-Nobel began her exploratory search on the mucous membranes of guinea pigs with no success. Then she broadened her search to strains of rats and mice used at the Lister Institute that were known to suffer from an unusual bronchopneumonia. Researchers had been unsuccessful in culturing the fluid released from the lesions of this bronchopneumonia, which would have provided the etiological cause of the disease. One day, Klieneberger-Nobel smeared culture plates containing a medium she had devised with the pus from the bronchopneumonia abscesses, thereby successfully growing the organisms responsible for the disease. She found that, after two or three days of incubation, minute colonies had grown that were similar in appearance to those of the pleuropneumonia and agalactia. Ledingham was astonished, and she continued to find similar organisms in strains of rats and mice brought to her from other locations.

On the strength of these early discoveries, Klieneberger-Nobel was awarded the Jenner Memorial Scholarship at the Lister Institute, which allowed her three years of uninterrupted time for research. Other scientists in the field of bacteriology were beginning to work successfully with her L-forms. It was found that L-forms existed not only in rodents but also in dogs. In addition, a saprophytic strain was located in sewage and soil.

In the late 1930s Klieneberger-Nobel read of a "rolling disease" described by Dr. Albert Sabin, a prominent American bacteriologist, that resulted from a toxoplasma infection of the brains of mice. Klieneberger-Nobel wrote to Sabin, who sent her freeze-dried brains from the infected animals. She was successful in growing cultures from these samples in her specially developed medium, and she shared her results with Sabin. Several weeks before her results were to be published in the *Lancet* (with Findlay et al. 1938), Sabin published his results in *Science* without mentioning the work of, or his correspondence with, Klieneberger-Nobel.

Just before the outbreak of World War II, she continued to work with Dr. George M. Findlay of the Welcome Bureau for Scientific Research, with whom she had identified a mycoplasma in the brains of mice exhibiting the "rolling disease." Their work examined a disease of rats called polyarthritis. When fluid from the joints of infected rats was smeared on her special nutrient agar, a large number of mycoplasma colonies grew. These colonies were identified as the same organisms (L-4) that Klieneberger-Nobel had previously cultivated from the swollen glands of rats. By injecting rats with an adjuvant made of the agar suspension and the L-4 organism, the severe arthritic condition was reproduced, demonstrating that the L-4 mycoplasma, *Mycoplasma arthritides*, was the responsible causative agent of polyarthritis in rats.

In 1954 and 1955 Klieneberger-Nobel made another discovery associated with the L-forms. She indicates in her *Memoirs* (1980, 121) that after reading a paper by Kwok-Kew Cheng, she recognized that the problem investigated therein, a bronchiectasis infection, was probably due to the L-3 *Mycoplasma pulmonis* that would later cause the rats to get bronchopneumonia. Klieneberger-Nobel had been unsuccessful originally in producing the disease experimentally. Then she was able to create ideal conditions for the multiplication and growth of the mycoplasma by tying off the bronchus, a procedure developed by Cheng. Klieneberger-Nobel was able to demonstrate that the etiological cause of rat bronchopneumonia was the *Mycoplasma pulmonis*.

Her research in identifying *Mycoplasma arthritides* from rats and *Mycoplasma neurolyticum* from mice contributed significantly to the understanding of these mycoplasmas in infectious diseases. Later isolation of *Mycoplasma pneumon iae*, a chief agent in human respiratory disease, is attributable to the work of Klienenberger-Nobel.

Her early discoveries opened the door for further research and clarification into mycoplasmas or L-forms. Thus, she initiated a whole new field, the study of mycoplasmas, as a distinct group of organisms.

BIBLIOGRAPHY

Works by Emmy Klieneberger-Nobel

Scientific Works

Space does not permit the listing of the complete works of Emmy Klieneberger-Nobel. Listed here are all works except those cited in Klieneberger-Nobel (1980). Included are her dissertation and all references cited in the text.

"Ueber die Grösse und Beschaffenheit der Zellkerne mit besonderer Berucksichtigung der Systematik." Diss., Goethe University Frankfurt am Main, 1917 (as E. Klieneberger).

"The colonial development of the organisms of pleuropneumonia and agalactia on serum agar and variations of the morphology under different conditions of growth." *Journal of Pathology and Bacteriology* 39 (1934): 409–420 (as E. Klieneberger).

(with G. M. Findley, F. O. MacCallum, et al.) "Rolling disease. New symptoms in mice associated with a pleuropneumonia-like organism." *Lancet* 235 (1938): 1511–1513 (as E. Klieneberger).

Memoirs. London: Academic Press, 1980.

Works about Emmy Klieneberger-Nobel

Chick, Harriette, et al. *War on Disease*. London: Deutsch, 1971.

Tully, Joseph G. "Foreword." In *Memoirs*, by Emmy Klieneberger-Nobel. London: Academic Press, 1980.

Other References

Obituary of Edmund Nobel. *Lancet* (Mar. 16, 1946): 402.
  Reproduced in *Memoirs* by Klieneberger-Nobel, 106–107.

Sabin, Albert B. "Identification of the filterable, transmissible, neurolytic agent isolated from toxoplasma-infected tissue, as a new pleuropneumonia-like microbe." *Science* 88 (2294) (1938): 575–576.

———. "Isolation of a filterable, transmissible agent with 'neurologic' properties from toxoplasma-infected tissues." *Science* 88 (2278) (1938): 189–191.

REBECCA CRAIGHILL LANCEFIELD
(1895–1981)

Teresa T. Antony

BIOGRAPHY

Rebecca Craighill was born January 5, 1895, at Fort Wadsworth in Staten Island, New York. Her father, William Edward Craighill, was a West Point engineering graduate and at the time of her birth was working on the army post at Fort Wadsworth. He was a native of Virginia, where his forefathers had settled in the early 1700s. Her mother, Mary Wortley Montague Byram, also came from Virginia, where her ancestors had settled. Later they moved to Mississippi.

The Craighills had six daughters, and Rebecca was their third one. The rigorous discipline instilled in the children by the father helped her face many difficult situations in her life. Even though the mother was a homemaker, she was a great admirer of the legendary Julia Tutwiler, a champion of education for women. Therefore, she cultivated in her daughters the importance of good education. Craighill said once that her mother could have become president of the United States if she did not have to raise six daughters.

Being an army officer, Craighill's father was transferred frequently from post to post, which often resulted in the interruption of the children's early education. However, this gave the children a certain degree of independence. Whenever there were satisfactory and convenient public schools near the army post, the Craighill children attended such schools, giving them a variety of geographic exposures and experiences. Rebecca Craighill was the first to venture into college education. Her sisters followed her example later. One of them graduated from medical school, an unusual accomplishment for a woman at that time.

Rebecca Craighill started attending Wellesley College (Boston) in the fall of 1912, with the vague idea of majoring in French and English. In her first year she had an English teacher who made her students memorize long lists of titles and authors, which Craighill found very boring. At the same time she noticed that her roommate, who was taking a course in zoology, appeared to be enjoying the subject. This aroused Craighill's curiosity and interest, and she started taking as many biology courses as could be fitted into the college requirement for the

B.A. degree. The biology courses included an elementary course in bacteriology. She realized that knowledge of chemistry was very important for the appreciation of bacteriology, and hence, she took as many chemistry courses as she could. She graduated from Wellesley in 1916.

She got a job teaching science and mathematics at a girls' boarding school, Hopkins Hall in Burlington, Vermont. She was asked to teach physical geography and proved to be a great teacher. In addition to a small salary, she received room and board and was able to save some money for future graduate studies. The following year Teachers College at Columbia University offered her a graduate scholarship. This grant had been established by the Daughters of the Cincinnati for the daughters of army or navy officers. By now Craighill had developed a great interest in bacteriology and genetics. Even though she was a student at Teachers College, she was permitted to take all the bacteriology and related courses in Professor Hans Zinsser's Department of Bacteriology at the College of Physicians and Surgeons of Columbia University. In 1918 she was awarded an M.A. degree. She then married Donald Lancefield, who was also at Columbia, pursuing graduate work. She could not continue work for a Ph.D. degree, because her husband was called into military service during World War I. He was stationed with the Sanitary Corps at the Rockefeller Institute for Medical Research (now Rockefeller University). Rebecca Lancefield wanted very much to work along with her husband and applied for a position at Rockefeller Institute. In June 1918 she was appointed a technician at the Institute to assist Oswald T. Avery and Alphonse R. Dochez in their study of the bacterium *Streptococcus*. This work was funded by a grant from the Surgeon General's office but was terminated when the funding ended.

The Lancefields returned to Columbia University, where Rebecca Lancefield worked as a research assistant for Professor C. W. Metz in his genetics laboratory. In 1921 Donald Lancefield finished his Ph.D. degree and was offered a position at the University of Oregon. She accompanied her husband and was able to obtain a position teaching bacteriology. At the end of the academic year, the Lancefields returned to Columbia University. Her husband continued his earlier work in Professor Thomas Hunt Morgan's laboratory, where he had completed his doctoral research. Rebecca Lancefield explored the possibility of doing her Ph.D. work in bacteriology under Zinsser at Columbia. Even though Zinsser was not very enthusiastic about women working in his laboratory, he made an exception in the case of Lancefield. He had been impressed by her earlier work at Rockefeller Institute and at Columbia University. In 1922 she obtained a position with Professor Homer Swift at Rockefeller Institute, assisting him in studies related to rheumatic fever. She worked on a bacterium called *Streptococcus viridans*. She was able to use her research with Swift for her Columbia doctorate. It was not easy, and often she had to carry her racks of test tubes between Columbia University and Rockefeller Institute. She received her Ph.D. degree in bacteriology in 1925, and her thesis was entitled ''The Immunological Relationships of *Streptococcus viridans* and Certain of Its Chemical Fractions.'' She remained at Rockefeller Institute for the rest of her professional life, and worked with dedication on the study of various aspects of streptococci.

At Rockefeller Institute Lancefield was promoted through the academic ranks from institute assistant (1925), through member and professor (1958). She became professor emeritus in 1965 but continued to go to her laboratory every day, driving her car from Douglaston, Queens to Manhattan until she fell and broke her hip in November 1980. She died on March 3, 1981, due to complications from the accident. Her husband, who had retired as chairman of the Biology Department at Queens College, City University of New York, died a few months after her death. She is survived by her daughter, Jane Lancefield Hersey, and by two grandsons, Donald and James Hersey.

Many of Lancefield's friends and colleagues felt that there was a great delay by the scientific community in recognizing her contributions. However, microbiologists had acknowledged her as the outstanding authority on streptococci. Many honors had been bestowed on her while she was still alive. She received the T. Duckett Jones Award of the Helen Hay Whitney Foundation (1960), the American Heart Association Achievement Award (1964), and the Medal of the New York Academy of Medicine (1973). Rockefeller University recognized her contributions with an honorary D.Sc. degree (1973). In 1976 her alma mater, Wellesley College, recognized her with an honorary doctorate on the occasion of the 60th anniversary of her graduation. She was elected to the National Academy of Sciences in 1970.

Lancefield enjoyed informal parties in her laboratory and initiated a pre-Thanksgiving eggnog party at Rockefeller Institute. The recipe she developed for making eggnog is still being used in the laboratory.

No name is so clearly written in the minds of bacteriology students, medical doctors, and research workers in bacteriology as the Lancefield classification of hemolytic streptococci. In spite of her success and the many honors she received, she was a very modest person. She preferred that her tributes go unnoticed. She was a great supporter of equal rights for women. She acknowledged that it is not easy to have a successful scientific career and raise a family without compromising one or the other. Her life serves as an example that it is possible, although it may be difficult, to be successful as a scientist as well as wife and mother.

WORK

In her early work as a technical assistant to Avery and Duchez at Rockefeller Institute, Lancefield got her first exposure to the study of the microorganism *Streptococcus*. These two scientists had collected 125 cultures from army personnel in a Texas camp, where there was a serious epidemic of streptococcal infection concurrent with a measles epidemic. Her group used the most reliable method (also the most difficult), known as the ''mouse protection test,'' to analyze the cultures. She was able to identify among those cultures four im-munologically specific types of streptococci. At this time, streptococcal classi-fication and related pathogenicity were in a state of disorganization. The microorganism *Streptococcus* consisted of a large, diverse number of species. Some were primary agents of disease in humans and animals, while others were harmless and without any medical significance.

Even though Lancefield was only a technician at that time, her intense interest and talents were primarily responsible for the completion of this work within a year. Her contributions in identifying 70% of the strains were acknowledged by Avery and Dochez, making her a coauthor of their extensive 34-page published paper ("Studies on . . ." 1919). This paper was also the first documented evi-dence of the specific types of hemolytic streptococci in infections. From her later studies she was able to suggest preventive measures for military personnel to combat pharyngitis and related problems. Her studies also showed that dom-inant strains of streptococci that cause infections vary from year to year in their characteristics. Complications in scarlet fever may be caused by a different strain from that which initiated the disease. Her observations were also responsible for interpreting the persistence of type-specific antibody in a host who still might be open to the possibility of reinfection with the same strain.

In her later investigations Lancefield used a simple, reliable method, called the "precipitin test," to identify various antigens that are the causative agents in streptococcal infections. This test is based on the principle that soluble protein antigens, when mixed with their corresponding antibodies, give visible precip-itates that are easy to detect. She analyzed sera from patients suffering from various types of streptococcal infections. She then identified the major chemical components (a polysaccharide) in the bacteria common to all streptococci ob-tained from acute infections in humans. This polysaccharide was named the C carbohydrate. All the strains of the bacteria that contained the C substance were given the designation of group A. On further study, she showed that group A streptococci can be subdivided into many specific types on the basis of serolog-ical analysis of other components of the organism. These components, called antigens, responsible for the type-specificity of group A streptococci, were shown to be proteins. She named these proteins M substances, and they proved to be responsible for the virulence of the organism. She further proposed that the antibodies developed in humans as a result of infection were responsible for the immunity of the host against further infections by the same strain of the organism.

This hypothesis, that the M protein was responsible for virulence, was con-trary to the belief prevailing at that time, that all type-specific substances are carbohydrates or polysaccharides. It took a lot of courage and confidence on the part of Lancefield to challenge the current theory. She remarked that many chemists believed that her method of preparing the antigen, by boiling the strains of bacteria in acid, would destroy the antigenic property of the M protein. By her meticulous work she was able to convince her critics that, as long as she maintained proper conditions in the preparations of the antigens, her specimens were viable.

During the studies Lancefield and her coworkers noticed that many patients had recurrent attacks of rheumatic fever. For some time it was believed that strains of different type-specific organisms initiated each new attack. However, in some cases, Lancefield and coworkers identified the same strain in recurrent attacks. Further studies showed that antibiotic therapy used for the patients would eliminate the infecting organism, thereby interfering with antibody production. This could result in having no immunity in the patients, and it explained the recurrent attacks by the same strain of the bacteria.

In 1928 Lancefield continued her studies of the complex antigens of *Streptococcus hemolyticus* and characterized the chemical and immunological properties of the substances: C polysaccharide and the M protein. She also isolated from the same organism a nucleoprotein and called it the P substance. She demonstrated that the P substance could stimulate the production of antibodies in rabbits, proving that it was a true antigen. She further showed that the C polysaccharide from the same source would not stimulate antibody production in guinea pigs, proving that the C substance is not a true antigen but only a hapten (i.e., a substance that by itself cannot elicit antibody production).

In 1933 Lancefield published one of her landmark papers, called, ''A serological differentiation of human and other groups of hemolytic streptococci.'' This extensive study, which she had undertaken single-handedly, utilized 106 strains of streptococci collected from various sources: humans; animals such as cows, guinea pigs, rabbits, horses, swine, chicken, and foxes; as well as cheese and other dairy products. Only a person familiar with bacteriological studies can appreciate the months and even years of work involved in the collection and identification of the strains, preparation of each antigen, and the prolonged process of inducing and testing the antibodies. She tested each antigen and antibody by all the methods available at that time and compared the results with those obtained using the ''precipitin'' method. The methods she used were change in the pH of dextrose broth, hydrolysis of sodium hippurate, reduction of methylene blue milk, growth of bile agar, fermentation of different types of sugars, and lysis with streptococcus bacteriophage. Out of the 106 strains studied, Lancefield was able to identify 104 strains. Each strain was further grouped according to their specific reactions, and they were given alphabetical letters. Thus, group A strains were mainly of human origin, while group B was found in cows and milk. Group C was found in pigs and lower animals, and group D in cheese products. The animal industry also benefited from her studies, since it was possible to prevent certain types of animal epidemics.

Lancefield next concentrated her studies on group B streptococci isolated from bovine (beef) sources as well as from patients of a London maternity hospital. These studies showed that, contrary to the earlier belief, group B streptococci could also infect humans. Further, it was shown that group B could cause septicemia and meningitis in newborn infants. Extending her studies of the polysaccharide present on the bacterial capsule of group B, she showed that the specific antibodies to this polysaccharide could give immunity against experimental infection in animals. She continued her studies on the complex biochemical and antigenic structure of these polysaccharides, thereby enhancing the understanding of their reactions.

At the time there was a misconception that rheumatic fever was caused by a microorganism called *Streptococcus viridans*. Lancefield's work showed that rheumatic fever was caused by group A streptococci. She further studied the immunological properties of another antigen in streptococci, called the R antigen, and showed that the R antigen and its corresponding antibody had no effect on the virulence of streptococci.

Lancefield was also able to corroborate her findings with the observations of another authority on streptococci, Dr. Fred Griffith of England. Even though Griffith had used a different method for the identification of the bacteria, the "slide agglutination" procedure (Griffith 1935), he and Lancefield were able to concur that they were both identifying the same M antigens characteristic of group A strains. They further confirmed their conclusions by exchanging with each other all the strains, sera, and data they had collected and analyzing the strains, using the established procedures. The agreement between the two scientists was so close (with a few exceptions) that Lancefield, for the sake of uniformity, adopted Griffith's numbering system to identify different strains of each group, instead of the patients' names, which she had been using. This numbering was further accepted by the nomenclature subcommittee of the International Congress of Microbiology. It designated the usage of the M antigen as the basis for the serological typing of all group A streptococci. This attempt by Lancefield to systematize the classification of streptococcal strains helped avoid confusion among bacteriologists and made international collaboration among them much easier.

During her study of Griffith's strains, Lancefield detected a discrepancy between her strains and some of his strains and noticed that he had included in his group A some strains that did not belong to that group. She named the antigen from these strains the T antigen. She also showed that the T antigen, unlike the M antigen, had no relationship to the virulence of the bacteria but could be present on the bacterial strains where M antigen is also present. Further, she demonstrated that the T antigen could be used as an additional marker in cases where the M protein is either absent or difficult to identify.

Studies of the M antigen's properties by Lancefield showed that when the M protein is present in an organism, the white blood cells of the human host are not able to engulf and destroy the streptococcal bacteria. However, if the antibodies to the M proteins are present in the blood, the positive effect of the M protein is neutralized, and white blood cells are capable of destroying the bacteria. She also showed that immunity to streptococcal disease is type-specific, and recovery from one type of infection does not protect the host from being infected again by a different strain of the same group A organism. Thus, she was able to explain how people get repeated streptococcal infections. After a study of the persistence of type-specific antibodies in humans, she was able to suggest that it is possible to have lasting immunity to the M antigen.

Lancefield presented review lectures at national and international conferences. Special mention should be made of the one she presented (1936) at the second International Congress of Microbiology in London. This lecture dealt with the innumerable problems encountered in typing (identifying) streptococci. In the Harvey Lecture she presented, she traced the history of streptococci (Lancefield 1941), starting with their earliest description in 1879 by Louis Pasteur, who observed streptococci as a chain of beads in the blood of patients that died of sepsis (Dubos 1950, 261–262). Included in the lecture was a complete review of her own work and its significance in the identification and classification of streptococci. She further pointed out the problems encountered in the identification due to cross-reactions and technical difficulties.

As president of the Society of American Bacteriologists during 1943–44 and later as president of the American Association of Immunologists, she provided leadership to those who worked in the field. She traveled frequently to visit microbiology laboratories in England and Europe and shared her knowledge and expertise with scientists working in that field. She invited scientists from abroad to work in her laboratory.

At the Rockefeller Institute, Lancefield supervised the large-scale production of the grouping and typing of sera to be given to the military services. The latter undertook intensive study of the epidemiology of streptococcal disease among military personnel, using the elegant, reliable methodology she had popularized. Lancefield helped other workers in the field find solutions to their research problems. She spent many hours trying to identify the unusual strains of bacteria sent to her from all over the world. She kept exhaustive and thorough records of all the strains she had collected and preserved lyophilized specimens for future reference. Thus, her laboratory became a reference center for streptococcal research.

BIBLIOGRAPHY

Works by Rebecca Craighill Lancefield

Scientific Works

Space does not permit the listing of the complete works of Rebecca Craighill Lancefield. Listed here are all works by Lancefield except those cited in McCarty (1987). Included here are all references cited in the text.

(with A. R. Dochez and O. T. Avery) "Antigenic relationship between strains of *Streptococcus hemolyticus*." *Transactions of the Association of American Physicians* 34 (1919): 63–67.

(———) "Studies on the biology of streptococcus. I. Antigenic relationship between strains of *Streptococcus haemolyticus.*" *Journal of Experimental Medicine* 30 (1919): 179–213.

"The immunological relationships of *Streptococcus viridans* and certain of its chemical fractions. I. Serological reactions obtained with antibacterial sera." *Journal of Experimental Medicine* 42 (1925): 377–395.

Results of Ph.D. dissertation research.

"The immunological relationships of *Streptococcus viridans* and certain of its chemical fractions. II. Serological reactions obtained with antinucleoprotein sera." *Journal of Experimental Medicine* 42 (1925): 397–412.

Results of Ph.D. dissertation research.

(with J. H. Quastel) "A note on the antigenicity of crystalline egg albumin." *Journal of Pathology and Bacteriology* 32 (1929): 771–773.

"Note on the susceptibility of certain strains of hemolytic streptococcus to a streptococcus bacteriophage." *Proceedings of the Society of Experimental Biology and Medicine* 30 (1932): 169–171.

"A serological differentiation of human and other groups of hemolytic streptococci." *Journal of Experimental Medicine* 57 (1933): 571–595.

"Specific relationship of cell composition to biological activity of hemolytic streptococci." *Harvey Lectures* 36 (1941): 251–290.

Works about Rebecca Craighill Lancefield

McCarty, M. *Biographical Memoirs of the National Academy of Sciences* 57 (1987): 227–246.

O'Hern, E. M. *Profiles of Pioneer Women Scientists*, 69–78. Washington, DC: Acropolis Books, 1985.

Other References

Dubos, Rene J. *Louis Pasteur: Free Lance of Science.* Boston: Little, Brown, 1950.

Griffith, F. "Serological classification of *Streptococcus pyogenes.*" *Journal of Hygiene* 34 (1934): 542–584.

———. "Aronson *Streptococcus.*" *Journal of Hygiene* 35 (1935): 23–37.

OL'GA BORISOVNA PROTOPOVA LEPESHINSKAIA (1871–1963)

John Konopak

BIOGRAPHY

Ol'ga Borisovna Protopova was born to a *petit bourgeois* family of small land-owners in Perm, Russia, on August 6 or 18, 1871. The discrepancy in birth dates can be explained by differences between the Russian Orthodox and the Gregorian calendars.

She became a revolutionary at age 23, in 1894. It required great courage and strength for a single woman to participate in scientific and social reforms. Little has been recorded concerning Protopova's early years. The first official notice of her activities appeared in 1894, when she declared herself for the Revolution. She joined the St. Petersburg Union of Struggle for the Emancipation of the Working Class. At this time she met and married her husband, Panteleimon N. Lepeshinskii (1864–1944), also a revolutionary organizer. He was a journalist and historian who was frequently arrested and later exiled because of his activities. She not only shared, but participated in, organizing revolutionary groups in Pskov, not far from Petrograd (1903). The Lepeshinskiis had at least one child, a daughter.

The life, career, and record of Ol'ga Protopova Lepeshinskaia offer important insights into Russian life at that time. During and after the Revolution there was civil unrest accompanied by social changes unparalleled in the twentieth century. Because of the turmoil, there are few records outside Russia about the life of Lepeshinskaia. There is no citation or reference to her in a recent comprehensive index to collective biographies of prominent scientists (Pelletier 1994); nor is she mentioned in an earlier biographical encyclopedia of scientists (Daintith, Mitchell, et al., 1981). In the history of science written in the West, her career merits little more than derisive or dismissive footnotes. Even in Soviet sources written before the fall of the USSR, Lepeshinskaia is remembered more for her contributions to the social revolution than to the scientific revolution.

In 1897 Lepeshinskaia graduated from the Rozhdestvensky Course in surgery. While in exile (1902), she studied medicine and graduated from the prestigious medical college of the University of Lausanne in Switzerland. Later, in 1915, she obtained the medical doctorate from the University of Moscow. She practiced medicine in Russia and in the Crimea, while also working for the Revolution.

For nearly 50 tumultuous years, Lepeshinskaia held top posts in the scientific and medical establishment in the USSR. From 1915 to 1919, she was Professor of Therapy at Moscow University. Following this (1919–20) she was a staff member of a laboratory at Tashkent. Subsequently (1920–26) she was on the staff of a laboratory in Moscow. She was staff biologist at the Histological Laboratory of the Timiriazev Biological Institute from 1926 to 1963. Concurrently (1936–63), she was a staff member of the Cytology Laboratory of the All-Union Institute for Experimental Medicine of the USSR Academy of Medical Sciences (serving as head from 1957 to 1963). From 1949 to 1963 she was on the staff of the Institute for Experimental Biology in Moscow (Debus 1968, 1027).

From her first taste of revolutionary politics in St. Petersburg to the final installation of Vladimir I. Lenin in October 1917, Lepeshinskaia was ceaselessly engaged in the struggle to overthrow the tsarist regime. She was one of the earliest members of the Communist Party, participating in its forerunner, the Russian Social Democratic Party (1898). Between 1897 and 1913 she was in exile, either in Siberia or in Europe. In recognition of her heroic revolutionary work, she was later awarded the Order of Lenin and the Order of the Red Banner of Labor. She was awarded the State Prize of the USSR (1950). She was also named an Academician, a full member of the Soviet Academy of Sciences, in 1950.

While exiled in Siberia with her husband, Lepeshinskaia met and became an early partisan of Lenin's faction. They both became trusted members of Lenin's inner circle (1897–1900). She was present on the platform of the Finland Station to greet Lenin on his return to Petrograd in October 1917.

Lepeshinskaia had stood in the forefront of the Revolution, as Cossacks or other imperial cavalry and police cut down, shot, and beat peasants and workers in demonstrations in the streets of St. Petersburg and elsewhere. Clearly, Lepeshinskaia was a political agent before she became a scientist. However, in the period before the success of the Revolution her interests in science began to awaken. She died in Moscow, at the age of 92, on October 2, 1963 (Obituary 1963).

WORK

In her time Lepeshinskaia was *the* authority on plant biology, cytology, histology, and a dozen other subjects in Soviet science. She was the definitive reference in the practice of Soviet agricultural science and controlled those studies from 1946 until about 1960. (At that time Nikita Khrushchev was in power.)

When this phase of Russian history ended, her influence was not only abolished but practically erased due to ideological, political, and scientific reasons.

Understanding the fall of Academician Lepeshinskaia requires examining the rise of her revolutionary alter ego, "Comrade Galia." This pseudonym, a nom de guerre, was a badge of revolutionary honor of great value and importance. It bespoke its bearer's intimate knowledge, connection to, and participation in, the bloody, dangerous work of rebellion and revolution. Her life serves as a reminder of the danger that continues to plague scientists; the too-close synergy of science with either technology or ideology.

It is unclear why Lepeshinskaia chose a career in biology. Perhaps she agreed with Lenin that there was a need for a "revolutionary science" to sustain and accelerate the social revolution. The key element in this strategy was the understanding, shared by Lenin, that an army marches on its belly. The Revolution could starve without adequate supplies of bread. Under the forced industrialization policies, necessitated to overcome Russia's longtime technological backwardness, millions of peasants were being recruited from the land to the cities. Of prime importance was the development of new agricultural practices that could supply bread and potatoes cheaply to the newly centralized economy. These needs provided the impetus toward the collectivization of the Soviet farm economy.

Lepeshinskaia was an adaptable, durable, and tough individual. She was rarely far from the centers of either power or controversy in revolutionary Russia. Under Joseph Stalin and Khrushchev she was always in a position to assert her ideas and opinions. For nearly ten years Lepeshinskaia was both source and purveyor of Soviet orthodoxy concerning the theories and practices of the biological sciences.

By 1946 her views had been imposed as "basic principles of all biology and medical science." By *dictat* from the highest levels of the State, Lepeshinskaia's theories were used to develop the new core curriculum in cytology, histology, embryology, biochemistry, microbiology, general pathology, and oncology throughout the postwar Soviet Union (Medvedev 1969, 182). To disagree with her publicly was tantamount to an attack on the State and was grounds for banishment, imprisonment in the gulags, or even death.

Trofim Denisovich Lysenko (1898–1976), a Soviet agronomist and geneticist, was best known for his theory that the heredity of plants can be altered if specific metabolic requirements are met at definite stages of development. That theory was rejected by most Western geneticists but was supported by the Communist Central Committee (Stein 1954). Lysenko's ideas were based on neo-Lamarckian theory (Wallace et al. 1981, 401). He exploited the State's needs and assumed control over Soviet agriculture and the biological sciences that supported it. Lepeshinskaia became an early and ardent Lysenkoist. This association and the work that came of it tarnished her reputation and probably guaranteed her exclusion from future discussions of central figures in biology.

Even Lysenko's detractors, by whom he is legitimately considered a "crank"

(Joravsky 1970), nowadays recognize him as a skilled agronomist, observer of horticulture, and lucky practitioner with a green thumb. He became influential because of the early, innovative, and especially inexpensive successes in improving grain production in winter wheat during the 1930s. He had gained renown in the Soviet scientific community in the early Stalin years by denouncing the genetics of Gregor Mendel, Thomas Hunt Morgan (Nobel laureate, 1933), and August Weissman as "bourgeois, reactionary Idealism" (Medvedev 1969). As early as 1927, Lepeshinskaia was championing those same views, and her work comported perfectly with Lysenko's ideas.

Lysenko won Stalin's approval with an early success at improving the Russian wheat harvest. His first and probably greatest coup was perfecting the process called "vernalization," by which winter wheat was induced to become more resistant to frost, by means of soaking and exposure to cold. This lowered the rate of "winter kill," the perpetual bane of Soviet grain production. Lysenko's stature also grew from the success of his recommendation to "mobilize" the chickens of the Ukraine to arrest an infestation of weevils on the crucial sugar beet crop during the Nazi invasion (1941).

Later Lysenko turned to fakery and quackery, encouraging these traits in others. The only caveat was that agricultural programs be cheap and promise great results. Lepeshinskaia was among his foremost and best-placed allies. She championed revolutionary views, denying the cytological explanation for the origin and organization of living matter. She was closely associated with the "revolutionary, anti-bourgeois, anti-intellectual science" promulgated by Lysenko at the behest of Stalin. On the basis of her involvement, some commentators (e.g., Medvedev 1978) have expressed doubt that Lepeshinskaia deserved the appellation "scientist."

Lepeshinskaia sublimated her apparently sound scientific instincts to the exigencies of communist doctrine, and she became known (and ultimately reviled) as one of the foremost disciples of, and apologists for, the infamous Lysenko. Her reputation inside and outside the Soviet Union suffered from this association. Nevertheless, Lepeshinskaia wielded immense power and influence, despite the fact that she was considered by many scientists a third-rate crank, an opportunistic, semieducated pseudoscientist who was the source of such "highly original . . . rubbish" as the rejuvenating effects of soda baths and enemas. She published works in which she "proved" spurious discoveries such as "mutual transformation of cells of plants and animals, spontaneous generation of *Infusoria* in broth of hay, formation of cells from egg albumen and of blood vessels from egg yolk" (Medvedev 1969, 182).

She served on the editorial boards of the Soviet scientific periodicals *Progress in Modern Biology* and the *Archives of Anatomy, Histology, and Epidemiology.*

Because Lepeshinskaia's science was so closely tied to the political climate, it must be understood in the context of those conditions. The life of Lepeshinskaia symbolizes the basic differences between communist ideology and prac-

tices and those accepted at that time in the West. Unfortunately, she lived to
see her work debunked and her reputation defamed.

BIBLIOGRAPHY

Works by Ol'ga Borisovna Protopova Lepeshinskaia

Scientific Works

The Origin of Cells in Living Matter and the Role of Living Matter in the Organism (in
    Russian). Moscow: Academia Nauk USSR, 1945, 1950, 1951, 1952. English
    translation. Moscow: Foreign Languages Publishing House, 1954.
Membranes of Living Cells and their Biological Significance (in Russian). Moscow: n.a.,
    1947, 1952, 1953.
The Development of Living Processes in the Pre-Cellular Period. London: Society for
    Cultural Relations with the USSR, 1951. Russian version. Moscow: Akademia
    Nauk USSR, 1952.
Solutions to the Problem of Living Matter (in Russian). Moscow: n.a., 1951.
Cells: Their Life and Origin (in Russian). Moscow: n.a., 1952.
The New Cell Theory and Its Factual Basis (in Russian). Moscow: n.a., 1955.

Works about Ol'ga Borisovna Protopova Lepeshinskaia

Debus, A. G. (ed.). *World Who's Who in Science.* Chicago: Marquis Who's Who, 1968.
Great Soviet Encyclopedia. Vol. 14, 427. New York: Macmillan, 1978.
Krushchov, G. "New developments in cell theory: A significant discovery of the Soviet
    biologist O. B. Lepeshinskaya." *Journal of Heredity* 42 (1951): 121–122.
Nachtsheim, Hans. "Biological phantasies: New developments in the case of Lysenko."
    Journal of Heredity 42 (1951): 122–123.
Obituary. *New York Times* (Oct. 4, 1963): 35.

Other References

Daintith, J., S. Mitchell, et al. (eds.). *A Biographical Encyclopedia of Scientists*, vol 2.
    New York: Facts on File, 1981.
Graham, L. R. *Science, Philosophy, and Human Behavior in the Soviet Union.* New
    York: Columbia University Press, 1987.
Joravsky, D. *The Lysenko Affair.* Cambridge, MA: Harvard University Press, 1970.
Medvedev, Z. A. *The Rise and Fall of T. D. Lysenko.* New York: Columbia University
    Press, 1969.
———. *Soviet Science.* New York: W. W. Norton, 1978.
Pelletier, P. A. (ed.). *Prominent Scientists: An Index to Collective Biographies.* 3d ed.
    New York: Neal-Schuman, 1994.
Stein, Jess (ed.). *Basic Everyday Encyclopedia.* New York: Random House, 1954.

Wallace, R. A., J. L. King, et al. *Biology: The Science of Life*. Glenview, IL: Scott, Foresman, 1981.

Wieczynski, J. L. (ed.). *The Modern Encyclopedia of Russian and Soviet History*, vol. 19. New York: Academic International Press, 1981.

RITA LEVI-MONTALCINI (1908–)

Mary Clarke Miksic

BIOGRAPHY

Rita Levi-Montalcini was born in Torino (Turin), Italy, on April 22, 1908. She shared her birthday with her fraternal twin sister, Paola. Their family included two older siblings, Gino seven years, and Anna (Nina) five years old. The family was prosperous and upper middle class, headed by Adamo Levi, an engineer and factory manager, and Adele Montalcini. The Sephardic Jewish families of both parents had been settled in the Piedmont region for many generations: the Levi family around Casale Monferrato and the Montalcini family in Asti. Education and community service were strong traditions in both families.

There were 18 children in Adamo Levi's family. His father, a lawyer, died when Adamo was just nine years old, and his father's brother, also a lawyer, took over the care of the family. This uncle established close, loving ties with all the children. The boys were given university educations, and two of the girls earned doctorates, very unusual at that time. Adamo Levi earned an engineering degree from the Turin Polytechnic. Though the family was observant of the Jewish religious traditions, the liberalism of the late nineteenth century strongly influenced Levi's thinking, and he moved away from religious practice, considering himself a "freethinker."

Adele Montalcini had a sister and three brothers. One brother became an ophthalmologist in Turin, and another brother, a chemist by training, devoted himself to the improvement of the rural countryside around Asti and became the Mayor and benefactor of Asti. Most of the Montalcinis maintained the traditional Jewish religious practices and invited the Levi-Montalcinis to share the rituals of Passover and Yom Kippur with them. Though Adamo Levi assumed an aloof and cynical attitude on these family occasions, his children were exposed to the meaning and ritual of the Jewish tradition.

In her autobiography (1988) Rita Levi-Montalcini recalls the sweetness and warmth of her mother and the quiet diplomacy with which she managed the household, as well as the more imperial style of her father. His devotion to rationality, innovation, and vision, as well as persistent hard work in the achievement of goals, set a high standard for the family. Regarding the education of his daughters, his goals were colored by his paternal concern for their happiness. His egalitarian impulses led him to insist that the children attend the local public elementary school in Turin, so as to expose them to children from all classes of society, rather than a private school, where they would meet only wealthy children. However, when it came time for middle and high schools, he sent his daughters to a girls' school that gave no training in mathematics, science, or the classics. This was despite the fact that Anna showed a strong talent for mathematics, and Rita and Paola were outstanding and enthusiastic students. Such deficient training precluded the possibility of the girls' attending the university. Gino, of course, was sent to schools that prepared him for the university, and he subsequently became one of Italy's prominent architects. This paternal decision reflects the attitude prevalent at the time about career choices for women. The father based his decision on the experience of his sister, who had achieved doctorate degrees in mathematics and literature but who struggled to balance academic careers and marriages in early twentieth-century Italian society. Like most parents, he wanted his daughters to be happy, and he assumed that he knew best how to accomplish this. Fortunately, in spite of the limitations imposed on them by his decisions, all three of his daughters succeeded in achieving productive, full lives in ways he probably never could have imagined. However, his example of living a full, passionately directed life very likely contributed to their success. Adamo Levi died of heart disease in 1932.

After graduating from the Girls' High School in Turin with an excellent record but with no preparation for the university and no vocational direction, Rita Levi-Montalcini lived at home, did a great deal of reading, and occupied herself with her large extended family. When she was 19 years old, her beloved governess, Giovanna Bruatto, became ill. Levi-Montalcini credits her experience with the illness and death of this woman from stomach cancer (1930) for her own determination to study medicine (Levi-Montalcini 1988, 38). Her mother supported her decision, but her father was more skeptical. When he was finally convinced that she was serious and understood the problems she would probably encounter, he agreed to provide the necessary tutors. She had to be prepared in Greek, Latin, mathematics, and science in order to sit for the university entrance examinations.

Levi-Montalcini, barely 20 years old at the time, convinced her cousin, Eugenia, to take on this challenge with her. Together the young women prepared for eight months, and when they took the examinations for entrance to the University of Turin in the fall, both women attained the highest scores on the examinations. They both entered the Faculty of Medicine in 1930.

At the Institute of Anatomy of the Turin Medical School, Levi-Montalcini came under the tutelage of Professor Guiseppe Levi. He was a teacher and researcher famous for his expertise, for the talented students he launched into the medical profession, and for his impatience and rage with incompetence or what he perceived as insufficient dedication. Levi-Montalcini passed her first-year examinations with honor and, along with other promising students, was invited to become an intern in the laboratory of this imposing professor. Thus began a fruitful scientific relationship that led her to the study of the nervous system and introduced her to other scientists who would make important contributions to her work. This relationship flourished through the difficult years of fascist repression, of the war, and beyond, until Professor Levi's death of stomach cancer in 1965 at the age of 92.

Levi-Montalcini earned a Doctorate in Medicine from the University of Turin in 1936. After graduation she worked as a research assistant at the Institute for Anatomy of the University of Turin and began the practice of medicine at the Turin Clinic for Nervous and Mental Disorders. She was dismissed from both positions, as were all of her Jewish colleagues, including Professor Levi, after the promulgation of the anti-Semitic racial laws in 1938. Jews could not hold teaching or government positions, could not publish their work, and were forbidden to practice medicine. At about this time, Levi-Montalcini became engaged to Germano R. Raising, a fellow student who graduated with her in 1936. However, his health declined, and he died in 1939 of tuberculosis (Levi-Montalcini 1988, 69). She never married.

Levi-Montalcini accepted an invitation to continue her research in the laboratory of Professor L. Laruelle, the Director of the Neurology Institute of Brussels, Belgium, where she worked from March 1939, until December 1939, when she returned to Turin. Back in Italy she tried to practice medicine clandestinely, helping those who would otherwise be without care. However, she was unable to write prescriptions and so could not give patients the treatments they often needed. In 1940, as Italy entered the war, life became more constrained. Reading was one of the few intellectual activities possible. An article by Viktor Hamburger, given to her two years before by Professor Levi on the development of the spinal cord and ganglia in chick embryos, impressed her for the rigor of its methodology and analysis.

She was familiar with the histologic techniques used by Santiago Ramon y Cajal for studying the nervous system, having used them in the work she did as Levi's intern. She decided that she could use those techniques to further investigate the development of the chick embryo and that she could do it at home. After conferring with her mother, brother, and sister, she was able to gather the required materials and instruments (some fashioned from household implements) necessary to begin work in microscopic surgery. She was joined in this research by Professor Levi, who returned to Turin in 1942 after the German occupation of Belgium made it impossible for him to continue to work there. Worsening conditions in Turin and the constant air raids directed at the city's industrial centers forced many residents into the countryside. At the end of 1942 Levi-Montalcini, with her mother and sister, retreated to a farmhouse in Asti, where the home laboratory was again set up in her small bedroom. Considerable ingenuity was required to find chick embryos and to convince farmers to sell them to her in the climate of wartime food shortages. In spite of the difficulties, this work laid the foundation for her future research.

In 1943 the resignation of Benito Mussolini and the invasion of the German army from the north threw the civilian population into chaos. This was especially true of the Jewish citizens, who were now subject to Nazi administrative laws as the Germans took over the cities. In this dangerous situation, some members of Levi-Montalcini's family attempted to flee over the Alps into Switzerland. The Allied armies were invading from the south. Levi-Montalcini, her mother, sister, brother Gino, and his new wife fled south, hoping to reach the Allied line. They were carrying forged identity papers and reached only as far as Florence. Here they were helped by a schoolmate of Levi-Montalcini who found them lodging with an Italian family. They lived under assumed identities in Florence from the fall of 1943 until May 1945 in conditions that grew progressively worse as Germans, partisans, and the British fought for control of the city.

When the British finally entered Florence, Levi-Montalcini volunteered her services to the Red Cross. Through the winter of 1944–45, she worked as a doctor in the refugee camp set up to receive the flood of refugees streaming into Florence. Patients were in very poor condition, and there were outbreaks of epidemic diseases. There was no effective treatment, and many people died. The helplessness she felt in the face of so much death and the emotional toll of watching the vulnerable die led her to move away from the practice of medicine and induced her to concentrate on research (Levi-Montalcini 1988, 108). After a period of recuperation, she returned to work in Professor Levi's laboratory at the Institute of Anatomy in Turin during the summer of 1945. She also enrolled in a course of biological studies, an area that her previous schooling had neglected.

Late in 1945 an invitation came from Viktor Hamburger, who was then at Washington University in St. Louis. He had seen an article she and Levi had published on the work they had done during the war (1942). He wrote to Levi and through him invited Levi-Montalcini to continue her work on spinal cord and ganglion development in his laboratory as a visiting research associate.

In September 1946, after finishing her biological studies, she left for St. Louis. A visit that was to last only six months led to an association of 30 years. During this time she became a full professor in the Zoology Department at Washington University. She made frequent trips to Italy, where she maintained a close relationship with her family (her mother died in 1963). In 1961 she established a research unit in Italy for the study of neurobiology. This Center for Neurobiology was supported by the National Science Foundation at first, but its support was supplemented after the first year by the Istituto Superiore Sanita (Institute of Health) and the Consiglio Nazionale delle Ricerche (National Research Council). Until she retired, Levi-Montalcini alternately spent six months directing the Center for Neurobiology in Rome and six months teaching and doing research in collaboration with Hamburger and numerous colleagues and graduate students at Washington University.

Important collaborative efforts contributed to the body of Levi-Montalcini's work. She perfected her cell culture technique, developing the tool that enabled her to study complex problems *in vitro* during a 1952 stay at the laboratory of Professor Carlos Chagas at the Institute of Biophysics of the University of Rio de Janeiro in Brazil. This exchange, supported by Hamburger, gave Levi-Montalcini the opportunity to work with Hertha Meyer, who, with Levi, had developed methods for studying nervous system tissues *in vitro*. The results attained in this laboratory led directly to the discovery of the nerve growth factor. In 1953, when she returned to St. Louis, Levi-Montalcini was able to persuade Stanley Cohen, a biochemist and a member of the Washington University research faculty, to isolate and purify the nerve growth factor, for which they both shared the Nobel Prize in physiology or medicine in 1986.

In 1979 Levi-Montalcini retired from active teaching and from the directorship of the Institute for Cell Biology in Rome, which had grown out of the original Center for Neurobiology. She continues to carry on research as a guest of the Institute and to follow and comment upon new advances in the understanding of neurogenesis.

In 1968 Rita Levi-Montalcini was elected a member of the National Academy of Sciences. In 1974 she was given the William Thompson Wakeman Award of the National Paraplegic Foundation. The Lewis S. Rosenstiel Award for Distinguished Work in Medical Research of Brandeis University was bestowed on her in 1982. She received the Louisa Gross Horowitz Prize of Columbia University in 1983. She and Stanley Cohen received the Albert Lasker Medical Research Awards in 1986. She received the National Medal of Science in June 1987. Honorary doctorates have been awarded to her by the University of Uppsala (Sweden), the Weizmann Institute (Israel), St. Mary's College, and Washington University School of Medicine.

WORK

The first student research project assigned to Rita Levi-Montalcini by Professor Levi was a problem in neurogenesis, the development of the nervous system. His major interest was nervous system development. In the late nineteenth century, histologic techniques had been developed that allowed the study of the microscopic structure of this complex system. Camillo Golgi invented a fixation-staining chrome-silver impregnation procedure (Golgi 1878). This technique was modified by S. Ramon y Cajal for use in the study of embryonic nervous systems (Ramon y Cajal 1889). He described the various regions and cell types in the nervous system, as well as the sequence of stages through which these regions and structures pass. This study of the metamorphosis from neural groove to brain and spinal cord opened this incredibly complex area of study to ongoing scientific research. A central question to be answered was whether neurogenesis is genetically programmed or whether the observed migration and differentiation are driven by environmental and, to some extent, chance encounters.

Levi-Montalcini, along with several other interns, was given the task of counting the number of neurons in the spinal ganglia of mice. Was the number of neurons the same in mice from the same litter? Was the number the same for mice from different litters? Was the number of ganglionic neurons genetically determined for all mice? The tedium of such a project and the dedicated focus required to bring any accuracy to the final numbers can be appreciated only by those who have studied serial microscopic sections of tissues and who understand that hundreds of thousand of cells must have been cataloged. This was Levi-Montalcini's introduction to silver-impregnated preparations and to the nervous system. Although she performed her counts as carefully as possible, she understood that human error is a factor inherent in such research. She found the lack of clarity and precision troubling, though Levi happily accepted the interns' results. She was less than ecstatic, therefore, when for her second year of internship Levi assigned her the task of following the development of convolutions in the brain of human fetuses to try to discover the underlying mechanisms responsible for this process. The assignment said more, perhaps, about the confidence Levi placed in his student than about his assessment of the feasibility of the project. Though she attacked the problem with determination and ingenuity, it was soon apparent that the specimens and techniques available to her would not be productive.

Annoyed by her lack of progress, Levi finally allowed her to take a new topic, the study of the development of collagenous reticular tissue in connective tissue, muscle, and epithelium. This project became the subject of her doctoral thesis. It involved the use of tissue culture, a new technique in which cells were grown in laboratory dishes on specially prepared media. This new technique was also enthusiastically embraced by Levi and his coworker, Hertha Meyer. The *in vitro* method enabled them to study sensory cell development and structure and to introduce experimental manipulations outside the complex environment of the nervous system.

During her student days as an intern in the laboratory of Professor Levi, Levi-Montalcini worked with dedication and thoroughness to master the areas that would dominate her intellectual life: neurogenesis and tissue culture.

At the Clinic for Nervous and Mental Diseases in Turin, Levi-Montalcini and Fabio Visintini conducted an electrophysiologic and histologic study of chick embryos. With implanted electrodes, Visintini was able to record the spontaneous activity and to stimulate specific neural centers of embryos at different stages of development. Levi-Montalcini studied the experimentally manipulated nervous systems using a modification of Ramon y Cajal's method. She was able to correlate specific stages of neurological development with the onset of specific kinds of neurological response. The precision of the experimental design, which allowed more objective analysis of results, appealed to her. The paper reporting these results was published in a Swiss journal (Levi-Montalcini and Visintini

1939) after being refused by an Italian publication for political reasons. Levi-Montalcini wrote in her autobiography that she still considers this paper one of her most rewarding (Levi-Montalcini 1988, 84).

Hamburger's paper (Hamburger 1934) on the degeneration of ganglia and nerve columns in chick embryos whose limbs had been experimentally removed stimulated Levi-Montalcini to set up her home laboratory in the war years, when her official career was thwarted by anti-Semitic political decrees. Despite the restrictive environment and lack of official position, Levi-Montalcini repeated Hamburger's experiments and studied histologic preparations of closely spaced stages of development in experimentally manipulated chick embryos. Careful study led her to propose a hypothesis, different from the one proposed by Hamburger, that gave new understanding to the relationship between the developing nervous system and its surroundings. Hamburger, following the lead of Hans Spemann (Nobel laureate, 1935), who worked with amphibia, proposed that in chick embryos, sensory and motor nerves innervate peripheral tissues by "induction." That is, the peripheral tissues send some signal (it was assumed to be chemical, though unspecified) that stimulates the outgrowth of nerves from the spinal cord and neural crest into the tissues. If the peripheral tissues are removed, no outgrowth occurs. If extra tissue is grafted onto the chick, extra outgrowth occurs. The failure of innervation, as well as excess innervation, affects the number and size of neurons in the spinal cord and neural crest.

Through patient study of closely spaced serial sections, Levi-Montalcini observed that in the absence of peripheral tissue (as when an embryonic limb has been amputated), the neurons develop and *do* grow out to the stump. However, if innervation is not successful (because the tissue is not there), the nerve fibers and cell bodies undergo a subsequent degeneration. Sympathetic neurons, as well as sensory and motor cells, are involved. Instead of "induction," she hypothesized that the peripheral tissue supplies a "trophic" factor that is necessary to maintain the integrity of the neuron. The relationship between the neuron and its field of innervated cells is one of mutual dependency. Without the neuron, the peripheral cell cannot function, but without the "trophic" factor supplied by the peripheral cell, the neuron will die.

In 1942 Levi-Montalcini invited Levi, who was also without laboratory or position, to join her in this project, and he became her enthusiastic collaborator. Together they published several papers on this research, one of which came to the attention of Hamburger (Levi-Montalcini and Levi 1942). His interest in their work and in the new hypothesis led him to invite Levi-Montalcini to St. Louis.

At Washington University, Levi-Montalcini continued her studies of the developing chick nervous system. After months of discouraging, painstaking study of hundreds of specimens, she "suddenly" noticed patterns in the migration of neurons in developing spinal cords and immediately realized that these "revealed the genetic programming at work" (Levi-Montalcini 1988, 141). Based on these observations, she and Hamburger published a paper in 1949 that changed the assumptions about the establishment of functional neuron groups. It had been assumed that neurons proliferated in nuclei that had established functional connections. This paper offered observations showing that neurons migrate around the developing central nervous system (CNS) according to patterns that are predictable from animal to animal of the same species and are, therefore, genetically programmed. After migration, functional connections are made, and if these fail to occur, the neurons degenerate.

In ten years Levi-Montalcini's work had demonstrated that peripheral or environmental factors were critical to the CNS and also that the genetic program played a major role in CNS development. Clarification of the contribution each factor made to development was difficult in such relatively large and slow-growing material as the embryo. She looked for a less complicated model than the intact chick nervous system and realized that tissue culture provided a tool where cells could be studied more rapidly, in a more controlled way. As a visiting researcher in Chagas's laboratory at the University of Rio de Janeiro, she conducted a series of *in vitro* experiments that showed that a substance produced by the cells of a mouse sarcoma exerted powerful growth-promoting effects on chick embryonic neurons. The extraction, purification, and structural identification of this nerve growth factor (NGF) was done at Washington University by her collaborator, Stanley Cohen. NGF proved to be a protein extractable from many tissues, with a strong effect on the development and survival of sympathetic ganglia. Cohen obtained antibodies to NGF and with Levi-Montalcini showed that treatment of cells and embryos with this antibody completely destroyed the ability of developing neurons to grow and to survive. NGF acted like the trophic factor that Levi-Montalcini had hypothesized. Further, in the presence of NGF, nerve fibers grew in profusion into many tissues, even if the fibers made no functional connection with the cells of the tissue. After NGF was injected into the brains of developing animals, sympathetic fibers grew into brain areas where they are not found normally. These results indicated that NGF exerted a distinct neurotropic effect, not unlike the "induction" hypothesized by Spemann and Hamburger. It became apparent that both induction and trophic relationships are important in neurogenesis and that NGF was a chemical signal that had both effects.

The discovery of NGF, the development of antibodies to it, and its use *in vitro* and *in vivo* provided scientists with tools with which to investigate and manipulate the development of the nervous system. In the early 1950s this discovery was an important stimulus to neurogenesists, those relatively few, die-hard scientists willing to wrestle with overwhelming complexity in the face of limited experimental technique and control. The Washington University laboratory and the Center for Neurobiology in Rome (later the Institute for Cell Biology) attracted numerous students and colleagues who have significantly advanced the understanding of how the nervous system develops. Neurogenesis is now just a part of a large and vital scientific entity called neuroscience, a dynamic collaboration among neurobiologists, biophysicists, biochemists, and neurologists. The discovery of NGF has been followed by the identification of other growth factors affecting the development, maintenance, and repair of other systems. These discoveries promise to improve medical scientists' understanding of many illnesses, such as cancer, birth defects, and senile dementia.

The answer to the central question as to the primacy of the genetic program or of the environment in development has been shown, by the work initiated by Levi-Montalcini and carried on by others, to be almost as complex as the nervous system itself.

Levi-Montalcini is a member of the Association for the Advancement of Science, the Society for Developmental Biology, the American Association of Anatomists, the Tissue Culture Association, and the Pontifical Academia della Scienza. She is also a member of the Harvey Society, the Belgian Royal Academy of Medicine, the European Academy of Science, Arts, and Letters, The National Academy of Science of Italy, and the Academy of Arts and Science of Florence.

BIBLIOGRAPHY

Works by Rita Levi-Montalcini

Scientific works

Space does not permit the listing of the complete works of Rita Levi-Montalcini. Listed here are all works except those cited in Levi-Montalcini, *Les Prix Nobels (1986)* (1987). Included here are all references cited in the text.

(with F. Visintini) "Relationship between the functional and structural differentiation of the nerve centers and pathways in the chick embryo" (in Italian and French). *Archives Suisses de Neurologie et de Psychiatrie* 43 (1939): 381–393; 44 (1939): 119–150.

(with G. Levi) "The consequences of destruction of a peripheral innervation territory on development of corresponding central nervous system neurons in the chick embryo" (in French). *Archives de Biologie* 53 (1942): 537–545.

(———) "Developmental correlations between the parts of the nervous system. The consequences of destruction of a limb bud on development of the corresponding region of the central nervous system in the chick embryo" (in Italian). *Commentaries, Pontifical Academy of Science* 8 (1944): 527–568.

(with R. Amprino) "Experimental studies of the origin of the ciliary ganglion in the chick embryo" (in French). *Archives de Biologie* 58 (1947): 265–288.

(with V. Hamburger) "Proliferation, differentiation and degeneration in the spinal ganglia of the chick embryo under normal and experimental conditions." *Journal of Experimental Zoology* 111 (1949): 457–501.

(with S. Cohen) "Purification and properties of a nerve-growth-promoting factor isolated from mouse sarcoma 180." *Cancer Research* 17 (1957): 15–20.

"Chemical stimulation of nerve growth." In *Symposium on the Chemical Basis of Development*, edited by H. B. Glass and W. D. McElroy, 645–664. McCollum-Pratt Institute Contribution 234. Baltimore: Johns Hopkins University Press, 1958.

"Destruction of the sympathetic ganglia in mammals by an antiserum to the nerve-growth-promoting factor." *Proceedings of the National Academy of Sciences* 46 (1960): 384–391.

(with P. U. Angeletti) "Biological properties of a nerve-growth-promoting protein and its antiserum." In *Regional Neurochemistry*, edited by S. S. Kety and J. Elkes, 362–377. New York: Pergamon, 1960.

(with S. Cohen) "Effect of the extract of the mouse submaxillary salivary glands on the sympathetic system of mammals." *Annals of the New York Academy of Sciences* 85 (1960): 324–341.

(with P. U. Angeletti) "Growth control of the sympathetic system by a specific protein factor." *Quarterly Review of Biology* 36 (1961): 99–108.

(———) "Noradrenaline and monoamine oxidase content in immunosympathectomized animals." *International Journal of Neuropharmacology* 1 (1962): 161–164.

(with F. Caramia and P. U. Angeletti) "Experimental analysis of the mouse submaxillary salivary gland in relationship to its nerve-growth factor content." *Endocrinology* 70 (1962): 915–922.

"Differentiation and growth control mechanisms in the nervous systems." In *Experiments in Biological Medicine*, vol. 1, edited by E. Hagen et al., 170–182. Basel and New York: Karger, 1967.

(with P. U. Angeletti) "Nerve growth factor." *Physiological Reviews* 48 (1968): 535–569.

(with J. S. Chen) "Axonal outgrowth and cell migration *in vitro* from nervous system of cockroach embryos." *Science* 166 (1969): 631–632.

(with K. R. Seshan) "Neuronal properties of nymphal and adult insect neurosecretory cells *in vitro*." *Science* 182 (1973): 291–293.

(with P. Calissano) "Nerve growth factor." *Scientific American* 240 (1979): 68–77.

(with L. Aloe) "Synthesis and release of the nerve growth factor from the mouse submaxillary glands: Hormonal and neuronal regulatory mechanisms." *Hormonal Cell Regulation* 5 (1981): 53–72.

(with P. Calissano, E. R. Kandel, et al., eds.) *Molecular Aspects of Neurobiology*. Berlin: Springer-Verlag, 1986.

"The nerve growth factor 35 years later." In *Les Prix Nobel (1986)*, 276–299. Stockholm: Nobel Foundation, 1987.

"The nerve growth factor 35 years later." *Science* 237 (1987): 1154–1162.

In Praise of Imperfection, My Life and Work. Translated by Luigi Attardi. New York: Basic Books, 1988.

Works about Rita Levi-Montalcini

Arehart-Treichel, J. "Levi-Montalcini, R., nerve growth factor may hold the key—but to what?" *Science News* 11 (1977): 330–331.

"Collaborators Cohen and Levi-Montalcini win the medicine Nobel Prize." *Science News* 130 (1986): 244.

Holloway, M. "Finding the good in the bad." *Scientific American* (Jan. 1993): 32, 36.

Liversidge, A. "Levi-Montalcini, R., an interview." *Omni* 10 (1988): 70–74.

Marx, J. L. "Nobel prize for physiology or medicine." *Science* 234 (1986): 543–544.

"The Nobel prizes: Physiology or medicine." *Scientific American* 255 (1986): 84–86.

Randall, F. "The heart and mind of a genius." *Vogue* 177 (1987): 480–481.

Wasson, T. (ed.). *Nobel Prize Winners*. New York: H. W. Wilson, 1987.

Other References

Golgi, Camillo. "A new microscopic technique method" (in Italian). *R. C. Istituto Lombardo Scientifica*, 2d s., 12 (1878): 5.

Hamburger, Viktor. "The effects of limb bud extirpation on the development of the central nervous system in chick embryos." *Journal of Experimental Zoology* 68 (1934): 449–494.

Ramon y Cajal, Santiago. "Staining of the central nervous system of chick embryos using Golgi's method" (in Spanish). *Gaceta Medica Catalana* 12 (1889): 6–8.

MARY FRANCES LYON (1925–   )

Judith A. Dilts

BIOGRAPHY

Mary Frances Lyon was born on May 15, 1925, in Norwich, England. Her parents were Clifford James and Louise Frances (Kirby) Lyon. Mary Lyon has a brother three years younger and a sister ten years younger. Her brother worked as a chartered accountant, while the sister was a schoolteacher and later a social worker. There was one other scientist in her family, her first cousin, Kenneth Blaxter, a nutritionist and Director of the Rowett Research Institute in Aberdeen. Lyon is not married and has no children.

Her father was the youngest of seven children. His father, George Hodgson Lyon, from Hull, Yorkshire, was a cooper by trade. He met his wife, Susanna Green, in King's Lynn, Norfolk, where she was a domestic servant. They moved to Norwich, where Mary Lyon's father was born. He received a basic state education and quit school to work in a solicitor's office at age 14. He enlisted at the outbreak of World War I in the Royal Army Medical Corps and served in Gallipoli, Egypt, and Palestine and later in the Royal Artillery in France. Following the war he was an inspector of taxes in the civil service.

Her mother was born in London, the second of four children of John Francis and Julia (Crowe) Kirby. John Kirby apprenticed as a shoemaker in Norwich and then went to work in London, where he married Julia Crowe, a bookkeeper. They went back to Norwich to settle and start a shoe factory. Lyon's mother spent her early life in Norwich and received a grammar school education. She trained and worked as a schoolteacher prior to her marriage.

Mary Lyon's family lived in several different towns, which disrupted her education and that of her siblings. At age five she spent a few months in a state school in Bradford, then a few months at a state school in Norwich, before moving to a small private school in Norwich. When she was ten, her family moved to Birmingham, where she attended the King Edward VI High School for Girls. There, she recalls, her interest in biology developed when she won an essay competition on "Children of the British Empire," held to mark the silver jubilee of King George V in 1935. The prize was a set of four books on wild flowers, birds, and trees. Her interest was strengthened by Miss Udall, a biology teacher at King Edward's School, who was "a very clear and precise thinker . . . who made the analytical aspects of biology very interesting" (Personal communication . . . 1996).

In 1939, when she was 14, her family moved to Woking, Surrey. There she attended Woking Grammar School for Girls. Because of World War II, school facilities were restricted, and she was unable to study biology in the sixth form. Instead, she traveled among three other local schools for lessons in physics, chemistry, and mathematics, preparing for university entrance. In 1943 she obtained entrance to Girton College, Cambridge. There, although she had not taken biology on the entrance examination, she was allowed to read biology. In addition to zoology as her main subject, she studied physiology, organic chemistry, and biochemistry. While at Girton College, she was named the Sophia Adelaide Turle Scholar (1944) and received the Gertrude Gwendolen Crewdson Prize (1945). She earned a B.A. degree in zoology from Cambridge (1946).

While at Cambridge, Lyon was fascinated by the major advances in experimental embryology of the 1930s and was much influenced by the writings of Conrad H. Waddington. It seemed to her that genes must underlie all embryological development, a relatively new idea at the time. Although genetics was not taught as a degree subject, Lyon took a course of lectures by R. A. Fisher. She recalls that while his lectures were "very mathematical and largely incomprehensible," he kept the interest of the undergraduates by engaging them in his research. The students tallied mice for mutant genes in an experiment on linkage and worked in his genetics garden in the summer, scoring plants for their flower type. Lyon began work on her Ph.D. degree with Fisher's mouse colony. As his student, she was able to develop her own project and thus decided to study the genetics of a balance defect that she had noticed in mice with the mutant gene pallid. She finished her work at the Institute of Animal Genetics in Edinburgh, where D. S. Falconer, a leading quantitative geneticist, was Lyon's supervisor on the project. Her thesis was entitled "Absence of Otoliths in a Mutant of the Mouse." She was awarded her M.A. and Ph.D. degrees by Cambridge in 1950.

Waddington, then head of the Institute of Animal Genetics in Edinburgh, offered her a post. He had grants from the Agricultural Research Council and the Medical Research Council (MRC) to study the genetic effects of atomic radiation. In the late 1940s scientists were concerned about the possible mutagenic effects of atomic radiation, whether from atmospheric weapons testing, possible nuclear wars, or peaceful uses of atomic energy. The MRC grant funded the mutagenesis experiments in mice in which Lyon participated.

In order that the work of the research group be extended and expanded, the group was transferred in 1955 to the MRC Radiobiology Unit at Harwell in Oxfordshire. Lyon worked in this group for the remainder of her career, except for a year at Girton College (1970–71), where she was Clothworkers Visiting

Research Fellow, working with Richard Gardner. From 1962 to 1986 Lyon was Head of Genetics Division, MRC Radiobiology Unit, Chilton, and from 1986 to 1990 she served as its Deputy Director.

Lyon was awarded a Sc.D. degree from Cambridge (1968) and was made a fellow of the Institute of Biology (1973). She became a fellow of the Royal Society (1973). Lyon has received numerous other honors and awards, including the Francis Amory Prize of the American Academy of Arts and Sciences (1977); she was named foreign associate of the National Academy of Sciences (USA) (1979); she became honorary member of the United Kingdom Environmental Mutagen Society (1982); she received the Royal Medal of the Royal Society (1984); she was named honorary fellow, Girton College (1985); she was awarded the San Remo International Prize for Genetics (1985), the Gairdner Foundation International Award (1985), and the William Allan Award of the American Society of Human Genetics (1986); and she was made an honorary member of the Genetics Society of Japan (1991).

WORK

Lyon spent the whole of her career with the Medical Research Council working on the estimation of the genetic hazards of radiation and other mutagenic agents. However, the project was not limited to the estimation of genetic hazards but was always interpreted to include the study of the nature of the mutations and formal mouse genetics relevant to medicine. Lyon's work focused on X-chromosome inactivation, the t-complex, and other aspects of mouse mutants and chromosome aberrations.

One of the outcomes of her mutagenesis work on mutant genes and chromosome aberrations was the Lyon hypothesis, the phenomenon of X-chromosome inactivation in mammals. It was known that the sex chromatin body in female mammals consisted of one condensed X-chromosome and that chromosomally XO mice were normal, fertile females. Work by Lyon on variegated female mice heterozygous for the X-linked genes controlling coat color led her to hypothesize that one of the X-chromosomes in the somatic cells of female mice is condensed during interphase and is thus genetically inactive. Shortly thereafter Lyon modified the hypothesis and extended it to all mammals. She noted that in humans who were sex chromosome aneuploids (e.g., XXX, XXY, XXXX), the number of inactive X-chromosomes was always one less than the number of X-chromosomes. Thus, rather than postulating that a single X-chromosome became inactive, Lyon proposed that a single X-chromosome remained active. However, because the majority of human XO females, unlike mice, show severe phenotypic abnormalities, Lyon suggested that a region of homology between the X and Y chromosomes does not undergo inactivation and thus does not require dosage compensation.

Lyon's work on coat color variegation in X-autosome translocations in mice suggested the idea of an X-inactivation center on the X-chromosome from which the inactivation spread. Since only one of the two segments of the translocation exhibited inactivation, she proposed the hypothesis that inactivation spread from the X into the attached autosome. Later work by Lyon and coworkers on translocations involving the locus for ornithine transcarbamylase (*Otc*) supported this hypothesis.

Lyon began her work on the *t*-complex in the mouse because *t*-haplotypes were thought to show very high mutation rates, a topic relevant to studies of mutagenesis. The putative mutations were discovered, however, to be crossovers. For a period of time she investigated the properties of these recombinant *t*-haplotypes. Lyon and her colleagues then returned to this research and localized the *t*-haplotypes by *in situ* hybridization to the proximal portion of chromosome 17. Further investigation on the *t*-complex dealt with the genetic basis of transmission-ratio distortion and male sterility produced by the *t*-complex. Lyon and coworkers showed that there are various distorter genes that act on a single responder gene. Homozygosity of any distorter in combination with heterozygosity of at least one other distorter gives rise to male sterility.

Other research by Lyon included work on mouse genetics with relevance to medicine and on germ-cell mutation in mammals. With her coworkers she showed that the pink-eyed dilution (*p*) locus of mouse chromosome 7 has a homolog in humans that is localized to chromosome 15, a region associated with the neurological disorders Prader-Willi and Angelman syndromes. Her work on germ-cell mutation in mammals indicated that only a small fraction of the mutation can be accounted for by the low doses and dose rates of environmental radiation.

Lyon is a member of a number of learned societies, including the Association for Radiation Research, the British Society for Developmental Biology, the Genetical Society, the Institute of Biology, the International Mammalian Genome Society, the Laboratory Animal Science Association, the Royal Society, and the United Kingdom Environmental Mutagen Society. She was honorary treasurer of the Genetical Society (1968–76) and its vice president (1976–79). From 1974 to 1980 she was convener, Mammalian Genetics Group Member, European Molecular Biology Organization.

She was chairman of the Committee on Standardized Genetic Nomenclature for Mice (1975–90) and chairman of Committee 4, International Commission for Protection against Environmental Mutagens and Carcinogens (1977–84). Lyon was a member of the Royal Society *ad hoc* Group of Human Fertilization and Embryology (1982–84) and was a member of the International Commission for Protection against Environmental Mutagens and Carcinogens (1985–90).

Lyon has served on the following editorial boards: *Cell, Cytogenetics, Development, Genetical Research, Genomics, Heredity, Journal of Embryology and Experimental Morphology, Journal of Genetics, Laboratory Animals, Mammalian Genome*, and *Mutation Research*. In addition, she was editor of the *Mouse News Letter* (1956–69).

BIBLIOGRAPHY

Works by Mary Frances Lyon

Scientific Works

Space does not permit the listing of the complete works of Mary Frances Lyon. Included here are her dissertation and all references by Lyon except those cited in Lyon (1972; "The William Allan . . ." 1988; "The genetic . . ." 1991; and "X-chromosome . . ." 1994).

(with R. A. Fisher and A.R.G. Owen) "The sex chromosome in the house mouse." *Heredity* 1 (1947): 355–365.

"Absence of Otoliths in a Mutant of the Mouse." Ph.D. diss., Cambridge University, 1950.

"Hereditary absence of otoliths in the house mouse." *Journal of Physiology—London* 114 (1951): 410–418.

"Absence of otoliths in the mouse: An effect of the pallid mutant." *Journal of Genetics* 51 (1953): 638–650.

"Stage of action of the litter-size effect on absence of otoliths in mice." *Zeitschrift für Induktive Abstammungs-und Vererbungslehre* 86 (1954): 289–292.

(with T. C. Carter and R.J.S. Phillips) "Induction of sterility in male mice by chronic gamma irradiation." *British Journal of Radiology* 27 (1954): 418–422.

"Ataxia: A new recessive mutant of the house mouse." *Journal of Heredity* 46 (1955): 77–80.

"The development of the otoliths of the mouse." *Journal of Embryology and Experimental Morphology* 3 (1955): 213–229.

"The developmental origin of hereditary absence of otoliths in mice." *Journal of Embryology and Experimental Morphology* 3 (1955): 230–241.

(with T. C. Carter and R.J.S. Phillips) "Gene-tagged chromosome translocations in eleven stocks of mice." *Journal of Genetics* 53 (1955): 154–166.

"Hereditary hair loss in the tufted mutant of the house mouse." *Journal of Heredity* 47 (1956): 101–103.

(with T. C. Carter and R.J.S. Phillips) "Further genetic studies of eleven translocations in the mouse." *Journal of Genetics* 54 (1956): 462–473.

(———) "Induction of mutations in mice by chronic gamma irradiation; interim report." *British Journal of Radiology* 29 (1956): 106–108.

"Twirler: A mutant affecting the inner ear of the house mouse." *Journal of Embryology and Experimental Morphology* 6 (1958): 105–116.

(with T. C. Carter and R.J.S. Phillips) "Genetic hazard of ionizing radiations." *Nature* 182 (1958): 409.

"A new dominant T-allele in the house mouse." *Journal of Heredity* 50 (1959): 140–143.

"Some evidence concerning the 'mutational load' in inbred strains of mice." *Heredity* 13 (1959): 341–352.

(with R.J.S. Phillips) "Crossing-over in mice heterozygous for t-alleles." *Heredity* 13 (1959): 23–32.

"Effect of X-rays on the mutation of t-alleles in the mouse." *Heredity* 14 (1960): 247–252.

"A further mutation of the mottled type in the house mouse." *Journal of Heredity* 51 (1960): 116–121.

"Zigzag: A genetic defect of the horizontal canals in the mouse." *Genetical Research* 1 (1960): 189–195.

(with T. C. Carter and R. J. Phillips) "The genetic sensitivity to X-rays of mouse fetal gonads." *Genetical Research* 1 (1960): 351–355.

(with G. D. Snell, J. Staats, et al.) "Standardized nomenclature for inbred strains of mice. 2nd listing." *Cancer Research* 20 (1960): 145–169.

"Gene action in the X-chromosome of the mouse (*Mus musculus* L.)." *Nature* 190 (1961): 372–373.

"Linkage relations and some pleiotropic effects of the dreher mutant." *Genetical Research* 2 (1961): 92–95.

(with T. C. Carter) "An attempt to estimate the induction by X-rays of recessive lethal and visible mutations in mice." *Genetical Research* 2 (1961): 296–305.

(with R.J.S. Phillips and A. G. Searle) "A test for mutagenicity of caffeine in mice." *Zeitschrift für Vererbungslehre* 93 (1962): 7–13.

"Lyonization of the X-chromosome." *Lancet* (2) (1963): 1120–1121.

(with M. C. Green, H. Gruneberg, et al.) "A revision of the standardized genetic nomenclature of mice." *Journal of Heredity* 54 (1963): 159–162.

"Genetics of the mouse." In *Animals for Research. Principles of Breeding and Management*, edited by W. Lane-Petter, 199–234. New York: Academic Press, 1964.

(with R. Meredith) "Investigations of the nature of t-alleles in the mouse. I. Genetic analysis of a series of mutants derived from a lethal allele." *Heredity* 19 (1964): 301–312.

(———) "Investigations of the nature of t-alleles in the mouse. II. Genetic analysis of unusual mutant allele and its derivatives." *Heredity* 19 (1964): 313–325.

(———) "Investigations of the nature of t-alleles in the mouse. III. Short tests of some further mutant alleles." *Heredity* 19 (1964): 327–330.

(with R.J.S. Phillips and A. G. Searle) "The overall rates of dominant and recessive lethal and visible mutation induced by spermatogonial X-irradiation of mice." *Genetical Research* 5 (1964): 448–467.

(with E. V. Hulse and C. E. Rowe) "Foam-cell reticulosis of mice: An inherited condition resembling Gaucher's and Niemann-Pick diseases." *Journal of Medical Genetics* 2 (1965): 99–106.

(with S. Ohno) "Cytological study of Searle's X-autosome translocation in *Mus musculus*." *Chromosoma* 16 (1965): 90–100.

"Lack of evidence that inactivation of the mouse X-chromosome is incomplete." *Genetical Research* 8 (1966): 197–203.

"Order of loci on the X-chromosome of the mouse." *Genetical Research* 7 (1966): 130–133.

"Sex chromatin and gene action in the X-chromosome of mammals." In *The Sex Chromatin*, edited by K. L. Moore, 370–386. Philadelphia: W. B. Saunders, 1966.

(with R. Meredith) "Autosomal translocations causing male sterility and viable aneuploidy in the mouse." *Cytogenetics* 5 (1966): 335–354.

(with T. Morris) "Mutation rates at a new set of specific loci in the mouse." *Genetical Research* 7 (1966): 12–17.

(with E. P. Evans and M. Daglish) "A mouse translocation giving a metacentric marker chromosome." *Cytogenetics* 6 (1967): 105–119.

(with T. Morris, A. G. Searle, et al.) "Occurrences and linkage relations of the mutant 'extra-toes' in the mouse." *Genetical Research* 9 (1967): 383–385.

(with J. M. Butler and R. Kemp) "The positions of the centromeres in linkage groups II and IX of the mouse." *Genetical Research* 11 (1968): 193–199.

"A true hermaphrodite mouse presumed to be an XO/XY mosaic." *Cytogenetics* 8 (1969): 326–331.

(with R. Meredith) "Muted, a new mutant affecting coat color and otoliths of the mouse, and its position in linkage group XIV." *Genetical Research* 14 (1969): 163–166.

(with T. Morris) "Gene and chromosome mutation after large fractionated or unfractionated radiation doses to mouse spermatogonia." *Mutation Research* 8 (1969): 191–198.

"The activity of the sex chromosomes in mammals." *Science Progress* 58 (1970): 117–130.

"X-ray-induced dominant lethal mutations in male guinea-pigs, hamsters, and rabbits." *Mutation Research* 10 (1970): 133–140.

(with S. G. Hawkes) "X-linked gene for testicular feminization in the mouse." *Nature* 227 (1970): 1217–1219.

(with T. Morris, P. Glenister, et al.) "Induction of translocations in mouse spermatogonia by X-ray doses divided into many small fractions." *Mutation Research* 9 (1970): 219–223.

(with S. Ohno) "X-linked testicular feminization in the mouse as a non-inducible regulatory mutation of the Jacob-Monod type." *Clinical Genetics* 1 (1970): 121–127.

(with R. J. Phillips and P. Glenister) "Dose-response curve for the yield of translocations in mouse spermatogonia after repeated small radiation doses." *Mutation Research* 10 (1970): 497–501.

(with N. V. Savkovic) "Dose-response curve for X-ray-induced translocations in mouse spermatogonia. I. Single doses." *Mutation Research* 9 (1970): 407–409.

(with E. V. Hulse) "An inherited kidney disease of mice resembling human nephrophthisis." *Journal of Medical Genetics* 8 (1971): 41–48.

(with B. D. Smith) "Species comparisons concerning radiation-induced dominant lethals and chromosome aberrations." *Mutation Research* 11 (1971): 45–58.

"X-chromosome inactivation and developmental patterns in mammals." *Biological Reviews of the Cambridge Philosophical Society* 47 (1972): 1–35.

(with B. M. Cattanach and H. G. Wolfe) "A comparative study of the coats of chimeric mice and those of heterozygotes for X-linked genes." *Genetical Research* 19 (1972): 213–228.

(with P. H. Glenister and S. G. Hawker) "Do the H-2 and T loci of the mouse have a function in the haploid phase of sperm?" *Nature* 240 (1972): 152–153.

(with D. G. Papworth and R.J.S. Phillips) "Dose-rate and mutation frequency after irradiation of mouse spermatogonia." *Nature, New Biology* 238 (1972): 101–104.

(with R.J.S. Phillips and H. J. Bailey) "Mutagenic effects of repeated small radiation doses to mouse spermatogonia. I. Specific-locus mutation rates." *Mutation Research* 15 (1972): 185–190.

(with R.J.S. Phillips and P. H. Glenister) "Mutagenic effects of repeated small radiation doses to mouse spermatogonia. II. Translocation yield at various dose intervals." *Mutation Research* 15 (1972): 191–195.

(with S. G. Hawker) "Reproductive lifespan in irradiated and unirradiated chromosomally XO mice." *Genetical Research* 21 (1973): 185–194.

(with I. Hendry and R. V. Short) "The submaxillary salivary glands as test organs for response to androgen in mice with testicular feminization." *Journal of Endocrinology* 58 (1973): 357–362.

(with R.J.S. Phillips and P. H. Glenister) "The mutagenic effect of repeated small radiation doses to mouse spermatogonia. III. Does repeated irradiation reduce translocation yield from a large radiation dose?" *Mutation Research* 17 (1973): 81–85.

"Evolution of X-chromosome inactivation in mammals." *Nature* 250 (1974): 651–653.

"Role of X- and Y-chromosomes in mammalian sex determination and differentiation." *Helvetica Paediatrica Acta Supplement* 34 (1974): 7–12.

"Sex chromosome activity in germ cells." *Basic Life Sciences* 4 (1974): 63–71.

"Symposium No. 6: Gene and chromosome inactivation. Introduction by the chairman." *Genetics* 78 (1974): 305–309.

(with P. H. Glenister) "Evidence from Tfm/O that androgen is inessential for reproduction in female mice." *Nature* 247 (1974): 366–367.

(with G. M. Simpson) "An investigation into the possible genetic hazards of ultrasound." *British Journal of Radiology* 47 (1974): 712–722.

"Implication of freezing for the preservation of genetic stocks." In *Basic Aspects of Freeze Preservation of Mouse Strains*, edited by O. Mühlbock, 57–65. Stuttgart: Gustav-Fischer Verlag, 1975.

(with D. Bennett, E. A. Boyse, et al.) "Expression of H-Y male antigen in phenotypically female Tfm/Y mice." *Nature* 257 (1975): 236–238.

(with B. D. Cox) "The induction by X-rays of chromosome aberrations in germ cells of male guinea-pigs, golden hamsters, and rabbits. I. Dose-response in post-meiotic stages." *Mutation Research* 29 (1975): 93–110.

(———) "The induction by X-rays of chromosome aberrations in male guinea-pigs, golden hamsters, and rabbits. II. Properties of translocations induced in post-meiotic stages." *Mutation Research* 29 (1975): 111–126.

(———) "The induction by X-rays of chromosome aberrations in male guinea-pigs, rabbits and golden hamsters. III. Dose-response relationship after single doses of X-rays to spermatogonia." *Mutation Research* 29 (1975): 407–422.

(———) "The mutagenic effect of triethylenemelamine (TEM) on germ cells of male golden hamsters and guinea-pigs." *Mutation Research* 30 (1975): 293–297.

(———) "The induction by X-rays of chromosome aberrations in male guinea-pigs and golden hamsters. IV. Dose-response for spermatogonia treated with fractionated doses." *Mutation Research* 30 (1975): 117–128.

(———) "X-ray induced dominant lethal mutations in mature and immature oocytes of guinea-pigs and golden hamsters." *Mutation Research* 28 (1975): 421–436.

(with P. H. Glenister and M. L. Lamoreux) "Normal spermatozoa from androgen-resistant germ cells of chimaeric mice and the role of androgen in spermatogenesis." *Nature* 258 (1975): 620–622.

(with R.J.S. Phillips) "Specific locus mutation rates after repeated small radiation doses to mouse oocytes." *Mutation Research* 30 (1975): 375–382.

(with H. C. Ward and G. M. Simpson) "A genetic method for measuring nondisjunction in mice with Robertsonian translocations." *Genetical Research* 26 (1975): 282–295.

"Chromosome condensation in relation to genetic activity." In *Organization and Expression of Chromosomes*, edited by V. G. Allfrey, 131–140. Berlin: Dahlem Konferenzen, 1976.

"Distribution of crossing-over in mouse chromosomes." *Genetical Research* 28 (1976): 291–299.

(with B. D. Cox and J. H. Marston) "Dose-response data for X-ray induced translocations in spermatogonia of Rhesus monkeys." *Mutation Research* 35 (1976): 429–436.

(with R. T. Schimke, W. J. Gehring, et al.) "Determination." In *Organization and Expression of Chromosomes*, edited by V. G. Allfrey, 53–69. Berlin: Dahlem Konferenzen, 1976.

"Genetic nomenclature and nomenclatorial rules in the mouse." *Immunogenetics* 5 (1977): 393–403.

(with K. B. Bechtol) "Derivation of mutant t-haplotypes of the mouse by presumed duplication or deletion." *Genetical Research* 30 (1977): 63–76.

(with A. Caine) "The induction of chromosome aberrations in mouse dictyate oocytes by X-rays and chemical mutagens." *Mutation Research* 45 (1977): 325–332.

(with E. P. Evans and C. E. Ford) "Direct evidence of the capacity of the XY germ cell in the mouse to become an oocyte." *Nature* 267 (1977): 430–431.

(with P. H. Glenister) "Factors affecting the observed number of young resulting from adjacent-2 disjunction in mice carrying a translocation." *Genetical Research* 29 (1977): 83–92.

(with H. Kacser, K. Mya Mya, et al.) "Maternal histidine metabolism and its effect on fetal development in the mouse." *Nature* 265 (1977): 262–266.

(with I. Mason) "Information on the nature of t-haplotypes from the interaction of mutant haplotypes in male fertility and segregation ratio." *Genetical Research* 29 (1977): 255–266.

(with D. G. Whittingham and P. H. Glenister) "Long-term storage of frozen mouse embryos under increased background irradiation." In *The Freezing of Mammalian Embryos in London, England*, edited by K. Elliott and J. Whelan, 273–290. Amsterdam: Elsevier/North-Holland, 1977.

(———) "Long-term storage of mouse embryos at −196° C: The effect of background radiation." *Genetical Research* 29 (1977): 171–181.

(———) "Re-establishment of breeding stocks of mutant and inbred strains of mice from embryos stored at −196° C for prolonged periods." *Genetical Research* 30 (1977): 287–300.

"Standardized genetic nomenclature for mice: Past, present, and future." In *Origins of Inbred Mice*, 3d ed., edited by H. C. Morse, 445–455. New York: Academic Press, 1978.

(with K. B. Bechtol) "H-2 typing of mutants of the t6 haplotype in the mouse." *Immunogenetics* 6 (1978): 571–584.

"Relative problems of assessing the genetic hazards of radiation and chemicals." *Mutation Research* 64 (1979): 107–108.

(with A. Caine) "Reproductive capacity and dominant lethal mutations in female guinea-pig and Djungarian hamsters following X-rays or chemical mutagens." *Mutation Research* 59 (1979): 231–244.

(with E. P. Evans, S. E. Jarvis, et al.) "t-Haplotypes of the mouse may involve a change in intercalary DNA." *Nature* 279 (1979): 38–42.

(with S. E. Jarvis, I. Sayers, et al.) "Complementation reactions of a lethal mouse t-haplotype believed to include a deletion.'' *Genetical Research* 33 (1979): 153–162.

(with R.J.S. Phillips and G. Fisher) "Dose-response curves for radiation-induced gene mutations in mouse oocytes and their interpretation.'' *Mutation Research* 63 (1979): 161–174.

(with P. H. Glenister) "Reduced reproductive performance in androgen-resistant Tfm/Tfm female mice.'' *Proceedings of the Royal Society of London—Series B: Biological Sciences* 208 (1980): 1–12.

(with S. Rastan, M. H. Kaufman, et al.) "X-chromosome inactivation in extraembryonic membranes of diploid parthenogenetic mouse embryos demonstrated by differential staining.'' *Nature* 288 (1980): 172–173.

"Nomenclature.'' In *The Mouse in Biomedical Research*, vol. 1, edited by H. L. Foster et al., 28–38. New York: Academic Press, 1981.

"Sensitivity of various germ-cell stages to environmental mutagens.'' *Mutation Research* 87 (1981): 323–345.

(with T. P. Dalton, J. H. Edwards, et al.) "Chromosome maps of man and mouse.'' *Clinical Genetics* 20 (1981): 407–415.

(with S. E. Jarvis, I. Sayers, et al.) "Lens opacity: A new gene for congenital cataract on chromosome 10 of the mouse.'' *Genetical Research* 38 (1981): 337–341.

"Problems of extrapolation from experimental data to human mutagenesis.'' In *Mutagens in Our Environment*, edited by M. Sorsa and H. Vainio, 127–136. New York: Alan R. Liss, 1982. Also in *Progress in Clinical and Biological Research* 109 (1982): 127–136.

(with P. H. Glenister) "A new allele sash (Wsh) at the W locus and a spontaneous recessive lethal in mice.'' *Genetical Research* 39 (1982): 315–322.

(with M. Kirk) "Induction of congenital anomalies in offspring of female mice exposed to varying doses of X-rays.'' *Mutation Research* 106 (1982): 73–84.

(with R.J.S. Phillips and G. Fisher) "Use of an inversion to test for induced X-linked lethals in mice.'' *Mutation Research* 92 (1982): 217–228.

"Sensitivity of various germ-cell stages to environmental mutagens.'' *Biologisches Zentralblatt* 102 (1983): 211–229.

"Problems in extrapolation of animal data to humans.'' In *Utilization of Mammalian Specific Locus Studies in Hazard Evaluation and Estimation of Genetic Risk*, edited by F. J. de Serres and W. Sheridan, 289–305. New York: Plenum, 1983.

(with I. B. Fritz and B. P. Setchell) "Evidence for a defective seminiferous tubule barrier in testes of Tfm and Sxr mice.'' *Journal of Reproduction and Fertility* 67 (1983): 359–363.

(with J. F. Loutit) "X-linked factor in acquired immunodeficiency syndrome?'' *Lancet* (1) (1983): 768.

(with E. Moustacchi, I. D. Adler, et al.) "Estimation of genetic risks and increased incidence of genetic disease due to environmental mutagens.'' *Mutation Research* 115 (1983): 255–292.

(with V. J. Buckle, J. H. Edwards, et al.) "Chromosome maps of man and mouse. 2.'' *Clinical Genetics* 26 (1984): 1–11.

(with K. Dudley, J. Potter, et al.) "Analysis of male sterile mutations in the mouse using haploid stage expressed cDNA probes.'' *Nucleic Acids Research* 12 (1984): 4281–4293.

(with P. H. Glenister, J. F. Loutit, et al.) "A presumed deletion covering the W and Ph loci of the mouse." *Genetical Research* 44 (1984): 161–168.

(with P. H. Glenister and D. G. Whittingham) "Further studies on the effect of radiation during the storage of frozen 8–cell mouse embryos at −196°C." *Journal of Reproduction and Fertility* 70 (1984): 229–234.

(with K. M. Kirk) "Induction of congenital malformations in the offspring of male mice treated with X-rays at pre-meiotic and post-meiotic stages." *Mutation Research* 125 (1984): 75–86.

(with J. D. West and J. Peters) "Genetic differences between two substrains of the inbred 101 mouse strain." *Genetical Research* 44 (1984): 343–346.

"Comparison of the dominant visible and other mutation tests in the mouse." In *Mutagenesis and Genetic Toxicology: Theoretical and Practical Results*, edited by P. Janiaud et al., 153–164. Paris: Editions Inserm, 1985.

"Measuring mutation in man." *Nature* 318 (1985): 315–316.

(with G. Fisher and P. H. Glenister) "A recessive allele of the mouse agouti locus showing lethality with yellow Ay." *Genetical Research* 46 (1985): 95–100.

(with J. F. Loutit and B. M. Cattanach) "The gene triplet Rw W Ph controls murine hematopoiesis." *British Journal of Haematology* 60 (1985): 219–232.

(with J. H. Rogers and K. R. Willison) "The arrangement of H-2 class I genes in mouse t- haplotypes." *Journal of Immunogenetics* 12 (1985): 151–166.

(with J. D. West, K. M. Kirk, et al.) "Discrimination between the effects of X-ray irradiation of the mouse oocyte and uterus on the induction of dominant lethals and congenital anomalies. 1. Embryo-transfer experiments." *Mutation Research* 149 (1985): 221–230.

(———) "Discrimination between the effects of X-ray irradiation of the mouse oocyte and uterus on the induction of dominant lethals and congenital anomalies. 2. Localized irradiation experiments." *Mutation Research* 149 (1985): 231–238.

(with J. D. West, J. Peters, et al.) "Genetic differences between substrains of the inbred mouse strain 101 and designation of a new strain 102." *Genetical Research* 46 (1985): 349–352.

(with P. H. Glenister) "Long-term storage of eight-cell mouse embryos at −196°C." *Journal of in Vitro Fertilization and Embryo Transfer* 3 (1986): 20–27.

(with B.L.M. Hogan, G. Horsburgh, et al.) "Small eyes (Sey): A homozygous lethal mutation on chromosome 2 which affects the differentiation of both lens and nasal placodes in the mouse." *Journal of Embryology and Experimental Morphology* 97 (1986): 95–110.

(with R. Renshaw) "Induction of congenital malformations in the offspring of mutagen treated mice." In *Genetic Toxicology of Environmental Chemicals, Part B: Genetic Effects and Applied Mutagenesis*, edited by C. Ramel et al., 449–458. New York: Alan R. Liss, 1986. Also in *Progress in Clinical and Biological Research* 209B (1986): 449–458.

(with J. Zenthon, E. P. Evans, et al.) "Location of the t-complex on mouse chromosome 17 by *in situ* hybridization with Tcp-1." *Immunogenetics* 24 (1986): 125–127.

"X-chromosome inactivation and sex determination." In *Meeting of the British Society for Developmental Biology on the Mammalian Y-Chromosome: Molecular Search for the Sex-Determining Factor*, 191. Cambridge, England: Company of Biologists, 1987.

(with N. Brockdorff, E.M.C. Fisher, et al.) "Construction of a detailed molecular map of the mouse X-chromosome by microcloning and interspecific crosses." *European Molecular Biology Organization Journal* 6 (1987): 3291–3298.

(with A. G. Searle, J. Peters, et al.) "Chromosome maps of man and mouse. III." *Genomics* 1 (1987): 3–18.

(with J. Zenthon, M. D. Burtenshaw, et al.) "Localization of the Hprt locus by *in situ* hybridization and distribution of loci on the mouse X-chromosome." *Cytogenetics and Cell Genetics* 44 (1987): 163–166.

"Clones and X-chromosomes." *Journal of Pathology* 155 (1988): 97–99.

"Experimental work on induced mutations." *Philosophical Transactions of the Royal Society of London—Series B: Biological Sciences* 319 (1988): 341–352.

"The William Allan Memorial Award address: X-chromosome inactivation and the location and expression of X-linked genes." *American Journal of Human Genetics* 42 (1988): 8–16.

(with J. E. Barker and R. A. Popp) "Mouse globin gene nomenclature." *Journal of Heredity* 79 (1988): 93–95.

(with N. Brockdorff, E.M.C. Fisher, et al.) "Localization of the human X-linked gene for chronic granulomatous disease to the mouse X-chromosome: Implications for X-chromosome evolution." *Cytogenetics and Cell Genetics* 48 (1988): 124–125.

(with S.D.M. Brown, N. Brockdorff, et al.) "The long-range mapping of mammalian chromosomes." In *Genetics of Immunological Diseases*, edited by B. Mock and M. Potter, 3–12. New York: Springer-Verlag, 1988. Also in *Current Topics in Microbiology and Immunology* 137 (1988): 3–12.

(with R. Renshaw) "Induction of congenital malformation in mice by parental irradiation: Transmission to later generations." *Mutation Research* 198 (1988): 277–284.

(with A. G. Searle, eds.) *Genetic Variants and Strains of the Laboratory Mouse.* 2d ed. Oxford, England: Oxford University Press, 1989.

(with A.G. Searle, J. Peters, et al.) "Chromosome maps of man and mouse. IV." *Annals of Human Genetics* 53 (1989): 89–140.

(with K. Willison, V. Lewis, et al.) "The t-complex polypeptide 1 (TCP-1) is associated with the cytoplasmic aspect of Golgi membranes." *Cell* 57 (1989): 621–632.

"Genetics. Evolution of the X-chromosome." *Nature* 348 (1990): 585–586.

"L. C. Dunn and mouse genetic mapping." *Genetics* 125 (1990): 231–236.

(with C. A. Howard, G. R. Gummere, et al.) "Genetic and molecular analysis of the proximal region of the mouse t-complex using new molecular probes and partial t-haplotypes." *Genetics* 126 (1990): 1103–1114.

(with J. Nasir, E. M. Fisher, et al.) "Unusual molecular characteristics of a repeat sequence island within a Giemsa-positive band on the mouse X-chromosome." *Proceedings of the National Academy of Sciences* 87 (1990): 399–403.

(with J. Peters, P. H. Glenister, et al.) "The scurfy mouse mutant has previously unrecognized hematological abnormalities and resembles Wiskott-Aldrich syndrome." *Proceedings of the National Academy of Sciences* 87 (1990): 2433–2437.

"The genetic basis of transmission-ratio distortion and male sterility due to the t-complex." *The American Naturalist* 137 (1991): 349–358.

"The quest for the X-inactivation center." *Trends in Genetics* 7 (1991): 69–70.

(with K. Artzt, D. Barlow, et al.) "Mouse chromosome 17." *Mammalian Genome* 1 (1991): S280–300.

(with K. Dudley, F. Shanahan, et al.) "Isolation and characterization of a complementary DNA clone corresponding to the mouse t-complex gene Tcp-1x." *Genetical Research* 57 (1991): 147–152.

(with P. H. Glenister) "A search for strain differences in response of mice to mutagenesis by thio-TEPA." *Mutation Research* 249 (1991): 317–322.

(with N. D. Mazarakis, D. Nelki, et al.) "Isolation and characterization of a testis-expressed developmentally regulated gene from the distal inversion of the mouse t-complex." *Development* 111 (1991): 561–572.

"Deletion of mouse t-complex distorter-1 produces an effect like that of the t-form of the distorter." *Genetical Research* 59 (1992): 27–33.

(with J. M. Gardner, Y. Nakatsu, et al.) "The mouse pink-eyed dilution gene: Association with human Prader-Willi and Angelman syndromes." *Science* 257 (1992): 1121–1124.

(with T. R. King, Y. Gondo, et al.) "Genetic and molecular analysis of recessive alleles at the pink-eyed dilution (p) locus of the mouse." *Proceedings of the National Academy of Sciences* 89 (1992): 6968–6972.

(with L. M. Silver, K. Artzt, et al.) "Mouse chromosome 17." *Mammalian Genome* 3 (1992): S241–260.

"X-inactivation controlling the X-chromosome." *Current Biology* 3 (1993): 242–244.

(with B. W. Ogunkolade, M. C. Brown, et al.) "A gene affecting Wallerian nerve degeneration maps distally on mouse chromosome 4." *Proceedings of the National Academy of Sciences* 90 (1993): 9717–9720.

(with L. M. Silver, K. Artzt, et al.) "Mouse chromosome 17." In *Genetic Maps: Locus Maps of Complex Genomes*, 6th ed., vol. 4, edited by S. J. O'Brien, 96–103. Plainview, NY: Cold Spring Harbor Laboratory Press, 1993.

"X-chromosome inactivation." In *Molecular Genetics of Sex Determination*, edited by S. S. Wachtel, 123–142. San Diego: Academic Press, 1994.

"The X-inactivation centre and X-chromosome imprinting." *European Journal of Human Genetics* 2 (1994): 255–261.

(with C. A. Everett, P. H. Glenister, et al.) "Mapping of six dominant cataract genes in the mouse." *Genomics* 20 (1994): 429–434.

(with J. Forejt, K. Artzt, et al.) "Mouse chromosome 17." *Mammalian Genome* 5 (1994): S238–258.

(with J. Loester, W. Pretsch, et al.) "Close linkage of the dominant cataract mutations (Cat-2) with Idh-1 and Cryge on mouse chromosome 1." *Genomics* 23 (1994): 240–242.

"The history of X-chromosome inactivation and relation of recent findings to understanding of human X-linked conditions." *Turkish Journal of Pediatrics* 37 (1995): 125–140.

"X-chromosome inactivation. Pinpointing the centre." *Nature* 379 (1996): 116–117.

Other Works

Personal communication to author, 1996.

Works about Mary Frances Lyon

Gillis, A.M. "Turning off the X-chromosome." *BioScience* 44 (3) (1994): 128–132.

Mix, M. C., P. Farber, et al. *Biology*. 2d ed., 365–366. New York: HarperCollins, 1996.

Ohno, S., L. N. Geller, et al. "The analysis of Lyon's hypothesis through preferential X-inactivation." *Cell* 1 (1974): 175–184.

HILDE PROESCHOLDT MANGOLD
(1898–1924)

Veronica Reardon Mondrinos

BIOGRAPHY

Hilde Proescholdt was born on October 20, 1898, in Gotha, Thuringia, a province in east-central Germany. There is little biographical information available about her parents and siblings. Her father, Ernest Proescholdt, was the owner of a soap factory in Gotha. Her mother was Gertrude (Bloedner) Proescholdt. Hilde was the middle child with an older and younger sister. Her family were well-to-do Protestants.

She attended the University at Jena in Germany for two semesters in 1918 and 1919, followed by two semesters at the University of Frankfurt. Here she heard a lecture by Hans Spemann on experimental embryology that contributed to her decision to pursue studies in that field. She enrolled at the Zoological Institute of Freiburg, Germany, in 1921 as a candidate for a Ph.D. degree. She studied under Spemann, who was the leader in embryological research.

In October 1921, she married Otto Mangold, Spemann's chief assistant. She completed her Ph.D. degree in zoology in 1923. Her dissertation was titled "On the Induction of Embryonic Transplants by Implantation of Organizers from Different Species."

In the spring of 1924 she moved with her husband and infant son to Berlin. Otto Mangold had been made Director of Experimental Embryology at the Kaiser Wilhelm Institute for Biology. Hilde Mangold died tragically of severe burns on September 4, 1924, when the gas heater in the kitchen of her apartment exploded. Her son was killed in World War II.

She was described by fellow student and friend Viktor Hamburger as unusually gifted, open, frank, cheerful, vivacious, and charming. He described her sense of humor and her intellect as penetrating and reflective. She had a wide range of interests that included the arts as well as the sciences.

WORK

Mangold's doctoral dissertation was the basis on which her adviser, Spemann, won the Nobel Prize in 1935. He had designed the experiments, and her task was to perform the detailed work. She was the codiscoverer of the "organizer," the chemical that directs the embryonic development of different organs and tissues. The experiments were performed during the spring of 1921 and 1922, during the short breeding season of amphibians.

The experiments were difficult to perform and required a great deal of skill, patience, and perseverance. Mangold transplanted a piece of the upper blastoporal lip of the gastrula of the newt, *Triturus cristatus*, to the flank of a gastrula of the common newt, *Triturus taeniatus*. During the early stages of her work she produced an embryo that displayed on its flanks a large secondary neural tube. Spemann included the results in a previously prepared paper by introducing the term "organizer." The embryonic organizer was composed of migratory cells that invaginated from the surface and induced the development of the neural tube.

Mangold was able to produce only six viable embryos of this transplantation. The transplanted tissue induced the formation of a second embryonic axis and included neural tubes, notochord, intestine, and kidney tubules. The embryo was composed of cells from the donor and the host. The experiments had been designed to help understand the determination of the axial organs in vertebrate embryos. Her work indicated that the organizer substance was responsible for the induced development of the neural tube.

Mangold did not live to see the great impact her experiments had on experimental embryology. She died just as her paper was being published. Her work would raise questions and give rise to numerous experiments in the field. She had continued to work on the organizer experiments in the spring of 1923. The study was with a species of salamander not used in the original experiments. In 1929 the results of the latter investigations were published by her husband in her name.

BIBLIOGRAPHY

Works by Hilde Proescholdt Mangold

Scientific Works

"Über Induktion von Embryonalanlagen durch Implantation artfremder Organisatoren." Ph.D. diss., Zoological Institute of Freiburg, 1923.
(with H. Spemann) "Über Induktion von Embryonalanlagen durch Implantation artfremder Organisatoren." *Archiv für Mikroskopische Anatomie und Entwicklungsmechanik* 100 (1924): 599–638.
"Organisatortransplantationen in verschiedenen Kombinationen bei Urodelen." *Archiv für Entwicklungsmechanik der Organismen* 117 (1929): 697–710.

Works about Hilde Proescholdt Mangold

Hamburger, V. *The Heritage of Experimental Embryology*, 173–180. Oxford, England: Oxford University Press, 1988.

———. Personal communication, 1994.

Oppenheimer, S. B. "The discovering of noggin." *American Biology Teacher* 57 (1995): 264–266.

ANNA MORANDI MANZOLINI (1716–1774)

Connie H. Nobles

BIOGRAPHY

Anna Morandi was born in 1716 in Bologna, Italy. Her parents were Rose (Giovanni) and Charles Morandi. She was raised in a traditional home where marriage, children, and a domestic lifestyle were natural choices for women.

Anna Morandi married her childhood sweetheart, Giovanni Manzolini, in 1736. She was 20 years old, and he was 24. After five years of marriage, Anna Morandi Manzolini was the mother of six children. Additional information concerning her children is unavailable.

Within only a few years after their marriage, Giovanni Manzolini fell victim to tuberculosis. Although poor, he was known for his hard work as a painter and an expert maker of anatomical wax models. In addition, he was Professor of Anatomy at the University of Bologna. However, when he became ill at the age of 30 and had trouble coping with the disease and life's circumstances, Anna Manzolini decided to help her husband with his work. This was a major turning point in her life.

In order to learn anatomy, Anna Manzolini had to dissect cadavers, which was extremely difficult for her. Fortunately, she was triumphant in overcoming her fears and learned anatomy and wax model-making. Her husband realized how indebted he was to his wife: "Elle m'a rendu a moi meme" (She restored me to myself) (Lipinska 1930, 97). Giovanni Manzolini was so encouraged by Anna Manzolini and her accomplishments that he again returned to his work.

In addition to making wax models, Anna Manzolini lectured in her husband's place. Soon wife and husband were recognized as a team by many artists, intellectuals, and anatomists in Europe. She surpassed Giovanni Manzolini in the scientific knowledge necessary to sculpt the models.

Upon her husband's death in 1755, Anna Morandi Manzolini was appointed Lecturer in Anatomy in her own name by the Institute of Bologna. In 1756 she received the title of Professor of Anatomy, and in 1760 she was given the added title of *Modellatrice*.

She was honored by numerous heads of state. Joseph II of Austria visited her and showed his appreciation of her skill and attainments by bestowing upon her gifts equivalent to those given to a sovereign. She was also invited by Catherine II to lecture in Russia.

Her wax models were highly prized while she was alive and long after her death. Some of her anatomical models were so skillfully molded that they were extremely difficult to distinguish from the actual body parts from which they were copied. After her death in 1774, the Medical Institute of Bologna acquired the collection of models that she had made and used. This collection is currently at the Institute of Science in Bologna.

WORK

Anna Manzolini's acute skill at dissection resulted in her discovery of several previously unknown anatomical parts, including the termination of the oblique muscle of the eye. She also held the distinction of having been the first person to reproduce, in wax, body parts of minute portions, including capillary vessels and nerves. One of her wax molds showed the stages of the fetus and how it was nourished.

Her collection of wax models was known throughout Europe as *Supellex Manzoliniana* and was eagerly sought after to aid in the study of anatomy. Her work became the archetype of such models as the exquisite Vassourie collection and the creations of Dr. Auzoux made from papier maché, which were the forerunners of those used in today's schools and colleges.

Anna Manzolini was invited to lecture to private classes by Giovanni Galli, surgeon and professor of gynecology, who opened a school of obstetrics in his house. In her lectures she clearly demonstrated the theoretical and practical expertise she possessed of the human anatomy. Not only was her lecture room frequented by students of many countries, but her reputation spread throughout Europe.

She was elected to membership in the Italian Royal Society, the Russian Royal Scientific Association, and the British Royal Society. The University of Milan sent her a blank contract offering her whatever conditions she requested in order to appoint her as chair of anatomy. However, Manzolini's love for Bologna was greater, and she would not leave.

After her death, a bust of Anna Morandi Manzolini was placed in the Pantheon in Rome. Another portrait in wax, which she modeled herself, was placed in the museum at the University of Bologna.

BIBLIOGRAPHY

Works about Anna Morandi Manzolini

Bolton, H. C. "The early practice of medicine by women." *Popular Science Monthly* 17 (1880): 191–202.

Boulding, E. *The Underside of History*, vol. 2, 160. Newbury Park, CA: Sage, 1992.

Fantuzzi, G. *Notizie Degli Scrittori Bolognesi, Raccolte da Giovanni Fantuzzi*, 113–116. Bologna: S. Tommasco d'Aquino, 1781–1794.

Lipinska, M. "Italie au XVIII siècle—école de Bologne." In *Les femmes et le progres des sciences medicales*, 96–105. Paris: Librairies de L'Académie de Mèdecine, 1930.

Petteys, C. *Dictionary of Women Artists*, 472. Boston: Hall, 1985.

Schiebinger, L. *The Mind Has No Sex?*, 16. Cambridge, MA: Harvard University Press, 1989.

BARBARA McCLINTOCK (1902–1992)

Virginia L. Buckner

BIOGRAPHY

Barbara McClintock was a maverick, an independent thinker, and a loner. From an early age she did what she wanted to and was not influenced by the conventions of the day. She was highly intelligent, had a sharp tongue, and often was less than patient with those who could not keep up with her mentally.

The third of four children born to Thomas Henry and Sara (Handy) Mc-Clintock, she was born June 16, 1902, in Hartford, Connecticut. Her siblings were Marjorie (b. 1898), Mignon (b. 1900), and Malcolm Rider, known as Tom (b. 1904).

Her father, Dr. McClintock, though born in Natick, Massachusetts, was considered a foreigner by her maternal grandfather, as the McClintocks had emigrated from the British Isles. The grandfather, Benjamin Franklin Handy, a Congregationist minister, did not consider McClintock a proper husband for his daughter, for both he and his wife, the former Sara Watson Rider, had ancestors who had come over on the *Mayflower*. Also, McClintock was still a student at Boston University Medical School. Despite these objections, McClintock and Sara Handy were married in 1898, and Sara's small inheritance (her own mother died soon after she was born) helped pay her husband's medical school debts.

Money continued to be scarce, and so her mother gave piano lessons to help out. Her life had not prepared her to take care of four children with little or no help. Therefore, soon after Tom was born, Barbara McClintock was sent to live with her paternal aunt and uncle for several years in rural Massachusetts. She remembered with enthusiasm the time she spent with her uncle. When McClintock returned home, relations with her mother were strained, but she enjoyed working on motors with her father.

In 1908 the family moved to Brooklyn, New York, where McClintock attended elementary school and then Erasmus Hall High School. During this period McClintock was an avid reader and enjoyed being by herself and thinking. The McClintock children were encouraged to do what they wanted to do. If the children did not want to go to school, they did not have to attend school. McClintock loved to ice-skate and played all sorts of sports with her brother and his friends. She persuaded her mother to have bloomers made for her so that she could climb trees with the boys. At Erasmus Hall High School Mc-Clintock discovered science. She found she enjoyed solving difficult problems, though often her solutions were not the ones the teacher expected.

Fearful both of the cost and that too much education would make her daughters unmarriageable, Sara McClintock persuaded the two older girls not to go to college. Her arguments did not sway her third daughter, but not until her father returned from the war in Europe and seemingly with her father's approval, did McClintock enroll in the College of Agriculture at Cornell University (1919).

Although McClintock had been different from other girls in high school, with her interests in sports and science, during her first year at Cornell she blossomed socially. She was elected president of the women's freshman class and was invited to join a sorority. When she became aware of their discriminatory policies, she declined to join. She continued to be strongly against honorary societies and rarely joined them unless it was necessary for her work. During the first two years of college, McClintock went out on dates, but then she decided that she needed no close personal attachment to anyone, including her family. Science became her passion. While a student at college, she bobbed her hair, long before this hairstyle became popular. In 1923 McClintock received her B.S. degree from the Cornell University School of Agriculture.

In 1927 McClintock received her Ph.D. degree in botany and genetics and was asked to remain at Cornell University as an instructor. From 1927 to 1931 she worked as a research assistant and instructor in botany. Her passion was the cytogenetics of maize (corn plant). Apparently, she was not a particularly good instructor but greatly preferred research. At Cornell she was supported by a fellowship from the National Research Council (1931–33). A faculty appointment was not possible. (Cornell appointed its first woman assistant professor in 1947.) In 1933 she received a Guggenheim Fellowship to go to Berlin, Germany. Here she worked with Richard B. Goldschmidt, head of the Kaiser Wilhelm Institute. Her year in Germany was traumatic, for she was politically naive and was upset by the rise of the Nazis. She had many Jewish friends and colleagues. Again in the United States, she returned to Cornell, a woman with an international reputation but no job. The Rockefeller Foundation funded a position for McClintock as a research associate at Cornell from 1934 to 1936. She divided her time between the University of Missouri at Columbia (UMC), California Institute of Technology (Cal Tech), and Cornell, doing the investigations with maize. She posed a difficult problem for her colleagues, because women could teach only at a women's college or work as assistants in laboratories. There were no openings for a research scientist who was a woman. McClintock wanted to be given the credit due her for her research work, and she felt this was not forthcoming because she was a female.

In the spring of 1936, Louis Stadler, a colleague from Cornell, persuaded the University of Missouri to offer McClintock a position as assistant professor at the genetics center he was opening with a grant from the Rockefeller Foundation. Although her research prospered, her time at UMC was not happy. Soon after she arrived, she felt that this job would not work out. The position had been especially created for her, but she was excluded from faculty meetings, and there was no chance for promotion. Nor was she made aware of job offers from other universities. Probably some of the problems were hers. She wanted consideration commensurate with her abilities as a research scientist and not as a "lady" scientist. If the corn, still grown at Cornell, matured late, she might stay at Cornell and arrive late for fall classes. She was often impatient with others who were less quick witted. However, some of the problems were undoubtedly due to the fact that UMC was in a small town in the middle of Missouri. Nobody was prepared for a woman research scientist who wore knickers, did what she wanted, and spoke her mind no matter what the situation. During this period at UMC, McClintock's national reputation continued to grow, and she was elected vice president of the Genetics Society of America (1939).

Although McClintock did not fit in with university faculties, she had the respect and support of her colleagues. A fellow geneticist and colleague from Cornell, Marcus Rhoades, suggested that she come to Cold Spring Harbor, a remote community on the North Shore of Long Island, which had started as the station for experimental evolution (1904). This became one of the early centers of genetic research in the United States. McClintock arrived as a guest at Cold Spring Harbor in June 1941 and stayed. That December Milislav Demerec became Director of the Department of Genetics at the Carnegie Institution of Washington at Cold Spring Harbor. One of his first acts was to give McClintock a one-year appointment, which became a permanent position almost immediately. Here McClintock had a place to grow her corn, a laboratory for her research, a salary, and a home with no teaching or administrative duties. She was to remain here until her death. In the early years at Cold Spring Harbor, McClintock was respected by her colleagues and was very productive.

In 1947 McClintock was awarded the American Association of University Women (AAUW) Achievement Award, the last award she would receive for some time. McClintock continued to work on the problems of maize chromosomes, but the world of genetics was changing. Interest shifted from chromosome cytology to their biochemistry (structure of DNA) and their physiology. Biology was changing from an observational science to a science concerned with the physical and chemical parts of cells. Also, the major emphasis of genetic studies shifted from maize and *Drosophila* to bacteria and bacteriophages (bacterial viruses). In this atmosphere McClintock was considered old-fashioned. She continued to work, using the microscope to visualize the chromosomes and being largely ignored by everyone else. In this period the chromosomes were considered to be unchanging, and the gene was thought to be a fixed unit of inheritance.

In her 1951 presentation to the summer symposium at Cold Spring Harbor, McClintock reported on the movement of genes, a process that she called transposition, but the concept was not accepted. She also reported that switches were the controlling elements in the chromosomes. McClintock had studied maize for 30 years and the process of transposition for 6 years. She knew maize intimately: the patterns of the leaves, the spots of color on the kernels, and the different structures on the chromosomes. Although she knew maize well, she failed to communicate her concepts to others who did not know the plant. Her 1951 report did not follow the accepted beliefs. It was hard for others to understand, as her proofs consisted of observational work on maize, work that the other geneticists did not know.

Another presentation to the summer symposium of 1956 was even less well received. McClintock did not get the deserved respect of her colleagues, for they could not see the logic in what she was reporting. She was ignored and even ridiculed for her unorthodox thinking. She became increasingly isolated as she continued her work, and she no longer participated in the summer symposia at Cold Spring Harbor. From this time on, she withdrew from her fellow scientists and published fewer papers, but she continued working on the genetics of maize. She carefully wrote up her data for the yearbook of the Carnegie Institution, which funded her work.

The National Academy of Sciences, concerned that the indigenous strains of maize in Central and South America could be lost because of modern agricultural methods, asked McClintock to train local cytologists. Between 1958 and 1960, she spent two winters in Central and South America. From examining the chromosomes McClintock could determine patterns of human migrations and trade, as people transported their precious seeds with them. While this information was interesting to anthropologists, it was not McClintock's passion, so she returned to her work at Cold Spring Harbor.

Recognition for this remarkable woman was long in coming. Between 1962 and 1969 McClintock served as a consultant in agricultural science to the Rockefeller Foundation. In 1965 she was appointed the Andrew White Professor-at-Large at Cornell University (a nonresident appointment). In 1967 the National Academy of Sciences awarded her the Kimber Genetics Award, and in 1970 she received the National Medal of Science for her discoveries concerning the chromosomes, work she had done in the 1930s. In the late 1970s, as molecular biologists also began to find transposable elements (first in bacteria and then in other organisms), McClintock's revolutionary work was rediscovered. Then she received the praise and awards that her work so richly deserved. She received an honorary doctorate from Harvard University (1979). In 1981 she won the Albert Lasker Basic Medical Research Award, Israel's Wolf Foundation Prize, and an annual fellowship for life from the MacArthur Foundation. In 1983 came the greatest recognition of all, when she was awarded the Nobel Prize in Physiology or Medicine for the discovery that genetic material is not fixed but instead is mobile, work accomplished in the early 1950s. She is the first woman to be the sole winner of the Nobel Prize in this category.

McClintock worked 12 hours a day, six days a week on her research, for research was her whole life. She died of natural causes on September 2, 1992, at Huntington Hospital, near her small apartment in Cold Spring Harbor. In her obituary in the *New York Times*, Dr. James Watson, Director of the Cold Spring Harbor Laboratory (Nobel laureate, 1962), is quoted as saying, "Dr. McClintock was one of the three most important figures in the history of genetics, one of the 'three M's.' The other two [were] Gregor Mendel and Thomas Hunt Morgan" (Kolata 1992).

WORK

McClintock's work involved the chromosomes and genes of maize. During the first year of her graduate studies, she found the ten chromosomes in maize and, using a light microscope, could identify them by their distinctive shapes and structures. She identified bands on chromosomes that related to observable traits in the maize, and this was the basis of her doctoral dissertation. In order to see the chromosomes more clearly, she developed new techniques of staining. She was unique because she both grew the maize and studied its chromosomes. In fact, plant breeding and genetics were in different departments when McClintock worked at Cornell. However, she felt that she must know each maize plant individually and intimately in order to understand fully what was happening to the chromosomes. She was acknowledged by her colleagues to be the expert in cytogenetics, the relationships between genes and the structure of the chromosome. In fact, George Beadle (Nobel laureate, 1958) asked McClintock for help on the cytogenetics of *Neurospora*, because the chromosomes of this fungus are very small, and nobody could distinguish them. In 1944 McClintock went to Stanford University to work on *Neurospora*. Within two months she had not only counted and identified the chromosomes but also worked out the meiotic cycle of this fungus.

In 1931 McClintock and her graduate student, Harriet Creighton, proved that genes of the chromosomes contain the information necessary for the visible traits. By using genetic markers, they showed that during early stages of meiosis actual exchanges of genetic material occur. They explained the process of crossing-over.

At the University of Missouri, McClintock was involved with mutations caused by X-rays. The radiation of the pollen caused breakage in the chromosomes, resulting in inversions, translocations, and deletions. These changes produced observable changes in the chromosomes and variation in the color, both in the plant foliage and in kernels of maize that developed from the pollen. McClintock determined that when a chromosome reformed, a part of the chromosome might be missing. She found distinctive rings formed from the deleted part of the chromosome. These rings did not get passed along during meiosis.

Missing part of the original chromosome caused the plant to produce kernels of maize with mixed colors in a given kernel. She became fascinated by the pattern of pigmentation in both the corn-husk tissue and the kernels. X-rays also may cause dicentric chromosomes, chromosomes with two centromeres. These chromosomes break and reform during meiosis in a breakage-fusion-bridge cycle. McClintock found that the breakage, followed by reformation of the chromosomes, led to many new mutations in endosperm color. During her close examination of maize chromosomes, McClintock found a chromosomal element that was necessary for the formation of the nucleolus. She called this element the "nucleolar organizer."

As McClintock continued to research maize genetics, she saw kernels of maize with spots and dots and husks with patches of white or yellow. She could see that the patterns were not random but were under some kind of control that appeared to act during the development of the plant. Therefore, she decided that some genetic elements acted as controlling elements or switches. She also determined that these controlling genes moved along the chromosome in response to signals from the cell during the development of the plant. She found that these controlling elements moved either along a chromosome or between chromosomes. McClintock named this process "transposition" and the movable controlling elements or transposable elements "transposons." This work was published in a report to the *Carnegie Yearbook* (1948). By 1951 McClintock had more evidence of traits that were controlled by the transposons. The controlling element that suppressed the pigment gene was called the "dissociator" (*Ds*), and the "activator" (*Ac*) blocked the action of the dissociator. McClintock found that the chromosome would break at a certain locus (*Ds*) and that this occurred only when the *Ac* was activated. The *Ac* element could control several different genes. The more *Ac* that was present in the genome, the lower the frequency of mutations. Since *Ac* could move, sometimes one sister chromatid had more *Ac* than the other sister chromatid, and therefore sister chromatids were not identical. This was the paper that she presented at the Cold Spring Harbor Symposium in 1951, a presentation that was ignored and even ridiculed by her colleagues. The scientists of the period thought that genes remained in the same place along the chromosome and could not comprehend her work. McClintock continued to work using only her observations of the maize plants, breeding experiments, and a light microscope for observation of the chromosomes. She found another system of control, the suppressor-mutator system (*Spm*). Again, there were two controlling elements. The first of these two elements could delete or cause a mutation in the second controlling element. These controlling elements could be found in several positions along the chromosome, another example of transposition. The reports of McClintock's research were published yearly in the *Carnegie Institution of Washington Yearbooks*, which nobody read. So her work was ignored.

When McClintock saw the paper on operons of bacteria published in 1961 by François Jacob (Nobel laureate, 1965) and Jacques Monod (Nobel laureate,

1965), she wrote a paper for the *American Naturalist* (1961), indicating that her system of dissociator and activator closely resembled the operator and regulator genes of Jacob and Monod. Still, no recognition came. In the years that followed, as more was learned about the genes along the chromosome, and as others found genes that move in different organisms from bacteria to nematodes, the work of McClintock received recognition. Throughout her life McClintock was considered by her contemporaries to be without peer in the observation of chromosomes. However, her research was ignored because she did not follow the leaders into molecular genetics but continued to work with her microscope, breeding experiments, and observation of the plants. Since her concepts have become known and understood, McClintock now is considered by many to have been a scientific visionary.

McClintock joined only the few organizations that were important to her professional life. In 1939 she was elected vice president of the Genetics Society of America and then president of that society in 1945. In 1944 McClintock was elected to the National Academy of Sciences, only the third woman to be so recognized.

BIBLIOGRAPHY

Works by Barbara McClintock

Scientific Works

Space does not permit the listing of the complete works of Barbara McClintock. Listed here are all works by McClintock except those cited in McClintock (1987) and Torrey Botanical Club (1886–1966). Included here are her dissertation and all references cited in the text.

"A Resume of Cytological Investigations of the Cereals with Particular Reference to Wheat." M.A. thesis, Cornell University, 1925.

(with L. F. Randolph) "Polyploidy in *Zea mays* L." *American Naturalist* 60 (1926): 99–102.

"A cytological and genetical study of triploid maize." Ph.D. diss., Cornell University, 1927.

(with G. W. Beadle) "A genic disturbance of meiosis in *Zea mays.*" *Science* 68 (1928): 433.

"A method for making acetocarmin smears permanent." *Stain Technology* 4 (1929): 53–56.

"A cytological demonstration of the location of an interchange between two non-homologous chromosomes of *Zea mays.*" *Proceedings of the National Academy of Sciences* 16 (1930): 791–796.

(with H. B. Creighton) "A correlation of cytological and genetical crossing-over in *Zea mays.*" *Proceedings of the National Academy of Sciences* 17 (1931): 492–497.

(———) "The correlation of cytological and genetical crossing-over in *Zea mays.* A corroboration." *Proceedings of the National Academy of Sciences* 21 (1935): 148–150.

(with M. M. Rhoades) "The cytogenetics of maize." *Botanical Review* 1 (1935): 292–325.

"Breakage-fusion-bridge cycle induced deficiencies in the short arm of chromosome 9." *Maize Genetics Cooperation News Letter* 18 (1944): 24–26.

"Mutable loci in maize." *Carnegie Institution of Washington Yearbook* 47 (1948): 155–169.

"Mutable loci in maize." *Carnegie Institution of Washington Yearbook* 53 (1954): 227–237.

"1. Spread of mutational change along the chromosome. 2. A case of *Ac*-induced instability at the bronze locus in chromosome 9. 3. Transposition sequences of *Ac*. 4. A suppressor-mutator system of control of gene action and mutational change. 5. System responsible for mutations at a_1-m2." *Maize Genetics Cooperation News Letter* 29 (1955): 9–13.

"1. Further study of the $a_1{}^{m-1}$ *Spm* system. 2. Further study of *Ac* control of mutation at the bronze locus in chromosome 9. 3. Degree of spread of mutation along the chromosome induced by *Ds*. 4. Studies of instability of chromosome behavior of components of a modified chromosome." *Maize Genetics Cooperation News Letter* 30 (1956): 12–20.

"1. Continued study of stability of location of *Spm*. 2. Continued study of a structurally modified chromosome 9." *Maize Genetics Cooperation News Letter* 31 (1957): 31–39.

"Some parallels between gene control systems in maize and in bacteria." *American Naturalist* 95 (1961): 265–277.

"(1) Restoration of A_1 gene action by crossing over. (2) Attempts to separate *Ds* from neighboring gene loci." *Maize Genetics Cooperation News Letter* 39 (1965): 42–51.

"Significance of chromosome constitutions in tracing the origin and migration of races of maize in the Americas." In *Maize Breeding and Genetics*, edited by W. D. Walden, 159–184. New York: John Wiley and Sons, 1978.

(with T. A. Kato Y. and A. Blumenschein) *Chromosome Constitution of Races of Maize. Its Significance in the Interpretation of Relationships between Races and Varieties in the Americas*. Chapingo, Mexico: Colegio de Postgraduados, Escuela National de Agricultura, 1981.

"Trauma as a means of initiating change in genome organization and expression." *In Vitro* 19 (1983): 283–284.

The Discovery and Characterization of Transposable Elements: The Collected Papers of Barbara McClintock. New York: Garland, 1987.

Works about Barbara McClintock

Craig, Patricia Parratt. *Jumping Genes*. Washington, DC: Carnegie Institution of Washington, 1994.

Fedoroff, N., and D. Botstein (eds.). *The Dynamic Genome*. Cold Spring Harbor, NY: Cold Spring Harbor Press, 1992.

Hammond, A. L. *A Passion to Know: Twenty Profiles in Science*. New York: Scribners, 1984.

Keller, E. F. *A Feeling for the Organism*. San Francisco: W. H. Freeman, 1983.

Kent, C. *Barbara McClintock*. New York: Chelsea House, 1988.

Kolata, G. ''Barbara McClintock, 90, gene research pioneer, dies.'' *New York Times* (Sept. 4, 1992).

Torrey Botanical Club. *Index to American Botanical Literature, 1886–1966*. Boston: G. K. Hall, 1969.

Other References

Jacob, F., and J. Monod. ''Gene regulatory mechanisms in the synthesis of proteins.'' *Journal of Molecular Biology* 3 (1961): 318–356.

BEATRICE MINTZ (1921–)

Maura C. Flannery

BIOGRAPHY

Beatrice Mintz, a member of the prestigious National Academy of Sciences, is noted for her work in developmental biology and genetics. She was born in Brooklyn, New York, on January 24, 1921, to Samuel and Janie (Stein) Mintz. She was educated in New York City schools and attended Hunter College in Manhattan, where she received her A.B. degree (1941) magna cum laude. After graduation she worked as a research assistant at the Guggenheim Dental Clinic in New York City and, in 1942, went to the University of Iowa to do graduate work, earning her M.S. degree (1944) and her Ph.D. degree in zoology (1946). During this time she served as a research assistant in developmental biology. At Iowa she worked with Emil Witschi, studying the effects of sex hormones on the development of the reproductive system in amphibians.

After completing her doctoral studies, she became an instructor in the Department of Biological Sciences at the University of Chicago. She remained at Chicago until 1960 and during this time rose from the rank of instructor to that of associate professor. Since 1960 she has worked at the Institute for Cancer Research at the Fox Chase Cancer Center in Philadelphia. She began there as an associate member and has been a senior member since 1965.

Because of the innovative nature of her work and the great creativity and energy she has brought to it, Mintz has received a number of prestigious awards. She has been a member of the National Academy of Sciences since 1973 and of the American Philosophical Society since 1982. In 1986 she was invited by the Vatican to become a member of the Pontifical Academy of Sciences, a body composed of distinguished scientists who meet to consider important questions of scientific and ethical significance (Holden 1987). Among the other honors Mintz has received are the Bertner Foundation Award (1977), the Award for Biological and Medical Science of the New York Academy of Sciences (1979), the Papanicolaou Award of the Papanicolaou Cancer Research Center (1979), the Medal of the Genetics Society of America (1981), the Amory Prize of the

American Academy of Arts and Sciences (1988), the Ernst Jung Medal (1990), and the first March of Dimes Prize in Developmental Biology (1996).

In November 1994 she was presented with Philadelphia's John Scott Award during a ceremony at the American Philosophical Society (Sankaran 1994). The award was established in the early 1800s by John Scott, a Scottish druggist living in Philadelphia. He entrusted the administration of the award to the city and intended it as a reward to individuals whose useful inventions contributed significantly to the "comfort, welfare and happiness" of mankind. Mintz sees the award as "a paeon in praise of crazy ideas," her most notable "crazy idea" being the fusing of embryos from mice with different genetic makeups (Sankaran 1994).

As an undergraduate at Hunter College, Mintz was elected to Phi Beta Kappa. Subsequently, she has been the recipient of several honorary doctorate degrees: from New York Medical College (1980), from the Medical College of Philadelphia (1980), from Northwestern University (1982), and from Hunter College (1986). She received an L.H.D. degree from Holy Family College (1988). Between 1973 and 1990 she also presented a number of named lectures, including the Harvey Society Lecture in 1976 (published in 1978) and a National Institutes of Health Lecture (1978). In 1951 Mintz was a Fulbright Research Scholar at the University of Paris and the University of Strasbourg. Other distinguished positions she holds are fellow of the American Association for the Advancement of Science; fellow of the American Academy of Arts and Sciences; and honorary fellow of the American Gynecological and Obstetrical Society. She was a member of the Board of Advisers of the Jane Coffin Children's Memorial Fund from 1977 to 1979. All of these honors indicate the esteem in which Mintz's work is held. She has uniquely combined research in developmental biology with work in genetics and cancer studies, thus illuminating all of these areas. Now in her 70s, she continues to be an active researcher, doing work on the skin cancer melanoma.

WORK

Beatrice Mintz's early work was on development of the reproductive system in amphibians. At the University of Iowa she studied the role of sex hormones in producing sex reversal in larvae of the frog *Rana clamitans* and of the salamander *Ambystoma mexicanum*. She continued this line of research for some time at the University of Chicago and expanded it to include work on animals that had undergone early hypophysectomy. This removal of the hypophysis (pituitary gland) did not alter the effects of hormones on sex development, thus showing that the hormonal effects were not mediated through the pituitary. In cases where complete sex reversal failed to occur when the hormone of the opposite sex was administered during development, microscopic examination of the tissues indicated that the tissue necessary for the development of the sex organs had already deteriorated to the point that it could not be recruited for this purpose. These early studies on amphibians display two hallmarks found in much of Mintz's work: ambitious research using a large number of individuals (work on 295 animals was reported in her doctoral study) and careful microscopic examination of tissues.

In the mid-1950s Mintz changed the focus of her research from amphibians to mice. She worked with mice that were heterozygous for the mutation W^j that involves a germ cell deficiency. Examining heterozygous animals during development, she found that germ cells formed in the eighth day of development were normal but that from the ninth day on, two populations of cells were apparent. One population proliferated and migrated normally, and the second population, presumably those cells where the defective gene was active, failed to increase in number or to move. Her observations also supported the view that the germ cells arose in nongonadal tissue and migrated to the gonads (or sex organs) during development.

After her move to Fox Chase, Mintz continued to work with the genetics of mouse embryos and developed the "crazy idea" that led to great strides in her own research and gained her so much esteem in the scientific community. Mintz reasoned that it would be valuable to be able to produce mice that had two genetically different populations of cells, for example, cells carrying a lethal genetic defect involving the blood combined with cells normal for blood development. This combination might produce an individual that would survive long enough to be studied yet still carry interesting effects of the defect. She achieved such mouse "mosaics" by fusing two embryos early in development.

While other researchers had attempted such fusions before, they had been unable to get the fused embryos to develop normally. One reason Mintz was successful was that she used a different method for removing the *zona pellucida*, the membrane enveloping the egg and early embryo. Others had mechanically stripped off the *zona*, but she removed it chemically by exposing the embryo to a solution of the protein-destroying enzyme pronase for only three minutes. With the *zona* removed, Mintz found that the embryos easily adhered to each other when they were each at the eight-to-ten-cell stage; at earlier stages, the cells of the two embryos were not sticky enough to adhere. She also found that maintaining a constant temperature of 37°C and a constant pH was essential to successful fusion. When the embryos successfully fused, they continued development to the blastocyst stage; they were then introduced into the uterine horn of a pseudopregnant mouse that had been mated with a vasectomized male. Using this procedure, over 25,000 offspring have been produced in Mintz's laboratory since 1967; this number includes over 500 morphologically normal adult animals. (Presumably, many other offspring would have reached adulthood but were sacrificed earlier in studies on development.)

Her 1971 article in *Methods in Mammalian Embryology* gives detailed descriptions of the painstaking methods she had developed to ensure a high degree of success in her embryo fusion studies. This paper has been cited in over 150 publications, and Mintz says that the technique is now so standard that "most people forget to cite the source'' (Sankaran 1994). While she called the earliest mice produced from fused blastocysts "mosaics," she came to find this term inadequate because it did not convey the order discernible in the mice or the fact that this order was interpretable in terms of developmental process and genetic information. She also saw problems with the term "chimera," which implies a monstrosity, both because the mice were normal and because there was a thorough integration of the two genetic components. These problems led her to coin a new term, "allophenic," to signify individuals with "a simultaneous, orderly manifestation of two (or more) allelic cellular phenotypes, or *allophenes*, each with a known and distinctive genetic basis. *Allopheny* is therefore the phenomenon of concurrent display of such allelic cellular phenotypes'' ("Gene control . . ." 1967, 349).

The number of citations of Mintz's 1971 article indicates how important allophenic mice became in the study of mammalian development and genetics. While genetic mosaics are easily produced in fruit flies, mammalian mosaics were almost unheard of before her work, except for a few cases of transplantation of fetal tissue that occurred much later in development than her fusions did and produced animals that had genetically different tissue only in the area of the transplant. The beauty of her technique was that it could produce individuals with combinations of cells from almost any two types of mice, and the two types of cells were often found together in the same organs.

One notable thing about Mintz's technique is that it was very labor-intensive; she and her coworkers produced exciting and valuable results, but they worked extremely hard to get them. Not only did each fusion involve the removal of embryos from two different individuals and the transfer of the resultant blastocyst into a pseudopregnant recipient, but all the tissues of each offspring resulting from this procedure had to be carefully examined to determine whether and to what extent each tissue possessed the genotypes of the two embryos used in the fusion. Sometimes only one genotype was present. This might be either because of damage to one of the embryos during the fusion or transfer or because only a few cells from the blastocyst form the individual, and the rest form the accessory tissues, such as the placenta.

In cases where both genotypes were present, the variety of combinations was impressive, and from these Mintz was able to decipher a great deal about the processes going on during development. For example, allophenic mice that possessed two different genotypes for coat color often had a striped appearance, though the patterns were not always regular. After examining a large number of these mice, Mintz concluded that, at most, there were 17 successive bands down each side of the animal: three on the head, six on the body, and eight on the tail. Each band suggests "progressive lateral, and also some anteroposterior, movement of cells during their proliferation, and each band has been interpreted as a single clone descended from one clonal initiator or primordial melanoblast cell'' ("Gene expression . . ." 1970, 22). Mintz uses the word "clone" here to refer to a group of cells that are all descended from one cell.

Similar studies on the skeletal systems of a large number of allophenic mice brought Mintz to the judgment that each vertebra is formed from four clones, while investigations of the eye led to the conclusion that the photoreceptor cells of each eye proliferate radially from a circlet of ten clonal-initiator cells. She noted that "from studies in allophenic mice, the clone appears to be the critical developmental unit in mammals by which fine-focus genetic control of differentiation is achieved" ("Gene expression . . ." 1970, 40). None of these findings were easily obtained. They involved not only the tedious examination under the microscope of the tissues of many animals but also insightful analysis of the data. All of Mintz's papers have these two characteristics. What makes her career truly remarkable is that she has been able to maintain this meticulous level of research throughout a variety of investigations.

Her development of allophenic mice was particularly significant because they could be used to tease out so much information about both normal and abnormal development. In terms of normal development, Mintz found support for the concept that immune tolerance is the result of nonreactivity of lymphoid cells, rather than of some blocking agent preventing their reactivity. She discovered, that no matter what strains of mice were combined to form allophenic individuals, the animals were permanently immunologically tolerant of their component cells. This was even the case when the two genotypes involved different histocompatibility antigens. Usually, when tissue was transplanted from an animal with one histocompatibility type into an animal with a different type, transplant rejection occurred. However, rejection did not occur when tissue of either contributing genotype was transplanted to an allophenic mouse, indicating the individual's tolerance for both of the donor tissue types.

In work on the blood-making or hematopoietic system in allophenic mice, Mintz found a high correlation between the ratio of each genotype in the erythrocytes and in lymphocytes, and the genotype ratio in circulating red and white blood cells. This supported the view that shared hematopoietic precursor cells gave rise to erythrocytes, granulocytes, and lymphocytes. In other work on the hematopoietic system, Mintz used allophenic mice to investigate abnormal development. She fused an embryo that carried the lethal genetic defect called macrocytic hypoplastic anemia with a genetically normal embryo. In the majority of cases involving these fusions that resulted in viable mice, both the red and white blood cells were genetically normal. This indicates that the initial action of the gene involved occurs very early in hematopoietic development, before differentiation into erythroid and lymphoid cell types has taken place. On the other hand, when embryos carrying the microcytic anemia gene were fused with normal embryos, the resulting mice carried only the normal gene in the erythrocytes, but both gene types were found in the white blood cells. This suggests that the gene was expressed or utilized only in the erythroid cell line.

In other work with genetic defects, Mintz found that when an embryo homozygous for the lethal t^{12} gene was fused with a normal embryo, the resulting blastocyst developed normally up to the midmorula stage but that the significant changes normally found in the late morula stage were limited or lacking. Mintz had selected the t^{12} mutation for study because it produced the earliest genetic lethal defect known in mammals. Using allophenic mice, she was able to pinpoint the stage at which the defect began to become apparent and to begin to isolate the cells in which problems first appeared. Again, careful microscopic observation was combined with an intellectually sophisticated research plan.

Mintz has spent the major part of her research career at an institution dedicated to cancer study, and while her work on the fundamentals of development bears on the question of cancer, it is not surprising that she has also done work that focuses directly on cancer. When she fused embryos of mice with a genetic susceptibility to cancer, with those lacking this susceptibility, she discovered that tumors found in the allophenic mice were almost always from the susceptible strain, and almost never did a tumor consist of cells of both strains. This result supports the premise that most tumors arise as a clone of cells; that is, they can be traced back to a single cell. While this idea is now commonly accepted, it was still controversial at the time Mintz did this work in the mid-1960s, and her results played a role in the final acceptance of the clonal model by cancer biologists.

In the 1970s Mintz continued her cancer research using teratocarcinoma cells, cells derived from tumors formed when embryonic cells from some mice strains are transplanted to sites such as the kidney capsule. In her initial experiments, Mintz used cells from a tumor line that had been maintained for eight years. She first tried fusing the teratocarcinoma cells with cells of a normal mouse embryo, but the cells would not adhere, so she injected a few teratocarcinoma cells into a normal mouse blastocyst. She found that, in many cases, development was normal, and the resulting mice were often mosaics, with cells from both the normal and teratocarcinoma cells in their tissues.

But these mice did not develop cancer. This was a rather astounding discovery: carcinoma cells, when placed in the right environment, could revert to normal behavior and could differentiate normally. This was the first clear indication that at least in this form of cancer, the abnormality was not genetic but was, rather, developmental. If the teratocarcinoma cells were placed in the right environment, that is, in a normal embryo, they could receive the normal cellular signals for differentiation and could become specialized in a normal way. Perhaps, even more amazingly, the teratocarcinoma cells, in the cases where they contributed the cells of the sex organs for the mosaic mouse, could produce normal offspring. In other words, what was a cancer cell in one environment became a normal germ cell in another environment: a normal germ cell from which a normal sex cell could arise. Mintz attempted the same experiment with a number of different teratocarcinoma cell lines and found that normal development occurred only when the teratocarcinoma cells used had a normal or close-to-normal chromosome number. From this Mintz concluded that a normal karyotype was related to normal development; her results also suggested that other tumors might derive from abnormal development rather than from genetic abnormalities.

Frequently in her research, when she discovered an interesting result such as the development of allophenic mice, Mintz looked for ways to use that work as a tool for further exploration of questions about development and genetics. This was the case with her work on teratocarcinomas. Since these cells were easily grown in culture, they could be readily screened for mutations. She subjected such cells to radiation and then screened them for a deficiency in hypoxanthine phosphoribosyltransferase (HPRT), the defect present in Lesch-Nyhan syndrome (a hereditary disorder of males in which there is excess uric acid production in blood and urine). The teratocarcinoma cells with the HPRT deficiency were then injected into a normal mouse blastocyst. Some of the individuals that resulted from this procedure were mosaics having both normal and HPRT-deficient cells. Though their contribution varied from animal to animal, the latter cells were found in all the tissues tested, except the blood. This might indicate that the activity of this gene is particularly necessary in blood cells, so those that lack it did not survive.

While much of her work involved *in vivo* studies, Mintz also used teratocarcinoma cells in *in vitro* experiments on genetic defects. She collaborated with Joseph Goldstein and Michael Brown in their research on familial hypercholesterolemia, a genetic disorder involved in vascular and coronary disease. Goldstein and Brown, who later received the Nobel Prize in physiology or medicine for their work (1985), had identified a cell surface receptor that normally binds low-density lipoprotein (LDL). Individuals with a genetic defect in this LDL receptor are susceptible to familial hypercholesterolemia at an early age. Mintz found that teratocarcinoma cells have LDL receptors and that mutant cells with defective receptors can easily be found with a fluorescence test for lipoprotein uptake by the cell.

As molecular biology became more sophisticated in the 1970s, and as individual genes could be cloned so that large numbers of copies of a gene in a pure form could be produced, Mintz took advantage of the availability of cloned deoxyribonucleic acid (DNA) and came up with new research strategies. While she continued to work with allophenic mice, they were no longer the primary focus of her research. Some of her investigations now involved microinjection of mouse eggs with foreign genes. In a 1974 study, she injected mouse blastocysts with simian virus 40 (SV40) DNA. She chose this virus because it is a tumor-causing agent. She found that SV40 DNA sequences were incorporated into several tissues of the healthy mice that developed from these blastocysts. Despite this, none of the mice had developed any form of cancer after a year. Mintz saw the long-term survival of these animals as presenting the possibility that such mice might be used for experimental investigation of how tumor viruses are transmitted and expressed.

In another series of experiments she used the tedious process of microinjection of DNA; she introduced two foreign genes, one encoding a tissue-specific protein, the blood protein human β-globin, and the other encoding an enzyme found in all tissues, thymidine kinase (TK). About 2,500 copies of a piece of DNA called a plasmid, containing both these genes, were injected into fertilized mouse eggs. The eggs that survived the microinjection procedure without overt injury were transferred into the oviducts of pseudopregnant females. Foreign DNA was found in 5 of the 33 fetuses that developed from the transferred eggs. From 3 to 50 copies of the β-globin DNA and 3 to 20 copies of the TK DNA were found in each of the 5 fetuses. The foreign DNA was incorporated into the mouse DNA and in some cases was transcribed into ribonucleic acid (RNA); thus, this donor DNA remained intact throughout development.

In another series of experiments on recombination of human and mouse genes, copies of the gene for human growth hormone (HGH) were injected into mouse eggs. Each of 229 eggs was injected with about 6,000 copies of a plasmid carrying the HGH gene. Of the injected eggs, 143 survived the injection, and ultimately 20 animals were born from the experimental eggs. Six of the 20 carried the HGH gene. Each of these animals became founders of new strains (named HUGH/1 to HUGH/6). In all the strains but HUGH/5, the DNA profile remained the same from generation to generation, indicating that the DNA was stably integrated into the host genome. In the homozygous form, both HUGH/3 and HUGH/4 were lethal. The assumption is that in these instances, the HGH gene was inserted into the mouse DNA in such a way that vital host sequences were disrupted, resulting in lethality, though the HGH gene was inserted into different chromosomes in HUGH/3 and HUGH/4. These are the first reported cases of early-stage lethalities resulting from insertional mutagenesis due to integration of recombinant DNA in the mouse germ line.

Experiments like these led Mintz to suggest that "insertional mutagenesis by exogenous DNA integration in the mammalian germ line may be expected to provide an important new tool for isolating and characterizing genes active in development. . . . The special promise of experimental insertional mutagenesis is that it may reveal some genes responsible for the very developmental phenomena that have remained most elusive" (Mintz et al. 1985, 451). This statement is very characteristic of Mintz's work. While she has developed many new techniques over the years and has used numerous techniques of others, to her a technique is not important in itself, but only as a means toward understanding the complexities of development and of gene expression. Mintz clearly states this over and over again in her papers. Her writings are always lucid, and her points are well made, but this leitmotif of experimental technique as a means to an end gives her papers their orientation. She makes clear how the work reported fits into the larger picture of developmental genetics.

Through the 1980s Mintz continued recombinant DNA studies on mice. She found several cases where the insertion of genes caused insertional mutations and rearrangements of host DNA, as the HGH gene had done. In other research that obviously had implications for human genetic engineering, she used handicapped retroviruses to introduce foreign genes into mouse hematopoietic cells.

She was then successful in reintroducing the retrovirally transduced cells back into mice.

In the 1990s Mintz has focused on malignant melanomas of both the eye (ocular) and the skin (cutaneous). She found that these melanomas arose in transgenic mice having an integrated recombinant gene made up of the tyrosinase promoter, which is expressed in pigment cells, and SV40 DNA transforming sequences. The tumors were similar to human melanomas. In a 1991 paper, Mintz and her coworkers note that "for both ocular and cutaneous melanomas, the transgenic mice offer numerous possibilities for experimental study of mechanisms underlying formation and spread of melanomas" ("Malignant melanoma..." 1991, 164). Since that time she has illustrated the validity of this statement in more recent reports. Further studies showed that these animals differ genetically only in the number of copies and chromosomal site of integration of the transgene or foreign DNA. Skin melanocytes from young mice with no apparent skin lesions were established in culture from donors with low, medium, and high numbers of transgene copies. All the cultures changed over time, but those with greater numbers of transgene copies became more abnormal in several different ways.

In 1991 Mintz also reported the results of studies involving the microinjection of albino mouse eggs with DNA for tyrosinase, the enzyme lacking in albinos. Five of the mice that developed from these cells had lightly pigmented patterns of stripes, indicating that the tyrosinase was functional in some of the cells. This patterning is reminiscent of that found in some allophenic mice. Mintz makes this comparison and argues that the similarity arises from the clonal theory of development that she elaborated in her work on allophenic mice. In the patterned transgenic mice, the potential for tumor susceptibility may be affected by clonal variation without further gene mutations or deletions.

Mintz has made a number of significant contributions to developmental biology and genetics over the past 50 years. What makes her work so striking is that she opened up whole areas of research that both she and many others have explored. She developed the first allophenic animals, which were used to elucidate a number of questions in normal mammalian development and in the effects of defective genes. Her experiments on teratocarcinoma cells in allophenic mice pioneered the concept that some teratocarcinomas were cancers arising from defective development rather than from defective genes. Her more recent work on melanoma has revealed aspects of this cancer that may make more effective treatments possible. In sum, her work has led her to be ranked among the most significant American women biologists of the twentieth century.

BIBLIOGRAPHY

Works by Beatrice Mintz

Scientific Works

Space does not permit the listing of the complete works of Beatrice Mintz. Listed here are all works by Mintz except those cited in "Allophenic mice . . ." (1971), "Gene control . . ." (1974), and "Putting genes . . ." (1983). Included here are her dissertation and all references cited in the text.

"Time and quantity relationships in hormonal sex reversal of *Rana clamitans* larvae." *Anatomical Record* 89 (4) (1944): 10.

(with C. Foote and E. Witschi) "Quantitative studies on response of sex characters of differentiated *Rana clamitans* larvae to injected androgens and estrogens." *Endocrinology* 37 (1945): 286–296.

"Effects of Testosterone Propionate on Sex Development in Female *Ambystoma* Larvae." Ph.D. diss., University of Iowa, 1946.

(with E. Witschi) "Determination of the threshold dose of testosterone propionate inducing testicular development in genetically female anurans." *Anatomical Record* 96 (4) (1946): 30–31.

"Effects of testosterone propionate on sex development in female *Ambystoma* larvae." *Physiological Zoology* 20 (1947): 355–372.

"Testosterone propionate minimum for induction of male development in anurans; comparative data from other vertebrates." *Proceedings of the Society for Experimental Biology and Medicine* 69 (1948): 358–361.

"Androgen-induced feminization of urodele larvae in the absence of the hypophysis." *Anatomical Record* 119 (4) (1954): 493–517.

"Embryological development of primordial germ cells in the mouse: Influence of a new mutation, W^j." *Journal of Embryological and Experimental Morphology* 5 (1957): 396–403.

(editor) *Environmental Influences on Prenatal Development.* Chicago: University of Chicago Press, 1958.

"Continuity of the female germ cell line from embryo to adult." *Archives d'Anatomie Microscopique et de Morphologie Experimentale* 155 (1959): 172.

"Embryological phases of mammalian gametogenesis." *Journal of Cellular and Comparative Physiology* 56 (Suppl. 1) (1960): 31–47.

(with R. L. Brinster) "A method for *in vitro* cultivation of mouse ova from two-cell to blastocyst." *Experimental Cell Research* 32 (1963): 205–208.

"Genetic mosaicism in adult mice of quadriparental lineage." *Science* 148 (1965): 1232–1233.

"Gene control of mammalian pigmentory differentiation. I." *Proceedings of the National Academy of Sciences* 58 (1967): 344–351.

"Gene expression in allophenic mice." *Symposia of the International Society for Cell Biology* 9 (1970): 15–42.

"Allophenic mice of multi-embryo origin." In *Methods in Mammalian Embryology,* edited by J. Daniel, 186–214. San Francisco: W. H. Freeman, 1971.

(with J. Gearhart and A. Guymount) "Phytohemagglutinin-mediated blastomere aggregation and development of allophenic mice." *Developmental Biology* 31 (1973): 195–199.

"Gene control of mammalian differentiation." *Annual Review of Genetics* 8 (1974): 411–470.

(with R. Jaenisch) "Simian Virus 40 DNA sequences in DNA of healthy adult mice derived from preimplantation blastocysts with viral DNA." *Proceedings of the National Academy of Sciences* 71 (1974): 1250–1254.

(with K. Ilmensee) "Normal genetically mosaic mice produced from malignant teratocarcinoma cells." *Proceedings of the National Academy of Sciences* 72 (1975): 3585–3589.

(with M. Dewey and A. Gervais) "Brain and ganglion development from 2 genotypic classes of cells in allophenic mice." *Developmental Biology* 50 (1976): 68–81.

"Teratocarcinoma cells as vehicles for mutant and foreign genes." *Brookhaven Symposia in Biology* 29 (1977): 82–95.

"Gene expression in neoplasia and differentiation." *Harvey Lectures* 71 (1978): 193–246.

(with M. Dewey) "Direct visualization by beta-galactosidase histochemistry of differentiated normal cells derived from malignant teratocarcinoma in allophenic mice." *Developmental Biology* 66 (1978): 550–559.

(———) "Genetic control of cell type specific levels of mouse beta-galactosidase." *Developmental Biology* 66 (1978): 560–563.

(with J. Goldstein, M. Brown, et al.) "Demonstration of low density lipoprotein receptors in mouse teratocarcinoma stem cells and description of a method for producing receptor-deficient mutant mice." *Proceedings of the National Academy of Sciences* 76 (1979): 2843–2847.

(with E. Wagner and T. Stewart) "The human β-globin gene and a functional viral thymidine kinase gene in developing mice." *Proceedings of the National Academy of Sciences* 78 (1981): 5016–5020.

(with J. Blanchet and R. Fleischman) "Murine adult hematopoietic cells produce adult erythrocytes in fetal recipients." *Developmental Genetics* 3 (1982): 197–206.

(with R. Fleischman and R. Custer) "Totipotent hematopoietic stem cells normal self-renewal and differentiation after transplantation between mouse fetuses." *Cell* 30 (1982): 351–360.

(with T. Stewart and E. Wagner) "Human beta-globin gene sequences injected into mouse eggs retained in adults and transmitted to progeny." *Science* 217 (1982): 1046–1048.

"Putting genes into mice." In *Perspectives on Genes and the Molecular Biology of Cancer*, edited by D. Robberson and G. Saunders, 207–220. New York: Raven Press, 1983.

(with E. Wagner, L. Covarrubias, et al.) "Prenatal lethalities in mice homozygous for human growth hormone gene sequences integrated in the germ line." *Cell* 35 (1983): 647–655.

(with K. Anthony and S. Litwin) "Monoclonal derivation of mouse myeloid and lymphoid lineages from totipotent hematopoietic stem cells experimentally engrafted in fetal hosts." *Proceedings of the National Academy of Sciences* 81 (1984): 7835–7839.

(with R. Fleischman) "Development of adult bone marrow stem cells in H-2 compatible and H-2 incompatible mouse fetuses." *Journal of Experimental Medicine* 159 (1984): 731–745.

(with L. Covarrubias and Y. Nishida) "Early developmental mutations due to DNA rearrangements in transgenic mouse embryos." *Cold Spring Harbor Symposia on Quantitative Biology* 50 (1985): 447–452.

(———) "Early postimplantation embryo lethality due to DNA rearrangements in a transgenic mouse strain." *Proceedings of the National Academy of Sciences* 83 (1986): 6020–6024.

(——— et al.) "Cellular DNA rearrangements and early developmental arrest caused by DNA insertion in transgenic mouse embryos." *Molecular and Cellular Biology* 7 (1987): 2243–2247.

(with R. Hawley, L. Covarrubias, et al.) "Handicapped retroviral vectors efficiently transduce foreign genes into hematopoietic stem cells." *Proceedings of the National Academy of Sciences* 84 (1987): 2406–2410.

(with M. Terao) "Cloning and characterization of a complementary DNA coding for mouse placental alkaline phosphatase." *Proceedings of the National Academy of Sciences* 84 (1987): 7051–7055.

(———, D. Pravtcheva, et al.) "Mapping of gene encoding mouse placental alkaline phosphatase to chromosome 4." *Somatic Cell Genetics* 14 (1988): 211–216.

(with B. Capel) "Neonatal W-mutant mice are favorable hosts for tracking development of marked hematopoietic stem cells." *Experimental Hematology* 17 (1989): 872–876.

(———, R. Hawley, et al.) "Clonal contributions of small numbers of retrovirally marked hematopoietic stem cells engrafted in unirradiated neonatal W-W-V mice." *Proceedings of the National Academy of Sciences* 86 (1989): 4564–4568.

(with B. Capel and R. Hawley) "Long and short-lived murine hematopoietic stem cell clones individually identified with retroviral integration markers." *Blood* 75 (1990): 2267–2270.

(with L. Larue) "Pigmented cell lines of mouse albino melanocytes containing a tyrosinase complementary DNA with an inducible promoter." *Somatic Cell Molecular Genetics* 16 (1990): 361–368.

(with M. Bradl) "Mosaic expression of a tyrosinase fusion gene in albino mice yields a heritable striped coat color pattern in transgenic homozygotes." *Proceedings of the National Academy of Sciences* 88 (1991): 9643–9647.

(———, A. Klein-Szanto, et al.) "Malignant melanoma in transgenic mice." *Proceedings of the National Academy of Sciences* 88 (1991): 164–168.

(with M. Bradl and L. Larue) "Clonal coat color variation due to a transforming gene expressed in melanocytes of transgenic mice." *Proceedings of the National Academy of Sciences* 88 (1991): 6447–6451.

(with A. Klein-Szanto, M. Bradl, et al.) "Melanosis and associated tumors in transgenic mice." *Proceedings of the National Academy of Sciences* 88 (1991): 169–173.

(with S. Porter) "Multiple alternatively spliced transcripts of the mouse tyrosinase-encoding gene." *Gene* 97 (1991): 277–282.

(——— and L. Larue) "Mosaicism of tyrosinase-locus transcription and chromatin structure in dark vs. light melanocyte clones of homozygous chinchilla-mottle mice." *Developmental Genetics* 12 (1991): 393–402.

(with A. Klein-Szanto) "Malignancy of eye melanomas originating in the retinal pigment

epithelium of transgenic mice after genetic ablation of choroidal melanocytes."
Proceedings of the National Academy of Sciences 89 (1992): 11421–11425.

(with L. Larue and N. Dougherty) "Genetic predisposition of transgenic mouse melanocytes to melanoma results in malignant melanoma after exposure to a low
ultraviolet B intensity nontumorigenic for normal melanocytes." *Proceedings of
the National Academy of Sciences* 90 (1992): 9534–9538.

(—— et al.) "Spontaneous malignant transformation of melanocytes explanted from
W^f/W^f mice with a kit kinase-domain mutation." *Proceedings of the National
Academy of Sciences* 89 (1992): 7816–7820.

(——) "Melanocyte culture lines from Tyr-SV40E transgenic mice models for the
molecular genetic evolution of malignant melanoma." *Oncogene* 8 (1993): 523–
531.

(with W. Silvers) "Transgenic mouse model of malignant skin melanoma." *Proceedings
of the National Academy of Sciences* 90 (1993): 8817–8821.

(—— and A. Klein-Szanto) "Histopathogenesis of malignant skin melanoma induced
in genetically susceptible transgenic mice." *Proceedings of the National Academy
of Sciences* 90 (1993): 8822–8826.

(with A. Klein-Szanto and W. Silvers) "Ultraviolet radiation-induced malignant skin
melanoma in melanoma-susceptible transgenic mice." *Cancer Research* 54
(1994): 4569–4572.

Works about Beatrice Mintz

Abelson, R. "Fox Chase Cancer Center thriving on recognition." *Philadelphia Business
Journal* 5 (50) (1995): 12.
Holden, C. "The Vatican weighs in." *Science* 235 (1987): 1455.
Marx, J. L. "Probing gene action during development." *Science* 236 (1987): 29–31.
——. "Tracking genes in developing mice." *Science* 215 (1982): 44–47.
Runkle, G., and A. J. Zaloznik. "Malignant melanoma." *American Family Physician* 49
(Jan. 1994): 91.
Sankaran, N. "City of Philadelphia's John Scott Award honors cancer researcher for
'crazy ideas.' " *The Scientist* 8 (24) (1994): 23.
Schmeck, H. M., Jr. "Cancer cells in experiment give rise to normal tissue." *New York
Times* (Feb. 13, 1979): C1, C2.
——. "Studying cloning is called aid to medical research." *New York Times* (June 1,
1978): 16.
"A transgenic first." *The Scientist* 10 (9) (1996): 30.

ANN HAVEN MORGAN (1882–1966)

Susan J. Wurtzburg

BIOGRAPHY

Ann (Anna) Haven Morgan was born on May 6, 1882, to Julia Alice (Douglass) and Stanley Griswold Morgan. She grew up in Waterford, Connecticut, with her younger sister, Christine, and a younger brother, Stanley.

Morgan completed her secondary schooling at Williams Memorial Institute in New London and in 1902 began study at Wellesley College, not the best location for a young, nonconformist woman student. Morgan's attendance at Boston boxing events and her general lack of compliance to Wellesley standards of behavior ensured a short tenure at the school. Within two years, she transferred to Cornell University, where she fared better, completing her A.B. degree in 1906. Upon graduation Morgan obtained an assistant teaching position in the Department of Zoology at Mount Holyoke College, South Hadley, Massachusetts (1906–7). Her enthusiasm for her work was soon rewarded by a promotion to instructor (1907–9).

During Morgan's first year at Mount Holyoke, she made the acquaintanceship of Cornelia M. Clapp (1849–1934), who was to become her good friend and mentor. Clapp, who had taught zoology and other subjects at Mount Holyoke since 1872, had benefited from the teaching of Lydia W. Shattuck,* who in turn had been educated in a curriculum strongly influenced by Mary Lyon. Morgan's biographical tribute to Clapp focuses on the power of female mentorship and on Clapp's determined quest for knowledge, recorded in her oft-repeated statement ''[I]f you want to do a thing there is no particular reason why you shouldn't do it'' (Morgan 1935, 2). Clapp had begun her Ph.D. studies late in life (completed 1896), and it was likely that Morgan was guided by her colleague's experiences, as she decided to obtain a doctorate, though at an earlier stage in her career.

In 1910, upon Morgan's return to graduate studies at Cornell, she was granted an instructorship (1909–11), later upgraded to a fellowship (1911–12). Her work at Cornell placed her in the midst of an enthusiastic and congenial group of young entomologists, many of them studying aquatic insects. Morgan's dissertation concentrated upon the biology of mayflies, resulting in the fitting nickname "Mayfly Morgan" (Bonta 1991, 245). She viewed mayfly biology as a unique means to study the "two problems which face every organism . . . [namely] maintaining its own life and continuing its race" (Morgan 1913, 371). The investigation of the mayfly life cycle was particularly relevant to the understanding of these two drives, since the "immature life of May flies is aquatic, and to it all adjustments concerned with food or safety are exclusively confined. The mature or adult life is aerial. It is solely devoted to reproduction. There is no provision for food or for other means of lengthening its life" (Morgan 1913, 371).

She completed her doctoral research in 1912 under the guidance of James G. Needham (Limnological Laboratory, Cornell University). Needham was highly supportive of his student, and in 1908 he helped her obtain membership in the prestigious Entomological Society of America. Several decades later, Needham contributed a foreword to Morgan's *Field Book of Ponds and Streams*, in which he applauded her ability to "make the knowledge of the whole range of life in ponds and streams a little more easy of access," since "with that knowledge will come appreciation, and a purpose to aid in keeping the waters free from pollution" (Morgan 1930, v). Needham's ongoing influence upon Morgan is indicated by her tribute to his "store of biological truth" in her *Field Book of Animals in Winter* (Morgan 1939, viii), written a decade later.

Upon the completion of her Ph.D. degree, Morgan returned to Mount Holyoke (as Ann rather than Anna). She was first employed as Instructor (1912–14), then Associate Professor (1914–17), and in 1918 she was promoted to full Professor. Morgan remained at Mount Holyoke and served as the Chair of the Zoology Department for more than 30 years (1916–47), leaving this influential position upon her retirement.

According to all accounts, Morgan was gifted with a powerful intellect and a wonderful vocabulary, which, combined with her public speaking ability, resulted in inspired teaching. "Students found her exacting but memorable—a short, trim woman with bobbed hair and blue eyes who always wore a tailored skirt, shirtwaist, tie, and white physician's coat, she lectured in clear, crisp tones" (Bonta 1991, 248).

Morgan's courses incorporated material from her own zoological studies, which focused on ecology (especially in the northeastern United States), the biology of water-dwelling animals, and animal adaptations to winter. Her teaching and writings also incorporated appeals for environmental preservation. Later in life she was renowned as a conservationist, and her writings indicate the strength of her belief that all living things are connected one to another. For example, Morgan's introduction to the *Kinships of Animals and Man* (1955, 1) states: "This book is about animals, those that are regularly called animals and others, the human animals. . . . It is about the relations of animals to one another, and to the plants upon which they depend, to water, to the sun, and to the earth about them. The organization and relationships inside and outside of animals are the keys to their existence.''

Throughout her tenure at Mount Holyoke, Morgan generally taught during the autumn and spring terms. When not teaching in South Hadley, her summers were devoted to research and teaching at various other locations: the University of Chicago (1916), the Marine Biological Laboratory, Woods Hole, Massachusetts (1918, 1919, 1921, 1923), Harvard University (1920), Yale University (1921), and the Tropical Laboratory in Kartabo, British Guiana (1926). Her choices of research settings testify to her enjoyment of the outdoors, and her earliest writings record her delight in a discipline that allowed her to explore the natural environment. For example, the site of her New York mayfly research is depicted through near lyrical prose: ''[T]hese streams flow along as quiet meadow brooks, or broadening out over stony beds are caught in a maze of ripply shallows'' (Morgan 1911, 95).

Research interests also developed into hobbies. Morgan's study of mayfly biology turned her attention to the insect lures used for fly-fishing. She became an accomplished manufacturer of feather-tied fishing lures, which she gave as gifts to friends and acquaintances. She also contributed to fishing lore through her publication on pond life, renowned as ''an angler's favorite'' (*Time* 1933). Morgan's zoological work also encouraged other artistic endeavors. She photographed and illustrated her research subjects to provide accompanying documentation for her articles and books. She also provided photographs for at least two exhibitions, one at Mount Holyoke and the other at Cornell University.

Morgan's dedication to zoological matters and to her students is apparent in both the quality and quantity of her publications. Much of this work was jointly authored with female colleagues and students from Mount Holyoke. It seems that Morgan's strongest friendships and working relationships were formed with other women, and indeed, she never married. Morgan's closest friend was A. Elizabeth Adams (1892–1962), a younger professor of zoology at Mount Holyoke. The two women shared a home for many years, and Morgan dedicated both her *Field Book of Ponds and Streams* (1930) and her *Kinships of Animals and Man* (1955) to Adams, acknowledging that ''Dr. Adams has been generous beyond my telling in giving her time, scholarship and keen critical sense'' (Morgan 1955, vi).

Based on her scientific achievements, Morgan was accorded the accolade of an entry in *American Men of Science* (1933). At that time, she, Ruth Benedict (an anthropologist) and Libbie Henrietta Hyman* (a zoologist) were the only three women to be included among 250 scientists.

From the 1940s onward, Morgan increased her efforts in environmental preservation and was involved in reforming the science curriculum through her membership on the National Committee on Policies in Conservation Education. She worked to increase the ecology and conservation content in general science courses and is considered one of the early forerunners of the ecology movement. In the mid-1940s she participated in a summer study session for teachers, the

Rhode Island Conservation Workshop. Morgan also increased her efforts at local conservation and for three summers (1944–46) worked for the Massachusetts State Biological Survey.

Retiring from Mount Holyoke in 1947, Morgan remained an active conservationist. She continued her workshops for teachers and, accompanied by Adams, traveled in the western United States and Canada. This experience impressed both women with the impetus of the western conservation movement and apparently inspired "her to support conservation efforts in her own Connecticut River Valley" (Bonta 1991, 249). Morgan continued her activities on behalf of an ecological, conservationist zoology until her death at age 84 of stomach cancer (June 5, 1966).

With her death, Ann Haven Morgan ensured her continued support of biological research. According to the terms of her will, she provided funds to the American Association of University Women for the establishment of a national fellowship, the Elizabeth Adams–Ann Morgan fellowship.

During her lifetime Morgan had received many honors, among them being a fellow of the American Association for the Advancement of Science, a Schuyler Fellow at Cornell University, a visiting fellow at Harvard University, and a visiting fellow at Yale University. She received research grants from Sigma Xi (1926, 1930) and the American Association for the Advancement of Science (1926, 1930).

WORK

Morgan's zoological interests covered a broad spectrum. Her research included investigations of the functions of gills in aquatic insects, studies of the sensory abilities of amphibians, the thyroid functions of these and other species, pond ecology, animal adaptations to winter, general ecology, and conservation. These research interests resulted in publications on mayflies, sponges, caddisflies, beetles, frogs and toads, newts, fowl, and general ecology (three books, published in 1930, 1939, and 1955). Many of these works include her own illustrations and photographs, displaying her artistic skills.

Her books demonstrate an ever-widening zoological focus, beginning with the biology of freshwater ponds (1930), then dealing with the adaptations of animals to winter (1939), and finally encompassing ecological aspects of animal biology (1955), stressing each animal's ecological niche. Morgan's views concerning animal and human interdependence place her among a select group of strong conservationists, of whom the most renowned is probably Rachel Carson,* the author of *Silent Spring*. An educational film to be used in the classroom, based on her work *Field Book of Animals in Winter*, was developed by Morgan (1949).

In addition to the value of her published contributions, Morgan served as a mentor and colleague to numerous women scientists. Twenty-four of her 45 scientific publications and presentations were coauthored with other women. Student recollections of her provide a fitting tribute to her abilities as a teacher and scientist. One former student wrote that "to have her for a teacher even one year was to have her live with you always" (Richardson 1966, 93). Another former student, an active professional in 1966, wrote, "I'm probably the way I am mostly because of her—a wonderful legacy and responsibility" (Richardson 1966, 93).

She held memberships in the Entomological Society of America, the American Society of Zoologists, the American Society of Naturalists, Sigma Xi, the American Limnological Society, the Ecological Society, the National Committee on Policies in Conservation, the American Association of Museums, the New York Herpetological Society, and the Association of Social Hygiene. She also served on the national Advisory Board of *Eugenics*.

NOTE

I would like to acknowledge the helpful assistance of Elaine D. Trehub, Archive Librarian; Sabine Cray at the Alumnae Office; and members of the Biology Department, all of Mount Holyoke College.

BIBLIOGRAPHY

Works by Ann Haven Morgan

Scientific Works

"May-flies of Fall Creek." *Annals of the Entomological Society of America* 4 (1911): 93–126.
"A Contribution to the Biology of May-flies." Ph.D. diss., Cornell University, 1912.
"Homologies in the wing-veins of may-flies." *Annals of the Entomological Society of America* 5 (1912): 89–111.
"A contribution to the biology of may-flies." *Annals of the Entomological Society of America* 6 (1913): 371–413.
"The wings of Ephemerida." In *The Wings of Insects*, edited by John H. Comstock, 214–223. Ithaca, NY: Comstock, 1918.
"The temperature senses in the frog's skin." *Journal of Experimental Zoology* 35 (1922): 83–114.
Common Water Insects, A Field Guide. Ithaca, NY: Slingerland-Comstock, 1929.
"Fresh-water sponges in winter." *Science Monthly* 28 (1929): 152–155.
"Illumination of anatomical preparations." *Science* 69 (Feb. 22, 1929): 255.
"The mating flight and the vestigial structures of the stump-legged mayfly, *Campsurus segnis* Needham." *Annals of the Entomological Society of America* 22 (1929): 61–69.
Field Book of Ponds and Streams: An Introduction to the Life of Fresh Water. New York: G. P. Putnam's Sons, 1930.
(with Margaret C. Grierson) "The effects of thymectomy on young fowls." *Anatomical Record* 47 (1930): 101–117.

(with Helen D. O'Neil) "The function of the tracheal gills in larvae of the caddisfly, *Macronema zebratum.*" *Bollettino di Zoologia* 5 (1930).

(———) "The function of the tracheal gills in larvae of the caddis fly, *Macronema zebratum*" Hagen. *Physiological Zoology* 4 (1931): 361–379.

(with S. Claire Sondheim) "Experiments in gill reduction in neotenous *Triturus virides-cens.*" *Proceedings of the Society for Experimental Biology and Medicine* 29 (1931): 299–301.

(with Margaret C. Grierson) "The functions of the gills in burrowing mayflies (*Hexagenia recurvata*)." *Physiological Zoology* 5 (1932): 230–245.

(———) "Winter habits and yearly food consumption of adult spotted newts, *Triturus viridescens.*" *Ecology* 13 (1932): 54–62.

(with S. Claire Sondheim) "Attempts to reduce the gills in neotenous newts, *Triturus viridescens.*" *Anatomical Record* 52 (1932): 7–29.

Laboratory Studies in General Zoology. Ann Arbor: Edwards Brothers, 1936.

(with Janet F. Wilder) "The oxygen consumption of *Hexagenia recurvata* during the winter and early spring." *Physiological Zoology* 9 (1936): 153–169.

Field Book of Animals in Winter. New York: G. P. Putnam's Sons, 1939.

(with Catherine H. Fales) "Seasonal conditions and effects of low temperature in the thyroid glands of amphibians. I. Adult *Triturus viridescens.*" *Journal of Morphology* 71 (1942): 357–389.

(with Barbara J. Johnson) "Seasonal conditions and effects of low temperature in the thyroid glands of amphibians. II. Terrestrial phase of *Triturus viridescens.*" *Journal of Morphology* 70 (1942): 301–321.

Animals in Winter. Encyclopedia Britannica Films, 1949.

Laboratory Studies in General Zoology, 4th ed., revised by A. Elizabeth Adams. Ann Arbor: Edwards Brothers, 1953.

Kinships of Animals and Man: A Textbook of Animal Biology. New York: McGraw-Hill, 1955.

Laboratory Studies to Accompany "Kinships of Animals and Man." New York: Mc-Graw-Hill, 1957.

"Plecoptera." In *McGraw-Hill Encyclopedia of Science and Technology,* vol. 10, 415–416. New York: McGraw-Hill, 1960.

(with L. Berner) "Ephemeroptera." In *McGraw-Hill Encyclopedia of Science and Technology,* 7th ed., vol. 6, edited by S. P. Parker, 445–446. New York: McGraw-Hill, 1992.

Other Works

"Zoology at Mount Holyoke." "Steps toward the new science building." *Mount Holyoke Alumnae Quarterly* 5 (1921): 17–23; 88–90.

"The new laboratory." *Mount Holyoke Alumnae Quarterly* 6 (1922): 121–127.

"Science and the new curriculum." *Mount Holyoke Alumnae Quarterly* 14 (1930): 14–15.

"Cornelia Maria Clapp: March 17, 1849–December 31, 1934: An adventure in teaching." *Mount Holyoke Alumnae Quarterly* 19 (1935): 1–3.

"Mount Holyoke's next president." *Mount Holyoke Alumnae Quarterly* 19 (1935): 28–29.

"A letter to the alumnae from Ann Haven Morgan." *Mount Holyoke Alumnae Quarterly* 31 (1947): 59–60.

Works about Ann Haven Morgan

Alexander, Charles P. "Ann Haven Morgan: 1882–1966." *Eatonia* 8 (1967): 1–2.
Bonta, Marcia Myers. *Women in the Field: America's Pioneering Women Naturalists.* College Station: Texas A & M University Press, 1991.
Holyoke Transcript-Telegram (June 6, 1966): 14, 18.
New York Times (June 6, 1966): 41.
Richardson, Dorothy. "In memoriam: Ann Haven Morgan 1882–1966." *Mount Holyoke Alumnae Quarterly* 50 (1966): 93.
Time 21 (12) (Mar. 20, 1933): 38.

LILIAN VAUGHAN SAMPSON MORGAN (1870–1952)

Katherine Keenan

BIOGRAPHY

Lilian Vaughan Morgan, the discoverer of the attached-X chromosome, was born Lilian Vaughan Sampson in 1870 in Hallowell, Maine. Her father, George Sampson, could trace his ancestry back to the *Mayflower*. Her mother, Isabella Merrick, was a descendant of the Vaughan family, early settlers of Hallowell. The family was prosperous and literate. Her mother's uncle, Samuel Vaughan Merrick, was the first president of the Pennsylvania Railroad. In the 1870s the Vaughan and Merrick families had the largest library north of Boston.

George and Isabella Sampson had three daughters: Edith, Lilian, and Grace. When Edith was five, and Lilian three, both parents and infant Grace died of tuberculosis. Edith and Lilian were raised by their maternal grandparents, Thomas B. and Elizabeth White Merrick in Germantown, Pennsylvania, in a large family.

The sisters were encouraged to study, and they both majored in biology at Bryn Mawr College. Lilian Sampson's undergraduate adviser was the progressive president of Bryn Mawr, M. Carey Thomas, and one of her biology professors was the preeminent cell biologist Edmund B. Wilson. Lilian Sampson graduated in 1891 at the top of her class and was awarded the European Fellowship.

After graduation, she spent the first of many summers at the newly established Marine Biological Laboratory in Woods Hole, Massachusetts. At the end of the summer she traveled to Switzerland on her fellowship to study anatomy at the University of Zurich with Arnold Lange, a student of Ernst Haeckel. She contined this work the following year at Bryn Mawr. She received an M.S. degree in biology (1894), and her adviser was Thomas Hunt Morgan.

In 1904 Lilian Sampson and Thomas Hunt Morgan were married. T. H. Morgan left Bryn Mawr to become Professor of Experimental Zoology with E. B. Wilson at Columbia University in New York. The first summer after their marriage they spent in California at Stanford University. Lilian Morgan published two papers from the work she completed there on planarian regeneration (L.V. Morgan 1905, 1906).

The Morgans had four children: Howard Key (b. 1906), Edith Sampson (b. 1907), Lilian Vaughn (b. 1910), and Isabel Merrick (b. 1911). She interrupted her research to stay at home for 16 years and raise her children. Losing her parents and a sister at age three gave her a lifetime commitment to her family. They always came first. Sadly, her older sister, Edith, died at a young age, also from tuberculosis. Lilian Morgan always had excellent health.

In the years she was not actively doing research, she never lost her interest in science. She was naturally very curious and explorative. During the years away from the laboratory, she was active in the education of her children. She was one of the founders of the Children's School of Science in Woods Hole, Massachusetts, and she also belonged to the Naples Table Association. This group sponsored investigators at the marine laboratory in Naples, Italy. She enjoyed photography, played the violin, and liked to travel, and she and her son, Howard, learned to drive an Overland vehicle from an instruction manual.

In the Morgan household she sought her husband's advice but made all the child-raising and housekeeping decisions herself. While she was not doing research, she clearly understood the work going on in her husband's laboratory on sex determination, sex-linkage, mutation, and crossing-over. It was one of the most exciting periods in genetics in the twentieth century. T. H. Morgan and his colleagues discovered the basic mechanisms of inheritance in *Drosophila*, the common fruit fly. Her husband received the Nobel Prize in Physiology or Medicine in 1933 for his work in genetics. Although Lilian Morgan was never part of the inner research circle, she discussed the daily results with her husband and was considered by many to be his intellectual equal.

In 1921 she returned to pursue research in genetics in T. H. Morgan's laboratory. At that time the youngest child, Isabel, was ten years old. Isabel Morgan Mountain (d. 1996) later carried on the family research tradition. Mountain developed a vaccine that immunized monkeys against polio. It was the forerunner of the Salk vaccine. She was one of ten scientists inducted into the Polio Hall of Fame established by the National Foundation for Infantile Paralysis at Warm Springs, Georgia. Lilian Morgan never assisted her husband in his work and went to great lengths not to appear intrusive in his laboratory. She always worked independently and was very absorbed in her work but undoubtedly she felt isolated (Schultz 1981). In 1922 she published the first of 12 papers in genetics.

The Morgans moved to Pasadena in 1928, when T. H. Morgan became head of the Division of Biology at the California Institute of Technology. Lilian Morgan could now spend more time in the laboratory. She again worked independently. In her career she never attended a scientific meeting or presented her work. Lilian Morgan did not have an academic appointment until one year after her husband's death. She became a research associate at the California

Institute of Technology in 1946. She died in Los Angeles, California, in 1952, at the age of 82.

WORK

Lilian Morgan began her research in anatomy and development. She studied biology when part of the discipline was about to take a dramatic turn away from natural history and toward the quantitative fields of genetics and physiology. Her early papers are in descriptive anatomy and embryology, while her later research was in genetics.

She carefully described the musculature of chitons (1894, 1895 as Sampson), anuran breeding and development (1900 as Sampson), and studied embryology and regeneration (1904 as Sampson; 1905, 1906). When she went back to the laboratory in the 1920s, she returned as a *Drosophila* geneticist. She had become thoroughly acquainted with the field. Lilian Morgan had her own stocks and worked independently of everyone else. On February 12, 1921, she discovered an unusual female fruit fly in a group she had anesthetized and was examining. She knew the discovery was important because it was a female fruit fly with markings that should have been in a male. In fact, it was a mosaic, with two different genotypes. The fruit fly woke up from the anesthesia and flew away. She searched for it and was able to recapture the tiny fly on a windowpane (Sturtevant 1965, 55).

This female, when crossed with a normal male, yielded very unusual flies. L.V. Morgan correctly concluded that the X chromosomes were transmitted not singly, but attached as XX. Undoubtedly, this conclusion was due, in part, to the recent discovery of nondisjunction by Calvin B. Bridges (Bridges 1913). In *Drosophila* the X chromosome is shaped like a rod. When two X chromosomes are attached, they look like the letter V. She confirmed her conclusion by chromosomal analysis. Her paper "Non-criss-cross inheritance in *Drosophila melanogaster*" is a classic in genetics (1922).

When the attached-X female is crossed with a normal male, the results are opposite to what one expects for sex-linked recessives. The X chromosome sex-linked recessive should pass from the mother to the sons, while the dominant allele from the father should go to the daughters. Exactly the reverse occurs with the attached-X strain. The mother passes the recessive to the daughters, and the sons receive the dominant from their fathers. This was a totally unexpected and startling result.

For example, if a normal male (XY) with a recessive allele for bar eye is crossed with an XXY female with normal round eyes, the results are bar-eyed males and round-eyed females. The sons are always like their father for the charcteristics on the X chromosomes. The males never inherit the single X from their mothers.

Lilian Morgan established the attached-X strain of *Drosophila*. It proved to be an excellent tool for studying crossing-over. Crossing-over, the exchange of chromosome segments at meiosis, had only recently been discovered by Alfred H. Sturtevant (1925). Sturtevant writes, "The chief advance in the understanding of the geometrical relations in crossing-over came as a result of the discovery of the attached-X's" (Sturtevant 1965, 55). In fact, E. G. Anderson (1925), using the attached-X strain, was able to show that crossing-over occurs at the four-stranded stage of meiosis, after chromatid formation.

The strain also confirmed both the theories of sex-linkage and the chromosome theory of sex determination. Although the results of an attached-X cross are opposite to the expected, it clearly and dramatically confirmed the theory of sex-linkage. While the bar-eyed cross showed an anomaly in the distribution of the X chromosome, it illustrated the same anomaly in the distribution of the bar-eyed trait. This discovery of the attached-X provided evidence for the newly discovered chromosome theory and confirmed the balance theory of sex determination in *Drosophila*. At the time very little was known about sex determination. Lilian Morgan's attached-X females are XXY females. In 1959 it was first recognized by P. A. Jacobs and J. A. Strong that in humans the XXY is a male (Klinefelter's syndrome). Furthermore, since the male's X chromosome always passes to the males (patroclinous inheritance), it never crosses over at meiosis. This provides a convenient way of maintaining certain traits or combinations of traits on the X chromosome in males. It is an excellent way of maintaining the purity of X chromosomes.

L.V. Morgan's second major contribution to contemporary genetics was the discovery of a closed X (ring) chromosome in *Drosophila*. When she crossed an attached-X yellow female with a scute broad apricot normal male, she would occasionally see a wild type female fly. These daughters had received one of their mother's attached-X chromosomes detached from its mate and an X from their father.

In one of her cultures she isolated one of these wild type females. When this female was mated with a regular forked-bar male, there were unexpected non-disjunctional sterile males and, more important, very low cross-over frequencies. She hypothesized that the low frequencies could be explained by an unusual chromosome. She was able to show that the wild type female daughter of the attached-X female had a single maternal X chromosome that was closed. The rod-shaped X had rounded up to form a ring (Morgan 1933). The ring X is another exceptional tool in genetic research. It is very useful for studying cross-over frequencies and generating gynandromorphs (an individual, part of whose body exhibits male sex characteristics, and part female characteristics). Ring Xs can be stable or unstable. Their instability can be turned to an advantage because, when they are eliminated very early in embryological development, they result in mosaic flies. Ring Xs also behave uniquely during mitosis, as reported by Barbara McClintock* (1938).

Lilian Morgan continued working on crossing-over, composites, the bar locus, polyploidy, nondisjunction, compound duplications, somatic crossing-over, intersexes, and the autosomal eye phenotype "sparkling." Her papers are thorough and are classic examples of observation, genetic deduction, and cytological confirmation. She was a first-rate scientist.

Lilian Morgan led a full life. She remained committed to her family, yet she made major contributions to contemporary genetics at a time when few women were engaged in biological research.

BIBLIOGRAPHY

Works by Lilian Vaughan Sampson Morgan

Scientific Works

"Die Muskulatur von Chiton." *Jenaischen Zeitschrift für Naturwissenschaft* 28 (1894): 460–468 (as L.V. Sampson).

"The musculature of chiton." *Journal of Morphology* 11 (1895): 595–628 (as L.V. Sampson).

"Unusual modes of breeding and development among anura." *American Naturalist* 34 (1900): 687–715 (as L.V. Sampson).

"A contribution to the embryology of *Hylodes martincensis*." *American Journal of Anatomy* 3 (1904): 473–504 (as L.V. Sampson).

"Incomplete anterior regeneration in the absence of the brain in *Letoplana littoralis*." *Biological Bulletin* 9 (1905): 187–193.

"Regeneration of grafted pieces of planarians." *Journal of Experimental Zoology* 3 (1906): 269–294.

"Non-criss-cross inheritance in *Drosophila melanogaster*." *Biological Bulletin* 42 (1922): 267–274.

"Polyploidy in *Drosophila melanogaster* with two attached X chromosomes." *Genetics* 10 (1925): 148–178.

"Correlation between shape and behavior of a chromosome." *Proceedings of the National Academy of Sciences* 12 (1926): 180–181.

"Composites of *Drosophila melanogaster*." *Carnegie Institution of Washington Publication* 399 (1929): 225–296.

"Proof that bar changes to not-bar by unequal crossing-over." *Proceedings of the National Academy of Sciences* 17 (1931): 270–272.

"A closed X chromosome in *Drosophila melanogaster*." *Genetics* 18 (1933): 250–283.

"Effects of a compound duplication of the X chromosome of *Drosophila melanogaster*." *Genetics* 23 (1938): 423–462.

"Origin of attached-X chromosomes in *Drosophila melanogaster* and the occurrence of non-disjunction of X's in the male." *American Naturalist* 72 (1938): 434–446.

"A spontaneous somatic exchange between non-homologous chromosomes in *Drosophila melanogaster*." *Genetics* 24 (1939): 747–752.

(with T. H. Morgan and H. Redfield) "Maintenance of a *Drosophila* stock center, in connection with investigations on the germinal material in relation to heredity." *Carnegie Institution of Washington Yearbook* 42 (1943): 171–174.

(with T. H. Morgan and A. H. Sturtevant) "Maintenance of a *Drosophila* stock center, in connection with investigations on the germinal material in relation to heredity." *Carnegie Institution of Washington Yearbook* 44 (1945): 157–160.

"A variable phenotype associated with the fourth chromosome of *Drosophila melano-gaster* and affected by heterochromatin." *Genetics* 32 (1947): 200–219.

Works about Lilian Vaughan Sampson Morgan

Keenan, K. "Lilian Vaughan Morgan (1870–1952): Her life and work." *American Zoologist* 23 (1983): 867–876.

Other References

Anderson, E. G. "Crossing over in a case of attached X chromosomes in *Drosophila melanogaster.*" *Genetics* 10 (1925): 403–417.

Bridges, C. B. "Non-disjunction of the sex chromosomes in *Drosophila.*" *Journal of Experimental Zoology* 15 (1913): 587–606.

Jacobs, P. A., and J. A. Strong. "A case of human intersexuality having a possible XXY sex-determining mechanism." *Nature* 183 (1959): 302–303.

McClintock, B. "The production of homozygous deficient tissues with mutant characteristics by means of the aberrant mitotic behavior of ring-shaped chromosomes." *Genetics* 23 (1938): 315–377.

Schultz, H.R. Letter to the author, 1981.

Sturtevant, A. H. "The effects of unequal crossing over at the bar locus of *Drosophila.*" *Genetics* 10 (1925): 117–147.

———. *A History of Genetics.* New York: Harper and Row, 1965.

ELIZABETH FONDAL NEUFELD (1928–   )

Mary Clarke Miksic

BIOGRAPHY

Elizabeth Fondal was born in Paris, France, in 1928. Her parents, Jacques and Elvira Fondal, were Russian refugees who came to France after the revolution in Russia. She went to primary school in France. In response to the increasing militarism and aggression of the fascist governments in Germany and Italy, she emigrated from Europe with her parents. Fondal was 12 years old when the family arrived in New York, where they were confronted with a new society, language, and school system. She finished high school in New York and in 1944 entered Queens College (later part of the City University of New York), graduating with a B.S. degree in 1948.

After college, Fondal worked as a research assistant at the Jackson Memorial Laboratory in Bar Harbor, Maine. There, with Dr. Elizabeth Russell, Fondal investigated the characteristic blood components of several inbred strains of mice. After a year at the Jackson Laboratory, she began her graduate studies at the University of Rochester, Department of Physiology, spending 1949–50 at this institution.

In 1951 she married Benjamin Neufeld and moved to Baltimore, where she was a research assistant at the McCollym-Pratt Institute of Johns Hopkins University. With Drs. Nathan O. Kaplan and Sidney P. Colowick she studied the enzymes and kinetics of electron transfer reactions involved in energy storage of bacteria and various animal tissue cells. Several papers were published based on this work that set Neufeld on her lifelong investigation into the intricacies of intracellular chemistry.

Neufeld entered the Ph.D. program in comparative biochemistry at the University of California, Berkeley, in 1952. As she completed her graduate studies, she worked in the laboratory of Dr. William Zev Hassid, primarily on plant cell enzyme reactions in complex carbohydrate chemistry, studying synthesis and degradation of starch and structural polymers. Her dissertation dealt with the role of enzymes in the formation of pentose nucleotides in mung beans. She was awarded the Ph.D. degree in comparative biochemistry in 1956.

From 1963 to 1984 Neufeld was associated with the National Institutes of Health (NIH) in Bethesda, Maryland. After ten years (1963–73) as a Research Biochemist at the National Institute of Arthritis, Metabolism and Digestive Diseases (NIAMDD), she became Chief of the Section of Human Biochemical Genetics. In 1979 she was appointed Chief, Genetics and Biochemistry Branch, National Institute of Arthritis, Diabetes, and Digestive and Kidney Diseases (NIADDK) of NIH and served from 1979 until 1983. From 1981 through 1983 she was also Deputy Director of the Division of Intramural Research at NIADDK. At NIH there is a strong spirit of collaboration between the basic and clinical scientists. In this dynamic atmosphere Neufeld's skills and experience in biochemistry were invaluable in tackling the classification and mechanisms of diseases caused by inherited abnormalities in intracellular chemistry.

In 1984 Neufeld returned to the University of California as Professor and Chair of the Department of Biological Chemistry at the School of Medicine, Los Angeles. She remains at this institution to the present, teaching, doing research, and fulfilling many administrative and public service responsibilities, in addition to her private roles as wife and mother of two children.

Many honors have been bestowed on Neufeld. She is an elected member of the National Academy of Sciences, the American Academy of Arts and Sciences, the Institute of Medicine, and the American Philosophical Society and an elected fellow of the American Association for the Advancement of Science (AAAS). In 1990 she was the California Scientist of the Year. She was the 1993 Distinguished Alumna of NIADDK and was also a National Medal of Science recipient in 1994. Neufeld has been given numerous awards, among which were the Dickson Prize of the University of Pittsburgh (1974), the Hillebrand Award of the Washington Area Chemical Society (1975), the Gairdner Foundation International Award (1981), the Albert Lasker Clinical Medical Research Award (1982), the Elliott Cresson Medal of Franklin Institute, Philadelphia (1984), and the Wolf Prize in Medicine (1988). Honorary doctorates have been conferred on Neufeld by the Université René Descartes, Paris (1978), by Russell Sage College, Troy, New York (1980), and by Hahnemann University School of Medicine, Philadelphia (1984). In 1979 Neufeld was invited to deliver a Harvey lecture on "Lessons from genetic disorders of lysosomes." She also delivered the keynote address at the 55th Annual Soma Weiss Student Research Day at Harvard University, giving a long-term view of her research into mucopolysaccharidoses. In her free time she enjoys hiking and traveling.

WORK

Neufeld's work has encompassed research, teaching, administration, and public service. Her contributions are so numerous and varied that a simple summary is difficult, but one could say that her research has been the scaffold upon which all the other contributions have been built. Neufeld's research career has had several phases, interdependent and progressing one to the other, starting from the fundamental base of rigorous training in biochemical theory and methods. In many different guises, the study of enzymes (their extraction, identification, and kinetics) has been the focus of her attention. At the time she began as a graduate student at Berkeley, most of the details of intracellular enzymatic pathways were unknown. During her career, the structure and function and even the genetics and synthesis of many of these important molecules have been described. Neufeld has contributed significantly to the development of this knowledge.

A valuable foundation was built while she worked in the laboratory of Dr. Kaplan at Johns Hopkins University. The behavior of a new enzyme, a pyridine nucleotide transhydrogenase, extracted in that laboratory from *Pseudomonas fluorescens*, was reported in a series of papers published between 1951 and 1953. After this bacterial enzyme study, Neufeld worked for several years on enzymes extracted from plant cells, such as sweet potato and mung bean. In the laboratory of Dr. Hassid she studied the enzymes that catalyzed the hydrolysis of polysaccharide polymers (such as amylose) and eventually focused on the synthesis of sugar nucleotides. Nucleotides were the subject of intense research because of the numerous roles they seemed to play in the metabolic life of cells. Neufeld and coworkers were able to extract, purify, and study enzymes (transhydrogenases) that could interconvert pentose nucleotides. It was hoped that these studies might lead to an understanding of the intracellular synthesis of pentose polymers, about which little was known at the time. Neufeld investigated the enzymes that catalyzed the synthesis of polysaccharides, in particular those active in the phosphorylation of glucuronic and galacturonic acid containing polymers in various plant cells.

When Neufeld moved in 1963 to NIAMDD in Bethesda, she had the opportunity to use her biochemical research skills in collaboration with clinical scientists and biologists who were studying human diseases caused by derangements in enzyme mechanics. Her work entered a new phase.

Within the institute, scientists were actively investigating a group of diseases characterized by abnormal accumulation of mucopolysaccharide (a complex carbohydrate polymer) within human cells, leading eventually to the degeneration of cellular function, deformity, and often death. Collectively, these diseases are called mucopolysaccharide (MPS) disorders. There are several different types of these disorders, which in the early 1960s were classified I–VI by V. A. McKusick et al. (1965), based on their clinical and inheritance patterns. However, since the study of deoxyribonucleic acid (DNA), ribonucleic acid (RNA), and the intracellular mechanisms of genetics was at a very early stage of development, the precise nature of the genetic mutation and its effects were unknown. What caused the accumulation of mucopolysaccharide was a mystery, in part because the details of intracellular metabolic pathways were poorly understood. It had been shown, however, by F. van Hoof and H. G. Hers (1964) that at least in one of the MPS disorders, Hurler syndrome (MPS I), the affected cells accumulated mucopolysaccharide in the lysosomes, intracellular organelles recently described by Christian De Duve. It had also been shown that fibroblasts, biopsied from the skin of patients with the MPS disorders, could be cultured in the laboratory and that these *in vitro* cells showed the functional and morphological changes characteristic of the disease. The processes and effects of the disease could be studied without disturbing or endangering patients.

Neufeld realized that her background in, and knowledge of, intracellular carbohydrate polymers could be applied to the biochemical problems underlying the MPS disorders. She probably did not imagine, though, that the study of MPS disorders and of other lysosomal storage diseases would occupy her for the next 30 years. These years, dedicated to work with many different coinvestigators and students, progressed along an ever-expanding pathway. Each stage of the work not only clarified some aspect of the diseases under study but also contributed to a more general understanding of the mechanisms at work in the genetics and intracellular life of all cells.

The first stage of Neufeld's work involved the identification of the chemical and morphological locus of the metabolic derangement. With her coworkers, she began a series of experiments on cultured cells from a patient with MPS I, as well as on normal cultured fibroblasts. Taking advantage of the fact that the intracellular mucopolysaccharide was a sulfated polymer, Neufeld used a radiolabeled sulfate as an identification tag to study its synthesis, storage, secretion, and decomposition *in vitro*. Analysis of the results of these experiments demonstrated that in normal cells there is a cyclical turnover of these molecules, so that a steady state is achieved, in which the rate of synthesis is matched by the rate of decomposition with minimal intracellular storage. The decomposition reactions take place in lysosomes and involve lysosomal enzymes. In the abnormal cells, however, there is a deficient rate of this lysosomal decomposition, and the excess mucopolysaccharide accumulate in the lysosomes. Neufeld and her coworkers had demonstrated that the basic problem was not one of synthesis or of secretion but involved deficient decomposition within the lysosomes.

An interesting paper by B. S. Danes and A. G. Bearn (1967) reported that fibroblasts, cultured from biopsies of individuals who were carriers (i.e., normal heterozygotes) for another MPS disorder, Hunter syndrome (MPS II), could be separated into two types. When the two types were independently cultured with radiolabeled sulfate, one type displayed the morphological changes characteristic of mucopolysaccharide storage. The other type remained normal. Since the carriers were normal and healthy and showed none of the symptoms of MPS disorder, Neufeld and her collaborator, Joseph Fratantoni, hypothesized that *in vivo* the normal cells must be secreting something that "helped" to normalize the function of the defective cells, thereby preventing the accumulation of destructive stored MPS. They initiated a series of experiments to test this hypothesis.

The first test was inadvertent when, in 1967, Fratantoni accidentally mixed a culture of fibroblasts from a patient with Hurler syndrome with a fibroblast culture derived from a patient with Hunter syndrome. This mixed culture was exposed to the radiolabeled sulfate, and when the sulfate uptake was analyzed, it was evident that all of the fibroblasts exhibited the normal steady-state pattern. That is, the rate of synthesis of mucopolysaccharide was matched to the rate of decomposition, such that mucopolysaccharide storage did not occur. Fratantoni and Neufeld realized that Hurler fibroblasts were secreting something that could normalize mucopolysaccharide decomposition in Hunter fibroblasts and that the Hunter fibroblasts secreted a different substance that improved mucopolysaccharide decomposition in the Hurler cells. They called these products "corrective factors," and further investigation showed that each type of MPS fibroblast secreted "corrective factors" that could normalize cells of the other types.

This finding provided a method for accurate differential diagnosis among these disorders, whose clinical pictures often overlapped. The "corrective factors" were found to be proteins and then were identified as enzymes. Normal fibroblasts, having normal enzyme profiles, when cultured along with MPS disorder cells, could provide the missing enzymes, "corrective factors," to normalize any of them. Ultimately, all of the missing enzymes were extracted and identified in Neufeld's laboratory and in others around the world and matched to the clinical types of MPS disorders. These exciting results implied that the clinical manifestations of the MPS disorders (to a large extent caused by accumulation of mucopolysaccharide within cells and their subsequent destruction) could be ameliorated by therapeutic replacement of the missing enzyme. The study of the possibility of enzyme replacement therapy opened a new phase of her research.

Neufeld's next investigations led to a more detailed understanding of lysosomal function and of the synthesis and transport within the cell of the lysosomal enzymes. Experiments with replacement enzyme preparations soon demonstrated that these preparations did not always enter the diseased cells as efficiently as did the "corrective factors." This problem was tackled in several laboratories. Neufeld and S. Hickman proposed (1972) that a functionally successful enzyme must have a "recognition site" that allowed uptake by the cell. The nature of this "recognition site" was finally identified by A. Kaplan et al. (1977) to be mannose-6-phosphate. The enzyme needed this mannose tag to get into the cell. This implied that there must also be a specific receptor on the cell membrane to receive this tag. Radioactive labeling of enzymes led to the location and subsequent identification of these membrane-bound receptors. From this work the process of lysosomal enzyme synthesis within the endoplasmic reticulum and insertion into the lysosome was demonstrated. It became evident that in the translocation of enzyme within the cell, during which it becomes attached to surface membranes, some is "leaked" into the surroundings. This "leaked" enzyme, with its recognition site intact, can be taken up by any other receptor-bearing cell and constituted the "corrective factor."

By the early 1980s the process at work in the correction of the defects in cultured MPS disorder fibroblasts was finally clear. The next development involved the study of many other inherited "lysosomal storage" diseases, in which cells are destroyed by abnormal accumulation of metabolites. In each case the identification of the deficient enzyme was an important step forward in understanding the disease, as in Gaucher disease (where because of a missing enzyme, glucocerebrosidase, glucosylceramide is stored and destroys cells) or in Tay-Sachs disease (where a missing hexosaminidase causes G_{m2} ganglioside storage and cell destruction). Cells from affected patients could be cultured *in vitro*, and the abnormality could be improved by uptake of the missing enzyme in its appropriate form from the culture medium. However, obstacles to successful therapeutic enzyme replacement *in vivo* remained. Very little functional enzyme was available in the 1980s, though the advances in recombinant DNA technology provided methods that might be used to develop enzyme supplies and thus another path for research.

Neufeld's laboratory and others have focused in the past decade on identifying the precise genes responsible for the synthesis of the deficient enzymes, as well as on cloning these genes and developing recombinant genetic vehicles so that functional enzymes can be synthesized in larger amounts. A promising line of genetically engineered Chinese hamster ovary cells was developed in Neufeld's laboratory. They are "over-expressors" of the recombinant gene for the missing enzyme in the Hurler syndrome, secreting significant excess enzyme. There is the hope that a retrovirus carrier can be developed to insert the engineered genetic sequence into cells of a patient *in vivo*, transforming them into "over-expressors." If even a small population of a patient's cells can be so modified, the excess enzyme taken up by deficient cells can normalize their function and alleviate the destructive intracellular accumulation, as did the "corrective factors" of Neufeld's early experiments.

There are still many unknowns ahead before a successful genetic replacement or enzyme replacement therapy becomes routine. It is still unclear how the patient's immune system would respond to cellular manipulation and whether long-term immunosuppressive drugs would be required, and this issue is under study. However, tremendous concrete benefit has already been derived from the work that Neufeld began almost 30 years ago. Synthesis of enzymes, their manipulation within the cell to make them recognizable and effective, and their transport to their functional locus within the lysosome have been clarified by Neufeld's contributions. Differential diagnosis, sensitive enzyme assay methods, and detailed knowledge about the genetic mutations involved (some of which have much more serious and life-threatening consequences than others) contribute to the physician's ability to counsel those who carry these mutations, giving them knowledge-based reproductive choices where none had existed.

Starting in the early 1970s, Neufeld has served on the editorial boards of professional journals, including the *Journal of Biological Chemistry*, the *Proceedings of the Society for Experimental Biology and Medicine*, *Biochemical and Biophysical Communications*, the *Archives of Biochemistry and Biophysics*, and *Biochemistry*. She also has been on the editorial boards of the *American*

Journal of Human Genetics, the *American Journal of Medical Genetics*, the *Journal of Clinical Investigation*, and *Human Mutation*.

Throughout her career Dr. Neufeld has led and contributed to many public service, educational, and research policy committees. She chaired the Gordon Conference on Lysosomes in 1984. At present she is a member of the Scientific Advisory Boards of the National MPS Society, Inc., the Tay-Sachs and Allied Diseases Association, and the Hereditary Disease Foundation and is a senior consultant to the Lucille P. Markey Charitable Trust. She is also a member of the California Council on Science and Technology and the Institute of Medicine Committee on Educating Dentists for the Future.

NOTE

I wish to thank Elizabeth Fondal Neufeld for her generosity in sending me reprints and bibliographic material, which gave me insight into her research.

BIBLIOGRAPHY

Works by Elizabeth Fondal Neufeld

Scientific Works

Space does not permit the listing of the complete works of Elizabeth Fondal Neufeld. Listed here are all works by Neufeld except those cited in Neufeld and Ginsburg (1965), Neufeld and Ginsburg (1969), Neufeld, Lim et al. (1975), Neufeld ("Lessons . . ." 1981), Neufeld (1991), and Neufeld and Muenzer (1995). Included here are her dissertation and all references cited in the text.

(with E. S. Russell) "Quantitative analysis of the normal and four alternative degrees of an inherited macrocytic anemia in the house mouse." *Blood* 6 (1951): 892–905.

(———— and C. T. Higgins) "Comparison of blood pictures of young adults from eighteen inbred strains of mice." *Proceedings of the Society for Experimental Biology and Medicine* 78 (1951): 761–766.

(with S. P. Colowick and N. O. Kaplan) "Pyridine nucleotide transhydrogenase. I." *Journal of Biological Chemistry* 195 (1952): 95–105.

(with N. O. Kaplan and S. P. Colowick) "Pyridine nucleotide transhydrogenase. II." *Journal of Biological Chemistry* 195 (1952): 107–119.

(————) "Pyridine nucleotide transhydrogenase. III." *Journal of Biological Chemistry* 205 (1953): 1–15.

(———— et al.) "Pyridine nucleotide transhydrogenase. IV." *Journal of Biological Chemistry* 205 (1953): 17–29.

(with W. Z. Hassid) "The hydrolysis of amylose by alpha-amylase and Z-enzyme." *Archives of Biochemistry and Biophysics* 59 (1955): 405–419.

(with N. O. Kaplan and S. P. Colowick) "Effect of adenine nucleotides on reactions involving triphosphopyridine nucleotide." *Biochimica et Biophysica Acta* 17 (1955): 525–535.

"Enzymatic Studies of Natural and Synthetic Amylose." Ph.D. diss., University of California, Berkeley, 1956.

(with V. Ginsburg and W. Z. Hassid) "Enzymatic synthesis of uridine diphosphate xylose and uridine diphosphate arabinose." *Proceedings of the National Academy of Sciences* 42 (1956): 333–335.

(with V. Ginsburg, E. W. Putman, et al.) "Formation and interconversion of nucleotides by plant extracts." *Archives of Biochemistry and Biophysics* 69 (1957): 602–615.

(with D. S. Feingold and W. Z. Hassid) "Enzymatic conversion of uridine diphosphate glucuronic acid to uridine diphosphate galacturonic acid, uridine diphosphate xylose and uridine diphosphate arabinose." *Journal of the American Chemical Society* 80 (1958): 4430–4431.

(———) "Enzymatic synthesis of uridine diphosphate glucuronic acid and uridine diphosphate galacturonic acid with extracts from *Phaseolus aureus* seedlings." *Archives of Biochemistry and Biophysics* 78 (1958): 401–406.

(with D. Mazia) "Non-protein sulfhydryl compounds in the division of eggs of *Strongylocentrotus purpuratus*." *Experimental Cell Research* 13 (1958): 622–624.

(with D. S. Feingold and W. Z. Hassid) "Enzymatic phosphorylation of D-glucuronic acid by extracts from seedlings of *Phaseolus aureus*." *Archives of Biochemistry and Biophysics* 83 (1959): 96–100.

(———) "Xylosyl transfer catalyzed by an asparagus extract." *Journal of Biological Chemistry* 234 (1959): 488–489.

(with W. Z. Hassid and D. S. Feingold) "Sugar nucleotides in the interconversion of carbohydrates in higher plants." *Proceedings of the National Academy of Sciences* 45 (1959): 905–915.

(with D. S. Feingold and W. Z. Hassid) "The 4-epimerization and decarboxylation of uridine diphosphate D-glucuronic acid by extracts of *Phaseolus aureus* seedlings." *Journal of Biological Chemistry* 235 (1960): 910–913.

(———) "Phosphorylation of D-galactose and L-arabinose by extracts from *Phaseolus aureus* seedlings." *Journal of Biological Chemistry* 235 (1960): 906–909.

(with G. A. Barber) "Rhamnosyl transfer from TDP-L-rhamnose catalyzed by a plant extract." *Biochemical and Biophysical Research Communications* 6 (1961): 44–48.

(with D. S. Feingold) "Isolation of UDP-D-galacturonic acid by extracts from germinating seeds of *Phaseolus aureus*." *Biochimica et Biophysica Acta* 53 (1961): 589–590.

(———, S. M. Ives, et al.) "Phosphorylation of D-galacturonic acid by extracts from germinating seeds of *Phaseolus aureus*." *Journal of Biological Chemistry* 236 (1961): 3102–3105.

(with G. Kessler, D. S. Feingold, et al.) "Metabolism of D-glucuronic and D-galacturonic acid by *Phaseolus aureus* seedlings." *Journal of Biological Chemistry* 236 (1961): 308–312.

"Formation and epimerization of TDP-D-galactose catalyzed by plant extracts." *Biochemical and Biophysical Research Communications* 7 (1962): 461–466.

(with B. Felenbok) "Hydrolysis of glucopyranosyl phosphate of the beta-D or alpha-L configuration of a plant phosphatase." *Archives of Biochemistry and Biophysics* 97 (1962): 116–212.

(with W. Z. Hassid) "Glycosidic bond exchange." In *The Enzymes*, 2d ed., vol. 6, edited by H. A. Lardy and P. Boyer, 278–315. New York: Academic Press, 1962.

(with F. Loewus and S. Kelly) "Metabolism of myoinositol in plants." *Proceedings of the National Academy of Sciences* 48 (1962): 421–425.

(with D. S. Feingold and W. Z. Hassid) "Preparation of UDP-D-xylose and UDP-L-arabinose." In *Methods in Enzymology*, vol. 6, edited by S. P. Colowick and N. O. Kaplan, 782–787. New York: Academic Press, 1963.

(with W. Z. Hassid) "Biosynthesis of saccharides from sugar nucleotides." *Advances in Carbohydrates Chemistry* 18 (1963): 309–356.

(with D. S. Feingold and W. Z. Hassid) "Enzymes of carbohydrate synthesis." In *Modern Methods of Plant Analysis*, vol. 7, edited by W. Ruthland, 474–519. Berlin: Springer-Verlag, 1964.

(with P. J. O'Brien) "A rapid procedure for the preparation of small quantities of UDP-N-[^{14}C]acetylglucosamine." *Biochimica et Biophysica Acta* 83 (1964): 352–354.

(with V. Ginsburg) "Carbohydrate metabolism." *Annual Review of Biochemistry* 34 (1965): 297–312.

(with C. W. Hall) "Inhibition of UDP-D-glucose dehydrogenase by UDP-D-xylose—a possible regulatory mechanism." *Biochemical and Biophysical Research Communications* 19 (1965): 456–461.

"UDP-D-galacturonic acid 4–epimerase from radish roots." In *Methods in Enzymology*, vol. 8, edited by E. F. Neufeld and V. Ginsburg, 276–277. New York: Academic Press, 1966.

(with V. Ginsburg, eds.) *Methods in Enzymology*, vol. 8. New York: Academic Press, 1966.

(with E. M. Goudsmit) "Isolation of GDP-L-galactose from the albumin gland of *Helix pomatia*." *Biochimica et Biophysica Acta* 121 (1966): 192–195.

(————) "Formation of GDP-L-galactose from GDP-D-mannose." *Biochemical and Biophysical Research Communications* 26 (1967): 730–735.

(with V. Ginsburg) "Complex carbohydrates of animals." *Annual Review of Biochemistry* 38 (1969): 371–388.

(with E. E. Grebner) "Stimulation of a protein glycosylation reaction by lysozyme." *Biochimica et Biophysica Acta* 142 (1969): 347–349.

(with M. Cantz and A. Chrambach) "Characterization of the factor deficient in the Hunter syndrome by polyacrylamide gel electrophoresis." *Biochemical and Biophysical Research Communications* 39 (1970): 936–942.

(with J. C. Fratantoni) "Inborn errors of mucopolysaccharide metabolism." *Science* 169 (1970): 141–146.

(with U. Weismann) "Metabolism of sulfated mucopolysaccharide in cultured fibroblasts from cystic fibrosis patients." *Journal of Pediatrics* 77 (1970): 685–690.

(————) "Scheie and Hurler syndromes: Apparent identity of the biochemical defect." *Science* 169 (1970): 72–74.

(with H. Kresse, U. Weismann, et al.) "Biochemical heterogeneity of the Sanfilippo syndrome: Preliminary characterization of two deficient factors." *Biochemical and Biophysical Research Communications* 42 (1971): 892–898.

"The biochemical basis of the inborn errors of mucopolysaccharide metabolism." In *Biochemistry of Glycosidic Linkage*, vol. 2, 711–715. New York: Academic Press, 1972.

"Mucopolysaccharidoses." In *Antenatal Diagnosis*, edited by A. Dorfman, 217–228. Chicago: University of Chicago Press, 1972.

"Mucopolysaccharidoses, the biochemical approach." *Hospital Practice* 7 (1972): 107–113.

(with R. W. Barton, M. Cantz, et al.) "Deficiency of specific proteins in the inborn errors of mucopolysaccharide metabolism." In *Sphingolipids, Sphingolipidoses and Allied Diseases,* edited by B. W. Volk and S. M. Aronson, 187–193. New York: Plenum, 1972.

(with M. Cantz, H. Kresse, et al.) "Corrective factors for inborn errors of mucopolysaccharide metabolism." In *Methods in Enzymology,* vol. 28, edited by V. Ginsburg, 884–897. New York: Academic Press, 1972.

(with S. Hickman) "A hypothesis for I-cell diseases: Defective hydrolases that do not enter lysosomes." *Biochemical and Biophysical Research Communications* 49 (1972): 992–999.

(with S. O. Lie and V. A. McKusick) "Simulation of genetic mucopolysaccharidoses in normal human fibroblasts by alteration of the pH of the medium." *Proceedings of the National Academy of Sciences* 69 (1972): 2361–2363.

(with P. J. O'Brien) "Biosynthesis of glycoproteins-glycosyltransferases." In *Glycoproteins, Their Composition, Structure and Function,* edited by A. Gottschalk, 1170–1184. New York: Elsevier, 1972.

"Mucopolysaccharidoses: The biochemical approach." In *Medical Genetics,* edited by V. A. McKusick and R. Claiborne, 141–147. New York: Hospital Practice, 1973.

(with M. Cantz) "The mucopolysaccharidoses studied in cell culture." In *Lysosomal Storage Diseases,* edited by F. van Hoof and H. G. Hers, 262–275. New York: Academic Press, 1973.

(with A. Milunsky) "The Hunter syndrome in a 46 XX girl." *New England Journal of Medicine* 288 (1973): 106.

"The biochemical basis of mucopolysaccharidoses and mucolipidoses." In *Progress in Medical Genetics,* edited by A. G. Steinberg and A. G. Bearn, 85–101. New York: Grune and Stratton, 1974.

"Uptake of lysosomal enzymes by cultured fibroblasts: Studies of mucopolysaccharidoses and I-cell diseases." In *Cell Communication,* edited by R. P. Cox, 217–231. New York: Wiley, 1974.

(with T. W. Lim, I. G. Bach, et al.) "An assay for iduronate sulfatase (Hunter corrective factor)." *Carbohydrate Research* 37 (1974): 103–109.

(with R. E. Stevenson and R. R. Howell) "Mucopolysaccharidoses." In *Current Diagnosis,* edited by H. F. Conn and R. B. Conn, Jr., 1065–1070. Philadelphia: W. B. Saunders, 1974.

(with T. W. Lim and L. J. Shapiro) "Inherited disorders of lysosomal metabolism." *Annual Review of Biochemistry* 44 (1975): 357–376.

(with M. Tondeur) "The mucopolysaccharidoses—biochemistry and ultrastructure." In *Molecular Pathology,* edited by S. B. Day and J. J. Yunis, 600–621. Springfield, IL: Charles Thomas, 1975.

(with C. J. Epstein, S. Yatsiv, et al.) "Genetic counselling for the Hunter syndrome." *Lancet* ii (1976): 737–738.

(with I. Liebaers) "Iduronate sulfatase activity in serum, lymphocytes and fibroblasts—simplified diagnosis of Hunter syndrome." *Pediatric Research* 10 (1976): 733–736.

(——— and T. W. Lim) "Iduronate sulfatase determination for the diagnosis of Hunter syndrome and the detection of the carrier state." In *Current Trends in Sphingo-*

lipidoses and Allied Disorders, edited by B. W. Volk and L. Schneck, 253–260. New York: Plenum, 1976.

(with L. J. Shapiro) "Genetic mucopolysaccharidoses and mucolipidoses." In *Basic Neurochemistry*, edited by G. J. Seifel et al., 569–580. Boston: Little, Brown, 1976.

(with R. E. Stevenson, R. R. Howell, et al.) "The iduronidase-deficient mucopolysaccharidoses: Clinical and roentgenographic features." *Pediatrics* 57 (1976): 111–112.

"The enzymology of inherited mucopolysaccharide storage disorders." *Trends in Biochemistry* 2 (1977): 25–26.

(with P. Di Natale and I. G. Leder) "A radioactive assay for alpha-L-iduronidase." *Clinica Chimica Acta* 77 (1977): 211.

(with I. Liebaers, C. J. Epstein, et al.) "The Hunter syndrome in females: Is there an autosomal recessive form of iduronate sulfatase deficiency?" *American Journal of Human Genetics* 29 (1977): 455–461.

(with B. R. Migeon, J. A. Sprenkle, et al.) "X-linked Hunter syndrome: The heterozygous phenotype in cell culture." *American Journal of Human Genetics* 29 (1977): 448–454.

(with V. A. McKusick and T. E. Kelly) "The mucopolysaccharide storage diseases." In *The Metabolic Basis of Inherited Disease*, 4th ed., edited by J. B. Stanbury et al., 513–514. New York: McGraw-Hill, 1978.

(with L. H. Rome and A. J. Garvin) "Human kidney alpha-L-iduronidase: Purification and characterization." *Archives of Biochemistry and Biophysics* 189 (1978): 344–353.

(with P. Di Natale) "The biochemical diagnosis of mucopolysaccharidoses, mucolipidoses and related disorders." In *Perspectives in Inherited Metabolic Diseases*, vol. 2, edited by B. Berra et al., 113–123. Milan, Italy: Ermes, 1979.

(with A. Frisch) "A rapid and sensitive assay for neuraminidase: Application to cultured fibroblasts." *Analytical Biochemistry* 95 (1979): 222–227.

(with L. H. Rome, A. J. Garvin, et al.) "Two species of lysosomal organelles in cultured human fibroblasts." *Cell* 17 (1979): 143–153.

(with G. N. Sando, P. Titus-Dillon, et al.) "Inhibition of receptor-mediated uptake of a lysosomal enzyme into fibroblasts by chloroquinone, procaine and ammonia." *Experimental Cell Research* 119 (1979): 359–364.

"Lessons from genetic disorders of lysosomes." *Harvey Lectures 1979–1980*, 41–60. New York: Academic Press, 1981.

"The receptor for lysosomal enzymes on the surface of cultured fibroblasts." In *Relationship between Toxins and Receptors*, edited by J. Middlebrook and L. Kohn, 11–14. New York: Academic Press, 1981.

"Recognition and processing of lysosomal enzymes in cultured fibroblasts." In *Lysosomes and Lysosomal Storage Diseases*, edited by J. W. Callahan and A. J. Lowden, 115–129. New York: Raven Press, 1981.

(with A. Frisch) "Limited proteolysis of the beta-hexosaminidase precursors in a cell-free system." *Journal of Biological Chemistry* 256 (1981): 8246–8248.

(with L. J. Shapiro) "Genetic mucopolysaccharidoses and mucolipidoses." In *Basic Neurochemistry*, 3d ed., edited by G. J. Siegel et al., 627–639. New York: Little, Brown, 1981.

(with M. C. Willingham, I. H. Pastan, et al.) "Morphologic study of internalization of a lysosomal enzyme by the mannose-6–phosphate receptor in cultured Chinese hamster ovary cells.'' *Proceedings of the National Academy of Sciences* 78 (1981): 6967–6971.

''Recent studies on the maturation of lysosomal enzymes.'' In *Inborn Errors of Metabolism in Humans*, edited by F. Cockburn and R. Gitzelmann, 103–110. New York: Alan Liss, 1982.

(with M. Owada) ''Is there a mechanism for introducing acid hydrolases into liver lysosomes that is independent of mannose-6–phosphate recognition? Evidence from I-cell disease.'' *Biochemical and Biophysical Research Communications* 105 (1982): 814–820.

(with R. L. Proia) ''Synthesis of beta-hexosaminidase in cell-free translation and in intact fibroblasts: An insoluble precursor alpha chain in a rare form of Tay-Sachs disease.'' *Proceedings of the National Academy of Sciences* 79 (1982): 6360–6364.

(with A. R. Robbins) ''Pleiotropic mutations of lysosomal function in human patients and in Chinese hamster ovary cells.'' In *Human Genetics*, Part B, *Medical Aspects*, edited by B. Bonne-Tamir, 177–185. New York: Alan Liss, 1982.

(with P. G. Robey) ''Defective phosphorylation and processing of beta-hexosaminidase by intact cultured fibroblasts from patients with mucolipidoses. III.'' *Archives of Biochemistry and Biophysics* 213 (1982): 251–257.

(with R. M. Shull, R. J. Munger, et al.) ''Canine alpha-L-iduronidase deficiency: A model of mucopolysaccharidoses. I.'' *American Journal of Pathology* 109 (1982): 244–248.

(with G. H. Thomas, S. Raghavavan, et al.) ''Nonuniform deficiency of hexosaminidase A in tissues and fluids of two unrelated individuals.'' *Pediatric Research* 16 (1982): 232–237.

(with G. Bach) ''Synthesis and maturation of cross-reactive glycoprotein in fibroblasts deficient in arylsulfatase A activity.'' *Biochemical and Biophysical Research Communications* 112 (1983): 198–205.

(with R. Myerowitz, A. R. Robbins, et al.) ''Studies of lysosomal enzyme biosynthesis in cultured cells.'' In *Methods in Enzymology*, vol. 96, edited by S. Fleischer and B. Fleischer, 729–736. New York: Academic Press, 1983.

(with G. G. Sahagian) ''Biosynthesis and turnover on the mannose-6–phosphate receptor in cultured Chinese hamster ovary cells.'' *Journal of Biological Chemistry* 258 (1983): 7121–7128.

''Mucopolysaccharide storage diseases.'' In *NIH: An Account of Research in Its Laboratories and Clinics*, edited by DeWitt Stetten and W. T. Carrigan, 330–336. New York: Academic Press, 1984.

(with A. d'Azzo and R. L. Proia) ''Defective synthesis or maturation of the alpha-chain of beta-hexosaminidase in classic and variant forms of Tay-Sachs disease.'' In *The Molecular Basis of Lysosomal Storage Disorders*, edited by R. O. Brady and J. A. Barranger, 251–256. New York: Academic Press, 1984.

(with R. L. Proia and A. d'Azzo) ''Association of alpha- and beta-subunits during the biosynthesis of beta-hexosaminidase in cultured human fibroblasts.'' *Journal of Biological Chemistry* 259 (1984): 3350–3354.

(with R. Myerowitz, R. Piekarz, et al.) ''Human beta-hexosaminidase alpha-chain: The coding sequence and homology with the beta-chain.'' *Proceedings of the National Academy of Sciences* 82 (1985): 7830–7834.

(with E. E. Grebner, D. A. Mansfield, et al.) "Two abnormalities of hexosaminidase A in clinically normal individuals." *American Journal of Human Genetics* 38 (1986): 505–515.

(with G. Zokaeem, G. Bayleran, et al.) "A shortened beta-hexosaminidase alpha-chain in an Italian patient with infantile Tay-Sachs disease." *American Journal of Human Genetics* 40 (1987): 537–547.

(with B. H. Paw) "Normal transcription of the beta-hexosaminidase alpha-chain gene in the Ashkenazi Tay-Sachs mutation." *Journal of Biological Chemistry* 263 (1988): 3012–3015.

"Natural history and inherited disorders of a lysosomal enzyme, beta-hexosaminidase (minireview)." *Journal of Biological Chemistry* 264 (1989): 10927–10930.

(with B. H. Paw and M. M. Kaback) "Molecular basis of adult-onset and chronic G_{m2} gangliosidoses in patients of Ashkenazi Jewish origin: Substitution of serine for glycine at position 269 of the alpha-subunit of beta-hexosaminidase." *Proceedings of the National Academy of Sciences* 86 (1989): 2413–2417.

(with L. C. Wood) "A cystic fibrosis phenotype in cells cultured from sweat gland secretory coil: Altered kinetics of Cl efflux." *Journal of Biological Chemistry* 265 (1990): 12796–12800.

"Lysosomal storage diseases." *Annual Review of Biochemistry* 60 (1991): 257–280.

(with B. H. Paw and L. C. Wood) "A third mutation at the CpG dinucleotide of codon 504 and a silent mutation at codon 506 of a HEX A gene." *American Journal of Human Genetics* 48 (1991): 1139–1146.

"The molecular basis of Tay-Sachs disease and related G_{m2} gangliosidoses among Jews and non-Jews." In *Genetic Diversity among Jews: Diseases and Markers at the DNA Level*, edited by B. Bonne-Tamir and A. Avinoam, 97–103. New York: Oxford University Press, 1992.

(with K. P. Menon and P. T. Tieu) "Architecture of the canine IDUA gene and mutation underlying canine mucopolysaccharidosis I." *Genomics* 14 (1992): 763–768.

(with E. D. Kakkis, A. Matynia, et al.) "Overexpression of the human lysosomal enzyme, alpha-L-iduronidase, in Chinese hamster ovary cells." *Protein Expression and Purification* 5 (1994): 225–232.

(with P. T. Tieu and K. Menon) "A mutant stop codon (TAG) in the IDUA gene is used as an acceptor splice site in a patient with Hurler syndrome (MPS IH)." *Human Mutation* 3 (1994): 333–336.

(with K. P. Menon) "Evidence for degradation of mRNA encoding alpha-L-iduronidase in Hurler fibroblasts with premature termination alleles." *Cellular and Molecular Biology* 40 (1994): 999–1005.

(with R. M. Shull, E. D. Kakkis, et al.) "Enzyme replacement in a canine model of Hurler syndrome." *Proceedings of the National Academy of Sciences* 91 (1994): 12937–12941.

(with J. Muenzer) "The mucopolysaccharidoses." In *The Metabolic and Molecular Bases of Inherited Disease*, 7th ed., edited by C. R. Scriver et al., 2465–2494. New York: McGraw-Hill, 1995.

"Research—a long-term view." In the *56th Annual Soma Weiss Medical and Dental Student Research Day: Book of Abstracts*, iv–vii. Boston: Harvard Medical School Office of Enrichment Programs, 1996.

Works about Elizabeth Fondal Neufeld

"National Medal of Science winners are named." *New York Times* (Sept. 9, 1994): 22.
Sankaran, N. "Eight to receive National Medals of Science." *The Scientist* 8 (1994): 1, 5, 8.
Schmeck, H. M., Jr. "7 medical experts win Lasker Prizes." *New York Times* (Nov. 18, 1982): 21.
———. "Two discoveries may help treat inherited diseases." *New York Times* (Mar. 27, 1979): III:1.

Other References

Danes, B. S., and A. G. Bearn. "A genetic study of clones in cell culture with particular reference to the Lyon hypothesis." *Journal of Experimental Medicine* 126 (1967): 509.
———. "Hurler's syndrome: Demonstration of an inherited disorder of connective tissue in cell culture." *Science* 149 (1965): 987–989.
De Duve, C. "Lysosomes." In *Subcellular Particles*, edited by T. Hayashi, 128–159. New York: Ronald Press, 1959.
———. "Principles of tissue fractionation." *Journal of Theoretical Biology* 6 (1964): 33–59.
Hoof, F. van, and H. G. Hers. "The ultrastructure of liver cells in Hurler's disease (gargoylism)" (in French). *Comptes Rendus Hebdomadaires des Seances de l'Academie des Sciences* 259 (1964): 1281.
Kaplan, A., D. T. Achord, et al. "Phosphohexosyl components of lysosomal enzyme are recognized by pinocytosis in human fibroblasts." *Proceedings of the National Academy of Sciences* 74 (1977): 2026–2030.
McKusick, V. A., D. Kaplan, et al. "The genetic mucopolysaccharidoses." *Medicine* 44 (1965): 445–483.

MARGARET MORSE NICE (1883–1974)

Alan Contreras and Milton Bernhard Trautman

BIOGRAPHY

Margaret Morse was born on December 6, 1883, in Amherst, Massachusetts. In her autobiography, *Research Is a Passion with Me* (1979), she notes that "I like to think that the same year saw the birth of the American Ornithologists' Union" (AOU). Her parents were Anson Daniel and Margaret Duncan (Ely) Morse. Her father was a professor of history at Amherst College; her mother was a graduate of Mount Holyoke College, who devoted most of her energies to raising Margaret and her six siblings (Trautman 1977).

Although her parents provided her with access to much of the intellectual stimulation available in a college community of the era, they held traditional views about their daughter's future as a wife and mother and did not encourage her to pursue a professional career. The Morse and Ely families had both lived in New England since the mid-1600s, and young Margaret Morse spent considerable time in her youth traveling to visit relatives, as well as traveling elsewhere in the United States and to Europe.

In her autobiography she relates that "the most cherished Christmas present of my life came in 1895, Mabel Osgood Wright's *Bird-Craft*." This book was one of the earliest guides to provide color illustrations of birds, some reprinted from the works of John James Audubon. She began keeping notes on local birds when she was eight years old, using for guidance *Bird-Craft* and a local annotated list.

Her notes show that even as a child she was keeping track of nests and taking pride in bringing the local list up-to-date with her own migration dates and other data. Because of this exceptional attention to detail, she was later able to compare her own notes on the rate of fledgling success of young American robins, chipping sparrows, and least flycatchers, written when she was 13 years of age, to records she collected 61 years later, surely, an accomplishment equaled by few scientists in any field. Her early data matched those of the mature ornithologist very closely (44.4% versus 46% success), a tribute to her remarkable powers of observation and commitment to recordkeeping.

Morse attended Amherst High School and, upon graduation, entered Mount Holyoke College in 1901. In addition to pursuing her interest in birds, she used her flair for languages to develop a working knowledge of German, French, and Italian. Her eventual fluency in German proved crucial in her major work, as it enabled her to publish her work of the song sparrow and to communicate easily with Dr. Konrad Lorenz and other European authorities.

At Mount Holyoke she studied under Mignon Talbot, Henrietta E. Hooker, Louise Wallace, Abby Turner, Mary Young, and Dr. Cornelia Clapp. Although she notes that this education proved very useful, her interest in ornithology was not predominant during this period. Both during her undergraduate years and after receiving her bachelor's degree (1906), she traveled extensively in Europe and North America with her family.

Dissatisfied with her lack of a direction for life, she entered the new Clark University in Worcester, Massachusetts (1906), one of only two women attending classes. She studied the food of the bobwhite. In her second year, the laboratory next to hers had a new instructor, Leonard Blaine Nice, another graduate student. They were married in 1909, and she remained at the college to complete the work that eventually led to her M.A. degree. This degree has been reported as being in zoology, but Clark University records show that her thesis was about environmental influences on the development of child vocabulary. In later years she complained, when identified as a housewife who studies birds, that "I am not just a housewife, I am a trained zoologist" (Bonta 1991, 222). Nice abandoned her plans to pursue a Ph.D. degree, but her study of the bobwhite was published (1910). Data derived from it were still in use as late as 1945, when she discovered her own information used in arguments for a bill on cat control before the Illinois legislature. The bobwhite was portrayed as a victim of cats, and its habit of eating insects injurious to crops was set forth as a reason to protect it (*Research Is a Passion with Me*, 36).

After four years, during which her husband completed his degree and taught at Harvard Medical School, the family moved to Norman, Oklahoma, in 1913, where he had accepted a faculty position. This move was the beginning of her prolific output of books and articles on birds. She received her M.A. degree from Clark in 1915. The family moved to Ohio in 1927, where her husband worked at Ohio State University for nine years. During this period she researched and published the monograph on song sparrows, for which she is best known. The family moved to Chicago in 1936, where they remained until Blaine Nice retired in 1952.

Unlike her parents, her husband was an enthusiastic supporter of her work. He not only helped prepare the first edition of her "The Birds of Oklahoma" (1924) but provided personal and financial support that allowed her to travel widely and meet the world's leading ornithologists. He sometimes traveled with her and is said to have taken great pleasure in his wife's professional reputation.

They had five children: Constance (b. 1910), Marjorie (b. 1912), Barbara (b. 1915), Eleanor (b. 1918), and Janet (b. 1923). Eleanor died at the age of nine. All of the children spent at least some time in the field with their mother, but Constance seems to have been the most inclined toward ornithology, spending up to five weeks at a time in the field with her mother (*Research Is a Passion with Me*, 1960) and working with her mother at the Delta Waterfowl Research Station in Manitoba. Her daughter, Constance, and her sister, Katharine, were sufficiently familiar with her work to serve as critics of drafts of some of her song sparrow publications, and the children sometimes accompanied their mother to the meetings of ornithological societies.

Nice was elected to honorary memberships in the British, Finnish, German, Dutch, and Swiss ornithological societies. She received the AOU's Brewster Medal (the highest award given by the organization) in 1942 for her studies of the song sparrow, the second woman honored, following Florence Merriam Bailey.* She received two honorary Doctor of Science degrees, one from Mount Holyoke College on the occasion of her 50th class reunion (1955) and another from Elmira College (1962). Ornithologist Robert Dickerman named a Mexican subspecies of song sparrow (*Melospiza melodia niceae*) to honor her (*Research Is a Passion with Me*).

Her last published article appeared in *Bird-Banding* when she was 87 years old. Margaret Morse Nice died in Chicago on June 26, 1974, of arteriosclerosis at the age of 90, a few months after the death of her husband.

WORK

Margaret Morse Nice followed the lead of Bailey and others in focusing on what birds *do* rather than what they *are*. This approach to ornithology (or zoology in general) had not attracted much attention from the male ornithologists of the day. She permanently broadened the field of ornithology through the song sparrow project that Deborah Strom describes in *Bird Watching with American Women* (1986) as "so vast and difficult that the mind boggles at the time and patience required."

First published in the German periodical *Journal für Ornithology* in 1933 and 1934 (because no American journal would print anything so long), Nice's work has been called by Konrad Lorenz "the first long-term field investigation of the individual life of any free-living wild animal." Lorenz, who later shared a Nobel Prize with Nikolaas Tinbergen (1973) for their work on the study of animal behavior, wrote the foreword to Nice's posthumous autobiography, *Research Is a Passion with Me*, and often credits her with being the founder of ethology (the study of animal behavior) as a science. Indeed, the journal now called *Ethology* was founded by Lorenz in Austria (1937), the same year that the Linnaean Society of New York first published "Studies in the Life History of the Song Sparrow" in the United States.

The late Dr. Milton B. Trautman recalls that Tinbergen wrote to Nice as follows: "In a long life you have found reward not only in the home circle for all your cares and sacrifices, but with remarkable creative power you have served science. Through your works you have become known to ornithologists throughout the entire world as the one who laid the foundation for the population studies now so zealously persecuted [*sic*]" (Trautman 1977).

Trautman's memorial in the *Auk* also mentions that the leading American ornithologist, Ernst Mayr, wrote to him stating that Nice "almost single-handedly initiated a new era in American ornithology and the only effective countermovement against the list-chasing movement." At the root of Margaret Morse Nice's success as a scientist lies an interest in *detail*. Never content to observe in a casual way, she followed a never-ending spiral of "why?" until she found what she wanted to know. Coupled with this commitment to discovering the entire pattern in a natural act was a relentless drive to *record* what she discovered.

For example, consider her interest in the vocal frequency and pattern of song sparrows. Not only did she develop a matrix of the different categories of song patterns, but she recorded how often certain patterns were used. Once, while sitting in one place tallying songs for an entire day from predawn till after dark, she noted that the sparrow sang 2,305 times.

She often expressed her commitment to see what was really there and to record it accurately. Unfortunately for some of the other ornithologists of her time (but good for the science itself), her thoroughness often unraveled earlier misconceptions. In *The Watcher at the Nest* (1939) she described her first attempt to get basic information about the song sparrow:

I went to the books and read that this species has two notes besides the song, and that incubation lasted ten to fourteen days and was performed by both sexes—meager enough information and all of it wrong. The men at the State Museum, fine field naturalists and well acquainted with the local birds, could not answer my questions; they did not know whether 4M's singing in late February meant that he had taken up his territory, nor could they tell me when the nesting song sparrows arrived.

Nice entitled her autobiography *Research Is a Passion with Me*, perhaps because it describes the tenacity with which she observed the song sparrows of Interpont, her Ohio home and its adjacent fields. Although she wrote several hundred papers and more than 3,000 book reviews and notes, her "Studies in the Life History of the Song Sparrow" (1937, 1943) will ensure her lasting impact on ornithology.

This work thoroughly covers every aspect of a song sparrow's life. It is based on extensive studies conducted on two banded pairs of birds and supplementary work with an additional 69 banded pairs. Beginning in 1929, she devoted eight years to studying these birds and focused on the way they interacted with each other within pairs, their breeding biology, their territoriality, innate and learned behavior, and song.

The first work she published, jointly with her husband, was "The birds of Oklahoma" (1924), which appeared under her name alone as a revised edition in 1931. Other book-length works were the two-volume "Studies in the life history of the song sparrow"; *The Watcher at the Nest* (1939), a semipopular account of her many field studies; "Development of behavior in precocial birds" (1962); and the posthumous autobiography *Research Is a Passion with Me.*

On the basis of her works Nice is considered one of the most significant ornithologists to have worked in North America and one of the most important in the world. Her research on breeding behavior and the ethology of young birds and also her work on territoriality remain basic to ornithology and continue to be valued. She also published a limited amount of material on child development.

Her early regional study, "The birds of Oklahoma," is now mainly of historical interest and has been superseded by more modern works. Her studies of birdsong have been outpaced by modern sound recording and analysis techniques. Nonetheless, *The Birds of Oklahoma* is worth examining as an archetype of what made Nice's work so distinctive. The book contained tables of Christmas bird count data and graphs such as "Composition of Bird Population by Species . . . According to the Eight Mile Censuses," material rarely included or merely summarized in other books of the period. Although she had as great an appreciation of nature as anyone, she was in essence a *quantifier* and thus a harbinger of today's ornithological norm.

Nice was neither a popularizer nor a transition figure, such as Bailey was, between the women who wrote for popular audiences and those who conducted research. Her work belongs essentially to the modern age of ornithology; indeed, she helped create and define one branch of it. She never attempted to reach a wide popular market with her work, except for the commercially unsuccessful *The Watcher at the Nest* in 1939. This book remains essentially unknown to the general public, perhaps because of its technical nature.

While most laboratory ornithologists of the day wrote in a rather dry manner (as they do today), Nice followed the pattern of the women who had come before her in combining a chatty style with a crisp array of facts. Witness this passage from the account of the Bewick's wren in *The Birds of Oklahoma* (rev. ed., 1931): "When thoroughly in the spirit of it, a bird may sing as often as ten times in a minute, but he does not keep up this rate for long. The largest number of songs I ever counted in one hour was 223." Although this informal, somewhat anthropomorphic style of writing is sometimes said to be evidence of an "unscientific" approach utilized by women ornithologists, the same style was in use by many prominent male ornithologists of the late 1800s and early 1900s.

She was six years old when Bailey's *Birds through an Opera Glass* was published (1889). Nice benefited from the groundbreaking work of Bailey and the other women who went before her. From the beginning of her career as a writer and researcher, she was a scientist, not a popularizer or a bird conservationist. Later she became more concerned with bird protection and wrote extensively about such issues.

During her long career Nice held a number of positions with ornithological organizations. In 1933–36 she served as vice president of the Columbus Audubon Society. She joined the AOU in 1907, rejoined in 1920 after a lapse, was elected a member in 1931, a fellow in 1937 (the second woman so honored after Bailey), and later became a life fellow. She served as a vice president of the Wilson Ornithological Society (one of the nation's three major ornithological societies, focusing on the eastern United States) from 1934 through 1937, then served two years as president, the only woman to have served as president of one of the "Big Three" at that time.

Nice was a founder of the Inland Bird-Banding Association (IBBA) and associate editor of its journal, *IBBA News*. She was also an associate editor of the *Wilson Bulletin* (1939–49). She was a member of the Cooper Ornithological Society (which focuses on the western United States). From 1940 to 1942 she served as president of the Chicago Ornithological Society.

Today ornithology employs many women, especially in state and national wildlife management agencies. Much of their work involves the study of wild birds in natural settings. Margaret Morse Nice has had a significant, lasting impact on ornithology and other sciences that involve the study of wild animals.

NOTE

Alan Contreras thanks Richard C. Hoyer, a 1994 graduate in zoology and German from Oregon State University, Shirley Williams of the Chemeketa Community College (Oregon) library, and Bruce Leach of the Ohio State University library for research assistance, as well as George Jobanek for the loan of research materials. Appreciation is expressed to Lona Contreras, Dr. Karen Garst, Dr. Sayre Greenfield, and Harriet Kofalk for reviewing and commenting on the manuscript of this chapter. The extensive memorial published in the *Auk* by the late Dr. Milton B. Trautman was the primary source for listing of Nice's many professional memberships and honors.

BIBLIOGRAPHY by Milton Bernhard Trautman

Works by Margaret Morse Nice

Space does not permit the listing of the complete works of Margaret Morse Nice. Listed here are all works by Nice except those cited in Nice, *The Birds of Oklahoma* (1931); Nice, "Studies" . . . (1937); Nice, "Studies" . . . (1943); and Nice, *Research Is a Passion with Me* (1979). Included here are all references cited in the text.

Scientific Works, Biology

"Food of the bobwhite." *Journal of Economic Entomology* 3 (1910): 295–313.
(with L. B. Nice) "A city kept awake by the honking of migrating geese." *Bird-Lore* 16 (1914): 119.

"A third Christmas census." *Proceedings of the Oklahoma Academy of Science* 2 (1922): 31–32.

"Nesting records from 1920 to 1922 from Norman, Oklahoma." *University of Oklahoma Bulletin* 3 (1923): 61–67.

"Extension of range of the robin and Arkansas kingbird in Oklahoma." *Auk* 41 (1924): 565–568.

(with L. B. Nice) "The birds of Oklahoma." *University of Oklahoma Bulletin* n.s. (286) (1924): 1–122.

"Observations on shorebirds in central Oklahoma in 1924." *Wilson Bulletin* 37 (1925): 199–203.

"The bird life of a forty acre tract in central Oklahoma." *University of Oklahoma Bulletin. Proceedings of the Oklahoma Academy of Science* 7 (1927): 75–93.

"New nesting records in Cleveland County in 1925 and 1926." *University of Oklahoma Bulletin. Proceedings of the Oklahoma Academy of Science* 7 (1927): 72–74.

"Pileated woodpeckers wintering in Cleveland County, Oklahoma." *Auk* 44 (1927): 103.

"Seasonal fluctuations in bird life in central Oklahoma." *Condor* 29 (1927): 144–149.

"Late nesting of indigo bunting and field sparrow in southeastern Ohio." *Auk* 45 (1928): 102.

"Magnolia warblers in Pelham, Massachusetts in 1928." *Wilson Bulletin* 40 (1928): 252–253.

"The morning twilight song of the crested flycatcher." *Wilson Bulletin* 40 (1928): 225.

"Adventures at a window shelf." *Oologist* 46 (1929): 161–163.

"Domestic pigeons nest hunting on a mountain top." *Auk* 46 (1929): 543–544.

"Eight-mile censuses in 1927." *Condor* 31 (1929): 79.

"The Harris sparrow in central Oklahoma." *Condor* 31 (1929): 57–61.

"A hawk census from Arizona to Massachusetts." *Wilson Bulletin* 41 (1929): 93–95.

"Vocal performances of the rock wren in Oklahoma." *Condor* 31 (1929): 248–249.

"American egret and anhinga nesting in Oklahoma." *Auk* 55 (1930): 121–122.

"A list of the birds of the campus of the University of Oklahoma." *Publications of the University of Oklahoma Biological Survey* 2 (1930): 195–207.

The Birds of Oklahoma. Rev. ed. Norman: University of Oklahoma Press, 1931.

"Notes on the twilight songs of the scissor-tailed and crested flycatchers." *Auk* 48 (1931): 123–124.

"Measurements of the white-throated and other sparrows to determine sex." *Bird-Banding* 3 (1932): 30–31.

"Female quail 'Bob-whiting.'" *Auk* 50 (1933): 97.

"Locating returned song sparrows as nestlings." *Bird-Banding* 4 (1933): 51–52.

"Robins and Caroline chickadees remating." *Bird-Banding* 4 (1933): 157.

"Zur Naturgeschichte des Singammers." *Journal für Ornithologie* 81 (1933): 552–595.

"Zur Naturgeschichte des Singammers." *Journal für Ornithologie* 82 (1934): 1–96.

"The Eighth International Ornithological Congress." *Bird-Banding* 6 (1935): 29–31.

"Some ornithological experiences in Europe." *Bird-Banding* 4 (1935): 147–154.

"Storks in trees." *Wilson Bulletin* 47 (1935): 270–271.

(with L. B. Nice and R. M. Kraft) "Erythrocytes and hemoglobin in the blood of some American birds." *Wilson Bulletin* 47 (1935): 120–124.

"Late nesting of myrtle and black-throated green warblers in Pelham, Massachusetts." *Auk* 53 (1936): 89.

"Curious ways of the cowbird." *Bird-Lore* 39 (1937): 196–201.

"Studies in the life history of the song sparrow. I. A population study of the song sparrow." *Transactions of the Linnaean Society of New York* 4 (1937): 1–247. Reprinted, New York: Dover, 1964.

   An expanded version of "Zur Naturgeschichte des Singammers" (1933, 1934).

"The Ninth International Ornithological Congress." *Bird-Banding* 9 (1938): 190–198.

The Watcher at the Nest. New York: Macmillan, 1939. Reprinted. New York: Dover, 1967.

"A spring trip to Louisiana." *Indiana Audubon Yearbook* 17 (1940): 6–13.

"Courtship feeding in various birds." *Auk* 58 (1941): 56.

"Spring and winter hawk censuses from Illinois to Oklahoma." *Auk* 58 (1941): 403–405.

"A bald song sparrow." *Wilson Bulletin* 55 (1943): 196.

"The laying rhythm of cowbirds." *Wilson Bulletin* 61 (1943): 231–234.

"Starlings and woodpeckers." *Auk* 60 (1943): 311.

"Studies in the life history of the song sparrow. II. The behavior of the song sparrow and other passerines." *Transactions of the Linnaean Society of New York* 6 (1943): 1–328. Reprinted, New York: Dover, 1964.

"New bird species recorded for Oklahoma since 1931." *Proceedings of the Oklahoma Academy of Science* 24 (1944): 14–16.

"Cowbirds anting." *Auk* 62 (1945): 302–303.

"How many times does a song sparrow sing one song?" *Auk* 62 (1945): 302.

"Phases de la croissance chez les Passereaux." *L'Oiseau* 16 (1946): 87–103.

"Weights of resident and winter visitant song sparrows in central Ohio." *Condor* 48 (1) (1946): 41–42.

"Obituary of James Schenk." *Auk* 64 (1947): 181.

"Desert and mountain in southern Arizona." *Illinois Audubon Bulletin* (67) (1948): 1–7.

"Late spring in Arkansas." *Chicago Naturalist* 10 (1948): 42–49.

(with R. Thomas) "A nesting of the Carolina wren." *Wilson Bulletin* 60 (1948): 139–158.

"California spring." *Illinois Audubon Bulletin* (70) (1949): 1–7.

"A plea for the pencil." *Passenger Pigeon* 11 (1949): 98–101.

"The question of sexual dominance." In *Festschrift zum 60 Geburtstag von Erwin Stresemann*, edited by E. Mayr, 158–161. Heidelberg: C. Winter, 1949.

"Development of a redwing (*Agelaius phoeniceus*)." *Wilson Bulletin* 62 (1950): 87–93.

"Obituary, Harold Michener, bander." *Auk* 50 (1950): 428–429.

"Red-eyed vireo in Jackson Park." *Illinois Audubon Bulletin* (73) (1950): 1–4.

(with C. T. Clark) *William Dreuth's Study of Bird Migration in Lincoln Park, Chicago.* Special Publication, no. 8. Chicago: Chicago Academy of Sciences, 1950.

(with C. Nice) "The appetite of a black and white warbler." *Wilson Bulletin* 62 (1950): 94–95.

"A first trip to Florida." *Passenger Pigeon* 13 (1951): 55–61.

(with C. Nice) "Potholes and prairies." *Illinois Audubon Bulletin* (80) (1951): 6–12.

"A baby sora rail." *Illinois Audubon Bulletin* (84) (1952): 1–2.

"Some letters of Althea Sherman." *Iowa Bird Life* 22 (4) (1952): 50–55.

"Song in hand-raised meadowlarks." *Condor* 54 (1952): 362–363.

(with R. W. Allen) "A study of the breeding biology of the purple martin (*Progne subis*)." *American Midland Naturalist* 47 (1952): 606–665.

(with C. Nice) "Riding mountain and delta marsh." *Passenger Pigeon* 14 (1952): 55–62.

"The earliest mention of territory." *Condor* 55 (1953): 316–317.

"On loving vultures." *The Florida Naturalist* 26 (1953): 1.

"Blue jay anting with hot chocolate and soap suds." *Wilson Bulletin* 67 (1955): 64.

"Eleven day incubation period of summer tanager." *Wilson Bulletin* 67 (1955): 309.

"Four generations of a song sparrow family." *Jack-Pine Warbler* 57 (1956): 241–242.

(with C. Nice and D. Ewers) "Comparison of behavioral development in snowshoe hares and red squirrels." *Journal of Mammalogy* 37 (1956): 64–74.

"Nesting success in altricial birds." *Auk* 74 (1957): 305–320.

(with W. E. Schantz) "Head-scratching in passerines." *Ibis* 101 (1959): 250–251.

(———) "Head-scratching movements in birds." *Auk* 76 (1959): 339–342.

"The belligerency of a pair of wood pewees." *Illinois Audubon Bulletin* (118) (1961): 1–7.

(with N. E. Collias) "A nesting of the least flycatcher." *Auk* 78 (1961): 145–149.

"Development of behavior in precocial birds." *Transactions of the Linnaean Society of New York* 8 (1962): 1–212.

"My debt to Konrad Lorenz." *Zeitschrift für Tierpsychologie* 20 (4) (1963): 461.

(with L. Kelso) "A Russian contribution to anting and feather mites." *Wilson Bulletin* 75 (1963): 23–26.

"Mississippi song sparrow." *U.S. National Museum Bulletin* 237, pt. 3 (1968): 1513–1523.

"Studies on the biology of the edible-nest swiftlets of Southeast Asia." *Bird-Banding* 42 (1971): 55.

Research Is a Passion with Me. Edited by Dorothy Spiers. Toronto, Canada: Consolidated Amethyst Communications, 1979.

Scientific Works, Child Psychology

"The Development of a Child's Vocabulary in Relation to Environment." Master's thesis, Clark University, 1915. Published in *Pedagogical Seminary* 22 (1915): 35–64.

"The speech of a left-handed child." *Psychological Clinic* 9 (1915): 115–117.

"The speech development of a child from eighteen months to six years." *Pedagogical Seminary* 24 (1917): 204–243.

"Ambidexterity and delayed speech development." *Pedagogical Seminary* 25 (1918): 141–162.

"Concerning all day conversations." *Pedagogical Seminary* 27 (1920): 166–177.

"A child and nature." *Pedagogical Seminary* 28 (1921): 22–39.

"A child that would not talk." *Proceedings of the Oklahoma Academy of Science* 2 (1922): 108–111.

"Handedness and speech." *Proceedings of the Oklahoma Academy of Science* 2 (1922): 10.

(with G. F. Miller and M. D. Miller) "A boy's vocabulary at eighteen months." *Proceedings of the Oklahoma Academy of Science* 3 (1923): 140–144.

"The speech development of a little girl." *Proceedings of the Oklahoma Academy of Science* 4 (1924): 147–168.

"Bird study and the children." *Mt. Holyoke Alumnae Quarterly* 9 (1925): 60–61. Reprinted. *Bulletin of the Massachusetts Audubon Society* 9 (6) (1925): 4–5.

"An analysis of the conversation of children and adults." *Child Development* 3 (1932):
240–246.

Works about Margaret Morse Nice

Bonta, M. M. *Women in the Field*. College Station: Texas A & M University Press,
1991.
Conway, J. K. (ed.). *Written by Herself*. New York: Vintage Books, 1991.
Johnston, D. W. (ed.). "Obituary: Margaret Morse Nice (1883–1974)." *Bird-Banding*
24 (1974): 360.
Taylor, R. "The legacy of Margaret Morse Nice." *Columbus Monthly* 20 (4) (1994):
127–128.
Trautman, M. B. "In memoriam: Margaret Morse Nice." *Auk* 94 (July 1977): 430–441.

Other References

Bailey, F. M. *Birds through an Opera Glass*. Boston: Houghton Mifflin, 1889.
Strom, D. (ed.). *Bird Watching with American Women*. New York: W. W. Norton, 1986.
Wright, M. O. *Bird-Craft*. New York: Macmillan, 1895.

RUTH PATRICK (1907–)

Barbara Mandula

BIOGRAPHY

Ruth Patrick was born in Topeka, Kansas, on November 26, 1907, and spent all but the first few months of her childhood in Kansas City, Missouri. Her father, Frank Patrick, a major influence on her life, was a lawyer and banker during working hours. His real interest, however, was natural science. He had graduated from Cornell College in Ithaca with a degree in botany and had studied law because his father maintained that he would never be able to support a family as a scientist. Her father spent most of his spare time studying diatoms, a diverse and beautiful group of algae. Ruth Patrick also fell in love with diatoms, using them as the basis of her long and distinguished research career.

Her mother, Myrtle Moriah (Jetmore) Patrick, was well educated but did not pursue a profession. Daughter of a Kansas judge, she attended Bethany College, a small school in Topeka. Her first love was music, and she knew some French. When Mr. Patrick asked permission to marry Myrtle, her father gave permission but added that his daughter did not know how to cook. Mr. Patrick responded, "I'm not marrying a cook, I'm marrying a wife." Ruth Patrick recalls that her mother had as little interest in housework as in cooking (Personal communications 1994). Mrs. Patrick's goal was for her daughter to marry well.

During Ruth Patrick's childhood, much of her mother's attention went to Ruth's older sister, Catherine, who was crippled by infantile paralysis. Ruth Patrick remembers that "I took care of her a lot, and managed to study nature on the side." Despite a lifelong limp, her sister graduated from Smith College, where she was one of the first to talk about women having equal rights with men. (Ruth Patrick, on the other hand, "never had any desire to be equal to anybody.") Catherine Patrick received a Ph.D. degree in psychology from Columbia University, wrote books on creative thought, and had an intense interest in poetry (Personal communications 1994).

Ruth Patrick's lifelong interest in streams began in her preschool years. Her father would take her and her sister on field trips every Sunday to collect specimens. If she had been good, her father would let her sit on his lap and look through his microscope as he studied and classified the specimens. She received her first microscope when she was seven years old (Holden 1975).

In common with many other women scientists of her generation, Ruth Patrick grew up as a "tomboy." "My father treated me the way he would have treated a son," she recalls. "When I was 8 or 10, he got me a pair of boots and a fishing rod and I would stand alongside of him and fish. I've always loved to fish" (Personal communications 1994).

Patrick's father was unusual in the early 1900s because he encouraged his daughters' interest in science from an early age. "I think most women who have had success in life have had fathers who believed in them. . . . The male symbol was very strong in my years of growing up. The male was the accomplisher. This may not be true for the next generation." Other scientists, such as G. Evelyn Hutchinson of Yale and Ivey F. Lewis, later echoed her father's confidence in her. "Ruth, you're doing well. Ruth you can do these things," they would say. Patrick observes that many women drop out of science careers because nobody encourages them or helps them focus on the rewards of being a scientist (Personal communications 1994).

There was family disagreement when it was time for Patrick to attend college. Patrick's mother had wanted her daughter educated in the social milieu provided by a college such as Vassar or Wellesley. Patrick avoided that fate by not taking the required entrance examinations. Instead, she spent her freshman year at the University of Kansas in Lawrence, which had an excellent biology department. One of the biology professors told her that to be a good biologist, she also needed to learn chemistry, mathematics, German, and French. She took the advice seriously (Personal communications 1994).

Patrick's mother did not want her to go back to Kansas because she did not know the men Patrick was dating. Patrick's father went along with his wife's wishes. So Patrick attended Coker College in South Carolina, largely because the school would give her credit for the courses she had already taken. She enjoyed Coker, graduating with a botany degree in 1929. To enhance her science education during college, her father arranged for her to take summer courses at first-rate laboratories such as the Woods Hole Oceanographic Institution in Massachusetts and the Cold Spring Harbor Laboratory on Long Island.

Allowed to choose her graduate school, Patrick chose the University of Virginia in Charlottesville because she could work with Lewis, the most knowledgeable algae expert in the country. She received her master's degree in 1931 and her Ph.D. degree in 1934.

In 1931, while still a graduate student, Patrick married Charles Hodge IV, an entomologist. Her father asked the husband-to-be if Ruth Patrick could keep her maiden name; the elder Patrick explained that he had always wanted to see the Patrick name amount to something in science. "Charles said he thought it would be a wonderful idea so people wouldn't get confused by two biologists publishing under the same name," Patrick remembers (Personal communications

1994). (Frank Patrick died in 1945, aware that his daughter had made a name for herself working in diatom taxonomy and stratigraphy.)

As a direct descendant of Benjamin Franklin, Hodge had deep family roots in the Philadelphia area. While he taught zoology at Temple University, Patrick completed her Ph.D. degree by taking courses at the University of Pennsylvania and then returning to the University of Virginia to pass examinations and write her thesis.

Ruth Patrick handled her family responsibilities as methodically and decisively as she did her science. The family must come first, she tells young women. "The first priority for money that you earn is to make sure that your husband has a good dinner and that your children are taken care of and well educated. This doesn't mean that the family takes all your time, but they are your first responsibility." When Patrick decided to have a child in her early 40s, she went to a child psychiatrist, a pediatrician, and a general practitioner and asked what they thought of her becoming a mother. She recounts, "They said that what's important is the quality of the time you spend with your child. When you're with your child, you should give him your full attention. And that's what I've always done" (Personal communications 1994). Her only child, Charles Hodge V, was born in 1951. At the age of seven he inherited the microscope Patrick's father had given her. When he grew up, her son became a physician.

In Philadelphia, Patrick was determined to continue her work, even in the absence of a suitable position. She began her more than 60-year association with the Academy of Natural Sciences (ANS) in 1933 as an unpaid researcher, later also serving as volunteer curator of the Leidy Microscopical Society (1937–47) and associate curator of the Microscopy Department (1939–47). At the same time she taught botany at the Pennsylvania School of Horticulture (1935–45). She joined the ANS full-time in 1945, when a grant provided funds for her to run the microscope section. In 1947 she founded the Limnology Department at ANS, which she chaired for the next 25 years. She remains curator of the department, which houses the world's most extensive collection of diatoms. In 1973 she was given the Francis Boyer Chair of Limnology, her current position. She was a lecturer in botany at the University of Pennsylvania from 1950 to 1970, when she became an adjunct professor. She was leader of a limnological expedition to Mexico in 1947 and leader of an expedition to Peru and Brazil in 1955. Since the death of her husband in 1985, Patrick has immersed herself in her work.

As a woman, Patrick encountered her share of difficulties, but she never allowed them to interfere with her work and individuality. For example, unlike many professional women of her time who tried to blend in by wearing pants as men did, she wore dresses when it suited her. When a senior male colleague disapprovingly observed that she was wearing lipstick, Patrick said that she did not see anything wrong with that. In a more telling instance, the president of the ANS insisted that an ANS board member oversee Patrick's handling of a large grant that was awarded to her. The president needed reassurance that the money would not be squandered, since "young women are well known for wasting money." That was the attitude of the day: that women really could not do anything worthwhile. Fortunately, says Patrick, that view has changed (Personal communications 1994).

In 1970 Patrick became the 12th woman elected to the National Academy of Sciences in its 100–year history. Following this recognition, other honors and awards were bestowed on her. In 1975 she received the Second International John and Alice Tyler Ecology Award, set up to compete in prestige and munificence with the Nobel Prize. Her numerous other honors and awards include Eminent Ecologist Award from the Ecological Society of America (1972), Philadelphia Award (1973), election to the American Academy of Arts and Sciences (1976), Founders Award of the Society of Environmental Toxicology and Chemistry (1982), Benjamin Franklin Award for Outstanding Scientific Achievement from the American Philosophical Society (1993), and the National Medal of Science (1996). In 1989, the groundbreaking ceremonies were held for the Ruth Patrick Science Education Center at the University of South Carolina–Aiken. She has received two dozen honorary degrees from various universities, including Princeton University, University of Pennsylvania, Drexel University, Swarthmore College, Rensselaer Polytechnic Institute, and the University of Massachusetts.

WORK

Patrick was drawn to applied science at a time when others thought that pure research was the only kind worth doing. "I was told by many of my professors that I would ruin my career by studying problems of the environment. I felt that if you used solid science to solve the problems of society, you weren't degrading yourself" (Personal communications 1994).

In her work, Patrick focused first on the taxonomy and ecology of diatoms. She then moved on to problems of aquatic pollution, expanded her studies to encompass many aspects of river systems, and then turned her attention to groundwater. Now she is integrating information on the structure and function of ecosystems in rivers.

Diatoms are among the most abundant and important organisms on Earth, consisting of thousands of species of one-celled algae. They are found in all freshwater environments and are also the major component of phytoplankton, the basis of marine food chains. Diatoms differ from other algae in having a silica shell, which gives them a variety of intricate and beautiful shapes—pinwheels, stars, spirals, rods, and so on (Woods and Forristal 1989). Throughout the 1930s and 1940s, Patrick studied and classified diatoms in river systems, an extremely laborious process.

Her big break came in the mid-1940s, when she was giving a paper about diatom communities. Having learned that different diatoms require different nutrients, she noted that diatoms might be useful for studying water constituents. This idea attracted William B. Hart, an executive of Atlantic Refining Company in Philadelphia, who immediately recognized that diatoms could help industry deal with the burgeoning problem of water pollution. He soon raised substantial funds so that Patrick could expand her studies. Although nobody at ANS thought that a woman could lead such a complex undertaking, Hart insisted that the ANS would get the money only if Patrick headed the project.

Patrick chose to study the Conestoga Basin, which encompasses a 475-square-mile area in Pennsylvania. The site was ideal because some tributaries were unpolluted, while others went past dairy farms, industrial plants, or sanitary facilities. The teams sampled 170 sites between June 15 and August 31, 1948. The resulting study of the Conestoga Basin probably represents Patrick's single most important contribution to science:

I was doing something very new. I didn't believe that one organism could tell you everything. I felt that you had to look at all the organisms in the stream—algae, bacteria, invertebrates, fish. I don't recall exactly where I got this idea. I had always looked at interrelationships. Pattern has always been the most important thing to me. (Personal communications 1994)

To carry out the study, she did other things that were new. She was the first scientist to use a multidisciplinary team to tackle an ecological water problem, and possibly the first scientist to use a team approach for the purpose. She recognized the necessity of measuring all relevant variables at the same time. In addition to monitoring many organisms, she planned to correlate the biotic findings with the physical characteristics and chemistry of the water. This had never been done before because characterization of rivers had been the domain of chemists and engineers, with no role apparent for biologists.

Patrick showed that the distribution of living organisms was the true measure of whether or not a body of water was healthy. Her results indicated that under healthy stream conditions, many species representing the various taxonomic groups should be present, but no single species should be represented by a great number of individuals. As pollution increases, the more sensitive species disappear, and more individuals of the remaining species are found.

Patrick found that natural aquatic communities have a structure, even though individual species come and go. She showed that a healthy riverain ecosystem has a certain distribution of different types and abundances of species representing many groups of organisms and that these distributions change in characteristic ways in response to pollution. In other words, biodiversity is the keystone on which healthy ecosystems are based. In her continuing work, she confirmed that similar patterns of species exist in the Amazon watershed and in other riverain communities of the world (Cairns, Personal communication 1994).

In addition to discovering that diatoms can provide information on pollution, Patrick found that they can provide answers to other scientific questions. While still a graduate student, she showed that identification of diatoms in sediments could help reconstruct ancient environments. For example, in the 750-square-mile Dismal Swamp between Virginia and North Carolina, she found freshwater diatom fossils below six feet of sediment, but saltwater species above that level. These data showed that an invasion of seawater had apparently been responsible for the death of an ancient forest. A similar analysis indicated that the Great Salt Lake in Utah had once been freshwater (Steinmann 1983). Patrick also showed how diatoms can provide information on animal behavior. Zoologists believed that a certain whale species migrated between the Arctic and the equator, but Patrick identified it with diatom species from the Ross Sea near Antarctica, indicating that the whales spent time there (Personal communications 1994).

Patrick has never lost her intellectual curiosity. She was over 70 years old when she became concerned about groundwater, a relatively neglected area in the early 1980s. She has since published numerous papers analyzing the characteristics of groundwater and showing the fragility of this water source and its interaction with surface water. In 1992 she published a book exploring the effects of environmental laws on three U.S. watersheds, the only three watersheds with enough data to attempt the analysis.

Between 1935 and 1945 Patrick taught botany and comparative anatomy at the Pennsylvania School of Horticulture. In the early 1950s she organized a course at the ANS called Fundamentals of Systematics, open to advanced students at the nearby colleges. As guest lecturers she brought in such famous biologists as Nobel laureate Barbara McClintock,* Theodore Dobzhansky, and Ernst Mayr. At the University of Pennsylvania she organized a course in her specialty, limnology, and again used outside scientists to supplement her own lectures. She still gives one or two lectures a year in the limnology course. She has trained about two dozen doctoral students from all over the world.

Patrick is concerned because researchers are now producing sloppy science as they rush to solve problems. People cannot understand how organisms interact in a stream without being able to identify the species, yet there is almost no money for training taxonomists, she observes. She is encouraged that many young people are motivated to learn about the environment and recognizes the need to give them basic training in chemistry, physiology, biochemistry, taxonomy, and other sciences. For example, understanding the chemistry of both water and soil is crucial to studying how species interact and transform detritus into nutrients and energy for other species (Personal communications 1994).

Unlike many environmental scientists, Ruth Patrick has not been at odds with industry. In the late 1940s the president of Du Pont became aware of her work on rivers and insisted that Patrick analyze the receiving waters before Du Pont built new plants. Patrick is also grateful for the industry funds that kept her department and her research going, especially before the days of the National Science Foundation. Because she is a firm believer that success in any endeavor depends on people working together, she sees nothing contradictory in working with industry. Only with full cooperation from all sectors will we be able to solve our pressing environmental problems, she says.

Patrick has served on a large number of environmental committees and advisory boards for public and private organizations. An abbreviated list of her memberships includes Advisory Committee for Research, National Science Foundation (1973–77); Executive Advisory Committee, Environmental Protection Agency (1974–79); Science Advisory Council of World Wildlife Fund–United States (1978–86); National Academy of Sciences chairperson of the Section on Population Biology, Evolution, and Ecology (1980–86); Energy Research Advisory Board, Department of Energy (1982–present). She also served on numerous panels convened by the state of Pennsylvania and by nearby academic institutions. She was the first woman and first environmentalist to serve on Du Pont's Board of Directors; she also has served on the Board of Directors of the Pennsylvania Power and Light Company. From 1973 to 1976 she was the first woman chair of the ANS Board of Trustees and has served as honorary chair since 1976.

Patrick has an impressive list of over 175 publications. She and Charles Reimer coauthored the definitive *Diatoms of the United States*, volume I in 1966, followed by volume II in 1975. In recent years Ruth Patrick has devoted almost all of her time to writing three comprehensive books. The first book describes the characteristics of the different types of estuaries, including how they function in terms of nutrients and energy transfer and the structure of their various ecosystems. It covers the organisms found in sand, on pilings, on vegetation, and in other habitats. The second book addresses the physical and chemical properties of freshwater. Patrick is now working to complete the third book, which will describe in detail the biological characteristics of the major watersheds of the United States. The task is difficult, partly because not all of the pertinent data are available. But Patrick, considered the leading limnologist in the world, is helping to fill in the missing pieces.

Asked about future research needs, Patrick emphasized basic knowledge and its integration. It is important to learn how air pollution moves through the ecosystem, getting into the soil, then into streams, and then into groundwater. She says that scientists often ignore the soil, but we are going to need to understand it and its organisms (protozoans, bacteria, fungi, and worms) before attempting to solve problems. Despite the severe constraints on her time, she is actively supporting efforts to increase government funding for integrated environmental research.

John Cairns, Jr., who worked closely with Patrick for two decades, has eloquently summarized her contributions:

Ruth was a scientist when women scientists were exceedingly uncommon; a leader in the development of the scientific team approach to problem solving when "lone-wolf" scientific specialists were dominant; a pioneer in the systems approach of looking at entire drainage basins; and, perhaps most importantly, living proof that theoretical and applied science cannot only co-exist, but are commonly synergistic. . . . By bringing theoretical and applied science together, Ruth Patrick forced the validation of ecological theory. (Cairns 1992)

NOTE

Special thanks go to Ruth Patrick, who took time out of her busy schedule to meet with me and then patiently corrected an early draft. I am grateful also to John Cairns, Jr., for his many insights and to David Blockstein for suggesting that Patrick be included in this volume.

BIBLIOGRAPHY

Works by Ruth Patrick

Scientific Works

Space does not permit the listing of the complete works of Ruth Patrick. Listed here are all works by Patrick except those cited in Patrick (''Diatom communities'' 1971; ''Stream communities'' 1975; ''Ecology . . .'' 1977; and ''What are . . .'' 1994). Included here are her dissertation and all references cited in the text.

(with I. F. Lewis and C. Zirkle) ''Algae of Charlottesville and vicinity.'' *Journal of the Elisha Mitchell Scientific Society* 48 (2) (1933): 207–223.

''The Diatoms of Siam and the Federated Malay States.'' Ph.D. diss., University of Virginia, 1934.

'' 'Karo' as a mounting medium.'' *Science* 83 (2143) (1936): 85–86.

''A taxonomic and distributional study of some diatoms from Siam and the Federated Malay States.'' *Proceedings of the Academy of Natural Sciences of Philadelphia* 88 (1936): 367–470.

The Culture of Plants in Nutrient Solutions. Pennsylvania School of Horticulture Bulletin 1. Philadelphia: Pennsylvania School of Horticulture, 1938.

''The occurrence of flints and extinct animals in Pluvial deposits near Clovis, New Mexico. V. Diatom evidence from the Mammoth Pit.'' *Proceedings of the Academy of Natural Sciences of Philadelphia* 90 (1938): 15–24.

''Nomenclatural changes in two genera of diatoms.'' *Notulae Naturae of the Academy of Natural Sciences of Philadelphia* (28) (Oct. 26, 1939): 1–11.

''Some new diatoms from Brazil.'' *Notulae Naturae of the Academy of Natural Sciences of Philadelphia* (59) (Nov. 25, 1940): 1–7.

''A suggested starting point for the nomenclature of diatoms.'' *Bulletin of the Torrey Botanical Club* 67 (7) (1940): 614–615.

''Diatoms.'' In *The Cochise Culture: Medallion Papers* no. 29, edited by E. B. Sayles and E. Antevs, 69–70. Globe, AZ: privately published, 1941.

''Diatoms of northeastern Brazil. Part I. Coscinodiscaceae, Fragilariaceae, and Eunotiaceae.'' *Proceedings of the Academy of Natural Sciences of Philadelphia* 42 (1941): 191–226.

''Diatoms.'' In ''Marine topography of the Cape May formation'' by P. MacClintock. *Journal of Geology* 51 (7) (1943): 458–472.

Estudo Limnologico e Biologica das Lagoas de Regiao Litoranea Sul-Riogradense. II. *Some New Diatoms from the Lagoa Dos Quadros. Boleton do Museu Nacional— nova serie. Rio de Janeiro—Brasil Botanica N2.* Rio de Janeiro, Brazil: Museu Nacional, 1944.

"Well-boring at Brandywine Lighthouse in Delaware Bay. II. Miocene diatoms." *Notulae Naturae of the Academy of Natural Sciences of Philadelphia* (133) (1944): 1–12.

"Factors affecting the distribution of diatoms." *Botanical Reviews* 14 (8) (1948): 473– 524.

"Biological measure of stream conditions." *Sewage and Industrial Wastes* 22 (7) (1950): 926–938.

"Some new methods for dealing with variation in diatoms." In *Proceedings of the 7th International Botanical Congress,* edited by H. Osvald and E. Aberg, 841. Waltham, MA: Chronica Botanica, 1950.

(with J. H. Wallace) "A consideration of *Gomphonema parvulum* Kutz." *Butler University Botanical Studies* 9 (1950): 227–234.

"A proposed biological measure of stream conditions." *Proceedings of the International Association of Theoretical and Applied Limnology* 11 (1951): 299–307.

(with P. Doudoroff, B. G. Anderson, et al.) "Bioassay methods for the evaluation of acute toxicity of industrial wastes to fish." *Sewage and Industrial Wastes* 23 (11) (1951): 1380–1397.

(with W. C. Starret) "Net plankton and bottom microflora of the Des Moines River, Iowa." *Proceedings of the Academy of Natural Sciences of Philadelphia* 104 (1952): 219–243.

"Aquatic organisms as an aid in solving waste disposal problems." *Sewage and Industrial Wastes* 25 (2) (1953): 210–214.

(with N. M. Wallace) "The effect of agar on the growth of *Nitzschia linearis.*" *American Journal of Botany* 40 (8) (1953): 600–602.

"Diatoms as an indicator of river change." In *Proceedings of the 9th Industrial Waste Conference, Purdue University Engineering Bulletin* 87, 325–330. West Lafayette, IN: Purdue University, 1954.

"Sexual reproduction in diatoms." In *Sex in Microorganisms,* edited by D. H. Wenrich et al., 82–99. Clinton, MA: Colonial Press, 1954.

"Diatoms as indicators of changes in environmental conditions." In *Biological Problems in Water Pollution,* edited by C. M. Tarzwell, 71–83. Cincinnati, OH: Robert A. Taft Sanitary Engineering Center, 1956.

"Some nomenclature problems and a new species and a new variety in the genus *Eunotia* (Bacillariophyceae)." *Notulae Naturae of the Academy of Natural Sciences of Philadelphia* (312) (Nov. 28, 1958): 1–15.

"Aquatic life in a new stream." *Water and Sewage Works* 106 (1959): 531–535.

"Bacillariophyceae." In *Fresh-Water Biology,* edited by H. B. Ward and G. C. Whipple, 171–189. New York: John Wiley and Sons, 1959.

"The development of the fauna and flora in a new stream bed, progress report." In *Proceedings of the 1959 Air and Water Pollution Abatement Conference,* 120– 136. Washington, DC: Manufacturing Chemists' Association, 1959.

"An evaluation of the contribution to the systematics of diatoms of electron microscope studies." *Proceedings of the 9th International Botanical Congress* 2 (1959): 296– 297.

"How is industry meeting its responsibilities in keeping streams clean?" In *7th Annual Pennsylvania Clean Streams Conference*, 24–28. Harrisburg: Pennsylvania State Chamber of Commerce, 1959.

"New species and nomenclatural changes in the genus *Navicula* (Bacillariophyceae)." *Proceedings of the Academy of Natural Sciences of Philadelphia* 111 (1959): 91–108.

"New subgenera and two new species of the genus *Navicula* (Bacillariophyceae)." *Notulae Naturae of the Academy of Natural Sciences of Philadelphia* (324) (1959): 1–11.

"Correct use and abuse of Philadelphia's rivers." In *Proceedings of the 88th Annual Meeting of the Fairmount Park Art Association*, 15–28. Philadelphia: Fairmount Park Association, 1960.

"Diatoms (Bacillariophyceae) from the alimentary tract of *Phoenicoparrus jamesii* (Sclater)." *Postilla* 49 (1961): 43–56.

"An evaluation of the contribution to the systematics of diatoms of the electron microscope studies." *Recent Advances in Botany* 1 (1961): 221–226.

"Use without abuse of our water resources." *Massachusetts Audubon Bulletin* (Jan.–Feb. 1961): 1–4.

(with L. R. Freese) "Diatoms (Bacillariophyceae) from Northern Alaska." *Proceedings of the Academy of Natural Sciences of Philadelphia* 112 (6) (1961): 129–293.

"Effects of river physical and chemical characteristics on aquatic life." *Journal of the American Water Works Association* 54 (5) (1962): 544–550.

"Limnology studies pollution." *Frontiers* 26 (3) (1962): 90–92.

"A discussion of the results of the Catherwood Expedition to the Peruvian headwaters of the Amazon." *Verhandlungen Internationale Vereinigung für Theoretische und Angewandte Limnologie* 15 (1964): 1084–1090.

"Use without abuse." *Frontiers* 29 (5) (1965): 133–137.

"The diatoms." *Memoirs of the Connecticut Academy of Arts and Sciences* 17 (1966): 74–77.

"Introduction and Part I—Limnological observations and discussion of results." In *The Catherwood Foundation Peruvian–Amazon Expedition*. Monograph 14, edited by C. W. Hart, Jr., and E. Peddicord, 1–40. Philadelphia: Academy of Natural Sciences, 1966.

"PEPCO's approach to use without abuse of the Patuxent River." In *Seminar Proceedings, Water Resources and Steam Generation of Electricity*, 46–63. College Park: University of Maryland, 1966.

(with C. W. Reimer) *Diatoms of the United States*, vol. 1. Monograph 13. Philadelphia: Academy of Natural Sciences, 1966.

"Diatom communities in estuaries." In *Estuaries*, edited by G. H. Lauff, 311–315. Washington, DC: American Association for the Advancement of Science, 1967.

"Natural and abnormal communities of aquatic life in rivers." *Bulletin South Carolina Academy of Sciences* 29 (1967): 19–28.

"The versatile diatom." *Frontiers* 31(3) (1967): 74–75.

(with R. S. Harvey) "Concentration of ^{137}Cs, ^{65}Zn, and ^{85}Sr by freshwater algae." *Biotechnology and Bioengineering* 9 (1967): 449–456.

"Aquatic communities and the problems of water quality." *General Systems* 13 (1968): 125–127.

"Effect of suspended solids, organic matter, and toxic materials on aquatic life in rivers."

In *Proceedings of the First Mid-Atlantic Industrial Waste Conference*, 275–283. Newark: University of Delaware, 1968.

"Effect of suspended solids, organic matter, and toxic materials on aquatic life in rivers." *Water and Sewage Works* (Feb. 1968): 89–92.

"Our freshwater environment." *Garden Journal* 18 (6) (1968): 174–176.

Water Research Programs—Aquatic Communities. Washington, DC: U.S. Department of the Interior. Office of Water Resources Research, 1968.

"The river and the watershed." *The Morton Arboretum Quarterly* 5 (4) (1969): 56–64.

"Aquatic communities of rivers." *New Jersey Nature News* 25 (3) (1970): 109–112.

"The diatom flora of some lakes of the Galapagos Islands." *Nova Hedwigia (Suppl.)* 31 (1970): 495–510.

"The diatoms: An account of the history and development of the Lago di Monterosi, Latium, Italy." *Transactions of the American Philosophical Society* 69 (4) (1970): 112–122.

"A healthy stream rests on a precarious web of life." *Better Living* 24 (5) (1970): 25–27.

"The river and the watershed." *Compost Science* 111 (4) (1970): 28–32.

(with J. Cairns, Jr., and R. L. Kaesler) "Occurrence and distribution of diatoms and other algae in the upper Potomac River." *Notulae Naturae of the Academy of Natural Sciences of Philadelphia* (436) (1970): 1–12.

"The diatom communities." In *The Structure and Function of Fresh-Water Microbial Communities*, Research Monograph 3, edited by J. Cairns, Jr., 151–164. Blacksburg: Virginia Polytechnic Institute and State University, 1971.

"The ecology of the river and watershed." *Technical Guidance Center Bulletin, University of Massachusetts* 3 (5) (1971): 1–3.

"Use without abuse of water resources." In *Environment, Man, Survival: Grand Canyon Symposium, 1970*, edited by L. H. Wullstein et al., 87–95. Salt Lake City: University of Utah, 1971.

(with R. R. Grant, Jr.) "Observations of stream fauna." In *Hydrology of Two Small River Basins in Pennsylvania before Urbanization*, edited by R. A. Miller et al., U.S. Geological Survey Professional Papers 701–A, 24–39. Washington, DC: Government Printing Office, 1971.

"Aquatic communities as indices of pollution." In *Indicators of Environmental Quality*, edited by W. A. Thomas, 93–100. New York: Plenum, 1972.

"The potential of various types of thermal effects on Chesapeake Bay." *Journal of the Washington Academy of Sciences* 62 (2) (1972): 140–144.

"Aquatic systems." In *Nitrogenous Compounds in the Environment*, EPA-SAB-73–001, 127–139. Washington, DC: Hazardous Materials Advisory Committee, U.S. Environmental Protection Agency, 1973.

"Diatoms as bioassay organisms." In *Bioassay Techniques and Environmental Chemistry*, edited by G. E. Glass, 139–151. Ann Arbor, MI: Ann Arbor Science, 1973.

"Ecological aspects of environmental impact statements." In *Proceedings of the Environmental Assessment and Impact Statements Conference, 1973*, 91–95. Philadelphia: Drexel University, 1973.

"The river and the watershed." *Morris Arboretum Bulletin* 24 (4) (1973): 63–66.

"Use of algae, especially diatoms, in the assessment of water quality." In *Biological Methods for the Assessment of Water Quality*, edited by J. Cairns, Jr., and K. L.

Dickson, Special Technical Publication 528, 76–95. Philadelphia: American Society for Testing and Materials, 1973.

(with T. Bott and R. L. Vannote) "The effects of natural temperature-variations on riverine communities." In *Report to the Congress by the Environmental Protection Agency*, Part III, Serial No. 93–14, 1181–1269. Washington, DC: Government Printing Office, 1973.

"A viewpoint from the life sciences." In *Water and the Environmental Crunch, Proceedings of the Princeton University Conference* 115, edited by W. E. Bonini, 51–58. Princeton: Princeton University Press, 1974.

(with D. M. Martin) *Biological Surveys and Biological Monitoring in Fresh Waters*. Academy of Natural Sciences of Philadelphia, no. 5. Philadelphia: Academy of Natural Sciences, 1974.

"Energy and the environment." *Frontiers* 39 (3) (1975): 5–9.

"Some thoughts concerning correct management of water quality." In *Urbanization and Water Quality Control, Proceedings* No. 20, edited by W. Whipple, Jr., 14–18. Minneapolis: American Water Resources Association, 1975.

"Stream communities." In *Ecology and Evolution of Communities*, edited by M. Cody and J. Diamond, 445–459. Cambridge, MA: Belknap Press of Harvard University Press, 1975.

(with C. W. Reimer) *Diatoms of the United States*, vol. 2, pt. 1, Monograph 13. Philadelphia: Academy of Natural Sciences, 1975.

"The effect of a stabilization pond on the Sabine estuary." In *Ponds as a Wastewater Treatment Alternative*, edited by E. F. Gloyna et al., 33–35. Austin: Center for Research in Water Resources, College of Engineering, University of Texas at Austin, 1976.

"Energy and the environment." In *A Change of Perspective*, 56–64. Auburn, GA: Auburn University, 1976.

"The formation and maintenance of benthic diatom communities." *Proceedings of the American Philosophical Society* 120 (6) (1976): 475–484.

"Overview of aquatic ecosystems." In *Biological Control of Water Pollution*, edited by J. Tourbier and R. W. Peirson, Jr., 39–40. Philadelphia: University of Pennsylvania Press, 1976.

"The role of aquatic plants in aquatic ecosystems." In *Biological Control of Water Pollution*, edited by J. Tourbier and R. W. Peirson, Jr., 53–59. Philadelphia: University of Pennsylvania Press, 1976.

(with P. Kiry) *Estuarine Surveys, Biomonitoring and Bioassays*. Academy of Natural Sciences of Philadelphia, no. 6. Philadelphia: Academy of Natural Sciences, 1976.

"The changing scene in aquatic ecology." In *The Changing Scenes in Natural Sciences 1776–1976*, edited by C. Goulden, Special Publications 12, 205–222. Philadelphia: Academy of Natural Sciences, 1977.

"Diatom communities." In *Aquatic Microbial Communities*, edited by J. Cairns, Jr., 141–159. New York: Garland, 1977.

"Ecology of freshwater diatoms and diatom communities." In *The Biology of Diatoms*, edited by D. Werner, *Botanical Monographs* 13, 284–332. Berkeley: University of California Press, 1977.

"The importance of monitoring change." In *Biological Monitoring of Water and Effluent Quality*, edited by J. Cairns, Jr. et al., Special Technical Publication 607, 157–189. Philadelphia: American Society for Testing and Materials, 1977.

"Effects of trace metals in the aquatic ecosystem." *American Scientist* 66 (2) (1978): 185–191.

"Identifying integrity through ecosystem study." In *Integrity of Water. Proceedings of the Symposium*, 155–162. Washington, DC: U.S. Environmental Protection Agency, Office of Water and Hazardous Materials, 1978.

"Some impacts of channelization on riverine systems." In *Environmental Quality through Wetlands Utilization*, edited by M. A. Drew, 15–28. Tallahassee, FL: Coordinating Council on the Restoration of the Kissimmee River Valley and Taylor Creek-Nubbin Slough Basin, 1978.

"Monitoring the condition of flowing waters by biological organisms." In *Proceedings of the First and Second USA–USSR Symposia on the Effects of Pollutants upon Aquatic Ecosystems*, vol. 1, section 7, EPA-600/3-78-076, 68–81. Washington, DC: U.S. Environmental Protection Agency, 1978.

"Pine barrens surface water." *Frontiers* 42 (2) (1978): 21.

(with W. C. Ackerman, D. J. Allee et al.) "Scientific and technological considerations in water resources policy." *EOS* 59 (6) (1978): 516–527.

(with B. Matson and L. Anderson) "Streams and lakes in the Pine Barrens." In *Pine Barrens, Ecosystems and Landscape*, edited by R. T. Foreman, 169–193. New York: Academic Press, 1979.

(with N. A. Roberts) "Diatom communities in the middle Atlantic states." *Nova Hedwigia* 64 (1979): 265–283.

"Freshwater ecosystems—acid rain." In *Environment, The Global Connection Summary Report*, 32–36. Washington, DC: Meridian House International, 1980.

(with T. Bott, R. Larson, et al.) *The Effect of Nitroacetic Acid (NTA) on the Structure and Functioning of Aquatic Communities in Streams*. Washington, DC: U.S. Environmental Protection Agency, 1980.

(with S. L. Friant and L. A. Lyons) "Effects of nonsettleable biosolids on stream organisms." *Journal of the Water Pollution Control Federation* 52 (2) (1980): 351–363.

"The world's wetlands." *EPA Journal* 7 (2) (1981): 13–16.

(with V. P. Binetti and S. G. Halterman) "Acid lakes from natural and anthropogenic causes." *Science* 211 (1981): 446–448.

(with J. W. Sherman) "The waters of Merom: A study of Lake Huleh. 7. Diatoms stratigraphy of the 54-M core." *Archiv für Hydrobiologie* 92 (2) (1981): 199–221.

"The energy dilemma." *Swarthmore College Bulletin* 79 (5) (1982): 2–4.

"Water: A critical problem." *The Garden Club of America Bulletin* 70 (4) (1982): 5–9.

"What is the condition of our surface water?" In *Proceedings of the National Water Conference*, edited by J. Wilson, 11–29. Philadelphia: Academy of Natural Sciences, 1982.

(with V. Pye) "Groundwater contamination in the United States." *Science* 221 (1983): 713–718.

(——— and J. Quarles) *Groundwater Contamination in the United States*. Philadelphia: University of Pennsylvania Press, 1983.

"Challenging problems in water management. Implications for the biological and environmental sciences." In *Proceedings of the University Council on Water Resources*, 40–45. Baton Rouge: Louisiana State University, 1984.

"A comparison of factors affecting the fate of toxics in surface and groundwater." In

Proceedings of the Second National Water Conference, edited by J. Wilson, 41–47. Philadelphia: Academy of Natural Sciences, 1984.

"Groundwater contamination." In *Proceedings of a Conference on Water Resources of Georgia*, edited by R. Arora and L. L. Gorday, 115–120. Atlanta: Department of Natural Resources—Environmental Protection Division and Georgia Geologic Survey, 1984.

"Some thoughts concerning the importance of patterns in diverse riverine systems." *Proceedings of the American Philosophical Society* 128 (1) (1984): 48–79.

"What should be the rationale for bioassays?" In *Multispecies Toxicity Testing*, edited by J. Cairns, Jr. New York: Pergamon Press, 1985.

"Diatoms as indicators of changes in water quality." In *Proceedings of the Eighth International Diatom Symposium*, edited by M. Ricard, 759–766. Koenigstein, Germany: Koeltz Scientific Books, 1986.

"Rural preservation in Pennsylvania. Environmental matters." In *Conference Proceedings*, 23–50. French Creek, PA: French and Pickering Creeks Conservation Trust, 1986.

"Water use and land management: An overview." In *Managing Water Resources*, edited by R. Patrick and J. Cairns, Jr., 1–5. Westport, CT: Praeger Press, 1986.

(with J. Cairns, Jr., eds.) *Managing Water Resources*. Westport, CT: Praeger Press, 1986.

"The Savannah River." In *Proceedings of a Symposium in Honor of Dr. Ruth Patrick*, 85–105. Aiken, SC: E. I. DuPont de Nemours Company, Savannah River Laboratory, 1987.

(with E. Ford and J. Quarles) *Groundwater Contamination in the United States*. 2d ed. Philadelphia: University of Pennsylvania Press, 1987.

"Changes in the chemical and biological characteristics of the upper Delaware River estuary in response to environmental laws." In *Ecology and Restoration of the Delaware River*, edited by S. K. Majumdar et al., 332–359. Philadelphia: Pennsylvania Academy of Sciences, 1988.

"Groundwater: Overview." In *Groundwater Contamination: Sources, Effects and Options to Deal with the Problem. Proceedings of the Third Water Conference of the Academy of Natural Sciences of Philadelphia, 1987*, edited by P. Churchill and R. Patrick, 13–18. Philadelphia: Academy of Natural Sciences, 1988.

(with P. Churchill, eds.) *Groundwater Contamination: Sources, Effects and Options to Deal with the Problem. Proceedings of the Third Water Conference of the Academy of Natural Sciences of Philadelphia, 1987*. Philadelphia: Academy of Natural Sciences, 1988.

"Past, present, and future of water use and management." In *Water Management in the 21st Century*, edited by A. I. Johnson et al., 15–19. Bethesda, MD: American Water Resources Association, 1989.

"Reevaluation of solutions to pollution problems." *Proceedings of the American Philosophical Society* 134 (2) (1989): 90–94.

"Groundwater and its contaminants." *United Nations Environment Programme for Industry and Environment* 13 (3–4) (1990): 29–32.

"What should be done to mitigate groundwater contamination?" *Environmental Health Perspectives* 86 (1990): 239–243.

(with D. D. Williams) "Aquatic biota in North America." In *Surface Water Hydrology*, vol. O-1, edited by M. Wolman and H. C. Riggs, 233–254. Boulder, CO: Geological Society of America, 1990.

"The importance of the technical characteristics of hazardous materials in considering environmental risk." *Risk Analysis* 11 (1) (1991): 27–29.

"The Savannah River—past, present, and future." In *Integrated Environmental Management*, edited by J. Cairns, Jr., and T. Crawford, 137–149. Chelsea, MI: Lewis, 1991.

(with H. Kunreuther) "Managing the risks of hazardous waste." *Environment* 33 (3) (1991): 12–16.

"Environmental risks—factors to consider in their control." In *Predicting Ecosystem Risk*, edited by J. Cairns, Jr., et al., 23–30. Princeton: Princeton Scientific, 1992.

Surface Water Quality: Have the Laws Been Successful? Princeton: Princeton University Press, 1992.

(with J. Hendrickson) "Factors to consider in interpreting diatom change." *Nova Hedwigia* 106 (1993): 361–377.

Rivers of the United States. 2 vols. New York: John Wiley and Sons, 1994–95.

"What are the requirements for an effective biomonitor?" In *Biological Monitoring of Aquatic Systems*, edited by S. L. Loeb and A. Spacie, 23–29. Boca Raton, FL: Lewis, 1994.

Other Works

Personal communications to the author, 1994.

Works about Ruth Patrick

Detjen, J. "In tiny plants, she discerns nature's warning on pollution." *Philadelphia Inquirer* (Feb. 19, 1989).

Holden, C. "Ruth Patrick: Hard work brings its own (and Tyler) reward." *Science* 188 (1975): 997–999.

Steinmann, M. "Rivers of America. The source is Ruth Patrick." *RF Illustrated* (The Rockefeller Foundation) (June 1983): 14–16.

———. "Ruth Patrick." In *Science Year 1986: The World Book Science Annual*, 351–362. Chicago: World Book, 1985.

Woods, M., and L. J. Forristal. "The river doctor." In *The World and I*, 340–347. Washington, DC: Washington Times, 1989.

Sources

Cairns, J., Jr. "Dedication." In *Advances in Modern Environmental Toxicology*, vol. 20, edited by J. Cairns, Jr., B. R. Niederlehner, et al., ix–xii. Princeton: Princeton Scientific, 1992.

Cairns, J., Jr. Personal communications, 1994.

LOUISE PEARCE (1885–1959)

Anne-Marie Scholer

BIOGRAPHY

Louise Pearce was born in Winchester, Massachusetts, on March 5, 1885. Her parents were Susan Elizabeth (Hoyt) and Charles Ellis Pearce, a cigar and tobacco dealer. Her only sibling was a younger brother, Robert Pearce, who became an attorney in New York City.

Some time after 1889 the Pearce family moved to California, where they lived on a ranch. She attended the Girls' Collegiate School in Los Angeles from 1900 to 1903. She then attended Stanford University, from which she received an A.B. degree in physiology and histology in 1907.

After college Louise Pearce moved to Boston to study for two years. During this time she worked as an instructor in embryology and assistant in histology for Boston University School of Medicine. She was offered a scholarship from the Woman's Medical College of Pennsylvania (now the Medical College of Pennsylvania at Philadelphia) in 1908 but declined due to financial constraints. In 1909 Pearce was admitted to Johns Hopkins University School of Medicine with advanced standing. She graduated third in her class in 1912 and was an intern at Johns Hopkins Hospital.

In 1913 Pearce had accepted a prestigious appointment to the staff of the Phipps Psychiatric Clinic at Johns Hopkins Hospital. However, she sought out a research position at the Rockefeller Institute for Medical Research (now Rockefeller University) instead, because she preferred investigative laboratory research. In a letter of application to Simon Flexner, she referred to her college and medical laboratory courses as those which she most enjoyed and in which she was most successful (Louise Pearce Papers, 7/7/1913).

Pearce was hired as an assistant fellow at the Rockefeller Institute to work under Flexner. He was married to Helen Thomas, sister of M. Carey Thomas, who was then President of Bryn Mawr College. The press release regarding Pearce's appointment refers to her intentions to study childhood infectious diseases ("Woman aid . . ." 1913). In actuality, her research addressed animal models for syphilis, gonorrhea, and trypanosomiasis (African sleeping sickness). Pearce continued at Rockefeller until her retirement in 1951, being promoted to associate fellow in 1916 and associate member in 1923. Limited funds due to the economic depression were cited by Flexner as the reason he was not able to promote her to full member. Such a promotion would have given her the freedom to pursue her research goals separately from those of her research colleague, Wade Hampton Brown.

She gained a boost from the visibility of her work on trypanosomiasis. As Brown was a family man, and Pearce was seen as having a more effective personality for the project, she was the one entrusted to travel to the Belgian Congo to test the developed drug on humans ill with the parasite (Flexner 1920). This trip resulted in Pearce's being awarded the Belgian Order of the Crown in 1921 and in international recognition of her work. Pearce was invited to give presentations, resulting in trips to Europe each summer from 1921 to 1939. She also was a medical consulting delegate for the International Conference on Christian Missions in Africa in 1926. On many of these trips Pearce also served as the Rockefeller Institute's representative to international pharmaceutical companies wishing to produce the drug.

Pearce also accepted a position as Visiting Professor of Syphilology at Peiping Union Medical College in what is now Beijing, China. She worked in China from November 1931 until May of 1932. During this time Japan invaded Manchuria, and 15 of Pearce's fellow visiting professors (presumably those of Japanese origin) were sent to Shanghai.

In addition to the Belgian Order of the Crown and the King Leopold II Prize for her work on trypanosomiasis (1953), Pearce received honorary doctorates from Wilson College, Beaver College, Bucknell College, Skidmore College, and the Woman's Medical College of Pennsylvania, where she served as President from 1946 to 1951. She also received the Elizabeth Blackwell Award from the New York Infirmary in 1951.

Louise Pearce left few personal papers or correspondence, so there is little information about her personal life. However, there is evidence that she had an active social life and that she numbered among her friends many of the professional women of the time. She never married or had children. In 1935 Pearce's department at Rockefeller Institute moved to Princeton, New Jersey. At this time two of Pearce's friends, Ida A. R. Wylie and Sara Josephine Baker, M.D., decided to move from their apartment in New York City to join Pearce in sharing a house. Wylie bought Trevanna Farm just outside Princeton, a house large enough to accommodate the three women, some maids, and Wylie's several dogs.

Wylie was a British novelist, best known for *The Daughter of Brahma* (1912) and her autobiography, *My Life with George* (published in 1940). She was also a very active suffragette. Baker was a physician and later the first Chief of the Department of Child Hygiene at the New York City Department of Health and a pioneer for public health education. Baker was also a suffragette. When Baker retired in 1923, she assumed responsibility for running the household, while writing her autobiography (published in 1939).

These three professional women apparently supported each other in their careers, dealing with Wylie's publisher or Pearce's department head in times of illness. With Baker and some servants to manage the household concerns, Wylie and Pearce were free to concentrate on work and business travel. Two or all three of the women would vacation together, either at Martha's Vineyard in Massachusetts or in Europe. They also attended parties given by Wylie's publisher. Baker died in 1945, and Wylie and Pearce continued to live at the farm. Pearce died on August 9, 1959, in a hospital in New York City. She had fallen ill on shipboard while returning from a trip to Europe. Wylie survived Pearce by only three months, and she left the bulk of her estate to the Woman's Medical College of Pennsylvania in Pearce's memory.

WORK

Louise Pearce's research career started as a medical student at Johns Hopkins University. She studied experimental jaundice and calcium metabolism under John King. As an intern she did a clinical study of Friedrich's ataxia and assisted with clinical microscopy.

When Pearce was hired by the Rockefeller Institute, she worked with Brown to develop animal models of trypanosomiasis and syphilis in rabbits. The courses of these diseases were characterized, and then arsenic-derived drugs were tested for effectiveness of treatment. Tryparsamide was found to be successful in clearing the blood and nervous system of the parasite, with no side effects.

In May 1920 Pearce was sent to the Belgian Congo to test the drug on humans, and in 1924 she returned to Africa to conduct larger trials. Tryparsamide was effective in humans, but large doses used for advanced cases resulted in damaged vision and occasional blindness. Pearce served as representative for the Rockefeller Institute and as a contact for the physicians conducting trials in her absence. During the second trip to the Belgian Congo, she urged her department head, Flexner, to end the trials so that widespread treatment of the afflicted could begin.

In 1928 Flexner suggested that Pearce produce a monograph on the tryparsamide trials, gathering together her results along with the data produced by her Rockefeller colleagues and by the physicians in Africa. The monograph was published in 1930 but also served to delay her other research.

From 1920 to 1928 Pearce and Brown also worked on describing an experimental model of syphilis in rabbits, complete with latency and lesions. The syphilitic infection of the nervous system was a clinically hopeless situation at that time. The hope was that tryparsamide, which was capable of penetrating the brain and spinal cord, would be able to kill syphilitic spirochetes in the nervous system. The drug did prove effective and was even more effective when combined with an artificial fever. This remained the accepted treatment of syphilis until the advent of penicillin.

Pearce continued to study syphilis, comparing normal blood components with those in syphilis-infected blood. She also studied the coinfection of syphilis and the vaccinia virus. She journeyed to China in 1932, taking along 125 rabbits, in order to compare the Western version of syphilis with an "Oriental" strain. What appeared to be two different strains of syphilis proved to be the differing effects of the same spirochete on two different species of rabbit.

In 1922, while characterizing syphilis-infected rabbits, Brown and Pearce found one rabbit with a malignant epithelial tumor of the scrotum. This tumor was successfully propagated to other rabbits. It was studied extensively to learn the biological laws affecting the susceptibility of the hosts to implanted cells. Pearce studied the effects of seasons and the exposure to light on normal and tumor-infected rabbits. She explored the effect of tumor presence on polio infection and on blood composition. With Thomas Rivers she did a series of experiments on the effect of an unknown coexisting virus on this tumor, trying to find a causal relationship. The effects of this tumor were so well characterized that the Brown-Pearce tumor became a popular test material for the study of cancer.

In 1932, 1933, and 1935, the Brown and Pearce rabbit colonies suffered epidemics of a deadly smallpoxlike disease. This damaged the breeding experiments that Brown was conducting but provided Pearce and her colleagues, Paul Rosahn, Harry Greene, and C. K. Hu, with another phenomenon to investigate. This disease proved to be viral in nature and a good model for human smallpox. The group was able to characterize the epidemiology, immune reactions, and susceptibility to the disease.

Around the time that Pearce returned from China, the Rockefeller Institute had elected to move Pearce, Brown, and Greene to Princeton in an attempt to provide enough room for the rabbit colonies and to decrease costs (Corner 1964). Shortly after this move, Brown became seriously ill. It was recommended that he cut back on new experimentation, downsize his breeding experiments, and push to finish those projects already started. This resulted in Pearce's experimental ideas being delayed, as she contributed to his work. Brown died in 1942, leaving Pearce to finish his experiments and to publish the data.

The majority of the breeding experiments were concerned with inherited disorders of constitution. Brown was interested in disorders that would make an animal more susceptible to disease. The rabbit colonies provided several inherited syndromes, such as a ricketslike forepaw malformation, excessively dense bones, a complex disorder that included premature aging and death, and achondroplasia. In 1947 the Institute voted to tell Pearce to stop experimentation, granting her three years' full pay to write up the results, followed by half pay until retirement. She continued to write articles until her death; and her assistant, Margaret Dunham, finished and submitted the last two articles posthumously. The remaining data were destroyed.

In addition to doing her research, Louise Pearce served as trustee of the New York Infirmary for Women and Children (1921–28) and of the Princeton Hospital (1940–44). At the urging of Florence Sabin,* she also was a member of the Board of Corporators of the Woman's Medical College of Pennsylvania from 1941 until her death.

Pearce was appointed to the Royal Society of Tropical Medicine and Hygiene in 1924. She was also appointed a member of the scientific advisory council of the American Social Hygiene Association (1921–41). She was a member of the National Research Council (1931–33) and was granted membership in the Pathology Society of Great Britain and Ireland in 1932.

Apparently, Pearce was supportive of woman's rights, because she became President of the Woman's Medical College of Pennsylvania in 1946, with the intention of holding this position open for a woman (Morantz-Sanchez 1985; Glazer and Slater 1987). She presided over the college's centenary celebration in 1950 and then retired in 1951. She was also active in the American Association for University Women, serving as Director (1945–51) and as Committee Chairman for International Relations (1946–51). Pearce was a member of the Executive Board of the American Medical Women's Association (1935–36) and an honorary member of the editorial board for its journal. Pearce also served on the Committee to Award International Fellowships for the International Federation of University Women (1938–46) and was a second Vice President (1950–53).

NOTE

I would like to thank Joy Henig and Sara Wolf at the Annenberg Library of Pine Manor College for their generous assistance. I would also like to thank Lisa LaPolt of the Rockefeller Archives Center; Teresa Taylor of the Medical College of Pennsylvania, Archives; and Robert Wolf of the Countway Library of Harvard Medical School.

BIBLIOGRAPHY

Works by Louise Pearce

Scientific Works

Space does not permit the listing of the complete works of Louise Pearce. Included here are all references cited in the text. Eliminated here are those references by Pearce included in Fay (1961).

(with J. H. King and J. E. Bigelow) "Experimental obstructive jaundice." *Journal of Experimental Medicine* 14 (1911): 159–178.

(with W. H. Brown) "Multiple infections with *Treponema pallidum* in the rabbit." *Proceedings of the Society for Experimental Biology and Medicine* 18 (1921): 258.

(with T. M. Rivers) "Effect of a filterable virus (Virus III) on the growth and malignancy of a transplantable neoplasm of the rabbit.'' *Journal of Experimental Medicine* 46 (1927): 81–99.

The Treatment of Human Trypanosomiasis with Tryparsamide. A Critical Review. Monograph no. 23. New York: Rockefeller Institute for Medical Research, 1930.

(with P. D. Rosahn and C. K. Hu) ''Experimental transmission of rabbit-pox by a filterable virus.'' *Proceedings of the Society for Experimental Biology and Medicine* 30 (1932–33): 894–896.

''Experimental syphilis: Transmission to animals and the clinical reaction to infection.'' In *Syphilis*, edited by F. R. Moulton. Publication no. 6, 58–71. Washington, DC: American Association for the Advancement of Science, 1938.

(with A. E. Casey and P. D. Rosahn) ''The association of blood cell factors with the transplantability of the Brown-Pearce tumor.'' *Cancer Research* 2 (1942): 284–289.

Other Works

(with D. W. Weeks) ''The problem of atomic energy.'' *Journal of the American Association of University Women* 39 (Winter 1946): 81–82.

''Implications of the Marshall Plan.'' *Journal of the American Association of University Women* 41 (Fall 1947): 67.

Works about Louise Pearce

Baker, S. J. *Fighting for Life*. New York: Macmillan, 1939.

Baumann, F. ''Memorial to Louise Pearce, M.D. (1885–1959).'' *Journal of the American Medical Women's Association* (Aug. 1960): 793.

Baumgartner, L. ''Baker, Sara Josephine.'' In *Notable American Women, 1607–1950. A Biographical Dictionary*, vol. 1, edited by E. J. James et al., 85–86. Cambridge, MA: Belknap Press of Harvard University Press, 1971.

Bendiner, E. ''Louise Pearce: A 'magic bullet' for African sleeping sickness.'' *Hospital Practice* (Jan.15, 1992): 207–221.

Corner, G. W. *A History of the Rockefeller Institute, 1901–1953. Origins and Growth.* New York: Rockefeller Institute Press, 1964.

————. *Pearce, Louise*. Accession #8. The Medical College of Pennsylvania, Archives and Special Collections on Women in Medicine. Philadelphia: N.p., 1977. Manuscript.

''Dr. Louise Pearce, physician, 74, dies.'' *New York Times* (Aug. 11, 1959), 27.

Fay, M. ''Louise Pearce. 5th March 1885–9th August 1959.'' *Journal of Pathology and Bacteriology* 82 (1961): 542–551. Includes almost complete bibliography.

Fleming, A. ''Good citizen of the world. The story of Louise Pearce.'' *Doctors in Petticoats*, 86–99. New York: J. B. Lippincott, 1964. Written for a young audience.

Flexner, S. *Reports to the Board of Scientific Directors of the Rockefeller Institute for Medical Research of the Directors of the Laboratories, the Hospital and the Department of Animal Pathology*. Vol. 8, 6–7. Rockefeller University Archives, Rockefeller Archives Center, North Tarrytown, NY, 1920.

Fulton, J. D. ''Dr. Louise Pearce.'' *Nature* 184 (Aug. 22, 1959): 588–589.

Glazer, P. M., and M. Slater. *Unequal Colleagues. The Entrance of Women into the Professions, 1890–1940*, 138–141, plus notes. New Brunswick, NJ: Rutgers University Press, 1987.

"Louise Pearce, M.D. Medical woman of the month." *Journal of the American Medical Women's Association* (Dec. 1948): 523.

Morantz-Sanchez, R. M. *Sympathy and Science. Women Physicians in American Medicine*, 347–348. New York: Oxford University Press, 1985.

"Named president of college." *Medical Women's Journal* 53 (1946): 8.

Rivers, T. *Tom Rivers: Reflections on a Life in Medicine and Science*. In an oral history prepared by Saul Benison, 82–85. Cambridge: MIT Press, 1967.
    Regarding work with Pearce on virus and tumor.

"Woman aid to Flexner. Dr. Louise Pearce gets important position at Rockefeller Institute." *New York Times* (Aug. 28, 1913): 6.

Wylie, I.A.R. *My Life with George*. New York: Random House, 1940.

Other References

American Association of University Women Archives, Washington, DC.
    Biography, bibliography, and vita.

Brown, Wade H. Papers. Box 450, Series B815. Rockefeller University Archives, Rockefeller Archives Center, North Tarrytown, NY.
    Administrative correspondence.

Medical College of Pennsylvania, Archives and Special Collections on Women in Medicine, Philadelphia.

Pearce, Louise. Papers. Box 450, Series P315. Rockefeller University Archives, Rockefeller Archives Center, North Tarrytown, NY.
    Also available microfilm of Simon Flexner's papers.

ALMIRA HART LINCOLN PHELPS (1793–1884)

Robert M. Hendrick

BIOGRAPHY

In 1859 Almira Hart Lincoln Phelps became the second woman elected to membership in the American Association for the Advancement of Science (AAAS). She was one of the most influential and widely read female educators in nineteenth-century America. In her obituary it was estimated that her various textbooks, self-improvement guides, and didactic novels had sold over 1 million copies ("Almira Hart . . ." 1884). Her *Familiar Lectures on Botany*, first published in 1829, went through at least 23 editions by the time the last one appeared in 1872; an abridged version of this textbook, *Botany for Beginners*, went through 13 editions (1833–54). Together they sold about 650,000 copies by the time they finally went out of print in the 1890s, an astounding figure for that period. She totally dominated the botany textbook market in the second third of the century (Keeney 1992, 61) and can be said to have done more than any other individual to spark the botanizing craze that gripped Americans in the middle of the century.

Almira Hart was born on July 15, 1793, in Berlin, Connecticut. Both sides of her family traced their antecedents back to important public figures in the early colonial years. Thomas Hooker and Stephen Hart, founders of towns in Connecticut, were ancestors on her father's side of the family while on her mother's side, Robert Hinsdale was a leader in early Massachusetts. Her mother, Lydia (Hinsdale), was the second wife of Samuel Hart, a public official holding a variety of posts in the local government. She raised the seven children of his first marriage and had 10 children of her own. Almira Hart was the last of these 17 children to be born. Four of the 17 had already died before her birth, and her oldest half sibling was 33 years older than she. Because of these age differences, she grew up especially close to her sister Emma, the second youngest in the family but still six years her senior. As she matured, in many ways she would pattern her own life after that of her sister.

Politically and religiously Samuel and Lydia Hart were a dissident family, making their home a center of Jeffersonianism, which put them at odds with the prevailing Federalism of New England. Samuel Hart also broke with the local religious establishment, Congregationalism, an act of some daring in the New England of that period. The family's progressive thinking was further expressed in their belief that girls should be educated in the same academic subjects as boys. Contrary to the prevalent views of the period, they were convinced that formal education would not unbalance the female mind or cause brain damage. After being taught at home in her early years, in 1804 Almira Hart entered the Berlin district school system, where her first teacher was her sister, Emma, who had decided on a career in teaching.

In 1807, although she was still only 20 years old, Emma Hart was hired to head a female academy in Middlebury, Vermont. Two years later she married a local physician, John Willard. Since their home was next to Middlebury College, they took in college students as boarders. In 1810 Emma Willard invited her sister to come live with her in order to be in a position to further her education. This "higher education" consisted of the sisters' discussing with the boarders the lectures the young men had attended at the College and borrowing their textbooks for their own study. These activities were fairly intense; after all, Hart was only a year or two younger than the male college students who were sharing their education with her. Although she subsequently spent a year at the Female Academy of Pittsfield, Massachusetts (1812–13), she apparently learned very little there because the curriculum focused on teaching the "accomplishments" young middle-class women were supposed to master: music, drawing, and so on. These were subjects that both Hart and Willard would later on play down in their own schools. The two years Almira Hart lived with her sister in Middlebury really constituted her "college education," consisting of borrowed textbooks, no laboratory training, and lectures filtered through undergraduate notes and discussions.

In 1814 Hart returned to Berlin and tried to run a boarding school out of her family's home. This effort failed, and in 1815 she returned to Middlebury, where her sister had established a successful school for females. Late that same year, Hart was hired to be the principal of the Sandy Hill Academy in Sandy Hill, New York. Here she began to institute reforms in female education that she and her sister agreed were necessary. These reforms consisted mainly in replacing emphasis on the "accomplishments" with solid academic subjects, particularly the sciences.

She left Sandy Hill less than two years later because of her marriage on October 4, 1817, to Simeon Lincoln, a newspaper editor and owner of the Federalist *Connecticut Mirror*. The couple lived first in Hartford, Connecticut, and then in nearby New Britain. They had three children: James Hart (b. April 1820 but died less than a year later), Jane Porter (b. 1821), and Emma Willard (b. 1822). Now a housewife, Almira Lincoln was apparently out of education for good. However, in October 1823, Simeon Lincoln died of yellow fever, leaving her a 30-year-old widow with two infant daughters and some debts. She had little choice but to return to teaching.

She was hired by her sister, Emma Willard, who by now had established a successful female school, the Troy Seminary, in Troy, New York. Willard invited her younger sister to teach at the school in 1824; in 1829 she made Lincoln the vice principal, and in October 1830 Willard left her in full charge, while she traveled through Europe for a year and a half. Almira Lincoln's association with the Troy Seminary completely changed her life. It allowed her to develop her ideas on female education, which she first presented in weekly lectures to her students during her sister's European grand tour and which were subsequently published as *Lectures to Young Ladies* (1833) and in revised versions as *The Female Student* (1836) and *The Fireside Friend* (1840).

More important, living in Troy allowed Lincoln to meet Amos Eaton, the leading scientific teacher at Rensselaer Institute. Eaton had an impact on Lincoln in a number of ways. He encouraged her to teach science to the girls at Troy Seminary. He was convinced that the female mind could master scientific subjects, and he often opened his lectures and laboratories to women on an informal basis. He encouraged Lincoln to study botany, and at his urging she began the lecture notes that became *Familiar Lectures on Botany* (1829), a textbook so successful that it almost drove Eaton's own *A Manual of Botany for the Northern States* (1817) out of the market. Most important, he provided Lincoln with the basic ideas that she incorporated into her botanical books.

Her teaching career at Troy ended in August 1831, when she married John Phelps, the father of one of the school's pupils. Phelps, a lawyer and politician from Vermont, already had six children, the youngest two of whom were raised by his new wife. The couple moved to Guilford, Vermont, and then to Brattleboro (1836). Once again a housewife, Almira Phelps became the mother of two more children, Charles Edward (1833) and Almira Lincoln (1836). Most of her free time in Vermont was spent in writing an impressively large series of science and educationally oriented books: *The Child's Geology* (1832), *Botany for Beginners* (1833), *Caroline Westerley*, a novel (1833), *Lectures to Young Ladies* (1833), *Chemistry for Beginners* (1834), *The Female Student* (1836), *Familiar Lectures on Natural Philosophy* (1837), *Natural Philosophy for Beginners* (1838), and *Familiar Lectures on Chemistry* (1838), as well as constant revisions of her most successful book, *Familiar Lectures on Botany*.

John Phelps's political and business careers took a severe downturn, largely because of his vocal antiabolitionist stance. He argued that slavery was protected by the Constitution and hence could not be abolished, a position also adopted by Almira Phelps and one that was extremely unpopular in Vermont in the late 1830s. Consequently, he urged his wife to accept the leadership of a new female seminary being established in West Chester, Pennsylvania. They left Vermont in October 1838, and until his death in April 1849, John Phelps served as a business manager for her at the various schools she headed. He encouraged his wife to exploit the success of her books, the ones on botany being the most remunerative. In fact, the success of her chemistry and natural philosophy efforts was mainly due to the reputation of the botanies. As late as 1900, her son was still receiving royalties from the sale of the botany texts (Bolzau 1936, 257, 264).

Almira Phelps's tenure at the West Chester Young Ladies' Seminary was extremely brief; the school went bankrupt in its first year of operation and was sold in November 1839. The new owner converted it to a school for boys. She then became the academic head of the Female Institute in Rahway, New Jersey. She reformed its curriculum by introducing more academic (especially scientific) courses and by stressing moral and religious instruction. The school soon had a national reputation, and within two years (November 1841) she was hired for the position that would, along with her writings, make her a national figure. The bishop of the Episcopal diocese of Maryland and the trustees of the Patapsco Female Institute in Ellicott's Mills (about ten miles outside Baltimore) hired John and Almira Phelps to convert that school to one "for Christian education." Her husband was to run the business end of the school, while Almira Phelps had full control over academic matters. However, John Phelps became paralyzed for the two years before his death in April 1849, so that, from early 1847 on, Almira Phelps had total control over the school.

She made it a remarkable success; from only 40 students when she first took over, by 1844 the school annually enrolled well over 100 girls, some from as far away as Canada. Most of its students, however, were from middle-class families in the southern states. In a brief autobiographical sketch, Almira Phelps referred to Patapsco as a "collegiate institution for the daughters of the South" (*Hours . . .* 1859, xxii). It was virtually the only progressive educational institution in the country that was not antislavery in outlook. Almira Phelps made no secret of her sentiments in this regard. In the second volume of her novel *Ida Norman* (written in 1854), she admonished abolitionists, "Let those who are expending their sympathies upon southern slaves . . . [the] pretended philanthropist who would benefit his country by throwing into her midst the torch of disunion, step aside from the log cabins of the negro, where resound the sounds of merriment coming from light hearts that feel no cares" (1855, 413). The reliance on southern students meant that the school would be closed with the outbreak of the Civil War in 1861, because most of the girls were immediately brought home by their worried families. By then, however, Phelps was connected to the school only in an advisory capacity. Already a widow for the second time, Almira Phelps resigned her position in August 1856, following the death of her daughter and assistant, Jane Lincoln, in a railroad accident. She moved to Baltimore, where she resided for the rest of her long life.

But in the 15 years she was in control of the Patapsco Female Institute, she made it a model school for young women that rivaled her sister's in Troy in terms of its national reputation. Her stress on mathematics and the natural sciences, her rejection of rote memory teaching techniques, and her reliance on the innovations of Johann Heinrich Pestalozzi made Patapsco one of the few progressive female schools in the country. Her goal was not so much to impart specific knowledge but rather to teach young women how to think. Her emphasis on high ethical and moral behavior, along with her deep religious faith, also gained her school favorable reviews in the popular press. Of equal importance, Almira Phelps trained teachers at Patapsco, sending out its graduates to all parts of the country to spread her gospel of the value of science courses for young women. Most important, in the words of her biographer, "Patapsco alumnae demonstrated the value of higher education for women, which had to be proved before women's colleges could be established successfully" (Bolzau 1936, 202).

The three decades of her retirement were very comfortable; according to her obituary in the *New York Times*, she had made over $1 million—a most impressive figure for that period—primarily through the sale of her various books ("Almira Hart . . ." 1884, 5). Short and obese, she had an elevator installed in her Baltimore home, which contemporaries referred to as "stately." In these later years, she remained active intellectually. In the words of the editors of a popular journal, the *National Quarterly Review*: "There is no female educational institution in America in which her text-books are not known, and there are few of the best similar European institutions to which their fame has not extended" ("Woman's duties . . ." 1874, 29–30).

With the outbreak of the Civil War, Phelps dropped her former Southern sympathies and became a strong Unionist supporter. Her son, Charles (who would later become a general, congressman, and judge on the Maryland Supreme Court), served as a colonel in the Seventh Maryland Infantry of the Union army and suffered severe wounds. In 1864 she edited a large collection of original essays and poems, *Our Country, in Its Relations to the Past, Present and Future*, as a fund-raising venture for the Maryland branch of the U.S. Sanitary Commission.

After the Civil War she turned out numerous articles on such wide-ranging topics as a history of the fine arts, the work of Johann Goethe, the life and times of Françoise d'Aubigné (Madame de Maintenon), England under the Stuarts, controversies in botany, the Cuban revolt against Spain, and Emma Willard's theory of circulation by respiration. But the topic that most kept her in the public eye in her later years was her active campaign against female suffrage.

Phelps had always held strongly conservative views on this subject. In her lectures to her students, she had insisted that the value of "female education should not be to lead woman from her own proper sphere, but to qualify her for the better discharge of those duties which lie within it" (*The Fireside Friend* 1840, 224). The value of teaching science and mathematics to young women was not to prepare them to be scientists or mathematicians. It was to provide them with the intellectual tools that would convince them to accept the role the Bible and social convention provided for them: to be dutiful children and wives and to be wise and loving mothers. Unless forced to do so by economic exigencies, she told her students not to seek careers: "The sphere of woman's duty is to be looked for in private and domestic life" (*The Female Student* 1836,

394). In her view, women should certainly not seek the vote. Those women who refuse to accept this situation, who want to leave "the sacred hearth, the domestic altar" were expressing "unnatural and depraved ambition"; they "deserve no longer to be recognized as women; they are monsters" (*The Female Student* 1836, 48).

Phelps continued to develop these themes in her later essays. In "Social Life in America," after criticizing American women's love of display, she urged women to be content to remain in the home, where their "influence is all-potent" (*Reviews and Essays* . . . 1873, 113, 115). In an essay on her sister's theory of circulation read to the AAAS in 1870, Phelps worked in an attack on the defenders of woman's rights as those advocating an overturn of God's providential order (*Reviews and Essays* . . . 1873, 320). In another essay Phelps used the Adam and Eve story as a lesson "full of instruction for woman." She argued that woman was created as a companion of man to cheer and assist him, and "a companion or assistant is in a secondary position." Woman's inherent weaknesses were symbolized by Eve's succumbing to temptation, while her ability to lead Adam astray represented woman's influence over man. Women must "submit to what God made us" and not seek the vote ("Woman's duties . . ." 1874, 32–33).

Even though she was in her late 70s, Phelps was a major organizing force in the Anti-Suffrage Committee of Washington, which functioned in 1870–71. Its purpose was to circulate a petition among prominent women in the various states. Phelps was very visible in this effort, serving as the Committee's corresponding secretary. After thousands of women's signatures were collected, the petition was submitted to Congress, asking that "no law extending suffrage to women may be passed." She also opposed female involvement in the "liquor crusade" because it would "intoxicate women with notoriety" and cause "the gross violation of womanly propriety and modesty" ("Woman's duties . . ." 1874, 51).

Not surprisingly, given the conservative tendencies that dominated her later writings, Phelps gave no public support for the movement to establish women's colleges. Certainly, these conservative tendencies, while gaining her public approval during her lifetime, were largely responsible for the posthumous decline of her reputation. Also, her continued adherence to Carolus Linnaeus's system of classification in botany and her anti-Darwinian position made her appear intellectually outmoded by the time of her death, on her birthday, July 15, 1884. She was buried in Baltimore.

WORK

In evaluating the significance of Almira Phelps's contributions to the biological sciences, one fact is obvious. In a very basic sense she was not a botanist at all, because she did not make a single addition to the body of botanical knowledge. All of her work, in botany as well as in chemistry and physics, was based on heavy borrowing from the work of others, even to the extent of using illustrations from other texts.

In her "Essay on Popular Science," which was read to the AAAS at its August 1869 meeting, she freely acknowledged this. She spoke of a "division of labor" in science between those who do science and those who "popularize science." She included herself in the latter category when she described herself as "an humble laborer in the field of education, with no claims to originality or discoveries" (*Reviews and Essays . . .* 1873, 284–285). She was, before all else, a popularizer of botany, not a botanist.

Yet her two botany texts were important. They helped introduce the science to several generations of Americans, including many who went on to become distinguished in that field. Even one of her opponents in the women's movement took time in her memoirs to remark that Phelps's "*Botany* was one of my most cherished companions on a Wisconsin farm" (Willard 1889, 562). Thousands of young Americans must have grown up with similar feelings.

Familiar Lectures on Botany and *Botany for Beginners* were unique in that period for several reasons. First of all, Phelps, as she did in all of her textbooks, applied the hands-on methods she had learned from Amos Eaton. Rather than have students learn botany solely from drawings in books, Phelps insisted that students have flowers or other plant forms at their desks for any botany lesson. As she stated, "In order to learn any part of Natural History, the student must *see much*, and exercise himself that he may *see clearly*" (*Familiar Lectures . . .* 1836, Appendix, 1). She also made certain that her botanies contained scores of illustrations. Unlike her competitors' books, hers were well illustrated with woodcuts that were carefully integrated with her text.

She extolled the study of botany because it introduced the student to the concepts of mental order and discipline: "The very logical systematic arrangement which prevails in Botanical science, has, without doubt, a tendency to induce in the mind the habit and love of order; which, when once established, will operate even in the minutest concerns." Botany, she argued, was especially beneficial to the female mind: "The study of Botany seems peculiarly adapted to females: the objects of its investigation are beautiful and delicate; its pursuits, leading to exercise in the open air, are conducive to health and cheerfulness" (*Familiar Lectures . . .* 1836, 14).

Most important, at least as far as the popular success of her botanies is concerned, Phelps insisted that "the study of Botany naturally leads to greater love and reverence for the Deity" (*Familiar Lectures . . .* 1836, 15). As she so often did in her books, she wrote a poem to make her point: "None can the life of plant or insect give / Save God alone / Each op'ning bud, and care perfected seed / Is as a page where we may read of God" (*Botany for Beginners* 1836, 12). Nor, she sternly lectured her readers, should they ignore the "presents" God has given them in the works of nature: "My dear children, it is your duty, as it should be your pleasure, to search into the wonders of created nature . . .

and to animate your pious feelings in thinking much upon the works of God''
(*Botany for Beginners* 1836, 22).

What made her botanies successful, then, were not so much her innovations
of copious illustrations and the urging of direct student involvement in the lessons. Rather, her introduction of a conservative ideology sprinkled throughout
her texts won her rave reviews in the popular press. She convinced her buyers
that botany would make children, especially females, more serious, more logical,
and more religious. These qualities attracted parents, teachers, and school boards
to her texts.

As far as the botanical content of her work was concerned, Phelps was a firm
adherent of the Linnaean, or ''artificial,'' system of classification. In her essay
on ''Popular Botany,'' she admitted that this created ''so many anomalies as
often to perplex the student,'' anomalies that the ''natural'' systems avoided.
Yet even as these newer systems were adopted by botanists, she insisted on
using the older system, because its ''beautiful simplicity'' made it ''the most
agreeable entrance into the temple of botanical science'' for the beginning student (*Reviews and Essays* . . . 1873, 247).

In fact, in her last essays, there is a sense that Phelps was very aware that
developments in botany had passed her by. She complained that the study of
botany had become too difficult for laypeople because of the emphasis on laboratory work: botany was in danger of becoming little understood outside the
sphere of learned professors. There was no longer any room for a self-taught
amateur like herself.

Nor did she accept the sweeping success of Darwin's work. Given her strong
religious beliefs and her adherence to the Linnaean system of classification, this
is not surprising. She attacked the ''modern infidelity'' of the belief that the
human race developed gradually, ''by fortuitous circumstances, from reptile or
monkeys, to the dignity of rational beings'' (''Woman's duties . . .'' 1874, 31).

Almira Hart Lincoln Phelps personified the development of botany in nineteenth-century America. Her emphasis on the intellectual and religious value of
the science, along with her clearly written texts, did more than any other single
contribution to make botany the people's science at midcentury. But like the
thousands who joined the botanizing clubs that were sprinkled across the country, Phelps was a self-taught amateur. Like them, she did not have the training
or the inclination to move the focus of her studies into the laboratory and under
a microscope. The natural systems, allied as they came to be with Darwinian
thought, were an anathema to her. And so the science passed her by.

BIBLIOGRAPHY

Works by Almira Hart Lincoln Phelps

Scientific Works

Familiar Lectures on Botany. Hartford: F. J. Huntington, 1829.
     Multiple editions and printings until 1872.

Dictionary of Chemistry. New York: Carvill, 1830.
    A translation of *Le Dictionnaire de Chimie* of Louis Vauquelin.
The Child's Geology. Brattleboro, VT: G. H. Peck, 1832.
Botany for Beginners. Hartford: F. J. Huntington, 1833.
    Multiple editions until 1854.
Chemistry for Beginners. Hartford: F. J. Huntington, 1834.
    Multiple editions until 1867.
Familiar Lectures on Natural Philosophy. New York: F. J. Huntington, 1837.
    Multiple editions to 1866.
Familiar Lectures on Chemistry. New York: F. J. Huntington, 1838.
Natural Philosophy for Beginners. New York: F. J. Huntington, 1838.
    Multiple editions until 1848.
"Essay on popular science." In *Reviews and Essays on Art, Literature, and Science*, by
    Almira H. L. Phelps. Philadelphia: Claxton, Remsen, and Haffelfinger, 1873.
"History and defence of Emma Willard's theory of circulation by respiration." In *Reviews and Essays on Art, Literature, and Science*, by Almira H. L. Phelps. Philadelphia: Claxton, Remsen, and Haffelfinger, 1873.
"Popular botany." In *Reviews and Essays on Art, Literature, and Science*, by Almira H. L. Phelps. Philadelphia: Claxton, Remsen, and Haffelfinger, 1873.

Other Works

Address on the Subject of Female Education in Greece. Troy, NY: Tuttle, 1833.
Caroline Westerley; or, The Young Traveller from Ohio. New York: Harper, 1833.
Lectures to Young Ladies, Comprising Outlines and Applications of the Different Branches of Female Education. Boston: Carter, Hendee, 1833.
"Observations upon an infant during its first year." An appendix to Madame Necker de Saussure's *Progressive Education.* Boston: Ticknor, 1835.
The Female Student; or, Lectures to Young Ladies on Female Education. New York: Leavitt, Lord, 1836.
"The influence of women on society." *American Ladies Magazine* 9 (1836): 558–563.
"Remarks on the education of girls." *Godey's Lady's Book* 18 (1839): 253–255.
The Fireside Friend or Female Student: Being Advice to Young Ladies on the Important Subject of Education. Boston: Marsh, Capen, Lyon, and Webb, 1840.
Ida Norman; or, Trials and Their Uses. New York: Sheldon, Lamport, and Blakeman, vol. 1, 1848; vol. 2, 1855.
Christian Households. New York: D. Dana, 1858.
Hours with My Pupils: or, Educational Addresses, etc. The Young Lady's Guide and Parent's and Teacher's Assistant. New York: Scribner, 1859.
(editor) *Our Country, in Its Relations to the Past, Present and Future.* Baltimore: J. D. Toy, 1864.
The Educator. New York: A. S. Barnes, 1868.
A Plea for Cuba; Addressed to the Senate and House of Representatives of the United States. Baltimore: n.p., 1870.
The Blue Ribbon Society; or, The School-Girls' Rebellion. Philadelphia: Charles Harvey, 1872.
Reviews and Essays on Art, Literature, and Science. Philadelphia: Claxton, Remsen, and Haffelfinger, 1873.
"Woman's duties and rights." *National Quarterly Review* 29 (1874): 29–54.

Works about Almira Hart Lincoln Phelps

"Almira Hart Lincoln Phelps." *New York Times* (July 16, 1884): 5.
Barnard, H. "Mrs. Almira Lincoln Phelps." *American Journal of Education* 17 (1868): 611–622.
Bolzau, E. L. *Almira Hart Lincoln Phelps: Her Life and Work.* Philadelphia: University of Pennsylvania, 1936.
    Dissertation.
Weeks, M. E., and F. B. Dains. "Mrs. A. H. Lincoln Phelps and her services to chemical education." *Journal of Chemical Education* 14 (1937): 53–57.

Other References

Eaton, A. *A Manual of Botany for the Northern States.* Albany, NY: Webster and Skinner, 1817.
Keeney, E. B. *The Botanizers: Amateur Scientists in Nineteenth-Century America.* Chapel Hill: University of North Carolina Press, 1992.
Willard, F. E. *Glimpses of Fifty Years: The Autobiography of an American Woman.* Chicago: H. J. Smith, 1889.

ELIZABETH (ELIZA) LUCAS PINCKNEY (1722–1793)

Virginia Pezalla

BIOGRAPHY

Elizabeth (Eliza) Lucas was born in the West Indies in December 1722. She was born into an upper-class English family that owned estates in both the West Indies and colonial South Carolina. Her father, George Lucas, was a lieutenant colonel in the British army and held a number of political posts in the West Indies, including that of Governor of the island of Antigua. Little is known about her mother, whose name was probably Ann Lucas (Pinckney 1972, 103). She and George Lucas had four children: Elizabeth (called Eliza), George, Thomas, and Mary (called Polly). In 1737, when Eliza Lucas was 16, her mother became ill, and George Lucas, hoping that a change of climate would improve his wife's health, moved the family to the plantation they owned on Wappoo Creek near Charleston, South Carolina. They had been there less than a year when war broke out between England and Spain, and George Lucas was ordered back to his military post in Antigua. He never returned to America and left the management of his plantation at Wappoo, as well as two other plantations, to his young daughter, Eliza Lucas.

Managing three plantations was an enormous responsibility for a 16-year-old girl. Fortunately, she found time to make copies in a letterbook of the numerous letters she wrote to her family and friends. Her letters reflect a lively interest in literature, music, and science. Like other children of her social class, she had been sent to England for schooling. The exact length of time she spent in England is not known, but it seems to have been a long period (Ravenel 1896, 3). At Wappoo she continued her education on her own, using books borrowed from neighbors and from her father's library. "I have," she wrote, "a little library well furnished (for my papa has left me most of his books) in which I spend part of my time" (Pinckney 1972, 7). Her letters contain frequent references to both classical writers, such as John Milton, Plutarch, and Virgil, and contemporary ones, such as the philosopher John Locke and the novelist Samuel

Richardson. She even read law books so that she could help her neighbors draft wills (Pinckney 1972, 41).

Music was also an important part of her life. She took weekly harpsichord lessons and was able to "tumble over one little tune" on the flute (Pinckney 1972, 63). Eliza Lucas also found time to tutor her young sister, Polly, and two slave girls. Her plan, an unusual one for the time, was to train these slave children to be teachers to other slaves: "I intend them for school mistres's for the rest of the Negro children" (Pinckney 1972, 34). She was equally interested in science, and in 1741 she described having seen a comet that she believed was one predicted by Isaac Newton (Pinckney 1972, 29). Her main scientific interest was agriculture. "I love the vegitable world extremly" she wrote to a friend (Pinckney 1972, 35). This interest seems to have come from her father, who was very unconventional. He actively encouraged his daughter's interest in agriculture and had, in her words, "an aversion" to the usual pursuits of young ladies of that time, such as embroidery (Pinckney 1972, 35). George Lucas supplied Eliza with the seeds she used in many of her early experiments.

Eliza Lucas found the management of the plantations a challenging task. She wrote to a friend: "I have the business of 3 plantations to transact, which requires much writing and more business and fatigue of other sorts than you can imagine" (Pinckney 1972, 7). She had little time for social life and was remarkably independent. Arranged marriages were common at this time, and, in 1740, her father sent her the names of two men he had selected as suitable husbands. In her return letter she replied that "the riches of Peru and Chili" would not be enough to get her to marry one of them and that "a single life is my only choice" (Pinckney 1972, 6). Not until four years later did she decide to marry a man of her own choice, Colonel Charles Pinckney, a widower who lived on a neighboring plantation. Since her first days at Wappoo, she had been friends with Pinckney and his first wife. She frequently visited them and borrowed books from Charles Pinckney. Mrs. Pinckney died in January 1744 after a long illness, and Charles Pinckney and Eliza Lucas were married in May of that year. Charles Pinckney was 25 years older, but it was a happy marriage. He was a lawyer and a prominent politician whose accomplishments included being a member of the Royal Provincial Council, chief justice of the colonial Supreme Court, and speaker of the South Carolina Assembly.

Elizabeth and Charles Pinckney had four children: Charles Cotesworth (b. 1746), George Lucas (b. and d. 1747), Harriott (b. 1748), and Thomas (b. 1750). Eliza Pinckney was a devoted mother and was very interested in the latest ideas about raising children. She sent to England for a new educational toy to help her baby, Charles, then four months old, "play himself into learning," and she had plans to teach him his letters before he could speak (Ravenel 1896, 113). During these busy years with young children, Eliza Pinckney was still able to find time to continue her interest in agriculture and experimented with growing hemp and flax and raising silkworms. In 1753 the family moved to London, where Charles Pinckney had been appointed colonial commissioner. The five years they spent in England were happy ones for the Pinckneys. They had an active social and cultural life. In 1758, however, the Pinckneys, fearing the outbreak of another war between France and England and worried about the safety of their plantations, returned to America. Unfortunately, soon after their return, Charles Pinckney contracted malaria and died on July 12, 1758. His death began a difficult time for Eliza Pinckney. She deeply missed him, "the most amiable, tender, and affectionate of husbands" (Pinckney 1972, 102). Her loneliness was increased by her separation from her two young sons, who had stayed in England to continue their education. "My heart bleeds at our separation," she wrote in 1758, "and I long more than I can express to be with them" (Pinckney 1972, 140). She did not see them again until they returned to America in their early 20s. However, she remained close to them through frequent letters that were filled with advice and affection.

After her husband's death, Pinckney, now 36 years old, took over the management of the family's seven plantations. She welcomed the distraction from her grief and loneliness: "A variety of imployment gives my thought a relief from melloncholy subjects" (Pinckney 1972, 144). During this time she wrote that her three pleasures were her garden, her books, and the education of her daughter, which she called "one of the greatest Businesses of my life" (Pinckney 1972, 81). She also found comfort in her strong faith in Christianity and was a member of the Episcopal Church.

Pinckney's busy and prosperous life as a plantation manager continued for the next 20 years until it was disrupted by the American Revolution. She chose to support the rebel cause despite her strong ties with England, and she paid heavily for this decision. When the British troops invaded South Carolina in 1779, they burned her houses, stole her possessions, and ruined her crops. Her sons were officers in the American army. Both were taken prisoner by the British, and one of them, Thomas, was badly wounded. By 1782 Pinckney's wealth was destroyed, and she wrote to a friend that she was unable to pay off even a small debt (Ravenel 1896, 301).

After the war, Pinckney went to live on the plantation of her widowed daughter, Harriott Horry. Her last years were happy ones. She helped raise her four granddaughters and took great pride in the accomplishments of her two sons, who became distinguished cultural and political leaders. In one of her last letters she wrote: "I now see my children grown up and, blessed be God! see them such as I hoped" (Ravenel 1896, 814). Pinckney remained a prominent figure in Southern society, and when President George Washington came to South Carolina in 1791, he stopped to visit her. In 1793, at the age of 71, Elizabeth Pinckney became ill with cancer. She died that May in Philadelphia, where she had gone for medical treatment. She was buried there in St. Peter's churchyard, with George Washington serving as one of the pallbearers at her funeral. His presence was a tribute to the influence and popularity of this remarkable woman.

WORK

Eliza Pinckney is most famous for introducing indigo as a cash crop into colonial South Carolina. Soon after taking over management of her father's plantations, she began experimenting with growing different types of crops. She left a record of these experiments in her letterbook, a book in which, as was a common practice of eighteenth-century businessmen, she made copies of her frequent letters to family and friends. Fortunately, this letterbook has been preserved and was published by the University of North Carolina (Pinckney 1972). Other letters, as well as family memories of Pinckney, have been published in a biography by her great-granddaughter, Harriet Ravenel (1896).

When Pinckney began her experiments, the main crop in the South was rice. Profits from this crop were low, however, due to wars with France and Spain. Hoping to find a new cash crop, she established large experimental gardens at her Wappoo plantation. In 1740 she wrote to her father about "the pains I had taken to bring Indigo, Ginger, Cotton and Lucerne and Casada to perfection" (Pinckney 1972, 8). Her father was very important to the success of these experiments, as he supplied her with seeds, including indigo, from the West Indies. Indigo, *Indigofera tinctoria*, is native to India and Africa and was highly valued for the beautiful blue dye that could be made from the leaves. It was the main dye used in blue fabrics. At the time she began growing indigo, most of the dye imported by England came from islands in the French West Indies, and England was eager to find a different source. "We please ourselves," she wrote, "with the prospect of exporting in a few years a good quantity from hence, and supplying our Mother Country with a manifature for which she has so great a demand, and which she is now supplyd from the French Collonys" (Ravenel 1896, 105).

First, however, it was necessary to learn how to grow indigo in a new climate. She had to discover, by experimentation, what type of soil to use and when to plant and harvest the indigo. She also selected plants that grew well by saving seed from the survivors of each year's crop. The first crop in 1740 failed because it was planted too late, and "the frost took it before it was dry" (Pinckney 1972, 16). Seed saved from this crop produced only a small yield the next year. In 1742 she again planted seed from her previous crop as well as some from the West Indies. The latter, however, did not grow, and she wrote to her father that "the last Indigo seed sent was not good. None of it came up. We shall save enough of our own to make a crop next year" (Pinckney 1972, 56). The next year's crop was also small, but, finally, in 1744, she writes of "a very good crop" (Ravenel 1896, 104). She generously distributed most of the seed from this crop to any neighbors who would agree to try it, and soon many planters in South Carolina were growing indigo.

Pinckney faced additional difficulties in learning how to make dye from the indigo leaves. This was a long and complicated process that involved harvesting the leaves at a certain age and fermenting them in large vats to extract their blue color. The liquid was then drained off, beaten until thick, and, after the addition of lime, allowed to sit until the dye particles settled out on the bottom. This residue was made into cakes that were dried and taken to market. An experienced French dyemaker, Nicholas Cromwell, had been hired to help with this process, but instead he tried to sabotage it to protect the dye industry of France. The dye he made was worthless, and he claimed that the climate was not suitable for growing indigo. She, however, was not easily fooled and, "by close watching and careful experimentation" (Ravenel 1896, 103), determined that Cromwell himself was to blame for the failure. Another dyemaker had to be found before she was able to produce, in 1744, dye of high enough quality to sell. The English government encouraged indigo planting by giving the colonists a bounty for every pound produced, and the export of indigo from the Southern colonies rapidly increased to over 1 million pounds a year (Ravenel 1896, 107). Indigo remained the basis of the Southern economy for the next 30 years until it was replaced by cotton after the American Revolution.

While Eliza Pinckney is most famous for the introduction of indigo into America, she also experimented with other plants. In 1745 she supervised the planting of two new crops, flax and hemp, on her father's plantation. It was hoped that these two crops could be used to make fabric for export. Her father again supplied her with the seeds and sent two indentured servants from Ireland to help with the spinning and weaving of the flax and hemp into cloth. These two crops turned out to be less profitable than cotton, and their cultivation soon stopped. The Irish weaving technique, however, was used on plantations for many years to weave cotton into cloth (Ravenel 1896, 130).

Pinckney also experimented with the culture of silkworms. She planted many mulberry trees and was very successful at raising the worms. She had three dresses woven from the silk she collected, and one of them can still be seen on display at the Smithsonian Institution in Washington, D.C. As with flax and hemp, however, the culture of silk was not profitable enough to be expanded into a larger industry, and the idea was eventually dropped.

References to many other agricultural experiments are found in her early letters. In 1742 she planted a fig orchard, "with design to dry and export them" (Pinckney 1972, 35). In another letter, she wrote of planting a cedar grove that, "shall be filled with all kind of flowers" (Pinckney 1972, 36), and soon after that, she wrote of planting a plantation of oak trees for future shipbuilding (Pinckney 1972, 38). With the assistance of Dr. Alexander Garden, a Carolinian botanist after whom the gardenia is named, she also planted many foreign species of trees, "trying which she could acclimatize" (Ravenel 1896, 102). Years later, in 1762, she wrote about sending American plants, sweet myrtle and palmetto, and seeds from a wild flowering shrub to friends in England (Ravenel 1896, 217). Most of Eliza Pinckney's experiments, however, occurred between the years of 1738, when she first arrived in America, and 1753, when she and her family moved to England for five years. Eliza Pinckney's letterbook ends abruptly in 1762, and she left no further accounts of agricultural experiments.

BIBLIOGRAPHY

Works by Elizabeth (Eliza) Lucas Pinckney

Scientific Works

The Letterbook of Eliza Lucas Pinckney, 1739–1762. Edited by Elise Pinckney. Chapel
Hill: University of North Carolina Press, 1972.

Works about Elizabeth (Eliza) Lucas Pinckney

Booth, Sally S. *The Women of '76.* New York: Hastings House, 1973.
Dexter, Elisabeth A. *Colonial Women of Affairs before 1776.* Boston: Houghton Mifflin,
1924.
Lee, Susan. *Eliza Pinckney.* Chicago: Children's Press, 1977.
Ravenel, Harriet H. *Eliza Pinckney.* New York: Charles Scribner's Sons, 1896. Reprint.
Spartanburg: Reprint, 1967.
Williams, Frances L. *Plantation Patriot.* New York: Harcourt, Brace, and World, 1967.

BEATRIX POTTER (1866–1943)

EdnaMay Duffy

BIOGRAPHY

Helen Beatrix Potter was born July 28, 1866, in Bolton Gardens (Kensington), London, where she lived for nearly 50 years with her parents, Rupert and Helen (Leech) Potter. She had one brother, Walter Bertram.

The Potters were a characteristic Victorian family of the moneyed middle class. Both of her parents were from wealthy families whose fortunes originated in the calico print and cotton industries of northern England. Beatrix's father was a lawyer who lived a life of leisure. He became a member of the Photographic Society of London and spent his days photographing landscapes and posed portraits of noteworthy contemporaries. Her mother had been an accomplished watercolor painter in her youth. Marriage afforded her a lifestyle of social calls, running a Victorian household, and perfecting embroidery skills.

Beatrix grew up shy and sickly, in reserved solitude for six years until her brother, Bertram, was born. Together they spent their days in the third-floor nursery, isolated from their parents. Her cloistered existence in Bolton Gardens was relieved by visits to Camfield Place, the country home of her paternal grandparents, spring holidays to the sea during the annual housecleaning, and summers in Scotland or the Lake District.

Victorian children were seen and not heard. Beatrix sharpened her powers of observation on the natural souvenirs she brought back to the nursery from her holiday travels. By the age of nine she had copious notes on her caterpillar observations and drawings to accompany them, along with a collection of toadstools, beetles, dead frogs, and minnows. In the country she drew everything, from the game brought home from a day of shooting, to the buttercups in the meadows. She caught rabbits, tamed them, drew them, and brought them back to London. She and Bertram skinned dead rabbits, boiled them until only bones were left, and then preserved, studied, and drew the skeletons.

Bertram was sent away to school, and Beatrix received a minimal education at home through her governess, as was typical for women of her times and status. Her studies included French, German, and needlework. From age 12 to age 17, she studied with a Miss Cameron, and in July 1881 she completed her only formal education qualification, the Art Student's Certification, second grade.

As she matured, her shyness increased. Her only companions were her menagerie and her diary. At the age of 14, she invented an elaborate secret code alphabet that she used in her journal until she was 30. Like her brother, she pursued painting. But while Bertram's canvases were consumed with the largeness of moors, sunsets, and misty glens, Beatrix painted the precise and the minute: the eye of a squirrel, the moss under a microscope. Her pet rabbit, Benjamin Bouncer, which she drew from every angle and toted everywhere in a basket, was the precursor of the infamous Peter Rabbit.

By the age of 19, Beatrix was permitted to visit the British Museum of Natural History, which was within walking distance of her home. Here she had free rein to draw, often sketching from the fossil cases. She then used her brother's microscope to draw mold spores, spider leg hairs, and butterfly scales collected on her country trips, which were still part of her life.

In her mid-20s her attention turned to the wild mushrooms and fungi found while on the family's holidays in Scotland. She continued to collect, study, and dissect her fungi specimens, amassing a collection of nearly 300 fungi drawings. She recorded her observations on their sporulation in her coded journal. In 1896, encouraged by her uncle, Sir Henry Roscoe, a chemist, Potter attempted to study at the Royal Botanic Gardens at Kew and to share her research on the germination of spores with the Linnaean Society of London.

As for her portfolios of fungi drawings, the Keeper of Botany at the Museum of Natural History informed her that her paintings lacked sufficient diagrammatic extension or detail to be scientifically useful. She made one more attempt to return to her fungi work but never received support or encouragement to continue. Instead, she turned to the picture letters she had sent to the children of her last governess, Annie Moore, and these became the basis of her children's books. In 1901, she published *The Tale of Peter Rabbit* on her own, following its rejection by a succession of publishers.

Frederick Warne Publishing persuaded her to publish in color, and a long relationship with that family and company began. Norman Warne became an admiring editor and Potter's fiancé, until his sudden death at age 37.

Living at Bolton Gardens and nursing her aging parents, Potter bought a farm, Hilltop House, in the Lake District and continued to entertain children with her menagerie of tame and wild pets in tales that grew out of an imagination that flourished from nearly three decades of self-entertainment.

The royalties from her children's books provided her with funds to purchase additional land in Sawrey, and in October 1913, at the age of 47, she married William Heelis, the solicitor in her land purchases. She treasured her privacy and withdrew from writing until 1916. At that time, Warne Publishing was experiencing financial problems, and *Appley Dapply's Nursery Rhymes* was published to help support the company in the war years.

Beatrix Potter spent the last 30 years of her life as a farmer and respected Herdwick sheep breeder in the Lake District. She had published over 25 children's books, and the royalties were used to buy 15 farms and numerous cottages to ensure that the countryside would not become urbanized. Her legacy includes not only the children's books and fungi portfolio but also 4,000 acres of land willed to the National Trust of the United Kingdom. She died December 22, 1943, at the age of 77 at Castle Cottage, Sawrey, of bronchitis and heart failure.

WORK

In 1867 the Swiss botanist Simon Schwendener described lichens as two organisms, but British botanists rejected his theories and continued to deny his hypotheses. By 1879 the German botanist Heinrich de Bary had coined the word "symbiosis," which Potter also adopted in her own research in 1894 (Gilpatrick 1972). Potter, a self-educated mycologist, was the first person in England to categorize the fungi of the British Isles and to affirm the relationship of algae and fungi in the lichen. She described this union ten years before it was accepted in botanical theory. Her manuscript indicating this scientific phenomenon was burned after her death. The carefully encoded notebooks were not deciphered until half a century later by Leslie Linder (*The Journal . . .* 1966).

On her trips to Scotland and the Lake District, Potter collected fungi to study and draw. While on holiday in Scotland, she met Charles McIntosh, who validated her botanical illustrations. McIntosh was a retired postman and amateur mycologist who was well versed on the mosses, ferns, and fungi of Perthshire. Potter's command of French and German from her tutorials allowed her to consult the natural historians in those languages, primarily Louis Pasteur and Oskar Brekfeld.

By the time she was in her 20s, Potter's observations and recordings on fungi became a prime focus of her life. Through her father she met Sir William Flower of the Museum of Natural History. However, in her father's absence, Sir William refused to acknowledge her, let alone support or mentor her scientific work.

Her uncle, Sir Henry Roscoe, was a chemist and supported her research by suggesting ways of testing algae for chlorophyll lines with a spectroscope. This work led Potter to determine that the fungus contributed minerals extracted from the air and water, which the algae needed.

Through her uncle, Potter finally met the Director of the Royal Botanic Gardens at Kew, Sir William Turner Thiselton-Dyer. She attempted to explain to him the symbiotic relationship of the algae and fungi through her drawings. However, he was indifferent and felt her drawings were too artistic to meet scientific criteria. He did, though, give her a student pass to Kew, which allowed her to study there, only to be greeted with either indifference (as she was female) or professional jealousy.

Discouraged by the indifference of the Kew scientists to her research, she attempted to continue her work without their assistance. As an amateur and woman, she shocked the scientific establishment with her independent experiments and theories. Her uncle encouraged Potter to put her theories on sporulation into writing, after a rejection of her theories by Thiselton-Dyer. The latter felt she needed formal education at the university to be accepted as a respected researcher.

Potter's uncle received a rude letter from Thiselton-Dyer. This prompted him to review Potter's work and encourage her to prepare a paper for the Linnaean Society, London's most prestigious scientific group. Sir Henry guided her through her written research, and on April 1, 1897, her paper "On the germination of the spores of Agaricineae" was read by George Massee of the Royal Society, as women were not allowed to attend these meetings. The paper discussed her experiments in trying to get spores to propagate outside their natural habitat. Before it could be published, Potter withdrew the paper in order to conduct additional research. However, she never received recognition of her work. At her death in 1943 or shortly thereafter, many of her voluminous research notes were burned, including her paper on spores.

She gave up her spore research two years after her paper was returned to her. The time and energy she had devoted to her research were channeled into writing and illustrating children's books, based on her careful drawings and love of nature, both of which were recognized and rewarded by her young readers. With the same precision she used to differentiate fungi, she perfected the animal illustrations of her small books. Later these same energies went into her conservation efforts in the Lake District. By 1901 she had completed 270 detailed watercolors of fungi. Her spore research remained unpublished, a frustration believed to be expressed in the foibles of some of the characters in over 25 of her children's books.

By her express wish, after her death in 1943 Potter's botanical drawings and watercolors were housed at the Armitt Trust Library in Ambleside, England, including the works she cited in her sporulation paper. As recently as 1986, these detailed drawings were cited to validate Potter's extensive knowledge of the Basidiomycetes, demonstrated through work she did 90 years earlier without present-day aids, such as time-lapse photography.

BIBLIOGRAPHY

Works by Beatrix Potter

Scientific Works

Space does not permit the listing of the complete works of Beatrix Potter. The *British Library General Catalog of Printed Books to 1975* lists over 25 publications by Potter. Listed below are works cited in the text.

"On the germination of the spores of Agaricineae." Paper read at the meeting of the Linnaean Society, London, April 1, 1897.

Other Works

The Tale of Peter Rabbit. Private ed., 1901. London: Frederick Warne, 1902.
Appley Dapply's Nursery Rhymes. London: Frederick Warne, 1917.

Works about Beatrix Potter

Battrick, Elizabeth. *The Real World of Beatrix Potter.* Norwich, England: Jarrold and Sons, 1987.
Crouch, Marcus. *Beatrix Potter—A Walck Monograph.* New York: Henry Z. Walck, 1961.
Findlay, W.P.K. *Wayside and Woodland Fungi.* London: Frederick Warne, 1967.
Gilpatrick, Naomi. "The Secret Life of Beatrix Potter." *Natural History* 81 (Oct. 1972): 38–41, 88–97.
Hobbs, Anne Stevenson. *Beatrix Potter's Art.* London: Frederick Warne, 1989.
———, and J. I. Whalley. *Beatrix Potter the V and A Collection.* London: Victoria and Albert Museum and Frederick Warne, 1985.
Hofer, Philip. *Beatrix Potter Letters to Children.* New York: Harvard College Library and Walker, 1966.
Holloway, Marguerite. "Trends in the sociology of science: A lab of her own." *Scientific American* (Nov. 1993): 94–103.
Jones, Chris. "An eye for nature." *New Scientist* (Dec. 24, 1987): 79.
Lane, Margaret. *The Tale of Beatrix Potter: A Biography.* London: Frederick Warne, 1946, 1960.
———. *The Magic Years of Beatrix Potter.* London: Frederick Warne, 1978.
Linder, Leslie. *Beatrix Potter 1866–1943: Centenary Catalog.* London: National Book League, 1966.
———. *A History of the Writing of Beatrix Potter.* London: Frederick Warne, 1971.
———. *The Journal of Beatrix Potter from 1881 to 1897.* London: Frederick Warne, 1966.
———, and E. Linder. *The Art of Beatrix Potter.* London: Frederick Warne, 1955.
Noble, Mary. "Beatrix Potter, naturalist and mycologist and Charles McIntosh, the Perthshire naturalist." *Notes from the Royal Botanic Garden Edinburgh* 44 (1987): 607–627.
———, and R. Davis. "Cup fungus or Basidiomycetes and Potterism." *Bulletin of the British Mycological Society* 20 (Oct. 1986): 145–147.
Parker, Ulla H. *Cousin Beatie.* London: Frederick Warne, 1981.
Peck, Robert M. "The tale before Peter Rabbit." *National Wildlife Federation* 20 (Jan.–Feb. 1990): 42–45.
Quinby, Jane. *Beatrix Potter—A Biographical Check List.* Portland, ME: Anthoensen Press, 1954.
Rea, Jeannette, J. Henderson, et al. "Lessons from women in the agricultural sciences." *The Agricultural Education Magazine* 62 (Aug. 1989): 20–21.
Taylor, Judy. *Beatrix Potter, Artist, Storyteller, Countrywoman.* London: Frederick Warne and Penguin Books, 1986.
———. *Letters to Children from Beatrix Potter.* London: Frederick Warne, 1992.

————. *So I Shall Tell You a Story . . . Encounters with Beatrix Potter*. London: Frederick Warne, 1993.

————, J. I. Whalley, et al. *Beatrix Potter 1866–1943—The Artist and Her World*. London: Frederick Warne with the National Trust, 1987.

Video

Beatrix Potter: Artist, Storyteller, and Countrywoman. Weston Woods, 1993.

EDITH SMAW HINKLEY QUIMBY (1891–1982)

Margaret (Betsy) G. Ott

BIOGRAPHY

Edith Smaw Hinkley was born in Rockford, Illinois, on July 10, 1891. The daughter of Arthur S. and Harriet (Hinkley) Hinkley, she traced her ancestry to old American lineage. She had one sister and one brother. Her father was trained as an architect but worked as a farmer. The family moved frequently, from Illinois, to Alabama, to California, then to Idaho, where Hinkley spent her teens in the mountains around Boise. She was active in sports. Hinkley described her family as middle-class and her home base in Boise as a "medium-sized city in the Middle West" (*Current Biography* 1950, 492).

Edith Hinkley graduated from Boise High School. She attended Whitman College in Walla Walla, Washington, on a full tuition scholarship for four years, earning her B.S. degree in 1912. At the time Whitman was one of the few colleges that welcomed women as students. Majoring in physics and mathematics, Hinkley intended to teach. In 1914, after a two-year appointment as a high school science teacher in Nyssa, Oregon, she was awarded a fellowship in physics from the University of California. She received her M.A. degree in 1915 at the age of 23.

There were several important influences during her education. Her natural aptitude for science was recognized and encouraged by a teacher in high school. In college, her mentors included B. H. Brown, professor of physics, and Walter Bratton, professor of mathematics, who later became president of Whitman College.

Hinkley met her future husband, Shirley Leon Quimby, while both were graduate students in physics at the University of California. They were married on June 9, 1915. Shirley Quimby's first position after graduate school was as a high school physics teacher in Antioch, California; at that time, Edith Quimby did not work outside the home. Shirley Quimby enlisted in the navy in 1918 during World War I, and Edith Quimby assumed his teaching position. After the war, in 1919, the Quimbys moved to New York City, where Shirley Quimby took a position as instructor of physics and became a doctoral student in physics at Columbia University. The Quimbys had no children.

Edith Quimby was an Episcopalian. As a Democrat, she was an active participant in the League of Women Voters. She liked detective stories, the theater, and playing bridge and enjoyed the domestic side of life, including being a hostess (*Current Biography* 1950, 493).

During her 40 active years in New York, Quimby became a biophysicist and educator whose life work was in clinical applications of radiation, particularly treatment of cancer and other tumors. In the early years she was the only woman in America engaged in medical research involving physics. One of her assistants, Rosalyn Yalow,* went on to win the Nobel Prize (1977).

Quimby's first position in New York was at the Memorial Hospital for Cancer and Allied Diseases as the assistant to Dr. Gioacchino Failla, chief physicist. Failla was so impressed by Quimby that he hired her in spite of the fact that he had not previously considered hiring a woman. She became his collaborator, and they worked together for more than 30 years. They executed lengthy studies of all aspects of therapeutic application of X-rays and radium, especially tumor treatment. Failla and Quimby were among the small group of scientists who founded the field of clinical radiation physics. She was highly regarded within that select group; her presentations at professional meetings always elicited warm praise.

Quimby held the position of assistant physicist at Memorial Hospital (1919–32). She was then promoted to associate physicist (1932–43). She was appointed concurrently as Assistant Professor of Radiology at Cornell Medical School (1938– 43). In 1943, she moved with Failla to the College of Physicians and Surgeons at Columbia University as Associate Professor of Radiology. While there, she engaged in biophysical research with X-rays, radioactive substances, and artificially radioactive isotopes, in cooperation with the College's clinical services members. Her responsibilities included creating and teaching a basic course in radiation physics for medical students, along with teaching future doctors how to use radiation safely for treatment of disease. In 1954 she was appointed full Professor, a position she held until 1960, when she was appointed Professor Emeritus. During World War II, from the time of her move to Columbia University until the end of the war, she held a half-time appointment with the Manhattan Project (working on the development of the atomic bomb). Her concern with peaceful uses of atomic energy manifested itself after the war, when she worked for the Atomic Energy Commission. She also served as a consultant to the Veteran's Administration on radiation therapy and as an examiner for the American Board of Radiology.

Edith Quimby was the first woman honored with the Janeway Medal of the American Radium Society, given to her in 1940 in recognition of 20 years of accomplishments. Following the tradition associated with this honor, she gave the keynote address, "The specification of dosage in radium therapy" (published in 1941), summarizing her investigations. In presenting her with the Janeway Medal, Dr. Zoe A. Johnston made the following remarks:

Her scientific accomplishments have been of the utmost importance. Her influence is far reaching. . . . Her straight thinking, precision as to detail, and an enthusiastic spirit are qualities which have enabled her to give such important contributions to the field of radiation therapy . . . her greatest influence is with the younger *men* [emphasis added] who have entered this field of endeavor. . . . It would be impossible to estimate the value of this fundamental training. (Quimby 1941)

In addition to the Janeway Medal, Quimby received the Gold Medal of the Radiological Society of North America (1941), a medal for achievement from the International Women's Exhibition (1947), and the American Design Award, given by Lord and Taylor (1949). She was honored by the Radiological Society of India (1952) and the American Cancer Society (1957). Alfred P. Sloan, Jr., founder of the Sloan-Kettering Institute for Cancer Research, personally honored Quimby in 1962 with the Katharine Berkan Judd Award, given to the scientist who made the "greatest advancement to the discovery of a cure for cancer" (Veglahn 1991, 54). She received the Coolidge Award for lifetime achievements (1978) from the American Association of Physicists in Medicine. In 1940 Whitman College honored her with the Sc.D. degree, and in 1956 Rutgers University awarded her a Sc.D. degree.

Edith Quimby died on October 11, 1982, at the age of 91, in New York City. Her husband died in May 1986 at the age of 93.

WORK

Quimby's work was under the direction of, and in collaboration with, Failla. Her research with Failla involved measuring radiation emission and penetration, so that exact dosages could be determined for radiotherapy. They investigated the amount of radiation emitted to air, on the skin, and in the body, in order to safeguard not only patients but also clinical workers. Prior to Quimby's groundbreaking work, doctors had to tailor the amount of exposure to each patient without precise knowledge of how much exposure a patient received or how deeply the radiation would penetrate in living tissue. Researchers at each institution developed their own measurements and standards, using filters of varying thicknesses and different metals. This created difficulties in sharing information and comparing results. Quimby published a methodology for determining individual exposure, as well as emissions into the work environment (Quimby 1942).

She contributed a great deal to the uniformity of measurements and standards. At the meeting of the American Radium Society in 1931, she presented a paper at the symposium on the biological effects of radium radiation (Quimby 1932). Quimby wrote approximately 75 technical articles for scientific journals, about 50 of them between 1920 and 1940, concerning experiments that established certain characteristics of radium and X-rays. By 1949 she had contributed chapters to five books; she also published lectures and presented poster sessions.

Quimby published determinations of the effect of size and shape of applicators for radium dosing, using different metals for filters and different thicknesses of filters. She explained her standardization of determining doses based on skin erythema, using any reddening of the skin (a threshold type of measurement) rather than any particular degree of redness or blistering. She meticulously explained the physical and mathematical rationale of her measurements of dosage and effects, often publishing hand-drawn graphs. Her papers contained detailed tables of the numerical values she measured and careful drawings of the placement of radium implants, reporting data needed by researchers and clinicians. In addition to publishing her work, she routinely attended national and international professional meetings to disseminate and explain her results and to interact with colleagues.

While the theoretical nature of her work was definitely within the realm of biophysics, its practical application was clearly of a clinical nature. She studied the effects of embedding a radium source (''seeds'' or ''needles'' covered in various metals) within tumors, using slabs of butter, discs of rubber, and layers of beeswax as models for the scattering of beta and gamma radiation through living tissue. A paper presented in 1932 at the meeting of the American Radium Society provided information on the placement of radium needles in treating breast tumors (Quimby 1934). Remarks after the presentation indicated the importance of this study for correcting inadequacies in current techniques.

In 1932 Quimby and George Pack published a paper that challenged the prevailing belief that total dosage, not intensity of dosage, was the only relevant factor in effectiveness of treating neoplasms. They discovered that long-term, low-intensity doses were more effective than the more commonly used short-term, high-intensity doses. They also showed that neoplasms had a slower recovery than normal tissue (Quimby and Pack 1932).

In the mid-1930s Quimby expanded her research to study rates of recovery of skin from the effects of irradiation. Little work had been done in this area before she began her studies. Once again, she made a major contribution in expanding the database for the clinical use of radiation (Quimby and MacComb 1936; Quimby and MacComb 1937). In 1939 Quimby authored an article on the need for precautions by researchers, particularly graduate students, when working with radiation. She delineated the general dangers, including those related to improper storage and use (Quimby, ''Radium protection'' 1939).

In 1940, along with Wilhelm Stenstrom and Eugene Pendergrass, Quimby reported the findings of the Research and Standardization Committee of the American Radium Society. The thrust of this study concerned the treatment of cervical carcinoma. Their report superseded a previous one, completed between 1926 and 1930 (Quimby, Stenstrom, and Pendergrass 1940).

While at Columbia, Quimby contributed to the synthetic development of a radioactive isotope of sodium. The availability of this isotope opened up new opportunities for medical applications. Quimby outlined the use of radioactive isotopes in a speech to the Medical Society of the State of New York (*Current Biography* 1950, 493). In the preface to *Safe Handling of Radioactive Isotopes*, she reviewed the use of radioactive isotopes, which became available after World War II (Quimby 1960, vii).

Quimby lectured in radiologic physics as assistant professor of radiology at Cornell University Medical College (1941–42). She continued to teach at Columbia University. Her many obituaries noted the "more than a thousand physicians" who owed their training to her care and diligence (Veglahn 1991, 54).

Quimby maintained memberships in many professional societies, as well as an appointment to the Advisory Committee on Isotope Usage of the Atomic Energy Commission, starting in 1950. Professional societies of which she was a member included the American Radium Society, of which she was president (1953–54); the American Roentgen Ray Society; the Radiological Society of North America; and the New York Academy of Sciences. She was also a member of Sigma Delta Epsilon.

BIBLIOGRAPHY

Works by Edith Smaw Hinkley Quimby

Scientific Works

Space does not permit the listing of the complete works of Edith Smaw Hinkley Quimby. Listed here are all works by Quimby except those cited in Quimby (1941, 1945). Included here are all references cited in the text.

"The effect of different filters on radium radiations." *American Journal of Roentgenology* 7 (1920): 492–501.

(with G. Failla and A. Dean) "Some problems of radiation therapy." *American Journal of Roentgenology* 9 (1922): 479–497.

"A simple nomogram for the determination of radium skin doses." *American Journal of Roentgenology and Radium Therapy* 10 (1923): 574–578.

"Comparison of different metallic filters used in radium therapy." *American Journal of Roentgenology and Radium Therapy* 13 (1925): 330–342.

"A method for the study of scattered and secondary radiation in X-ray and radium laboratories." *Radiology* 7 (1926): 211–217.

(with G. Failla, F. Adair, et al.) "Dosage study relative to the therapeutic use of unfiltered radon." *American Journal of Roentgenology and Radium Therapy* 15 (1926): 1–35.

"The skin erythema dose with a combination of two types of radiation." *American Journal of Roentgenology and Radium Therapy* 17 (1927): 621–625.

(with W. G. Sargent) "Influence of diaphragm on applicability of inverse square law to X-ray dosage." *Radiology* 10 (1928): 1–9.

"A comparison of radium and radon needles and permanent radon implants." *American Journal of Roentgenology and Radium Therapy* 23 (1930): 49–56.

(with H. R. Downes) "A chemical method for the measurement of quantity of radiation." *Radiology* 14 (1930): 468–481.

(with H. E. Martin) "Calculations of tissue dosage in radiation therapy: A preliminary report." *American Journal of Roentgenology and Radium Therapy* 23 (1930): 173–196.

(———) "Fixed pre-calculated irradiation dosage of intra-oral epidermoid carcinoma." *American Journal of Surgery* 8 (1930): 136–142.

(with F. W. Stewart) "Comparison of various sources of interstitial radiation." *Radiology* 17 (1931): 449–470.

"The grouping of radium tubes in packs or plaques to produce the desired distribution of radiation." *American Journal of Roentgenology and Radium Therapy* 27 (1932): 18–39.

(with B. J. Lea, G. T. Pack, et al.) "Irradiation of mammary cancer, with special reference to measured tissue dosage." *Archives of Surgery* 24 (1932): 339–410.

(with G. T. Pack) "The time-intensity factor in irradiation." *American Journal of Roentgenology and Radium Therapy* 28 (1932): 650–667.

(with L. D. Marinelli) "The influence of filtration on surface and depth intensities of 200 KV X-rays." *Radiology* 21 (1933): 21–29.

"Determination of dosage for long radium or radon needles." *American Journal of Roentgenology and Radium Therapy* 31 (1934): 74–91.

(with M. M. Copeland and R. C. Woods) "The distribution of Roentgen rays within the human body." *American Journal of Roentgenology and Radium Therapy* 32 (1934): 534–551.

(with C. D. Lucas, A. N. Arneson, et al.) "A study of back-scatter for several qualities of Roentgen rays." *Radiology* 23 (1934): 743–750.

(with A. N. Arneson) "Distribution of Roentgen radiation within average female pelvis for different physical factors of irradiation." *Radiology* 25 (1935): 182–197.

"The use of long target-skin distances in Roentgen therapy." *American Journal of Roentgenology and Radium Therapy* 36 (1936): 343–349.

(with W. S. MacComb) "The rate of recovery of human skin from the effects of hard or soft Roentgen rays or gamma rays." *Radiology* 27 (1936): 196–207.

(with L. D. Marinelli) "Study of cones or other collimating devices used in Roentgen therapy." *Radiology* 26 (1936): 16–26.

(with A. N. Arneson) "A comparison of paraffin and water phantoms for Roentgen-ray depth dose measurements." *American Journal of Roentgenology and Radium Therapy* 37 (1937): 93–97.

(with W. S. MacComb) "Further studies on the rate of recovery of human skin from the effects of Roentgen- or gamma-ray irradiation." *Radiology* 29 (1937): 305–312.

"A dosage chart for X-ray therapy." *Radiology* 31 (1938): 308–311.

(with F. E. Adair and E. L. Frazell) "Study of tissue dosage and radiation effect in cases of operable cancer of breast treated by combination of pre-operative irradiation and radical mastectomy." *Radiology* 30 (1938): 588–597.

"Some considerations regarding supervoltage X-rays." *Medical Woman's Journal* 46 (1939): 278–280.

"Dosage specifications in X-ray therapy." *New York State Journal of Medicine* 39 (1939): 509–515.

"Measurement of tissue dosage in radiation therapy." *Radiology* 32 (1939): 583–590.

"Radium." In *Nelson Looseleaf Text on Radiation Therapy.* New York: T. Nelson, 1939.

"Radium protection." *Journal of Applied Physics* 10 (1939): 604–608.

Syllabus of Lectures on the Physical Basis of Radiation Therapy. Ann Arbor, MI: Edwards Brothers, 1939.

(with A. U. Desjardins) "Radon seeds. Physical considerations. Do they leak and do they irritate tissues?" *Journal of the American Medical Association* 112 (1939): 1822–1829.

(with G. C. Laurence) "Radiological Society of North America Standardization Committee. Technical Bulletin no. 1." *Radiology* 35 (1940): 138–159.

(with W. Stenstrom and E. P. Pendergrass) "Report of the research and standardization committee of the American Radium Society." *American Journal of Roentgenology and Radium Therapy* 43 (1940): 118–126.

"The specification of dosage in radium therapy." *American Journal of Roentgenology and Radium Therapy* 45 (1941): 1–17.

"Some practical considerations regarding the employment of various qualities of Roentgen rays in therapy." *Radiology* 38 (1942): 261–273.

"Some aspects of the teaching of radiological physics." *Radiology* 41 (1943): 315–329. Spanish translation. *Revista de radiologia y fisioterapia* 11 (1944): 101–113.

(with J. F. Nolan) "Dosage calculation for various combinations of parametrial needles and intracervical tandems." *Radiology* 40 (1943): 391–402.

"Dosage table for linear radium sources." *Radiology* 43 (1944): 572–577.

"Radiation therapy: Tissue dosage of Roentgen rays and gamma rays." *Medical Physics* (1944): 1165–1180.

"Safety in diagnostic radiology (some facts for general practitioners)." *Medical Woman's Journal* 51 (1944): 23–25.

(with O. Glasser, L. S. Taylor, et al.) *Physical Foundations of Radiology.* New York: Paul B. Hoeber, 1944; 2d ed. 1952; 3d ed. 1961.

(with B. C. Smith) "Tracer studies with radioactive sodium in patients with peripheral vascular disease." *Science* 100 (Aug. 25, 1944): 175–177.

(———) "The use of radioactive sodium in studies of circulation in patients with peripheral vascular disease: A preliminary report." *Surgery, Gynecology and Obstetrics* 79 (1944): 142–147.

"The history of dosimetry in Roentgen therapy." *American Journal of Roentgenology and Radium Therapy* 54 (1945): 688–703.

(with B. C. Smith) "The use of radioactive sodium as a tracer in the study of peripheral vascular disease." *Radiology* 45 (1945): 335–346.

"Basic facts regarding X-ray and radium therapy." *Journal of the American Medical Women's Association* 1 (1946): 4–7.

"Protection against X-rays and gamma rays; tolerance dose or tolerance intensity." *Radiology* 46 (1946): 57–59.

(with T. C. Evans) "Studies on the effects of radioactive sodium and of Roentgen rays on normal and leukemic mice." *American Journal of Roentgenology and Radium Therapy* 55 (1946): 55–66.

(with S. A. Thompson and B. C. Smith) "The effect of pulmonary resuscitative procedures upon the circulation as demonstrated by use of radioactive sodium." *Surgery, Gynecology and Obstetrics* 83 (1946): 387–391.

"Radioactive sodium as a tool in medical research." *American Journal of Roentgenology and Radium Therapy* 58 (1947): 741–753.

"Radiosodium as a tool in biological and medical research." *Nucleonics* 1 (1947): 2–10.

(with D. J. McCune) "Uptake of radioactive iodine by the normal and disordered thyroid gland in children." *Radiology* 49 (1947): 201–205.

(with B. C. Smith) "The use of radioactive sodium in the study of peripheral vascular disease." *Annals of Surgery* 125 (1947): 360–371.

(with T. R. Talbot, Jr., and A. L. Barach) "Method of determining site of retention of aerosols within respiratory tract of man by use of radioactive sodium; preliminary report." *American Journal of Medical Science* 214 (1947): 585–592.

"Calculation of dosage in radioiodine therapy." In *Radioactive Iodine—Its Use in Studying Certain Functions of Thyroid Tumors*, edited by R. W. Rawson, 43–50. BNL-C-5. Upton, NY: Brookhaven National Laboratory, 1948.

"Fifty years of radium." *American Journal of Roentgenology and Radium Therapy* 60 (1948): 723–730.

(with V. K. Frantz and T. C. Evans) "Radioactive iodine studies of functional thyroid carcinoma." *Radiology* 51 (1948): 532–551.

(with L. D. Marinelli and G. J. Hine) "Dosage determination with radioactive isotopes. I. Fundamental dosage formulas." *Nucleonics* 2 (4) (1948): 56–66.

(———) "Dosage determination with radioactive isotopes. II. Practical considerations in therapy and protection." *American Journal of Roentgenology and Radium Therapy* 59 (1948): 260–281.

(with I. Mufson and B. C. Smith) "Use of radioactive sodium as a guide to the efficacy of drugs used in treatment of diseases of the peripheral vascular system. Preliminary report." *American Journal of Medicine* 4 (1948): 73–82.

(with B. C. Smith) "Radioactive sodium in peripheral vascular disease studies." *Surgical Clinics of North America* 28 (1948): 304–323.

(with S. C. Werner and C. Schmidt) "Clinical experience in diagnosis and treatment of thyroid disorders with radioactive iodine (eight-day half-life)." *Radiology* 51 (1948): 564–581.

(———) "The clinical use of radioactive iodine." *Bulletin of the New York Academy of Medicine* 24 (1948): 549–560.

(with L. D. Marinelli and G. J. Hine) "Dosisbestimmung bei radioaktiven Isotopen; fundamentale Dosierungsformeln. I." *Strahlentherapie* 80 (1949): 453–466.

(with E. B. Schlesinger) "The usefulness of radioactive substances in studying certain aspects of circulation in the brain." *Transactions of the American Neurological Association* 74 (1949): 94–98.

(with S. C. Werner) "Late radiation effects in Roentgen therapy for hyperthyroidism." *Journal of the American Medical Association* 140 (1949): 1046–1047.

(with S. C. Werner and C. Schmidt) "Radioactive iodine, I-131, in the treatment of hyperthyroidism." *American Journal of Medicine* 7 (1949): 731–740.

(———) "The use of tracer doses of radioactive iodine, I[131], in the study of normal and disordered thyroid function in man." *Journal of Clinical Endocrinology* 9 (1949): 342–354.

"Radiation-detection instruments." *New York Journal of Dentistry* 20 (1950): 303–306.

"Therapeutic use of radioisotopes." *New York Journal of Dentistry* 20 (1950): 301–302.

(with C. B. Braestrup) "Planning the radioisotope program in the hospital." *American Journal of Roentgenology and Radium Therapy* 63 (1950): 6–12.

(with L. D. Marinelli and G. J. Hine) "Dosisbestimmung bei radioaktiven Isotopen. II.

Biologische Betrachtungen und praktische Anwendungen." *Strahlentherapie* 81 (1950): 587–594.

(with S. C. Werner, L. D. Goodwin, et al.) "Some results from the use of radioactive iodine, I^{131}, in the diagnosis and treatment of toxic goiter." *American Journal of Roentgenology and Radium Therapy* 63 (1950): 889–894.

(with S. C. Werner and C. Schmidt) "Influence of age, sex, and season upon radioiodine uptake by the human thyroid." *Proceedings of the Society for Experimental Biology and Medicine* 75 (1950): 537–540.

"Achievement in radiation dosimetry, 1937–1950." *British Journal of Radiology* 24 (1951): 2–5.

"Dosimetry of internally administered radioactive isotopes." In *A Manual of Artificial Radioisotope Therapy*, edited by P. F. Hahn, 46. New York: Academic Press, 1951.

"Radioactive isotopes in clinical diagnosis." In *Advances in Biological and Medical Physics*, edited by H. J. Curtis, 243–268. New York: Academic Press, 1951.

"Safety in use of radioactive isotopes." *American Journal of Nursing* 51 (1951): 240–243.

"The teaching of radiological physics as a part of training in radiology." *American Journal of Roentgenology and Radium Therapy* 66 (1951): 453–458.

(with H. Speert and S. C. Werner) "Radioiodine uptake by the fetal mouse thyroid and resultant effects in later life." *Surgery, Gynecology and Obstetrics* 93 (1951): 230–242.

"Disposal of radioactive waste from hospital laboratories." *Laboratory Investigation* 2 (1953): 49–56.

(with V. Castro) "The calculation of dosage in interstitial radium therapy." *American Journal of Roentgenology, Radium Therapy and Nuclear Medicine* 70 (1953): 739–749.

"The clinical radiation physicist—a specialist within the field of radiology." *American Journal of Roentgenology, Radium Therapy and Nuclear Medicine* 72 (1954): 733–739.

"Interaction of ionizing radiation with matter." *Proceedings of the Second National Cancer Conference* 1952 (1954): 955–962.

"Radioisotope program in the general hospital." *Journal of the American Medical Association* 154 (1954): 499–501.

(with V. Castro and C. Soifer) "Dosage determination for rotation therapy in the horizontal plane." *Radiology* 63 (1954): 201–219.

(with M. M. Kligerman, E. G. Rosen, et al.) "Rotation therapy techniques applicable to standard deep X-ray machines." *Radiology* 62 (1954): 183–193.

(with V. Castro and C. Soifer) "Calculation of dosage in vertical rotation therapy using standard isodose charts." *American Journal of Roentgenology, Radium Therapy and Nuclear Medicine* 73 (1955): 815–826.

"The background of radium therapy in the United States, 1906–1956." *American Journal of Roentgenology, Radium Therapy and Nuclear Medicine* 75 (1956): 443–450.

"Isotope studies of blood flow and blood cells." *American Journal of Roentgenology, Radium Therapy and Nuclear Medicine* 75 (1956): 1068–1081.

"Radioactive isotopes as aids in medical diagnosis." *New England Journal of Medicine* 252 (1956): 1–9.

"The use of radioactive isotopes as an adjunct to surgery." *Surgical Clinics of North America* 36 (1956): 345–359.

(with V. Castro and C. Soifer) "Applications of Mayneord contour projector to various dosage problems in radiotherapy." *Radiology* 66 (1956): 362–372.

(with B. S. Cohen, V. Castro, et al.) "Calculation of tissue doses and data for the production of isodose charts, using standard depth-dose tables." *Radiology* 66 (1956): 667–685.

(with S. H. Madell, M. M. Kligerman, et al.) "Statistical appraisal of the use of radioactive iodinated human serum albumin for detection of liver metastases." *Radiology* 67 (1956): 210–217.

(with B. S. Cohen) "Effects of radiation quality, target-axis distance, and field size on dose distribution in rotation therapy." *American Journal of Roentgenology, Radium Therapy and Nuclear Medicine* 78 (1957): 819–830.

(with V. Kneeland, F. Morton, et al.) "A comparison of the carcinogenic effect of internal and external irradiation on the thyroid gland of the male Long-Evans rat." *Endocrinology* 61 (1957): 574–581.

(with S. C. Werner) "Acute leukemia after radioactive iodine (I-131) therapy for hyperthyroidism." *Journal of the American Medical Association* 165 (1957): 1558–1559.

(——— and B. Coelho) "Ten year results of I-131 therapy of hyperthyroidism." *Bulletin of the New York Academy of Medicine* 33 (1957): 783–806.

"Training programs in clinical use of radioactive isotopes." *American Journal of Roentgenology, Radium Therapy and Nuclear Medicine* 79 (1958): 138–141.

(with S. Feitelberg and S. Silver) *Radioactive Isotopes in Clinical Practice*. Philadelphia: Lea and Febiger, 1958.

Safe Handling of Radioactive Isotopes. New York: Macmillan, 1960.

"Medical radiation physics in the United States." *Radiology* 78 (1962): 518–522.

(with J. M. Hanford and V. K. Frantz) "Cancer arising many years after radiation therapy. Incidence after irradiation of benign lesions in the neck." *Journal of the American Medical Association* 181 (1962): 404–410.

(with S. Feitelberg) *Radioactive Isotopes in Medicine and Biology*, vol. 1, 2d ed. Philadelphia: Lea and Febiger, 1963.

"What nurses should know about radiation hazards." *International Nursing Review* 11 (1964): 18–22.

(with E. Hiza) "Evaluation of the resin uptake of I-131 triiodothyronine as a test of thyroid function." *Journal of Nuclear Medicine* 5 (1964): 489–499.

"Gioacchino Failla (1891–1961) and the development of radiation biophysics." *Journal of Nuclear Medicine* 6 (1965): 377–382.

(with E. F. Focht and M. Gershowitz) "Revised average geometric factors for cylinders in isotope dosage. I." *Radiology* 85 (1965): 151–152.

"Douglas Quick 1891–1966." *Radiology* 87 (1966): 142–143.

"Historical events leading to the clinical application of radium." *Cancer Journal for Clinicians* 16 (1966): 165–166.

"The fear of radiation." *American Association of Industrial Nurses Journal* 15 (1967): 19.

"Some historical notes about the American Radium Society." *American Journal of Roentgenology, Radium Therapy and Nuclear Medicine* 99 (1967): 499–502.

(with S. Feitelberg and W. Gross) *Radioactive Nuclides in Medicine and Biology.* 3d ed. Philadelphia: Lea and Febiger, 1970.

Other Works

The First Fifty Years of the American Radium Society, Inc., 1916–1966. n.p., 1967? [*sic*]

Works about Edith Smaw Hinkley Quimby

"Coolidge Award to Edith Quimby." *Physics Today* 31 (Jan. 1978): 89.
Chronicle of lifetime achievements.
Current Biography: Who's News and Why 1949, edited by A. Rothe, 492–493. New York: H. W. Wilson, 1950.
Desmond, P. "Edith H(inkley) Quimby." In *The Annual Obituary 1982*, edited by J. Podell. New York: St. Martin's Press, 1983.
"Edith Hinkley Quimby." *Journal of Nuclear Medicine* 6 (1965): 383–385.
Obituary. *New York Times* (Oct. 13, 1982): A28.
Veglahn, N. J. *American Profiles: Women Scientists*, 48–56. New York: Facts on File, 1991.

DIXY LEE RAY (1914–1994)

Janet Newlan Bower

BIOGRAPHY

Dixy Lee Ray was born in Tacoma, Washington, on September 3, 1914. Her parents were Alvis Marion and Frances (Adams) Ray. Her father's family, originally from Tennessee, were actively involved in the Southern Baptist Convention in addition to being commercial printers. Her paternal grandmother, Marguerite, was a direct descendant of Hugh Williamson, a signer of the Constitution. Her mother's family was originally from the central states but moved to Tacoma.

Ray was the second-born of five daughters. Her father had hoped for a son. As a result, there was no name planned for a girl (Moritz 1973, 345). She was finally given the name Margaret, which she legally changed to Dixy Lee when she was 16 years old.

While growing up, Ray was nicknamed "Dicks," shortened from "the little Dickens" (Guzzo 1980, 15). This gave an indication early in her life of the energy, enthusiasm, and strong opinions that were to characterize her career.

Usually, when asked why she changed her name, her response was, "Nobody's damned business" (Guzzo 1980, 22). However, she did relate that her grandmother was a Southern Lee from North Carolina, and she was an admirer of Robert E. Lee, so she chose as her name Dixy Lee Ray (Moritz 1973, 345).

Alvis and Frances eloped while in high school at the ages of 19 and 17, respectively. Although Alvis Ray had only a few months left before graduation, he chose to enter the printing field. Dixy Lee Ray felt that her father was a frustrated electrical engineer (Guzzo 1980, 23); however, he remained in commercial printing throughout his life. Her mother was a homemaker, busy raising five girls.

The Ray family lived the usual middle-class lifestyle. Girl Scouts, camping, and summers spent at their acreage on Fox Island (Puget Sound) were Dixy Lee Ray's favorite pastimes. Athletic at 12 years of age, she was the youngest girl to climb Mount Rainier.

Ray graduated from Stadium High School, Tacoma, in 1933. She attended Mills College, Oakland, California, on a merit scholarship. Her decision to attend Mills came after careful consideration of how she could earn enough money to cover her room and board. Ray's father directed her to do many of the chores a son would be expected to do, such as cutting wood and moving heavy objects. "Having been endowed by nature with a physique one could hardly call willowy, I saw greater opportunity for part-time employment at a woman's college" (Russell 1974, 490).

After studying with Dr. Alexander Pringle Jameson, Ray decided to take her B.A. degree in zoology. Because of a love for the stage, she minored in drama. She graduated from Mills College in 1937 with membership in Phi Beta Kappa. The following year she received an M.A. degree in zoology and a teacher's certificate from Mills.

Ray taught biology at Oakland High School for the next four years, in part to help pay off her educational debt. During the last two years of her high school teaching career, she worked on weekends at the Hopkins Marine Biological Station, Stanford University, in Pacific Grove, California. While there she worked in serology and developed a substance that trapped bacterial flagella (Poole and Poole 1963, 20). She also worked with soil amoebae, developing a technique to bring them out of the dry stage and cultivate them at any time.

Awards of a John Switzler fellowship (1942) and a Van Sicklen fellowship (1943–45) allowed Ray to complete her doctoral work at Stanford University. She graduated in 1945 with a Ph.D. degree in biology.

Ray never married. At one time she explained that, after suffering the emotional abuse of her father's disappointment that she was not a boy and the subsequent harsh attitude he took toward her, she never wanted to become dependent on anyone (Guzzo 1980, 33). At another point she said simply, "I'm too ugly." She did become a devoted aunt to eight nieces and nephews.

Ray attributed success to her efforts to prove to her father that she was someone to be proud of, even if she was not the son he had always hoped for (Guzzo 1980, 24). She had never known her father's affection. She felt that she had to work and work hard to gain his love and respect. Her father died in 1947. He never did tell her he loved her or that he was proud of her.

Although she was raised in a Baptist family, in adulthood Ray simply described herself as Protestant. She felt religion was a very private matter, and it was not necessary to elaborate any further. However, when asked by a reporter what she thought of born-again Christians, she replied she thought being born into a strict Baptist family once was enough (Boren and Harrell 1994).

She ardently claimed her political preference as Independent; however, she ran for governor of Washington state on the Democratic ticket (1977) and won. When Ray ran for a second term as governor, she was defeated in the Democratic primary. She then went into semiretirement. Living on the Fox Island acreage her family held for so long, she concentrated on farming, raising fruits, vegetables, and poultry. She was a student of American Indian culture, with particular interest in the Kwikseutanik tribe. Her expert wood carvings reflected this interest.

Among the 21 schools that conferred honorary degrees upon Dr. Ray are Rensselaer Polytechnic Institute (D.Sc., 1978), University of Missouri (D.Sc., 1976), University of Alaska (D.Sc., 1975), University of Maryland (LL.D., 1975), Smith College (D.Sc., 1975), and Michigan State College (LL.D., 1974).

Her more than 40 honors included membership in the Danish Royal Society of Natural History (1965); recipient of the William Clapp Award, American Society of Corrosion Engineers (1958); the Axel-Axelson Johnson Award, Royal Academy of Science and Engineers, Stockholm, Sweden (1974); the Walter H. Zinn Award, American Nuclear Society (1977); the Centennial Medallion Award, American Society of Mechanical Engineers (1980); and the Centennial Medal, Institute of Electrical and Electronic Engineers (1984).

Dixy Lee Ray died January 2, 1994, on Foxtrot Farm, Fox Island, of bronchial infection and viral pneumonia. She was 79 years old.

WORK

From 1945 to 1976 Ray was on the faculty of the University of Washington, first as assistant, then as an associate professor of zoology, concentrating on invertebrate marine biology. In addition, she was a member of the executive committee of the Friday Harbor Laboratories, the university's marine science center on San Juan Island, located in the San Juan Archipelago (1945–60).

In 1952 Ray received a Guggenheim grant for a year's study of amoeba substance and its relation to serums at Hopkins Marine Biological Station (Noble 1979, 96). She also described a new species of soil amoeba, *Hartmannella astronyxis* (Ray and Hayes 1954).

Ray was awarded a grant from the Office of Naval Research (1955) to compare and contrast the *Limnoria* (life and phenomena) from the warm waters of the Mediterranean Sea and from the cold waters of Washington's coast. Her work on marine organisms that digest wood submerged in seawater proved that these organisms actually consumed the wood. She obtained a patent for a treatment that she developed to protect wood submerged in seawater. The Office of Naval Research asked her to plan a symposium on the problem. The symposium was organized by Ray (1957), sponsored by the University of Washington and held at Friday Harbor. She not only conducted the proceedings but compiled and edited the papers in *Marine Boring and Fouling Organisms* (1959).

In 1958 Ray was asked to develop 15 half-hour educational television programs on marine biology. "Doorways to Science" provided an opportunity for her to use not only her background in science but her love of theater as well. The program and Ray became immensely popular in Washington state.

The National Science Foundation invited Ray to act as special assistant to the Assistant Director for the Division of Biological and Medical Science (1960–62). Two of her research studies resulted in the establishment of the National

Center for Atmospheric Research and the National Radio Astronomy Center (Boren and Harrell 1994). In her two years in Washington, D.C., she learned there were two perceptions of science, that of the scientist and that of the public. "I made up my mind that if there was ever anything I could do to build bridges between scientists and the public, I would do it" (Russell 1974).

She was appointed to the Committee on Oceanography, National Academy of Sciences, as well as its Committee on Postdoctoral Fellowships (1960–63). In 1962 Ray was invited to serve on the Subcommittee on International Biological Stations (1962–65). From 1965 to 1967 she served on the Subcommittee on Productivity of Marine Communities.

Ray became director of the Pacific Science Center in Seattle in 1963 and remained there until 1972. Her goal was to make the Center a place of learning for the public. She wanted people from all walks of life to understand and love science rather than fear it. Her ultimate concern was that only a small group of elites would eventually control the nation's scientific future. Under her direction the institution became a center for displays, conferences, experimentation, and improved teaching techniques. The Center validated her belief that science should be understood and appreciated by the public at large, not just by trained scientists. While she was director, she was invited to participate on the American Association for the Advancement of Science's (AAAS) Committee on Public Understanding of Science (Moritz 1973, 346). Ray was elected a fellow of the AAAS in 1962.

In 1964 Ray was appointed to the visiting faculty at Stanford University and named Chief Scientist for the International Indian Ocean Expedition, the first long-term, multinational cooperative effort to investigate this sea. She took charge of scientific activities on the university's research vessel, *Te Vega*.

Ray served as a member of numerous national committees, including the Smithsonian Institution's Advisory Council for the National Museum Act and the Smithsonian Research Awards Committee (Moritz 1973, 346). President Richard Nixon appointed Ray to the Atomic Energy Commission (AEC) in 1972. She explained her acceptance by saying, "[E]nergy is at the heart of Western Civilization, and that seemed like an interesting place to be" (Stead and Waugh 1993, x). Her agency responsibilities included minority hiring, communications, public information, and improved science education (Moritz 1973, 347; Guzzo 1980, 5).

When the vacancy arose, Nixon appointed Ray Chairman of the AEC in February 1973. She was the first woman appointed to a full five-year term as Commissioner of the AEC, as well as the first to sit as its chair.

As Chairman of the AEC she set three goals: the encouragement of further research in nuclear medicine, improved education to give the public a better understanding of nuclear energy, and the creation of a moderate governmental policy toward nuclear energy and the environment (Stead and Waugh 1993, x). It was her belief that nuclear energy was safe but that the public needed a better understanding of its potential, and the government needed better safety regulation of its production. She explained that her top priority was "to clean up public misconceptions about nuclear energy. It is ridiculous to associate bombs with nuclear power plants" (Russell 1974, 489).

That attitude was underscored with the Rasmussen Report, issued by the AEC while she was Chairman. The report emphasized the safety of nuclear energy by arguing that a person "has about as much chance of dying from an atomic reactor accident as of being struck by a meteor" ("Dixy Lee Ray . . ." 1994).

She was concerned that the same agency responsible for the development of nuclear power was also responsible for its safety regulation. Ray preferred to have two separate agencies that could act as a check against each other (Noble 1979, 100). In 1973 she reorganized the AEC by setting up a Division of Reactor Safety Research, which was separate from, and independent of, the Division of Reactor Development and Technology.

While Ray was Chairman, Nixon asked her to prepare a plan to develop new sources of energy. Her plan, proposed in "The Nation's Energy Future," was a $10 billion program that included the development of new forms of nuclear energy production as a supplement, and eventually a replacement, for finite fossil fuels (Ray, "A Biologist . . ." 1974, 495).

Ray was known nationally by this time for her ardent support of nuclear energy with proper safety standards. She was equally well known for her unusual lifestyle, which included living in a mobile home especially designed for her. Her constant companions were her two dogs. They accompanied her to her office daily, were admitted to her office with their individual "Q-tags" (security photo tags), and remained with her through conferences and interviews.

When the AEC was disbanded, Nixon appointed Ray Assistant Secretary of State for Oceans and International Environmental and Scientific Affairs (1975). Her responsibilities included coordination of all international scientific, technological, and environmental affairs, from space technology, to oceans and fisheries. In the six months she held her office, she assembled all the scientists at the State Department under her division, started a survey of all treaties and agreements in effect at that time, and demanded that her department be consulted before any commitment by the United States was made (Noble 1979, 103). When Secretary of State Henry Kissinger ignored her request for consultation, she angrily resigned. Publicly she criticized the Secretary and the Department for not putting enough emphasis on scientific considerations.

In 1977, when Ray ran for governor of Washington state, she campaigned for "honest government, open government, and responsible government" (Stead and Waugh 1993). Once elected as the state's first female governor, she was criticized for administering it like a conservative Republican, despite the fact that she ran as a Democrat.

The two major scientific problems of her term were nuclear energy and the eruption of Mount St. Helens. While concerned about environmental issues, she rejected popular no-growth policies and encouraged the development of nuclear energy. At the same time she crossed paths with the federal government's lax disposal policies at the Hanford nuclear waste disposal dump. Ray established the Washington State Hazardous Materials Commission (1978) and closed the Hanford site until the federal government agreed to safe packing, transportation, and disposal of the nuclear waste deposited in Washington state.

In April 1980 Ray formed the Mount St. Helens Watch Group. This was a combination of eight state agencies involved in various aspects of the potential eruption. Red and blue zones were created on the volcano to determine who would be allowed to enter. The volcano erupted in May, killing approximately 60 people.

After retirement she was a consultant on KVI-radio and a board member/ treasurer for the Greater Tacoma Community Foundation. Ray was also a consultant for the U.S. Department of Energy, the Los Alamos National Laboratory, and the Lawrence Livermore National Laboratory. She was a founding board member and consultant to the Washington Institute for Policy Studies, as well as a columnist for its publication, *CounterPoint*.

Ray sat on the Board of Directors of Brookhaven National Laboratories; the Science Research Advisory Committee of the U.S. Coast Guard; the Defense Science Board; the Visiting Committee for Nuclear Engineering, Massachusetts Institute of Technology; the Board of Directors, Americans for Energy Independence; and the Board of Directors, Associated Universities, Inc.

Her work and effort in science covered an array of fields: marine biology, education, nuclear energy, and governmental administration. As early as 1969, she warned, "We're heading for an elite class which will have knowledge while the rest remain in ignorance" (Boren and Harrell 1994). Her numerous awards and recognitions reflect the appreciation of the American public.

Her concerns about "knee-jerk" environmentalism and its negative effects on conservation, as well as the effects of such a scientific approach on citizens' everyday life, were emphasized in her two books, *Trashing the Planet . . .* (Ray and Guzzo 1990) and *Environmental Overkill* (Ray and Guzzo 1993).

BIBLIOGRAPHY

Works by Dixy Lee Ray

Scientific Works

"The Peripheral Nervous System of *Lampanyctus leucopsarus*: With Comparative Notes on Other *Iniomi*." Ph.D. diss., Stanford University, 1945.
"Peripheral nervous system of *Lampanyctus leucopsarus*." *Journal of Morphology* 87 (1950): 61–178.
"Agglutination of bacteria: A feeding method in the soil ameba." *Journal of Experimental Zoology* 118 (1951): 443–458.
"The occurrence of cellulase in *Limnoria lignorum*." In *Report Marine Borer Conference*, K-1–K-5, L-1–L-4. Pt. Hueneme, CA: U.S. Naval Civil Engineering Research and Evaluation, 1951.

(with Jean R. Julian) "Occurrence of cellulase in *Limnoria*." *Nature* 169 (1952): 32–33.

"Digestion of wood by *Limnoria lignorum* (Rathke)." *Proceedings XIV International Congress of Zoology* [Copenhagen] (1953): 279.

(with R. E. Hayes) "*Hartmannella astronyxis*: A new species of free-living amoeba." *Journal of Morphology* 95 (1954): 159–188.

"Recent research on the biology of marine wood borers." *American Wood Preservation Association Proceedings* 54 (1958): 120–128.

"An integrated approach to some problems of marine biological deterioration and destruction of wood in sea water." *Marine Biology, Oregon State College Biology Colloquium*, 1959.

(editor) *Marine Boring and Fouling Organisms*. Friday Harbor Symposia, University of Washington, Department of Zoology. Seattle: University of Washington Press, 1959.

"Marine fungi and wood borer attack." *American Wood Preservation Association Proceedings* 55 (1959): 1–7.

"Nutritional physiology of *Limnoria*." In *Marine Boring and Fouling Organisms*, Friday Harbor Symposia, University of Washington, Department of Zoology, edited by D. L. Ray, 46–60. Seattle: University of Washington Press, 1959.

"Some properties of cellulase from *Limnoria*." In *Marine Boring and Fouling Organisms*, Friday Harbor Symposia, University of Washington, Department of Zoology, edited by D. L. Ray, 372–397. Seattle: University of Washington Press, 1959.

"Trends in marine biology." *Marine Biology, Oregon State College Biology Colloquium*, 1959.

(with D. E. Stuntz) "Possible relation between marine fungi and *Limnoria* attack on submerged wood." *Science* 129 (1959): 93–94.

"Some marine invertebrates useful for genetic research." In *Perspectives in Marine Biology*, edited by A. A. Buzzati-Traverso, 497–512. Los Angeles: University of California Press, 1960.

(with Frank Johnson) "Simple method of harvesting *Limnoria* from nature." *Science* 135 (1962): 795.

"Miracle of water: Excerpts from address, April 1968." *American Forests* 74 (1968): 8.

"AEC and national energy goals." *Edison Electric Institute Bulletin* 41 (3) (1973): 129–132.

The Nation's Energy Future; A Report to Richard M. Nixon, President of the United States. Washington, DC: Atomic Energy Agency, 1973.

"What's holding up nuclear power? Interview." *U.S. News and World Report* 75 (1973): 23–25.

(with I. McManus) "Nuclear energy and the environment; interview." *American Forests* 79 (1973): 12–15.

"A biologist looks at the energy crisis." *BioScience* 24 (9) (1974): 495–497.

"Department of outraged virtue; letter." *Environment* 16 (1974): 4–5.

(with R. Gillette) "A conversation with Dixy Lee Ray, interview." *Science* 189 (1975): 124–127.

(with Anibal L. Taboas) *Waste Management: The Missing Link*. Washington, DC: DOE Defense Waste and Byproducts Management Conference, 1983.

"Nuclear waste: What good is it?" *Tennessee Law Review* 53 (3) (1986): 475–479.

"Acid rain: What to do?" *Environmental Science and Technology* 22 (4) (1988): 348.

"Who speaks for science?" *Imprimis* 17 (8) (1988).

"Why doesn't the public see it the way we do?" In *Proceedings of International Conference on Enhanced Safety of Nuclear Reactors*. Washington, DC: School of Engineering and Applied Science, George Washington University, 1988.

"Global warming: What's known and what is not known." *Washington Times* (Apr. 22, 1990).

"Greenhouse earth: The facts about global warming." *Our Land* 11 (1) (1990).

"Why we'll need more nuclear power." *Fortune* 121 (7) (1990): 88.

(with Louis R. Guzzo) *Trashing the Planet, How Science Can Help Us Deal with Acid Rain, Depletion of the Ozone, and Nuclear Waste (among Other Things)*. Washington, DC: Regnery Gateway, 1990.

"Global warming." *Flower and Garden* 35 (1991): 38–39.

"Global warming and CO_2." *CounterPoint* 1 (3) (1993): 13.

"Ozone depletion." *CounterPoint* 1 (1) (1993): 11.

"A real environmental disaster: The President's climate control plan." *CounterPoint* 1 (4) (1993): 13.

"The timber plan." *CounterPoint* 1 (2) (1993): 11.

(with Doug Esser) "Radiation reports raise needless fears, says Dixy Lee Ray; interview." *Daily World* [Aberdeen, WA] (Dec. 31, 1993): A-1, A-9.

   The last interview before her death.

(with Louis R. Guzzo) *Environmental Overkill: Whatever Happened to Common Sense?* Washington, DC: Regnery Gateway, 1993.

(with Patricia Poore, Albert Gore, Jr., et al.) "Trees." *GARBAGE the Practical Journal for the Environment* 5 (3) (1993): 30–36.

Works about Dixy Lee Ray

"AEC former head runs for governor." *New York Times* (Mar. 16, 1976): 26.

Boehm, G.A.W. "Extraordinary first lady of the AEC." *Reader's Digest* 105 (1974): 81–85.

Boren, Rebecca, and D. C. Harrell. "Former Gov. Dixy Lee Ray dies." *Seattle Post-Intelligencer* (Jan. 3, 1994): A-4.

Cher, D., and W. J. Cook. "Whistling Dixy." *Newsweek* 88 (1976): 47.

"Defeat for Dixy Lee Ray." *Time* 116 (1980): 25.

"Dixy Lee Ray, 79, Ex-Governor; led Atomic Energy Commission." *New York Times* (Jan. 3, 1994): A24.

"Dixy rocks the Northwest." *Time* 110 (1977): 26–29.

Emery, Parris. *America's Atomic Sweetheart: A Composite Biographical Profile of Washington's Maverick Lady Governor, Dixy Lee Ray*. Seattle: Century, 1978.

"50,000 warned of volcano flood threat." *New York Times* (May 21, 1980): 1.

Gillette, R. "A conversation with Dixy Lee Ray." *Science* 189 (1975): 124–127.

———. "Ray nominated to AEC." *Science* 177 (1972): 246.

———. "Ray's shift to State Department will test Kissinger's interest in science." *Science* 186 (1974): 612–613.

———. "Ray to be AEC chairperson." *Science* 179 (1973): 667.

"G.O.P. wins Illinois governorship, Jay Rockefeller elected in W. Virginia, Dixy Lee Ray, Democrat, elected in Washington." *New York Times* (Nov. 3, 1976): 21.

"Gov. Ray may reopen nuclear waste site if enforcement is stricter." *New York Times* (Nov. 7, 1979): 21.

Guzzo, Louis R. *Is It True What They Say about Dixy?: A Biography of Dixy Lee Ray.* Mercer Island, WA: Writing Works, 1980.

Gwertzman, Bernard. "Dr. Ray quits State Dept., critical of Kissinger policy." *New York Times* (June 21, 1975): 1.

"House panel rebuffs Carter on energy tax." *New York Times* (June 14, 1977): 51.

Jacobs, M. "Ray: Toward a more effective international science policy." *Physics Today* 28 (1975): 61.

Keerdoja, E., and P. Abremson. "Dixy Lee Ray is still speaking out." *Newsweek* 97 (1981): 16.

Marshall, E. "Governor Ray relents, opens waste site." *Science* 206 (1979): 1165.

Mathews, T., and P. S. Greenberg. "Washington: Lady with a chainsaw." *Newsweek* 89 (1977): 45.

Moritz, Charles (ed.). *Current Biography Yearbook 1973.* New York: H. W. Wilson, 1973.

"New U.S. atomic energy chief Dixy Lee Ray." *New York Times* (Feb. 7, 1973): 18.

Noble, Iris. *Contemporary Women Scientists of America.* New York: Julian Messner, 1979.

"Offshore oil plan draws criticism." *New York Times* (May 13, 1987): I-25.

"Plea slated on nuclear dumping." *New York Times* (Oct. 26, 1979): IV-9.

Poole, Lynn, and Gray Poole. *Scientists Who Work Outdoors.* New York: Dodd, Mead, 1963.

Russell, Christine. "Profiles." *BioScience* 24 (9) (1974): 489–492.

"Search ends for victims of volcano with toll at 22." *New York Times* (June 2, 1980): 14.

Stead, Elizabeth, and Kathleen Waugh. *Dixy Lee Ray 1977–1981, Guide to the Governor's Papers*, vol. 6. Olympia, WA: Office of the Secretary of State, 1993.

"Surprises from nation's two women governors." *U.S. News and World Report* 83 (1977): 45.

"Technology court is urged by Dr. Ray." *New York Times* (June 5, 1975): 12.

"3 Governors urge states to play atom-waste role." *New York Times* (Oct. 5, 1979): 21.

"Town accessible after eruption, Dixy Lee Ray signed order allowing residents to reenter the 'red zone.' " *New York Times* (Oct. 2, 1980): 18.

"U.S. plans 1.3 billion barrel oil reserve." *New York Times* (May 17, 1975): 33.

"U.S. to ease some energy saving rules." *New York Times* (Feb. 20, 1981): IV-2.

"U.S. wins safeguards in German deal with Brazil." *New York Times* (June 4, 1975): 16.

Verheyden-Hilliard, Mary Ellen. *Scientist and Governor, Dixy Lee Ray.* Bethesda, MD: Equity Institute, 1985.
     An illustrated biography for young readers.

Williams, D. A., and M. Reese. "Can Dixy rise again?" *Newsweek* 96 (1980): 28.

JANET DAVISON ROWLEY (1925–   )

Thomas A. Firak

BIOGRAPHY

Janet Davison was born in New York City on April 5, 1925. She was the only child of Hurford and Ethel (Ballantyne) Davison. Her American-born parents were of English and Scottish descent. Her father obtained a Bachelor's degree from the University of Chicago (1922) and an M.B.A. degree from the Harvard Business School (1924). He became a college educator in areas relating to retail store management. Her mother earned a Bachelor's degree from the University of Chicago (1923) and a Master's degree in education from Columbia University (1936). She became a high school English teacher and then a high school librarian. The family was middle-class and lived primarily in the Midwest. Her father was an undefined Protestant, and her mother was an Episcopalian. Both were Republicans.

Davison's parents were very supportive of her and her interests. Her mother felt that she was a gifted child who was not sufficiently challenged at her grammar school. She transferred her to a junior high school in New Jersey, where Davison began to apply herself more seriously. Davison's interest in science soon became apparent and continued throughout high school. She read a great deal, especially enjoying mysteries by Dorothy L. Sayers.

Davison received a scholarship to a program set up by Robert Maynard Hutchins at the University of Chicago called "The Four Year College." She entered the program in 1940 at the age of 15. Here she completed her last two years of high school and first two years of college. She was treated like a college student and was expected to display the independence and responsibility of one. There were 65 people in her entering class; class size averaged 20 to 25 students. Lectures were given but, more important, discussions abounded between the students and the dedicated faculty. She was encouraged to learn, think, and defend a particular idea. She considers her education there extraordinary and very important to her development.

Davison went on to earn a Ph.B. degree (1944), a B.S. degree (1946), and an

M.D. degree (1948), all from the University of Chicago. On December 18, the day after her graduation from medical school, she married Donald Adams Rowley. Her husband, also a medical doctor, is presently Professor Emeritus in the Department of Pathology at the University of Chicago.

The Rowleys had four sons: Donald Adams, Jr. (b. 1952), David Ballantyne (b. 1954), Robert Davison (b. 1960), and Roger Henry (b. 1963). There are also three grandchildren.

Balancing family life and career was extremely important to Rowley. She was home with the children during their early years and spent time with them on outings to museums and parks. Therefore, Rowley worked only two to three days per week for over 20 years after obtaining her medical license in 1951. Her critical discoveries in 1972 led her to devote more time to research, and by 1975, the year in which her youngest child turned 12, she was working full-time.

As Rowley grew more successful in her work, the number of travel requests increased. Scientific meetings and administrative responsibilities on boards of directors demanded that she be out of town 20% of the time. This situation created problems at home, especially since her husband did not travel a great deal. The two are still adjusting to the strain.

Rowley is an avid gardener who has turned her city lot near the campus in Hyde Park (Chicago) into an attractive, restful, and productive garden. Besides flowers, shrubs, and trees, the garden has an abundance of produce to provide ingredients for fresh meals in the summer. She has confided that some of her best thinking is done while working in the garden. Her other interests include the opera and traveling worldwide to hike, camp, observe wildlife, and become acquainted with the indigenous peoples. The Rowleys also enjoy lakeside activities with the family at their cottage on Lake Michigan.

After obtaining her medical license, Rowley became an attending physician (1953–54) at the Infant and Prenatal Clinics in the Department of Public Health in Montgomery County, Maryland. From 1955 through 1961 Rowley worked part-time as a research fellow in a clinical setting with retarded children at the Dr. Julian Levinson Foundation in Chicago. She also worked as a clinical instructor in neurology at the University of Illinois School of Medicine. During the years 1961 and 1962, Rowley studied chromosomes as a National Institutes of Health special trainee in the radiobiology laboratory of the Churchill Hospital at Oxford, England. Upon returning to the United States in 1962, Rowley began her long-standing professional association with the University of Chicago. Starting part-time in the Department of Hematology as a research associate and an assistant professor with Dr. Leon Jacobson, she progressed through the ranks to her present position as the Blum-Riese Distinguished Service Professor in the Department of Medicine and the Department of Molecular Genetics and Cell Biology at the Franklin McLean Memorial Research Institute.

Rowley has been the recipient of numerous honors and awards. Some of these accolades include the Dameshek Prize from the American Society of Hematology, the King Faisal International Prize in Medicine, and the William Proctor Prize for Scientific Achievement from Sigma Xi. She has also received the G.H.A. Clowes Memorial Award from the American Association for Cancer Research, the Charles S. Mott Prize from the General Motors Cancer Research Foundation, the Allen Award from the American Society of Human Genetics, and the de Villiers Award from the Leukemia Society of America.

Rowley has been a visiting scientist at Oxford and Distinguished Visiting Professor at Sloan-Kettering and the Mount Sinai School of Medicine. She has received honorary degrees from the universities of Arizona, Pennsylvania, and Southern California and from Knox College.

Rowley's career was very much affected by her parents' early encouragement and her experience at the "Four Year College." While in medical school, she had several female professors who served as role models. Women were then still a relative rarity in medical school. In 1944 she was one of three women accepted into a class of 65 students. (In the 1990s at the University of Chicago, women make up approximately 40% of entering classes.)

Dr. Jacobson, the director of the Argonne Cancer Research Hospital at the University of Chicago, was an important influence upon Rowley. After returning from Oxford, where she had gained her first research experience, she was hired by Jacobson (1962) as a part-time research cytogeneticist. He took the chance of hiring her, even though she as yet had no research credentials. She was given three feet of bench space and a microscope. Jacobson encouraged her for several years, and then the results began to materialize.

Dr. Charles B. Huggins, who won the Nobel Prize for his work with hormones and prostate cancer (1966), was another influence upon Rowley. He was one of her teachers in medical school and was always happy to talk with her about cancer research.

Rowley's husband also played an important role when she was writing her early research papers. As a research scientist, he had already accumulated much experience and reviewed her working drafts. He often made substantial contributions and improvements in the style of the presentation.

Rowley has given much thought to the important issue of women in science. She has never felt deterred because of her gender. She believes that this is largely due to the supportive atmosphere at the University of Chicago. Her female colleagues at other academic institutions have felt very beleaguered. Their male colleagues either did not pay attention to them or refused to give them credit for their intellectual ability.

There has not been a proportionate increase of females in the middle ranks of academe, even though more females are entering graduate school. Rowley feels that several factors contribute to this dilemma. To attain the rank of professor requires long hours and often fierce competition, and there may be gender bias among colleagues. Perhaps this situation proves distasteful to some women. Others may decide that their families and other opportunities and responsibilities are more important to them.

WORK

Rowley's research work began at Oxford, England. In 1961 her husband went on sabbatical to England, and so she applied to the National Institutes of Health for a Public Health Service Fellowship to work at Oxford, studying chromosomes. She had previously worked in a clinical setting with retarded children. In 1959 it was discovered that Down's syndrome was caused by a trisomy (an extra copy) of chromosome 21 (Lejeune et al. 1959). Rowley knew that chromosomes played an important role in other clinical abnormalities, and she wanted to analyze them. At Oxford she collaborated with Laslo Lajtha in studying the pattern of deoxyribonucleic acid (DNA) replication in normal and abnormal human chromosomes.

The study of human chromosomes was a relatively new field in 1962. The correct number of human chromosomes had been discovered only six years earlier (Tijo and Levan 1956). Methods for studying chromosomes were still rather primitive. A particularly important discovery by Peter Nowell and David Hungerford (1960) had linked a specific abnormal chromosome, the Philadelphia chromosome, with a particular type of leukemia. The Philadelphia chromosome was found to be missing a piece in patients with the chronic form of bone-marrow or myeloid leukemia. Because of the lack of more sophisticated techniques, it could not be determined whether the Philadelphia chromosome belonged to pair number 21 or 22. Neither could the fate of the missing piece be determined.

In 1970 Rowley returned to Oxford on sabbatical to learn the newly developed chromosome banding techniques. Once back in Chicago, she applied her new skills to bone-marrow samples from leukemic patients. Karyotypes of the chromosomes were prepared. This involved photographing metaphase chromosomes, cutting them out, and ordering them in their respective pairs. Rowley remembers the excitement she felt when noticing that there appeared to be extra material on one of the copies of chromosome 9 in the leukemic cells of patients with chronic myeloid leukemia and whose cells had the Philadelphia chromosome. This happened to be the case not just for a single patient but for several of them. As more cases were analyzed, nearly all of the patients displayed this cytogenetic abnormality.

Rowley developed the use of quinacrine fluorescence and Giemsa staining to identify chromosomes in cloned cells (Rowley 1973). While working in Sweden and Scotland, she demonstrated that the Philadelphia chromosome was involved in a translocation with chromosome 9. A translocation occurs when a piece of one chromosome breaks off and becomes associated with another chromosome. Translocations may involve a reciprocal exchange of material as a result of breakages in two chromosomes. In the type of leukemia she was studying, the break occurs near the end of the long arm of chromosome 9 (band 9q34) and in the proximal half of the long arm of chromosome 22 (band 22q11). This was the first consistent translocation to be identified for any disease in any species.

At the same time, Rowley identified another recurring translocation involving chromosomes 8 and 21 in patients with acute myeloblastic leukemia. By 1990 more than 70 recurring translocations had been detected in malignant cells.

These discoveries eventually altered the paradigm concerning cancer and chromosomal abnormalities. Before the development of chromosome banding, some leukemias displayed a bewildering array of chromosomal changes, whereas others had an apparently normal karyotype. This variability led to the view that chromosomal changes were merely epiphenomena and had little significance. Rowley's work, on the other hand, laid the foundation for a new paradigm. She demonstrated that several translocations were specifically associated with particular subtypes of leukemias. In other words, a specific chromosomal change is associated with a specific malignancy, a view that met with much resistance at the time. In the mid-1970s Rowley went to the Sunday morning hematology education meetings and "preached" to her colleagues the importance of chromosomal changes in human leukemia and lymphoma. Years later prominent scientists revealed to her that her work and ideas had very much influenced their careers.

The fact that the breakpoints in the translocations consistently occurred in the same regions led Rowley to believe that there must be an abnormal juxtaposition of two specific genes. These genes would normally be activated only in appropriate cell types or stages of differentiation. However, when a translocation occurred, the genes were inappropriately expressed and played a role in establishing a cancer. The next logical step was to clone (at the molecular level) the junctions at which the translocations occurred. By this time Rowley's discoveries had been accepted, and many research institutions joined in the hunt for these breakpoints.

Concurrently, there was rapid progress in the field of oncogenes (cancer-causing genes). Earlier in the century, work done by Francis Peyton Rous (Nobel laureate, 1966) identified viruses that caused tumors in animals (Rous 1911). It was found that these viruses had special genes that caused tumors. In the early 1970s it was shown that these viral oncogenes had counterparts in human cells. The counterparts were dubbed proto-oncogenes. They turned out to be the progenitors of viral oncogenes. The function of proto-oncogenes in normal cells is to mediate cell growth and differentiation. When proto-oncogenes are altered, they become oncogenic.

Later, in the 1980s, it was demonstrated that the chromosomal location of several proto-oncogenes corresponded to breakpoint junctions in various translocations. The cloning of these breakpoints verified that, indeed, proto-oncogenes were the genes that were in some way altered as a result of a translocation. In the case of chronic myeloid leukemia, the translocation involving the Philadelphia chromosome results in a fusion protein. A large distal portion of a proto-oncogene, called ABL, on chromosome 9 is joined to the proximal region of the BCR gene on chromosome 22. The two gene segments are fused and ultimately produce a chimeric protein that is larger than the normal ABL protein. A specific malignant state is a consequence of this process.

Rowley and her colleagues continued to identify other translocations associated with different types of leukemias. Recently they identified a gene, MLL (myeloid-lymphoid, or mixed lineage leukemia), that is located in the breakpoints of various translocations involving chromosome 11 band q23. One of the important features of this gene is that it is frequently associated with two types of leukemia: the acute lymphoid and acute myeloid leukemias. Lymphoid cells grow and mature in lymphatic tissue, such as the lymph nodes and spleen. Myeloid cells mature in the bone marrow. The alteration of the MLL gene, therefore, seems to be associated with an early progenitor cell capable of both lymphoid and myeloid differentiation. The MLL gene is involved in nearly 30 translocations. Rowley has analyzed hematological malignancies associated with deletions in chromosomes.

In addition, Rowley is a forerunner in applying the polymerase chain reaction (PCR), fluorescence *in situ* hybridization (FISH), and other molecular techniques to clinical work. The precision and sensitivity of such tools have brought about a revolution in clinical oncology. This is especially important in establishing an accurate diagnosis and prognosis, along with a reasonable treatment plan. Using molecular methods to monitor patients who are achieving remission has proven to be extremely sensitive. A precise knowledge of what genes and gene products are involved in each particular malignancy will allow for specific therapies. Eventually, there is hope that these targeted treatments will replace the more general and often extremely toxic therapies, such as radiation and chemotherapy.

As an educator, Rowley does not teach any formal courses at the University of Chicago but speaks to groups of students, whether in high school or medical school, when requested. She spends a great deal of time helping graduate students and postdoctoral fellows in her laboratory. She believes that researchers, especially those supported by public funds, have a responsibility to communicate with people in order to educate them, so that they may understand the excitement of research and its implications for improving their health.

In conjunction with many of her preeminent colleagues, Rowley has worked as a co-organizer of important conferences and symposia. She initiated and helped organize the first five International Workshops on Chromosomes in Leukemia. She also has organized many symposia involving genes, chromosomes, and cancer.

Rowley has served on numerous boards, such as the National Cancer Advisory Board, National Cancer Institute; the Medical Advisory Board, Leukemia Society of America; and the American Board of Medical Genetics. She was the president of the American Society of Human Genetics and is presently on both the Scientific and Medical Advisory Boards of the Howard Hughes Medical Institute. She is also a member of the National Academy of Sciences, the Institute of Medicine, and the American Philosophical Society and a fellow of the American Academy of Arts and Sciences.

Rowley is the cofounder and coeditor of the journal *Genes, Chromosomes and Cancer*. She currently serves on the editorial boards of the following journals: *Cancer Genetics and Cytogenetics, Genomics, International Journal of Hematology, Leukemia*, and *Oncology Research*. Throughout her career Rowley has been a groundbreaker in the field of cytogenetics and, specifically, cancer genetics. She has also been an active leader in the scientific and medical community. Her work and ideas have opened new frontiers that are revolutionizing medicine. Rowley is fortunate to see that her work is benefiting humanity in her lifetime, as well as establishing a legacy for the future. On October 23, 1995, a symposium was held at the University of Chicago in honor of her 70th birthday. World-renowned scientists lectured on the impact of her discoveries on cancer biology. A warm and generous individual, Rowley richly deserves such recognition.

BIBLIOGRAPGHY

Works by Janet Davison Rowley

Scientific Works

Space does not permit the listing of the complete works of Janet Davison Rowley. Listed here are all works by Rowley except those cited in Sandberg (1980, 1990).

(with C. W. Gilbert, S. Muldal, et al.) "Time sequence of human chromosome duplication." *Nature* 195 (1962): 869–873.

(with S. Muldal, C. W. Gilbert, et al.) "Synthesis of deoxyribonucleic acid on X-chromosomes of an XXXXY male." *Nature* 197 (1963): 251–252.

(———) "Tritiated thymidine incorporation in an isochromosome for the long arm of the X chromosome in man." *Lancet* 1 (1963): 861–863.

(with S. Muldal, L. Lindsten, et al.) "Thymidine uptake by a ring X chromosome in a human female." *Proceedings of the National Academy of Sciences* 51 (1964): 779–786.

"Some current problems in population cytogenetics." *Proceedings of the Institute of Medicine of Chicago* 27 (1968): 149–151.

"Uses and pitfalls of chromosomal labeling." In *Radioisotopes in Medicine: In vitro Studies 13, AEC Symposium Series*, edited by G. A. Andrews, R. M. Kniseley, et al., 679–694. Oak Ridge, TN: U.S. Atomic Energy Commission (USAEC), 1968.

"Clinical aspects of chromosome studies." *Cancer Journal for Clinicians* 19 (1969): 299–305.

(with E. Pergament) "Possible non-random selection of D group chromosomes involved in centric-fusion translocations." *Annales de Genetique* (1969): 177–183.

"Chromosomes in myelodysplasias." In *Myeloproliferative Disorders of Animals and Man*, edited by W. J. Clarke et al., 556–569. Washington, DC: USAEC, 1970.

"Sex chromosome abnormalities and neuropsychiatric disorders." *Illinois Medical Journal* 137 (1970): 528–530.

"Identification of a translocation with quinacrine fluorescence in a patient with acute leukemia." *Annales de Genetique* 16 (1973): 109–112.

"Identification of human chromosomes." In *Human Chromosome Methodology*, 2d ed., edited by J. J. Unis, 17–46. New York: National Academy Press, 1974.

"The significance of nonrandom chromosomal changes in myeloproliferative disorders." *Proceedings of the 16th International Congress on Hematology. Excerpta Medica* 415 (1976): 915–917.

"A possible role for nonrandom chromosomal changes in human hematologic malignancies." In *Chromosome Today*, vol. 6, edited by A. De La Chapelle and M. Sorsa, 345–355. North Holland: Elsevier, 1977.

"Chromosomes in acute non-lymphocytic leukemia." *British Journal of Haematology* 39 (1978): 311–316.

"Chromosomes in Ph1 positive chronic myeloid leukemia." *British Journal of Haematology* 39 (1978): 305–309.

"Nonrandom involvement of chromosome segments in human hematologic malignancies." In *Differentiation of Normal and Neoplastic Hematopoietic Cells*, edited by B. Clarkson et al., 709–722. New York: Cold Spring Harbor Laboratory, 1978.

(with G. J. Elfenbein, D. S. Borgaonkar, et al.) "Cytogenetic evidence for recurrence of acute myelogenous leukemia after allogenic bone marrow transplantation in hematopoietic cells." *Blood* 52 (1978): 627–636.

(with A. V. Carrano, B. H. Mayall, et al.) "Chromosomal DNA cytophotometry in 20q-nonspecific myeloid disorders." *Cancer Research* 39 (1979): 2984–2987.

(with P. K. Martin and J. M. Baron) "The use of bone core biopsies for cytogenetic analysis." *American Journal of Human Genetics* 51 (1979): 163–166.

(with V. L. Swain and A. V. Carrano) "Transcription and hybridization of 125 I-cRNA from flow sorted chromosomes." *Chromosoma* 70 (1979): 293–304.

"Chromosome studies in children and adults with leukemia." In *Modern Trends in Human Leukemia IV*, edited by R. Neth et al., 27–33. Berlin: Springer-Verlag, 1980.

"The implications of nonrandom chromosome changes for malignant transformation." In *Modern Trends in Human Leukemia IV*, edited by R. Neth et al., 151–155. Berlin: Springer-Verlag, 1980.

"Second International Workshop on Chromosomes in Leukemia." *Cancer Research* 40 (1980): 4826–4827.

(with G. H. Borgstrom, L. Teerenhovi, et al.) "Clinical implications of monosomy 7 in acute non-lymphocytic leukemia." *Cancer Genetics and Cytogenetics* 2 (1980): 115–126.

(with Y. Kaneko, I. Check, et al.) "The 14q+ chromosome on pre-B ALL." *Blood* 56 (1980): 782–785.

(with R. A. Streuli, J. R. Testa, et al.) "Dysmyelopoietic syndrome: Sequential clinical and cytogenetic studies." *Blood* 55 (1980): 636–644.

(with J. R. Testa, C. Hawkins, et al.) "A balanced translocation (17;22) and a pericentric inversion of chromosome 5." *Cancer Genetics and Cytogenetics* 2 (1980): 270–275.

"Nonrandom chromosome changes in hematologic diseases." In *Atlas of Blood Cells*, edited by D. Zucker-Franklin et al., 605–635. Philadelphia: Lea and Febiger, 1981.

"Third International Workshop on Chromosomes in Leukemia." *Cancer Genetics and Cytogenetics* 4 (1981): 95–142.

(with M. C. Egues, M. Waghray, et al.) "Cytogenetic analysis of bone marrow and peripheral blood samples stored in fixative for several years." *Stain Technology* 56 (1981): 109–112.

(with M.E.B. Gaeke, J. W. Vardiman, et al.) "Human T-cell lymphoma with suppressor effects on the mixed lymphocyte reaction (MLR)." *Blood* 57 (1981): 634–641.

(with Y. Kaneko and M. C. Egues) "Interstitial deletion of 11p limited to Wilms' tumor cells in a patient without aniridia." *Cancer Research* 41 (1981): 4577–4578.

(with Y. Kaneko, J. D. Variakojis, et al.) "Chromosome abnormalities in Down's syndrome patients with acute leukemia." *Blood* 58 (1981): 459–466.

(with R. A. Streuli, Y. Kaneko, et al.) "Lymphoblastic lymphoma in adults." *Cancer* 47 (1981): 2510–2516.

(ith M. Waghray, M. C. Egues, et al.) "Methods of processing marrow samples may affect the frequency of detectable aneuploid cells." *American Journal of Hematology* 11 (1981): 409–415.

"Chromosome gains in malignant disease." In *Gene Amplification*, edited by R. T. Schimke, 291–296. New York: Cold Spring Harbor Laboratory, 1982.

"Consistent translocations in human leukemia." In *Human Genetics, part B: Medical Aspects*, edited by B. Bonne-Tamir, 233–240. New York: Alan R. Liss, 1982.

(with Y. Kaneko) "Clinical significance of chromosome abnormalities in childhood and adult leukemia." In *Pediatric Oncology, Pancreatic Tumors in Children*, edited by G. B. Humphrey et al., 57–78. Boston: Kluwer, 1982.

(————, R. A. Larson, et al.) "Nonrandom chromosome abnormalities in angio-immunoblastic lymphadenopathy." *Blood* 60 (1982): 877–887.

(with Y. Kaneko, H. S. Maurer, et al.) "Chromosome pattern in childhood acute non-lymphocytic leukemia (ANLL)." *Blood* 60 (1982): 389–399.

(with Y. Kaneko, D. Variokojis, et al.) "Correlation of karyotype with clinical features in acute lymphoblastic leukemia (ALL)." *Cancer Research* 42 (1982): 2918–2929.

(————) "Lymphoblastic lymphoma: Cytogenetic, pathologic, and immunologic studies." *International Journal of Cancer* 30 (1982): 273–279.

(with M. Otter) "Leukemia, lymphomas, and related disorders." In *The Principles and Practice of Medical Genetics*, edited by A.E.H. Emery and D. L. Rimoin, 1076–1090. Edinburgh: Churchill Livingstone, 1982.

(with C. M. Richman) "Correlation of *in vitro* culture pattern and Q-banded karyotype in acute nonlymphocytic leukemia." *American Journal of Hematology* 14 (1982): 37–48.

(with S. Z. Salahuddin, P. D. Markham, et al.) "Establishment, characterization, and differentiation induction of a new human diploid myelo-monocytic cell line (HL 92) derived from a patient with acute myelo-monocytic leukemia." *Leukemia Research* 6 (1982): 729–741.

(with K. K. Wong, H. M. Golomb, et al.) "Prognostic significance of post-transfusion hepatitis and chromosomal abnormalities in adult acute nonlymphocytic leukemia." *Cancer Genetics and Cytogenetics* 5 (1982): 281–292.

"Chromosome changes in leukemic cells as indicators of mutagenic exposure." In *Chromosomes and Cancer*, edited by J. D. Rowley et al., 140–159. New York: National Academy Press, 1983.

"Correlation of karyotype and oncogenes in human leukemia and lymphoma." In *Normal and Neoplastic Hematopoiesis*, edited by D. W. Golde et al., 225–246. New York: Alan R. Liss, 1983.

(with K. S. Albain, M. M. Le Beau, et al.) "Development of a dysmyelopoietic syndrome in a hairy cell leukemia patient treated with chlorambucil: Cytogenetic and morphologic evaluation." *Cancer Genetics and Cytogenetics* 8 (1983): 107–115.

(with Y. Kaneko, K. Kondo, et al.) "Further chromosome studies on Wilms' tumor cells of patients without aniridia." *Cancer Genetics and Cytogenetics* 10 (1983): 191–197.

(with Y. Ueshima, G. Alimena, et al.) "Cytogenetic studies in patients with hairy cell leukemia." *Hematology/Oncology* 1 (1983): 215–226.

(with J. E. Ultmann, eds.) *Chromosomes and Cancer*. New York: National Academy Press, 1983.

"Consistent chromosome rearrangements in human malignant disease and oncogene location." In *The Cancer Cell*, edited by G. F. van de Woude et al., 221–226. New York: Cold Spring Harbor Laboratory, 1984.

"Fourth International Workshop on Chromosomes in Leukemia." *Cancer Genetics and Cytogenetics* 11 (1984): 249–360.

"Implications of consistent chromosome rearrangements." In *Genes and Cancer*, edited by J. M. Bishop, J. D. Rowley, et al., 503–524. New York: Alan R. Liss, 1984.

"Significance of chromosome rearrangements in leukemia and lymphoma." In *Leukemia*, edited by I. L. Weissman et al., 179–202. Berlin: Springer-Verlag, 1984.

(with J. M. Bishop et al., eds.) *Genes and Cancer*. New York: Alan R. Liss, 1984.

(with J. M. Haren, F. Wong-Staal, et al.) "Chromosome pattern in cells from patients positive for human T-cell leukemia virus." In *Human T-Cell Leukemia-Lymphoma Viruses*, edited by R. C. Gall et al., 85–89. New York: Cold Spring Harbor Laboratory, 1984.

(with R. H. Jacobs, R. A. Larson, et al.) "Hypercalcemia and lytic bone lesions in a patient with B-cell non-Hodgkin's lymphoma." *New England Journal of Medicine* 310 (1984): 263–264.

(with Y. Kaneko, D. Variakojis, et al.) "Prognostic implications of karyotype and morphology in patients with non-Hodgkin's lymphoma." *International Journal of Cancer* 32 (1984): 683–692.

(with H. P. Koeffler) "Therapy related acute nonlymphocytic leukemia." In *Neoplastic Diseases of the Blood*, edited by P. H. Wiernik et al., 357–381. Edinburgh: Churchill Livingstone, 1984.

(with K. Kondo, R. R. Chilcote, et al.) "Chromosome abnormalities from patients with sporadic Wilms' tumor." *Cancer Research* 44 (1984): 5376–5381.

(with P. K. Martin) "An improved technique for sequential R-, Q- and C-banding of bone marrow chromosomes." *Stain Technology* 58 (1984): 7–12.

(with J. B. Miller, J. R. Testa, et al.) "The pattern and clinical significance of karyotypic abnormalities in patients with idiopathic and postpolycythemic myelofibrosis." *Cancer* 55 (1984): 582–591.

(with C. M. Richman and H. M. Golomb) "Chronic granulocytic leukemia." In *Leukemias*, edited by J. M. Goldman et al., 208–238. Boston: Butterworths, 1984.

(with Y. Ueshima) "Chromosome studies in patients with multiple myeloma and related paraproteinemias." In *Neoplastic Diseases of the Blood*, edited by P. H. Wiernik et al., 495–512. Edinburgh: Churchill Livingstone, 1984.

(———, D. Variakojis, et al.) "Cytogenetic studies in patients with chronic T-cell leukemia/lymphoma." *Blood* 63 (1984): 1028–1038.

"Chromosomes in mutagen-induced leukemia." In *The Role of Chemicals and Radiation in the Etiology of Cancer*, edited by E. Huberman, 409–418. New York: Raven Press, 1985.

"Report of children's cancer study group." *Journal of Clinical Oncology* 1 (1985): 1–2.

(with R. R. Chilcote and E. Brown) "Lymphoblastic leukemia with lymphomatous features (L-ALL) is associated with abnormalities of the short arm of chromosome No. 9." *New England Journal of Medicine* 313 (1985): 286–291.

(with M. Dean, M. Park, et al.) "The human met oncogene is related to the tyrosine kinase oncogenes." *Nature* 318 (1985): 385–388.

(with M. O. Diaz, M. M. Le Beau, et al.) "The role of the c-mos gene in the 8;21 translocation in human acute myeloblastic leukemia." *Science* 229 (1985): 767–769.

(———) "Trisomy 8 in hematologic neoplasia and the c-myc and c-mos oncogenes." *Leukemia Research* 9 (1985): 1437–1442.

(with H. A. Drabkin, M. O. Diaz, et al.) "Isolation and analysis of the 21q+ chromosome in the acute myelogenous leukemia 8;21 translocation: Evidence that c-mos is not translocated." *Proceedings of the National Academy of Sciences* 82 (1985): 464–468.

(with M. M. Le Beau, M. O. Diaz, et al.) "Chromosomal localization of the human T-cell receptor beta-chain genes." *Cell* 41 (1985): 335–350.

(with M. G. Pearson) "The relationship of oncogenesis and cytogenetics in leukemia and lymphoma." *Annual Review of Medicine* 36 (1985): 471–483.

(———, J. W. Vardiman, et al.) "Increased numbers of marrow basophils may be associated with a t(6;9) in ANLL." *American Journal of Hematology* 18 (1985): 393–403.

(with W. C. Pugh, M. G. Pearson, et al.) "Philadelphia chromosome-negative chronic myelogenous leukemia: A morphologic reassessment." *British Journal of Haemotology* 60 (1985): 457–467.

(with D. Sheer, D. M. Sheppard, et al.) "Localization of the c-erbA1 immediately proximal to the APL breakpoint of chromosome 17." *Annals of Human Genetics* 49 (1985): 167–171.

(with C. R. Suarez, M. M. Le Beau, et al.) "Acute megakaryoblastic leukemia in Down's syndrome: Report of a case and review of cytogenetic findings." *Medical and Pediatric Oncology* 13 (1985): 225–231.

(with Y. Ueshima, M. L. Bird, et al.) "A 14;19 translocation in B-cell chronic lymphocytic leukemia: A new recurring chromosome aberration." *International Journal of Cancer* 36 (1985): 287–290.

(with C. A. Westbrook, M. M. Le Beau, et al.) "Chromosomal localization and characterization of c-abl in the t(6;9) of acute nonlymphocytic leukemia." *Proceedings of the National Academy of Sciences* 82 (1985): 8742–8746.

"The Philadelphia chromosome translocation." In *Genetic Rearrangements in Leukemia and Lymphoma*, edited by J. M. Goldman et al., 82–99. Edinburgh: Churchill Livingstone, 1986.

(with C. D. Bloomfield, A. I. Goldman, et al.) "Chromosomal abnormalities identify high-risk and low-risk patients with acute lymphoblastic leukemia." *Blood* 67 (1986): 415–420.

(with M. O. Diaz, M. M. Le Beau, et al.) "Interferon and c-ets-1 genes in the translocation (9;11)(p22;q23) in human acute monocytic leukemia." *Science* 231 (1986): 165–167.

(with R. H. Jacobs, M. A. Cornbleet, et al.) "Prognostic implications of morphology and karyotype in primary myelodysplastic syndromes." *Blood* 67 (1986): 1765–1772.

(with T. W. McKeithan, E. A. Shima, et al.) "Molecular cloning of the breakpoint junction of a human chromosomal 8;18 translocation involving the T-cell receptor alpha-chain gene and sequences on the 3' side of c-myc." *Proceedings of the National Academy of Sciences* 83 (1986): 6636–6640.

(with E. A. Shima, M. M. Le Beau, et al.) "Gene encoding the alpha chain of the T-cell receptor is moved immediately downstream of c-myc in a chromosomal 8;14 translocation in a cell line from a human T-cell leukemia." *Proceedings of the National Academy of Sciences* 83 (1986): 3439–3443.

"Chromosomal abnormalities in childhood tumors." In *Hematology of Infancy and Childhood*, 3d ed., edited by D. G. Nathan et al., 1179–1494. Philadelphia: W. B. Saunders, 1987.

"Correlation of chromosome abnormalities with histologic and immunologic characteristics in non-Hodgkin's lymphoma and adult T-cell lymphocytic leukemia." *Blood* 70 (1987): 1554–1564.

(with R. R. Chilcote, M. M. Le Beau, et al.) "Association of red cell spherocytosis with deletion of the short arm of chromosome 8." *Blood* 69 (1987): 156–159.

(with M. M. Le Beau, R. S. Lemons, et al.) "Chromosomal localization of the human G-CSF gene to 17q11–12 proximal to the breakpoint of the t(15;17) in acute promyelocytic leukemia." *Leukemia* 1 (1987): 795–799.

(with T. W. McKeithan, T. B. Shows, et al.) "Cloning of the chromosome translocation breakpoint junction of the t(14;19) in chronic lymphocytic leukemia." *Proceedings of the National Academy of Sciences* 84 (1987): 9257–9260.

(with M. J. Pettenati, M. M. Le Beau, et al.) "Assignment of CSF-1 to 5q33.1: Evidence of clustering of genes regulating hematopoiesis and for their involvement in the del(5q) in myeloid disorders." *Proceedings of the National Academy of Sciences* 84 (1987): 2970–2974.

(with N. Quintrell, R. Lebo, et al.) "Identification of a human gene (HCK) that encodes a protein-tyrosine kinase and is expressed in hemopoietic cells." *Molecular and Cellular Biology* 7 (1987): 2267–2275.

(with M. J. Ratain, L. S. Kaminer, et al.) "Acute nonlymphocytic leukemia following etoposide and cisplatin combination chemotherapy for advanced non-small cell carcinoma of the lung." *Blood* 70 (1987): 1412–1417.

(with C. M. Rubin, R. A. Larson, et al.) "Association of a chromosomal 3;21 translocation with the blast phase of a chronic myelogenous leukemia." *Blood* 70 (1987): 1338–1342.

(with C. M. Rubin, C. A. Westbrook, et al.) "Philadelphia chromosome-positive acute lymphoblastic leukemia: Detection of a DNA rearrangement 50–250 kilobases proximal to BCR." In *Recent Advances in Leukemia and Lymphoma*, edited by R. P. Gale et al., 125–131. New York: Alan R. Liss, 1987.

(with C. A. Westbrook, C. M. Rubin, et al.) "Molecular analysis of TCRB and ABL in a human T-cell leukemia cell line (SUP-T3) with a chromosomal 7;9 translocation." *Proceedings of the National Academy of Sciences* 84 (1987): 251–255.

"The specificity of chromosomal abnormalities in leukemia and lymphoma." In *Genetic*

Targeting in Leukemia, edited by D. Pinkel et al., 8–21. Philadelphia: Lippincott, 1988.

(with M. O. Diaz, S. Ziemin-van der Poel, et al.) "Homozygous deletion of the a- and 61 interferon genes in human leukemia and derived cell lines." *Proceedings of the National Academy of Sciences* 85 (1988): 5259–5263.

(with B. N. Harris, E. M. Davis, et al.) "Variant 9;11 translocations: Identification of the critical genetic rearrangement." *Cancer Genetics and Cytogenetics* 30 (1988): 171–175.

(with L. J. Joseph, M. M. Le Beau, et al.) "Molecular cloning, sequencing and mapping of EGR2: A human early growth response gene encoding a protein with zinc fingers (cell growth/transcriptional regulatory/multigene family)." *Proceedings of the National Academy of Sciences* 85 (1988): 7164–7168.

(with R. A. Larson, M. Wernli, et al.) "Short remission durations in therapy-related leukemia despite cytogenetic complete responses to high-dose cytarabine." *Blood* 72 (1988): 1333–1339.

(with Y. S. Li, J. Anastasi, et al.) "A recurring chromosome rearrangement, dic(16;22), in acute nonlymphocytic leukemia." *Cancer Genetics and Cytogenetics* 35 (1988): 143–150.

(with B. L. Samuels, R. A. Larson, et al.) "Specific chromosomal abnormalities in acute nonlymphocytic leukemia correlate with drug susceptibility *in vivo*." *Leukemia* 2 (1988): 79–83.

(with C. A. Westbrook, C. M. Rubin, et al.) "Long-range mapping of the Philadelphia chromosome by pulsed-field gel electrophoresis." *Blood* 79 (1988): 697–702.

"Finding order in chaos." *Perspectives in Biology and Medicine* 32 (1989): 371–384.

"Molecular analysis of rearrangement in Philadelphia (Ph1) chromosome-positive leukemia." In *Modern Trends in Human Leukemia*, vol. 8, edited by R. D. Neth, 3–10. Berlin: Springer-Verlag, 1989.

"Principles of molecular cell biology of cancer: Chromosomal abnormalities." In *Cancer: Principles and Practice of Oncology*, 3d ed., edited by V. T. DeVita et al., 81–97. Philadelphia: Lippincott, 1989.

"The scientific revolution in medicine." In *High School Biology: Today and Tomorrow*, edited by W. G. Rosen, 30–36. New York: National Academy Press, 1989.

"Sixth International Workshop on Chromosomes in Leukemia-Lymphoma." *Cancer Genetics and Cytogenetics* 40 (1989): 149–216.

(with A. L. Hooberman, J. J. Carrino, et al.) "Unexpected heterogeneity of BCR-ABL fusion mRNA detected by polymerase chain reaction in Philadelphia chromosome-positive acute lymphoblastic leukemia." *Proceedings of the National Academy of Sciences* 86 (1989): 4259–4263.

(with M. M. Le Beau) "Cytogenetic and molecular analysis of therapy-related leukemia." In *Viral Oncogenesis and Cell Differentiation*, edited by L. Diamond et al., 130–140. New York: New York Academy of Sciences, 1989.

(with T. W. McKeithan, M. O. Diaz, et al.) "Cloning the 14;19 translocation breakpoint in chronic lymphocytic leukemia." In *The Regulation of Proliferation and Differentiation in Normal and Neoplastic Cells*, edited by E. Frei III, 125–142. New York: National Academy Press, 1989.

(with J. K. Park, T. W. McKeithan, et al.) "An (8;14)(q24;q11) translocation involving the T-cell receptor a and the MYC oncogene 3' region in a B-cell lymphoma." *Genes, Chromosomes and Cancer* 1 (1989): 15–22.

(with S. C. Raimondi, I. D. Dube, et al.) "Clinicopathologic manifestations and breakpoints of the t(3;5) in patients with acute nonlymphocytic leukemia." *Leukemia* 3 (1989): 42–47.

(with J. L. Schwartz, T. Karrison, et al.) "Chromosomal sensitivity of lymphocytes from individuals with therapy-related acute non-lymphocytic leukemia." *Mutation Research* 216 (1989): 119–126.

(with Y. Ueshima, J. M. Haren, et al.) "Culture conditions in chronic lymphocytic leukemia: Relation to karyotype." *Leukemia* 3 (1989): 192–194.

"Chromosomal abnormalities in human cancer." In *The Biology of Human Leukemia*, edited by A. M. Mauer, 177–199. Baltimore: Johns Hopkins University Press, 1990.

"Leukemias, lymphomas, and related disorders." In *Principles and Practice of Medical Genetics*, 2d ed., edited by A.E.H. Emery et al., 1391–1410. Edinburgh: Churchill-Livingstone, 1990.

"Molecular cytogenetics: Rosetta stone for understanding cancer." *Cancer Research* 50 (1990): 3816–3825.

"The Philadelphia chromosome translocation: A paradigm for understanding leukemia." *Cancer* 65 (1990): 2178–2184.

"Recurring chromosome abnormalities in leukemia and lymphoma." In *Seminars in Hematology*, edited by E. R. Jaffe et al., 122–136. Philadelphia: W. B. Saunders, 1990.

(with M. O. Diaz, R. Espinosa III, et al.) "Mapping chromosome band 11q23 in human acute leukemia with biotinylated probes: Identification of 11q23 translocation breakpoints with a yeast artificial chromosome." *Proceedings of the National Academy of Sciences* 87 (1990): 9358–9362.

(with M. O. Diaz, C. M. Rubin, et al.) "Deletions of interferon genes in acute lymphoblastic leukemia." *New England Journal of Medicine* 322 (1990): 77–82.

(with M. M. Le Beau) "Cytogenetics." In *Hematology*, 4th ed., edited by W. J. Williams et al., 78–89. New York: McGraw-Hill, 1990.

(with C. M. Rubin, R. A. Larson, et al.) "T(3;21)(q26;q22): A recurring chromosomal abnormality in therapy-related myelodysplastic syndrome and acute myeloid leukemia." *Blood* 76 (1990): 2594–2598.

(with M. Thangavelu, O. I. Olopade, et al.) "Clinical, morphologic, and cytogenetic characteristics of patients with lymphoic malignancies characterized by both the t(14;18)(q32;q21) and the t(8;14)(q24;q32) or the t(8;22)(q24;q11)." *Genes, Chromosomes and Cancer* 2 (1990): 147–148.

"Cytogenetics: Past, present and future." In *Molecular Foundations of Oncology*, edited by S. Broder, 3–16. Baltimore: Williams and Wilkins, 1991.

(with R. C. Burnett, J. C. David, et al.) "The LCK gene is involved in the t(1;7)(p34; q34) in the T-cell acute lymphoblastic leukemia derived cell line, HSB-2." *Genes, Chromosomes and Cancer* 3 (1991): 461–467.

(with J. Gao, P. Erickson, et al.) "Isolation of a yeast artificial chromosome spanning the 8;21 translocation breakpoint, t(8;21)(q22;q22.3), in acute myelogenous leukemia." *Proceedings of the National Academy of Sciences* 88 (1991): 4882–4886.

(with Y. S. Li, M. M. Le Beau, et al.) "The proportion of abnormal karyotypes in acute leukemia samples related to the method of preparation." *Cancer Genetics and Cytogenetics* 52 (1991): 93–100.

(with C. M. Rubin, D. C. Arthur, et al.) "Therapy-related myelodysplastic syndrome and acute myeloid leukemia in children: Abnormalities of chromosomes 5 and 7 are common." *Blood* 78 (1991): 2982–2988.

(with C. M. Rubin, M. M. Le Beau, et al.) "Impact of chromosomal translocation on prognosis in childhood acute lymphoblastic leukemia." *Journal of Clinical Oncology* 9 (1991): 2183–2192.

(with S. Ziemin-van der Poel, N. R. McCabe, et al.) "Identification of a gene (MLL) which spans the breakpoint in 11q23 translocations associated with human leukemias." *Proceedings of the National Academy of Sciences* 88 (1991): 10735–10739.

"The der(11) chromosome contains the critical breakpoint junction in the 4;11, 9;11, and 11;19 translocations in acute leukemia." *Genes, Chromosomes and Cancer* 5 (1992): 264–266.

"Human leukemia genes: Search for the villains." In *Haematology and Blood Transfusion*, vol. 35, edited by R. Neth et al., lvii–lxii. Berlin: Springer-Verlag, 1992.

(editor) *Advances in Understanding Genetic Changes in Cancer.* Washington, DC: National Academy Press, 1992.

(with J. Anastasi, M. M. Le Beau, et al.) "Detection of trisomy 12 in chronic lymphocytic leukemia by fluorescence *in situ* hybridization to interphase cells: A simple and sensitive method." *Blood* 79 (1992): 1796–1801.

(with S. K. Bohlander, R. Espinosa III, et al.) "A method for the rapid sequence-independent amplification of microdissected chromosomal material." *Genomics* 13 (1992): 1322–1324.

(with P. Erickson, J. Gao, et al.) "Detection of AML t(8;21) breakpoints and a fusion transcript with similarity to *Drosophila* segmentation gene, runt." *Blood* 80 (1992): 1825–1831.

(with R. A. Larson, M. M. Le Beau, et al.) "Balanced translocations involving chromosome bands 11q23 and 21q22 in therapy-related leukemia." *Blood* 79 (1992): 1892–1893.

(with N. R. McCabe, R. C. Burnett, et al.) "Cloning of cDNAs of the MLL gene that detect DNA rearrangements and altered RNA transcripts in human leukemic cells with 11q23 translocations." *Proceedings of the National Academy of Sciences* 89 (1992): 11794–11798.

(with W. L. Neuman, C. M. Rubin, et al.) "Chromosomal loss and deletion are the most common mechanisms for loss of heterozygosity from chromosomes 5 and 7 in malignant myeloid disorders." *Blood* 79 (1992): 1501–1510.

(with O.I. Olopade, R. B. Jenkins, et al.) "Molecular analysis of deletion of the short arm of chromosome 9 in human gliomas." *Cancer Research* 2 (1992): 2523–2529.

(with O. I. Olopade, M. Thangavelu, et al.) "Clinical, morphologic, and cytogenetic characteristics of 26 patients with acute erythroblastic leukemia." *Blood* 80 (1992): 1–9.

(with N. Onodera, N. R. McCabe, et al.) "Hyperdiploidy arising from near-haploidy in childhood acute lymphoblastic leukemia." *Genes, Chromosomes and Cancer* 2 (1992): 331–336.

(with J. K. Park, M. M. Le Beau, et al.) "A complex genetic rearrangement in a t(10; 14)(q24;q11) associated with a T-cell acute lymphoblastic leukemia." *Genes, Chromosomes and Cancer* 4 (1992): 32–40.

(with M. J. Ratain) "Therapy-related acute myeloid leukemia secondary to inhibitors of topoisomerase II: From the bedside to the target genes." *Annals of Oncology* 3 (1992): 107–111.

(with M. Thangavelu, L. Snyder, et al.) "Cytogenetic characterization of B-cell lymphomas from SCID mice injected with lymphocytes from EBV-positive donors." *Cancer Research* 52 (1992): 4678–4681.

(with J. Anastasi, J. Feng, et al.) "Cytogenetic clonality in myelodysplastic syndromes studied with fluorescence *in situ* hybridization: Lineage, response to growth factor therapy, and clone expansion." *Blood* 81 (1993): 1580–1585.

(with J. C. Aster and J. Sklar) "The clinical applications of new DNA diagnostic technology on the management of cancer patients." *Journal of the American Medical Association* 270 (1993): 2331–2337.

(————) "The impact of new DNA diagnostic technology on the management of cancer patients: Survey of diagnostic techniques." *Archives of Pathology and Laboratory Medicine* 117 (1993): 1104–1109.

(with R. C. Burnett, R. Espinosa III, et al.) "Molecular analysis of a t(11;14)(q23;q11) from a patient with a null-cell acute lymphoblastic leukemia." *Genes, Chromosomes and Cancer* 7 (1993): 38–46.

(with M. O. Diaz, N. R. McCabe, et al.) "Analysis of 11q23 chromosome translocation breakpoints associated with human leukemia." In *Cancer Chemotherapy: Challenges for the Future*, edited by K. Kimura et al., 95–103. Tokyo: Excerpta Medica, 1993.

(with H. Kobayashi, R. Espinosa, et al.) "Analysis of deletions of the long arm of chromosome 11 in hematologic malignancies with fluorescence *in situ* hybridization." *Genes, Chromosomes and Cancer* 8 (1993): 246–252.

(————) "Do terminal deletions of 11q23 exist? Identification of undetected translocations with fluorescence *in situ* hybridization." *Genes, Chromosomes and Cancer* 7 (1993): 204–208.

(————) "Heterogeneity of breakpoints of 11q23 rearrangements in hematologic malignancies identified with fluorescence *in situ* hybridization." *Blood* 82 (1993): 547–551.

(————) "Variability of 11q23 rearrangements in hematopoietic cell lines identified with fluorescence *in situ* hybridization." *Blood* 11 (1993): 3027–3033.

(with F. Mitelman) "Chromosomal abnormalities in human cancer and leukemia." In *Cancer: Principles and Practices of Oncology*, 4th ed., edited by V. T. DeVita et al., 67–91. Philadelphia: Lippincott, 1993.

(with G. Nucifora, C. R. Begy, et al.) "The 3;21 translocation on myelodysplasia results in a fusion transcript between the AML1 gene and the gene for EAP, a highly conserved protein associated with the Epstein-Barr virus small RNA EBER1." *Proceedings of the National Academy of Sciences* 90 (1993): 7784–7788.

(with G. Nucifora, D. J. Birn, et al.) "Detection of DNA rearrangements in the AML1 and ETO loci and of an AML1/ETO fusion mRNA in patients with t(8;21) AML." *Blood* 81 (1993): 883–888.

(————) "Involvement of the AML1 gene in the t(3;21) in therapy-related leukemia and in chronic myeloid leukemia in blast crisis." *Blood* 81 (1993): 2728–2734.

(with G. Nucifora and R. A. Larson) "Persistence of the 8;21 translocation in patients with AML-M2 in long-term remission." *Blood* 82 (1993): 712–715.

(with O. I. Olopade) "Recurring chromosome rearrangements in human cancer." In

Cancer Medicine, 3d ed., edited by J. F. Holland et al., 99–120. Philadelphia: Lea and Febiger, 1993.

(with B. W. Porterfield, H. Pomykala, et al.) "The use of methylthioadenosine phosphorylase activity to select for human chromosome 9 in interspecies and intraspecies hybrid cells." *Somatic Cell and Molecular Genetics* 19 (1993): 469–477.

(with C. M. Rubin) "Chromosomal abnormalities in childhood malignant diseases." In *Hematology of Infancy and Childhood*, 4th ed., edited by D. G. Nathan et al., 1179–1206. Philadelphia: W. B. Saunders, 1993.

(with H. G. Super, N. R. McCabe, et al.) "Rearrangements of the MLL gene in therapy-related acute myeloid leukemia in patients previously treated with agents targeting DNA-topoisomerase II." *Blood* 82 (1993): 3705–3711.

(with M. J. Thirman, H. J. Gill, et al.) "A cDNA probe detects all rearrangements of the MLL gene in leukemias with common and rare 11q23 translocations." *New England Journal of Medicine* 329 (1993): 909–914.

"Can we meet the challenge?" *American Journal of Human Genetics* 54 (1994): 403–413.

"Chromosome translocations: Dangerous liaisons." *Leukemia* 8 (1994): 1–6.

"Cytogenetic and molecular analysis of pediatric neoplasms: Diagnostic and clinical implications." *Pediatric Pathology* 14 (1994): 167–176.

"DNA diagnosis in oncology." *Journal of Technology Assessment in Health Care* 10 (1994): 644–654.

"Genetics: Transforming medicine from art to science." In *Science and Medicine in the 21st Century—A Global Perspective*, 13–28. London: Royal Society, 1994.

"Rearrangements involving chromosome 11q23 in acute leukemia." In *Seminars in Cancer Biology*, vol. 4, edited by T. H. Rabbitts, 377–385. London: Academic Press, 1994.

(with S. K. Bohlander, R. Espinosa III, et al.) "Sequence-independent implication and labeling of yeast artificial chromosome for fluorescence *in situ* hybridization." *Cytogenetics and Cellular Genetics* 65 (1994): 108–110.

(with D. F. Claxton, P. Liu, et al.) "Detection of fusion transcripts generated by the inversion 16 chromosome in acute myelogenous leukemia." *Blood* 83 (1994): 1750–1756.

(with H. J. Gill, P. G. Rothberg, et al.) "Clonal, non-constitutional rearrangements of the MLL gene in infant twins with ALL: *In utero* chromosome rearrangements of 11q23." *Blood* 83 (1994): 641–644.

(with N. R. McCabe, M. Kipiniak, et al.) "DNA rearrangements and altered transcripts of the MLL gene in a human T-ALL cell line Karpas 45 with an X;11 (q13;q23) chromosome translocation." *Genes, Chromosomes and Cancer* 9 (1994): 221–224.

(with G. Nucifora) "The AML1 and ETO genes in acute myeloid leukemia with a t(8; 21)." *Leukemia and Lymphoma* 14 (1994): 353–362.

(with O. I. Olopade) "Human chromosome maps: Relevance to identifying genetic changes in tumors of the nervous system." In *Molecular Genetics of the Nervous System*, edited by A. J. Levine et al., 179–194. New York: Wiley-Liss, 1994.

(with J. Pedersen-Bjergaard) "The balanced and the unbalanced chromosome aberrations of acute myeloid leukemia may develop in different ways and may contribute to malignant transformation." *Blood* 83 (1994): 2780–2786.

(with G. J. Swansbury, S. D. Lawler, et al.) "Long-term survival in acute myelogenous leukemia: A second follow-up of the Fourth International Workshop on Chromosomes in Leukemia." *Cancer, Genetics and Cytogenetics* 73 (1994): 1–7.

(with M. M. Le Beau) "Cytogenetics." In *Williams Hematology*, 5th ed., edited by E. Beutler et al. New York: McGraw-Hill, 1995.

"The role of chromosome aberrations in human hematologic malignant diseases." In *Some Aspects of Dynamic DNA*, edited by K. S. McCarty et al. New York: Plenum, in press.

(with G. Nucifora, C. R. Begy, et al.) "Consistent intergenic splicing and production of multiple transcripts between AML1 at 21q22 and unrelated genes at 3q26 in (3; 21)(q26;q22) translocations." *Proceedings of the National Academy of Sciences* (in press).

Other Works

Personal communications, 1994.

Works about Janet Davison Rowley

Hollmann, M. R. "With a little bit of lab space." *University of Chicago Magazine* (Winter 1988).

Other References

Lejeune, J., et al. "Study of somatic chromosomes in 9 infants with mongolism." *Comptes Rendus Hebdomadaires des Seances de l'Academie des Sciences* 248 (1959): 1721–1722.

Nowell, P. C., and D. A. Hungerford. "A minute chromosome in human granulocytic leukemia." *Science* 132 (1960): 1497.

Rous, P. "A sarcoma of the fowl transmissible by an agent separable from the tumor cells." *Journal of Experimental Medicine* 13 (1911): 347–411.

Sandberg, A. A. *The Chromosomes in Human Cancer and Leukemia*. New York: Elsevier, 1980, 1990.

Tijo, J., and A. Levan. "The chromosome number of man." *Hereditas* 42 (1956): 1–6.

JANE ANNE RUSSELL (1911–1967)

David R. Stronck

BIOGRAPHY

Jane Anne Russell was born on February 9, 1911, in Los Angeles County, California, in the area now known as Watts. Both of her parents were of English descent: Josiah Howard and Mary Ann (Phillips) Russell. At the turn of the century her parents moved to California, probably from Maryland. Jane Anne Russell was the youngest of five children; she had two sisters and two brothers.

Her family was poor; her father worked as a rancher and later as a deputy sheriff. He built a homestead near Los Angeles, while his wife cared for the large family. Russell learned how to sew; later she became an expert knitter and made many beautiful sweaters. She developed a love for gardening that became an important part of her later life. Also, she enjoyed origami (Japanese paper folding).

Russell was an outstanding student in high school, especially in mathematics. When she graduated from Polytechnic High School, Long Beach, California (1928), she was ranked second in her class. At age 17 she entered the University of California, Berkeley. After four years, in 1932, she graduated with the degree of Bachelor of Arts and was ranked first in her class. Her academic achievements were recognized by the Kraft Prize, a Phi Beta Kappa key, a Stewart Scholarship, and the University Gold Medal.

She remained at the University of California, Berkeley, to continue graduate studies. She was supported by her work as a technician for the chairman of the Department of Biochemistry, Edward S. Sundstroem. Russell held the California Fellowship in Biochemistry (1934) and the Rosenberg Fellowship (1935). She was accepted as a Ph.D. candidate in the Institute of Experimental Biology at the University of California. The director of the Institute, Herbert M. Evans, organized research on vitamins and hormones. Russell worked under the supervision of Leslie L. Bennett, investigating the role of pituitary hormones in carbohydrate metabolism.

For three months in 1936 she was a research associate at the Department of

Pharmacology, Washington University School of Medicine in St. Louis, Missouri, with Nobel laureates (1947) Carl and Gerty Cori.* By the time she completed her doctorate (1937), six of her papers on pituitary hormones in carbohydrate metabolism were published, and several more had been accepted for publication. Dr. Russell remained at the Institute for another year, supported by a Porter Fellowship from the American Physiological Society (1937–38).

In 1938 Russell moved to Yale University, New Haven, Connecticut, where she served as a postdoctoral fellow in the Department of Physiology (1938–41). From 1941 to 1950 she was an instructor at Yale. Under C.N.H. Long at Yale's Department of Physiology, she continued her research for 12 years (1938–50). On August 26, 1940, Russell married Alfred Ellis Wilhelmi, a close research collaborator, but continued to use Russell as her professional name.

During her years at Yale, Russell received many awards. She was given the CIBA Award of the Endocrine Society (1945). In 1949 she served on the National Institutes of Health (NIH) peer review committee for research grant applications. She served as the vice president of the Endocrine Society (1950–51).

Russell and Wilhelmi moved to Emory University in Atlanta, Georgia, in 1950. He was appointed Professor and Chairman of the Department of Biochemistry, while she was appointed an Assistant Professor. From 1950 to 1954 she was supported by the NIH, while investigating how growth hormone acts on metabolism and the endocrine system.

Russell was well recognized as a national leader. From 1954 to 1957 she served on a committee of the National Research Council for the evaluation of postdoctoral fellowships. She was a member of the Board of the National Science Foundation (1958–64). In 1961 she and her husband were honored by sharing the Upjohn Award of the Endocrine Society. She was elected to membership in Sigma Xi.

Despite her extensive research publications, her many national honors, and her excellent teaching, she was not promoted to the rank of associate professor until 1953. She spent 12 years at this rank before becoming professor in 1965. Unfortunately, she developed mammary cancer in 1962. During the last years of her life, she cared for her aged mother, who lived with her. Prior to Russell's death she was seen by a friend editing manuscripts for the *American Journal of Physiology*. She worked up to the end of her life without complaint. Russell died on March 12, 1967.

WORK

Prior to receiving her Ph.D. degree, Russell had already published three papers. They indicated that fasting rats lose muscle glycogen following pituitary removal, while injections of anterior pituitary extract prevented this loss. Hypophysectomized rats (with their pituitary surgically removed) injected with pituitary extract exhibited a depressed carbohydrate oxidation following an acceleration. With R. E. Fisher and C. F. Cori she demonstrated that pituitary extracts given to normal animals affected their glucose utilization. Her most important early article was on the relationship between the anterior lobe pituitary and carbohydrates (Russell, "The relation . . ." 1938). Between 1936 and 1939 Russell's work produced about 16 papers concerned with pituitary extract and carbohydrate metabolism. She also published a review article on the subject for *Physiological Reviews* (1938), an honor at such a young age.

When Russell arrived at Yale University in 1938, she was already recognized as a known authority in her field. She found that during carbohydrate deprivation, some unknown pituitary factor was needed to maintain adequate levels of tissue carbohydrate, as well as blood glucose. This factor was determined to be growth hormone (somatotropin). Later research on the mechanisms of body growth and the maintenance and breakdown of carbohydrates was based on her early studies.

Russell designed relatively simple experiments that were capable of providing decisive answers to important questions. Her work showed that the carbohydrate regulating activity of the pituitary extracts is found in the growth hormone that was discovered, purified, and isolated by Evans et al. (1943, 1944).

In 1930 Russell reported that the effects of the adrenal cortex hormones on carbohydrate metabolism were different from those of the anterior pituitary hormones. She supported these data in 1941 by demonstrating that animals without an anterior pituitary exhibited the same "glycostatic" effects whether they were eviscerated or not. Later, in 1941, Russell's group at Yale University studied metabolic changes associated with hemorrhagic shock. She and coworkers developed a micromethod for determining amino nitrogen in the blood and used the technique to show the inhibitory effect that insulin and anterior pituitary extract had upon eviscerated animals whose blood nitrogen had risen. Russell and collaborators developed a method to prepare enough pure crystalline growth hormone to perform further investigations on carbohydrate metabolism (1947).

When Russell and her husband moved to Emory University, they continued to study growth hormone and its effect upon reducing urea formation in nephrectomized animals. Growth hormone was shown to decrease protein catabolism by aiding the entry of amino acids into tissues and incorporation of them into proteins. It was further indicated that growth hormone could affect cardiac and skeletal muscle. She also studied the influences of epinephrine and norepinephrine on glycogenolysis. She proposed that fluctuating growth hormone may have a role in starvation and the hypoglycemic state. She theorized that growth hormone is needed to prevent the breakdown of essential structural proteins. Her theories were later confirmed by Solomon Berson, Rosalyn Yalow* and others. Russell also pioneered the use of statistical analyses for analyzing biological data.

She was a member of the American Physiological Society and served on its editorial board. In 1962 she served as a section editor of metabolism and endocrinology for the *American Journal of Physiology* and the *Journal of Applied Physiology*.

Russell was an excellent lecturer, with great clarity and style. She cared for her students and was proud of their successes. Her students remember her as a wonderful professor and leader (Long 1967).

BIBLIOGRAPHY

Works by Jane Anne Russell

Scientific Works

"Carbohydrate levels in fasted and fed hypophysectomized rats." *Proceedings of the Society for Experimental Biology and Medicine* 34 (1936): 279–281.

(with L. L. Bennett) "Maintenance of carbohydrate levels in fasted hypophysectomized rats treated with anterior pituitary extracts." *Proceedings of the Society for Experimental Biology and Medicine* 34 (1936): 406–409.

(with R. E. Fisher and C. F. Cori) "Glycogen disappearance and carbohydrate oxidation in hypophysectomized rats." *Journal of Biological Chemistry* 115 (1936): 627–634.

(with R. I. Pencharz and C. F. Cori) "Relation of anterior and posterior lobe of the hypophysis to insulin sensitivity in the rat." *Proceedings of the Society for Experimental Biology and Medicine* 35 (1936): 32–35.

"Carbohydrate Metabolism in the Hypophysectomized Rat." Ph.D. diss., University of California, Berkeley, 1937.

"Effects of hypophysectomy and of anterior pituitary extracts on disposition of fed carbohydrate in rats." *Proceedings of the Society for Experimental Biology and Medicine* 37 (1937): 31–33.

"Production of refractoriness to action of anterior pituitary extracts in depressing oxidation of fed carbohydrate." *Proceedings of the Society for Experimental Biology and Medicine* 37 (1937): 33–34.

(with L. L. Bennett) "Carbohydrate storage and maintenance in the hypophysectomized rat." *American Journal of Physiology* 118 (1937): 196–205.

(with G. T. Cori) "A comparison of the metabolic effects of subcutaneous and intravenous epinephrine infusions in normal and hypophysectomized rats." *American Journal of Physiology* 119 (1937): 167–174.

"The action of insulin and anterior pituitary extracts in normal and hypophysectomized rats." *American Journal of Physiology* 124 (1938): 774–790.

"The anterior pituitary factor which maintains muscle glycogen in fasted hypophysectomized rats." *Endocrinology* 22 (1938): 80.

"The effects of hypophysectomy and anterior pituitary extracts on the disposition of fed carbohydrate in rats." *American Journal of Physiology* 121 (1938): 755–764.

"The effects of thyroxin on the carbohydrate metabolism of hypophysectomized rats." *American Journal of Physiology* 122 (1938): 547–550.

"The relation of the anterior pituitary to carbohydrate metabolism." *Physiological Reviews* 18 (1938): 1–27.

(with J. M. Craig) "Adrenal cortical hormone and ant. pituitary extract on carbohydrate levels in fasted hypophysectomized rats." *Proceedings of the Society for Experimental Biology and Medicine* 39 (1938): 59–62.

"Effects of anterior pituitary and adrenal cortical extracts on metabolism of adrenalectomized rats fed glucose." *Proceedings of the Society for Experimental Biology and Medicine* 41 (1939): 626–628.

"The relationship of the anterior pituitary and the adrenal cortex in the metabolism of carbohydrate." *American Journal of Physiology* 128 (1940): 552–561.

(with A. E. Wilhelmi) "The metabolism of amino-acids and keto-acids in the adrenalectomized rat kidney slices." *American Journal of Physiology* 129 (1940): 453–454.

"Carbohydrate metabolism in the eviscerated rat." *American Journal of Physiology* 133 (1941): 434.

(with A. E. Wilhelmi) "Glyconeogenesis in kidney tissue of the adrenalectomized rat." *Journal of Biological Chemistry* 140 (1941): 747–754.

(———) "The metabolism of kidney tissue in the adrenalectomized rat." *Journal of Biological Chemistry* 137 (1941): 713–725.

"The anterior pituitary in the carbohydrate metabolism of the eviscerated rat." *American Journal of Physiology* 136 (1942): 95–104.

(with J. Tepperman) "Carbohydrate metabolism in the eviscerated rat. II." *Federation Proceedings* 1 (1942): 76–77.

"The adrenals and hypophysis in the carbohydrate metabolism of the eviscerated rat." *American Journal of Physiology* 140 (1943): 98–106.

"The relationship of the anterior pituitary to the thyroid and the adrenal cortex in the control of carbohydrate metabolism." In *Essays in Biology in Honor of Herbert M. Evans*, 507–527. Berkeley: University of California Press, 1943.

(with B. O. Frame and A. H. Wilhelmi) "Colorimetric estimation of amino nitrogen in blood." *Journal of Biological Chemistry* 149 (1943): 255–270.

"The colorimetric estimation of small amounts of ammonia by the phenol-hypochlorite reaction." *Journal of Biological Chemistry* 156 (1944): 457–461.

"Interference with the colorimetric determination of lactic acid (Barker-Summerson method) by nitrate and nitrite ions." *Journal of Biological Chemistry* 156 (1944): 463–465.

"Note on the colorimetric determination of amino nitrogen." *Journal of Biological Chemistry* 156 (1944): 467–468.

(with C.N.H. Long and F. Engel) "The role of the peripheral tissues in the metabolism of protein and carbohydrate during hemorrhagic shock in the rat." *Journal of Experimental Medicine* 79 (1944): 1–7.

(with C.N.H. Long and A. E. Wilhelmi) "Oxygen consumption of liver and kidney tissue from rats in hemorrhagic shock." *Journal of Experimental Medicine* 79 (1944): 23–33.

"Carbohydrate metabolism." *Annual Review of Biochemistry* 14 (1945): 309–332.

(with A. E. Wilhelmi, H. G. Engel, et al.) "Some aspects of the nitrogen metabolism of liver tissue from rats in hemorrhagic shock." *American Journal of Physiology* 144 (1945): 674–682.

(with A. E. Wilhelmi and C.N.H. Long) "The effects of anoxia and of hemorrhage upon the metabolism of the cerebral cortex in the rat." *American Journal of Physiology* 144 (1945): 683–692.

(———) "The effects of hepatic anoxia on the respiration of liver slices *in vitro*." *American Journal of Physiology* 144 (1945): 669–673.

(with E. G. Frame) "Effects of insulin and anterior pituitary extracts on the blood amino nitrogen in eviscerated rats." *Endocrinology* 39 (1946): 420–429.

(with C.N.H. Long) "Amino nitrogen in the liver and muscle of rats in shock after hemorrhage." *American Journal of Physiology* 147 (1946): 175–180.

(with J. B. Fishman and A. E. Wilhelmi) "A crystalline pituitary protein with high growth activity." *Science* 106 (1947): 402.

(with A. E. Wilhelmi and J. B. Fishman) "A new preparation of crystalline anterior pituitary growth hormone." *Journal of Biological Chemistry* 176 (1948): 735–745.

"The endocrine system." In *Textbook of Physiology* by W. H. Howell, 16th ed., edited by J. F. Fulton, 1091–1159. Philadelphia: Saunders, 1949.

(with M. Cappiello) "The effects of pituitary growth hormone on the metabolism of administered amino acids in nephrectomized rats." *Endocrinology* 44 (1949): 333–344.

(———) "The relationship of temperature and insulin and dosage to the rise in plasma amino nitrogen in the eviscerated rat." *Endocrinology* 44 (1949): 127–133.

(with A. E. Milman) "Effects of growth hormone on the blood sugar and insulin sensitivity of the rat." *Federation Proceedings* 8 (1949): 111.

"Physiology of the pituitary-adrenal system." *Bulletin of the New York Academy of Medicine* 26 (1950): 240–250.

(with B. A. Illingworth) "Effect of purified growth hormone on the glycogen content of tissues of the rat." *Federation Proceedings* 9 (1950): 65–66.

(with A. E. Milman) "Some effects of purified pituitary growth hormone on carbohydrate metabolism in the rat." *Endocrinology* 47 (1950): 114–125.

(with A. E. Wilhelmi) "The glycostatic action of purified growth hormone." *Endocrinology* 47 (1950): 26–29.

"The effect of growth hormone on glycosuria in the diabetic rat." *Endocrinology* 48 (1951): 462–470.

"The effect of purified growth hormone on urea formation in nephrectomized rats." *Endocrinology* 49 (1951): 99–104.

"Metabolic functions of the endocrine glands." *Annual Review of Physiology* 13 (1951): 327–366.

(with B. A. Illingworth) "The effects of growth hormone on glycogen in tissues of the rat." *Endocrinology* 48 (1951): 423–434.

"Effects of growth hormone on metabolism of nitrogen." In *Protein Metabolism, Hormones and Growth*, edited by W. H. Cole et al., 46–61. New Brunswick, NJ: Rutgers University Press, 1953.

"Hormonal control of glycogen storage." In *Hormonal Factors in Carbohydrate Metabolism. Ciba Foundation Colloquia on Endocrinology*, vol. 6, edited by G.E.W. Wolstenholme et al., 193–206. Boston: Little, Brown, 1953.

"Symposium on amino acid metabolism; hormonal control of amino acid metabolism." *Federation Proceedings* 14 (Sept. 1955): 696–705.

(with W. L. Bloom) "Effects of epinephrine and of norepinephrine on carbohydrate metabolism in rat." *American Journal of Physiology* 183 (Dec. 1955): 356–364.

(———) "Extractable and residual glycogen in tissues of the rat." *American Journal of Physiology* 183 (Dec. 1955): 345–355.

(with G. A. Adrouny) "Effects of growth hormone and nutritional status on cardiac glycogen in rat." *Endocrinology* 59 (1956): 241–251.

(with W. L. Bloom) "Hormonal control of glycogen in heart and other tissues in rats."
   Endocrinology 58 (1956): 83–94.
"Effects of growth hormone on protein and carbohydrate metabolism." *American Journal of Clinical Nutrition* 5 (4) (1957): 404–416.
"The use of isotopic tracers in estimating rates of metabolic reactions." *Perspectives in
   Biology and Medicine* 1 (2) (1958): 138–173.
(with A. E. Wilhelmi) "Growth (hormonal regulation)." *Annual Review of Physiology*
   20 (May 1958): 43–66.
(———) "Endocrines and muscle." In *Structure and Function of Muscle*, vol. 2, edited
   by G. H. Bourne, 142–198. New York: Academic Press, 1960.
(with F. W. Fales and J. N. Fain) "Some applications and limitations of the enzymic,
   reducing (Somogyi), and anthrone methods for estimating sugars." *Clinical
   Chemistry* 7 (Aug. 1961): 389–403.
(with L. Crook) "Comparison of metabolic responses of rats to hypoxic stress produced
   by two methods." *American Journal of Physiology* 214 (5) (1968): 1113–1116.
(———) "Effects of growth hormone and cortisol on endocrine-deficient rats subjected
   to hypoxia." *American Journal of Physiology* 214 (5) (1968): 1117–1121.
(with A. McKee) "Effect of acute hypophysectomy and growth hormone on FFA mo-
   bilization, nitrogen excretion and cardiac glycogen in fasting rats." *Endocrinology*
   83 (1968): 1162–1165.

Works about Jane Anne Russell

Emory University Biographical Files. Special Collections, Robert W. Woodruff Library,
   Emory University, Atlanta, Georgia.
Long, C.N.H. "In memoriam: Jane A. Russell." *Endocrinology* 81 (1967): 689–692.
Tepperman, Jay. "Jane Anne Russell (Wilhelmi)." *Physiologist* 10 (1967): 443–444.

Other References

Evans, Herbert M., Walter Marx, et al. "Purification of the growth hormone of the
   anterior pituitary." *Journal of Biological Chemistry* 147 (1943): 77–89.
Evans, Herbert M., and Choh Hao Li. "The isolation of pituitary growth hormone."
   Science 99 (1944): 183–184.

FLORENCE RENA SABIN (1871–1953)

Linda H. Keller

BIOGRAPHY

Florence Rena Sabin was born on November 9, 1871, in a frame house in Central City, a Colorado mining town. The patriarch of the Sabin family, William, fled from France to England to escape persecution as a Huguenot by Cardinal Richelieu and then came to settle in Rehoboth, Massachusetts, in 1643. Florence Sabin's great-grandfather, Levi, had studied at Dartmouth Medical School to become a physician. Her paternal uncle, Robert, was also a physician.

Florence Sabin's father, George Kimball Sabin, had wanted to become a physician but had succumbed to gold fever and moved west to Colorado. Her mother, Serena Miner, a native of Vermont, went to Colorado by stagecoach in 1867 and taught school until she met George Sabin. They were married in Black Hawk, Colorado. Florence Sabin had a sister, Mary, who was two years older then she, and the girls were very close friends. Two brothers were subsequently born, but both died within a year, Richman in 1877 and Albert Hall in 1879. Her father moved the family to Denver, where Florence Sabin began school at the age of five.

While innate intelligence and a natural curiosity certainly contributed to Florence Sabin's many successes throughout her career, many people influenced her life and career. After her mother died in 1878, Florence and Mary went to Wolfe Hall, an Episcopal boarding school in Denver, for a year and then moved to Chicago to live with their uncle Albert, who was a schoolteacher. During the four years she lived there, Sabin learned to know and love music and developed a love of all things natural. She learned the rewards of self-motivation and activity. Uncle Albert taught her the enjoyment of good reading, and through him, she developed the characteristics of sympathy and understanding.

When she was 12, the sisters moved to Vermont to live with their grandparents. After their grandmother died in 1885, they began school at the Vermont Academy. Sabin's exceptional intelligence was recognized and appreciated there. She began to learn the importance of developing good work and study habits. Sabin was discouraged from a career in music but turned to her natural interest in the sciences.

In 1889 she followed her sister to Smith College, where again her exceptional intelligence and enthusiasm brought her to the attention of the faculty. After her sister had graduated, Sabin felt lonely, being on her own for the first time. Through the encouragement of several faculty members at Smith College, she decided to become a physician. She was encouraged to consider Johns Hopkins University in Baltimore, where efforts to establish a medical school were under way. She was advised that, as a doctor, she would have to become involved in the women's movement for equal rights and status.

Sabin graduated Phi Beta Kappa from Smith College in 1893 but could not immediately apply to Johns Hopkins due to lack of money. She went back to Wolfe Hall in Denver to teach mathematics. There she organized Saturday morning nature walks with her students. During the following summer, she and her sister were hired by Mrs. Ella Strong Denison to spend the summer with her three children on Lake Geneva in Northwoodside, Wisconsin. At the lake Sabin guided the children to discover the natural world for themselves. Mrs. Denison filled the void of her mother's death, and they maintained a lifelong relationship.

Two years later Sabin returned to Smith College as an assistant in zoology for one year, then received a stipend for summer study at the Marine Biological Laboratory in Woods Hole, Massachusetts. Having saved enough money, at the age of 25 she entered the Johns Hopkins University as a medical student in the class of 1900. She was one of 14 women in a class of 44 students. The requirements for entrance into this medical school were the highest in the country, as were the qualifications of the faculty. At the time, Johns Hopkins Medical School was the only one in the country that would accept women. Its opening embodies a historic victory for women. An endowment of $500,000 was bestowed on the Medical School by Mary E. Garrett of Baltimore. Its terms stipulated that successful applicants must have a college degree, including a knowledge of the sciences and a proficiency in French and German, and that women students must be admitted on the same terms as men.

Sabin had no trouble passing the entrance requirements. She immediately felt an intellectual attraction to the Anatomy Department of Dr. Franklin Paine Mall, who also sensed in her the promise of a research investigator. In her sophomore year he assigned her a research project on the origin of the lymphatic system, which initiated some of her most significant scientific work. Her close association with Dr. Mall continued until his death in 1917, at which time Sabin was given the honor of writing his life history (Sabin 1934). Thus, she was able to acknowledge fully his singular influence over her chosen career in medical research.

Sabin received her M.D. degree in 1900 and remained associated with Johns Hopkins Medical School through 1925. She fulfilled her internship (1900–1), later was a research fellow (1901–2), then joined Dr. Mall's staff as an assistant in anatomy (1902). She was promoted through the academic ranks to the position of professor of histology (1917), becoming the first woman full professor at the university.

The Rockefeller Institute for Medical Research had opened in 1904 under the directorship of Dr. Simon Flexner, who knew Sabin and invited her to join his staff. While it was not an easy decision to leave Johns Hopkins, Sabin accepted Flexner's offer, setting up the Department of Cellular Studies at the Institute in 1924. Soon she joined an interinstitutional research program, which involved 21 universities and other research institutions, all working toward the advancement of knowledge about tuberculosis. After her retirement from the Institute in 1938, she remained a member emeritus until her death. Although she moved back to Colorado to live with her sister, she frequently traveled east to attend board meetings.

After World War II ended, the Governor of Colorado established the Post War Planning Committee to facilitate the re-entry of soldiers into civilian life. In 1944 Sabin accepted the appointment of Chairman of its Sub-Committee on Health. She continued to work on state health matters until her eightieth birthday. Sabin retired from active participation in medical service at the age of 81 in order to take care of her sister, Mary. She died of a heart attack on October 3, 1953, in her home in Denver.

Sabin's work brought her honors and awards throughout her life. In the period from 1924 to 1926 she was the first woman president of the American Association of Anatomists, and in 1925 was elected to membership in the National Academy of Sciences, the first woman to achieve that distinction. She received numerous honorary Sc.D. degrees, including those from Smith College (1910), the University of Michigan (1926), Mount Holyoke College (1929), the University of Colorado (1935), the University of Pennsylvania (1937), and the Woman's Medical College of Pennsylvania (1950). She received many prestigious awards, including the National Achievement Award sponsored by Chi Omega Sorority in 1932, the Trudeau Medal from the National Tuberculosis Association in 1945, and the Lasker Award from the American Public Health Association in 1951. Several buildings at the University of Colorado were named for her, and the Sabin Award in Public Health was established in her honor. A bronze statue of Sabin by Joy Buba was placed in Statuary Hall in the national Capitol in 1958.

WORK

Sabin's first research problem, undertaken as a medical student, was to study the fiber tracts of the brain stem, which resulted in her construction of a three-dimensional model showing the tracts and nuclei of the medulla oblongata of a newborn. The "Sabin model" became a teaching aid used worldwide for studying the anatomy of the nervous system. In 1901 she published her *Atlas of the*

Medulla and the Midbrain, which served as the classic textbook in medical schools for many years.

The study of the development of the lymphatic system was assigned to her by Dr. Mall in her second year of medical school. Through her observations, she concluded that the lymphatic vessels are modified veins, closed at their collecting ends, with fluid entering them by seepage. She thought that they arose from a cell differentiation of a vasoformative type, which would become the endothelial lining of the vessels. Her findings were controversial. The opposing theory stated that the lymphatic vessels arose independently of the vascular system. While the earliest stages of development were still being debated in the late 1950s, her work suggested the origin of the lymphatic system from angioblasts and permitted the distinct separation of lymph spaces from tissue spaces. She won the $1,000 prize from the Naples Table Association, and on December 15, 1915, she summarized her research in the prestigious Harvey Lecture (Sabin 1915–16).

Sabin was one of the first histologists to work with fresh tissue and living cells by using vital staining, a laborious but rewarding practice. Among other things, she was able to study the origin of blood cells, blood vasculature, the structure and growth of bone marrow, and the functions of monocytes as phagocytic cells. As part of the collaborative research on tuberculosis, Sabin observed the cellular reactions to various tubercle components (lipids, polysaccharides, and proteins), and formulated the theory that antibodies were formed by the phagocytic cells that engulfed these antigens. Again, while the validity of some of her interpretations may have come into question, the accuracy of her observations remains sound. Her research findings on tuberculosis laid much of the foundation for an understanding of inflammation and other pathologic processes. In general, her accuracy and devotion to truth led to a better understanding of the anatomy, physiology, and pathology of the body in health and disease.

As a teacher Sabin was outstanding. During her first experiences with teaching, she developed her talents for inspiring her students to love what they were learning and doing. Sabin was able to win the confidence and respect of her students. She was able to recognize gifted students and stimulate them to independent work.

Sabin was an energetic member of many scientific committees and influential boards. She and her Sub-Committee on Health worked hard to develop and enact the "Sabin health bills," which were aimed at protecting the citizens of Colorado from preventable and controllable diseases. Although in her 70s, she tirelessly campaigned for these health bills until the new state health laws went through the Colorado legislature in 1947. Her life's philosophy can be summed up with the words of Leonardo da Vinci, which appear on her personally designed bookplate: "Thou, O God, dost sell unto us all good things at the price of labour" (MacCurdy 1938).

BIBLIOGRAPHY

Works by Florence Rena Sabin

Scientific Works

Space does not permit the listing of the complete works of Florence Rena Sabin. Listed here are all works by Sabin, except those cited in McMaster and Heidelberger (1960). Included here are all references cited in the text.

An Atlas of the Medulla and the Midbrain. Baltimore: Friedenwald, 1901.

"Tuberculous pericarditis with effusion; repeated tappings; bacilli in the exudate; recovery." *American Medicine* 3 (1902): 388–389.

"The method of growth of the lymphatic system." *Harvey Lectures, Series XI* (1915–16): 124–145.

Franklin Paine Mall: The Story of a Mind. Baltimore: Johns Hopkins University Press, 1934.

Works about Florence Rena Sabin

Best, Katherine, and Katharine Hillyer. "Colorado's little doctor." *Coronet* 25 (Mar. 1949): 99–103.

Bluemel, Elinor. *Florence Sabin: Colorado Woman of the Century.* Boulder: University of Colorado Press, 1959.

Booth, Alice. "America's twelve greatest women." *Good Housekeeping* 92 (June 1931): 50–51.

"Florence Rena Sabin, M.D." *Journal of the American Medical Women's Association* 5 (1950): 466–467.

Maisel, Alfred. "Colorado's lady dynamo." *Survey Graphic* 36 (Feb. 1938): 138–140.

McMaster, P. D., and M. Heidelberger. "Florence Rena Sabin." *Biographical Memoirs of the National Academy of Sciences* 34 (1960): 271–319.

Melville, Mildred. "Woman with two careers." *Today's Health* 31 (Feb. 1953): 42–43.

Obituary. *British Medical Journal* (Oct. 31, 1953): 997–998.

Obituary. *New York Times* (Oct. 4, 1953).

Other References

MacCurdy, Edward (ed.). *The Notebooks of Leonardo da Vinci.* New York: Reynal and Hitchcock, 1938.

RUTH SAGER (1918–1997)

Carol A. Biermann

BIOGRAPHY

Ruth Sager was born on February 7, 1918, in Chicago, Illinois, one of three daughters of Leon B. Sager, an advertising executive, and Deborah (Borovik) Sager. Sager's family life was intellectually stimulating, as her parents immersed the children in literature and the arts. The youngsters were encouraged to achieve academically. In 1934, at the age of 16, Sager graduated from New Trier High School in Winnetka, Illinois. She then attended the University of Chicago.

At that time the university promoted a broad liberal arts education. Sager's educational experiences there exposed her to many fields in which previously she had little interest. Sager stated that her four years as an undergraduate really changed her life (Personal communications 1994). She switched from a liberal arts major to biology because she found all the science courses exciting. At first she thought that she might become a doctor, and she took all the courses preliminary to medical school. However, her enthusiasm for research became paramount. Sager graduated Phi Beta Kappa from the University of Chicago in 1938, with a B.S. degree in physiology.

She supported herself for a few years as a secretary. Then she attended Rutgers University in New Jersey, where she obtained an M.S. degree in plant physiology (1944). Her unpublished master's thesis, "Nutritional Status of the Tomato Seedling in Relation to Successful Transplanting," was part of a wartime research project that was never completed. While at Rutgers she was a research and teaching assistant in plant physiology and a research assistant for a peach-breeding project (Hunter College Archives 1948).

Sager has been described as a woman with a ready smile and a voice that "carries a merry lilt." Apparently, she respected logic and order, because she chose the field of genetics, which possesses these attributes (*Current Biography* 1967, 367–370). She has always traveled extensively to gather information as well as to disseminate her knowledge to the scientific community. In 1944 Sager married Seymour Melman, an industrial economist and social critic.

In 1945 Sager matriculated for a Ph.D. degree at Columbia University. From 1946 to 1947, she received a fellowship for research in corn genetics. Professor Marcus M. Rhoades, Sager's mentor, describes her as being "very articulate, has a thorough knowledge of modern genetics and a pleasing personality" (Hunter College Archives 1948). Both he and Barbara McClintock* were very strong influences upon her development as a scientist (Personal communications 1994). At Columbia Sager was elected to Sigma Xi, and she graduated in 1948 with a Ph.D. degree in genetics. Her dissertation was entitled, "Mutability of Waxy Locus in Maize." Of her dissertation research Dr. Louis J. Stadler, a specialist in the field, stated, "[I]t was a very beautiful piece of work . . . and a very valuable contribution" (Stadler 1951).

Following her graduation Sager became a Merck postdoctoral fellow for the National Research Council (1949–51) under the supervision of Dr. Sam Granick of Rockefeller Institute for Medical Research (now Rockefeller University). She spent the summer after receiving her Ph.D. degree at the Hopkins Marine Station, Pacific Grove, California. She took a microbiology course with Professor Cornelius B. Van Niel to increase her genetics and biochemical background and to perfect her microbial techniques. Sager now concentrated upon the problem of understanding the chloroplast, the organelle of photosynthesis, from all aspects of its morphology and physiology.

At this juncture a crucial choice was made with the aid of Professor Van Niel and her principal adviser, Gilbert Morgan Smith, to focus her research on *Chlamydomonas reinhardi*, a soil organism, affectionately called "Clammy" by Sager. Smith had carried out the essential experiments showing that this organism was best suited for the experiments Sager wanted to perform (Personal communications 1994). Her relationship with the algae is reminiscent of Barbara McClintock's "feelings" for her organism, the corn plant. Granick stated that "Dr. Sager has demonstrated skill and persistence in this task" and "[m]y estimate of Dr. Sager is that she is an excellent investigator, a careful planner of experiments, for which [she] has demonstrated a considerable ingenuity and insight" (Granick 1950).

Sager was a Merck fellow (1949–51) at the Rockefeller Institute for Medical Research. Then she became a staff member at Rockefeller (1951–55). From 1955 to 1960 she served as a research associate at Columbia University and then as a senior research scientist (1960–66) at Columbia University's Department of Zoology in the laboratory of Professor Francis Ryan, supported solely by her grants. Together with Ryan, Sager wrote the successful textbook *Cell Heredity* (1961). The book began with a description of deoxyribonucleic acid (DNA) instead of with the studies of Gregor Mendel. In 1962 Sager was a nonresident fellow at the University of Edinburgh, Scotland Institute of Animal Genetics (Debus 1968, 1466). During these years Sager published more than 50 pioneering scientific papers on cytoplasmic genes. In addition to her research she has enjoyed collecting modern art. After Ryan's death Sager received her appointment to Hunter College (now part of the City University of New York) as Professor of Biology (1966–75).

In the early 1970s there were several crucial changes in Sager's life. She married Dr. Arthur Pardee, a biochemist and Professor of Biochemistry, then at Princeton University. He remained her supportive life partner and significant influence upon her life. In addition, Sager wrote her second book, *Cytoplasmic Genes and Organelles* (1972).

In the 18 years prior to becoming full Professor, Sager's only other positions were those of research fellows. The practice of assigning many scientists to positions of low stature was common during this period, particularly for women.

Sager held a Guggenheim research fellowship at the Imperial Cancer Research Fund Laboratory in London, England (1972–73). There she was fortunate to be guided by the Director of the laboratory, Dr. Michael Stoker (now Sir Michael Stoker). This facilitated her switch from chloroplast genetics to cancer research (Personal communications 1994).

From 1975 to 1997 Sager served as Professor of Cellular Genetics in the Department of Microbiology and Molecular Genetics at the Harvard Medical School. She was also Chief of the Division of Cancer Genetics, Dana-Farber Cancer Institute. Sager was passionately concerned with adequate funding for vital cancer research. In 1977 she expressed concern that governmental National Institutes of Health funds were not being allocated properly. She stressed the need for basic cancer research (*New York Times* 1977, 26).

In 1977 Sager was elected to the National Academy of Sciences in recognition for her work on cytoplasmic genes. The following year she was elected to the American Academy of Arts and Sciences. The National Cancer Institute honored Sager with the Outstanding Investigator Award in 1986. This enabled her to devote her laboratory to full-time investigations of the genetic basis of cancer. The Gilbert Morgan Smith Medal of the National Academy of Sciences was presented to her (1988). In 1990 she received the Schneider Memorial Lecturer Award, University of Texas, Galveston, and the Prince Takamatsku Cancer Research Fund Lecturer Award, Japan. Sager was elected to the Institute of Medicine (1992).

Sager was known as "the high priestess of the second genetic system," referring to cytoplasmic genetics, because her investigations were of primary importance in altering biologists' ideas concerning inheritance and evolution (*New York Times* 1965).

On March 29, 1997, Sager died of bladder cancer at her home in Brookline, Massachusetts.

WORK

"Genetics is the core science of biology. Its ultimate subject matter is a class of chemical compounds, the hereditary determinants, which are the prime movers of cellular metabolism and, thereby, of the even more complex processes of development and evolution," stated Sager in her classic textbook *Cell Heredity* (Sager and Ryan 1961). Sager's fascination with genetics dates back to her years as a graduate student at Columbia University, when she engaged in corn genetics. At this time nobody could anticipate that the work that Sager was to perform would modify the basic ideas of genetics.

While at Rockefeller University, Sager began to question the prevailing dogma of genetics that genes in eukaryotic cells occur only in the chromosomes of the nucleus and that these were the only mechanisms of inheritance. Actually, several reports had alluded to the possible existence of nonchromosomal genetics. Non-Mendelian genes (not obeying the principles set up by Mendel) were first proposed by Carl Correns in 1909. He had shown that mutations in color patterns (i.e., white-striped offspring of all-green plants) were always inherited from the maternal parent. This type of inheritance persisted through generations and did not revert to Mendelian principles. Sager was familiar with the work of Correns and others.

In a 1949 letter to Granick, Sager discussed her impending visit to Hopkins Marine Station in order to ''find *the* organism'' for her Merck fellowship research. By August of that year, Sager had started working with *Chlamydomonas* and was investigating methods to screen for mutants. The problem was to study the origin, growth, and development of the chloroplast, using an integration of many techniques, including electron microscopy, genetics, and biochemistry. *Chlamydomonas reinhardi*, isolated and characterized by G. M. Smith, was a good selection for chloroplast research since its chloroplast was of large size. The organism could grow and reproduce rapidly on various media and could be handled by microbial techniques. Above all, *Chlamydomonas* had two mating types, and classical genetic analysis could easily be carried out. Biochemical studies could also be applied to the organism. *Chlamydomonas*, a photosynthetic green alga, could also be grown in the dark on reduced carbon source. This would allow mutants with a damaged chloroplast to survive.

Sager and Tsubo (1961) had developed a system for producing mutations in nonchromosomal genes. Mutant strains were isolated. Some were unable to grow because they could not photosynthesize, some were temperature-sensitive, and some were resistant to various antibiotics. The organisms may reproduce sexually, as there are two mating types (plus and minus), which allow crossing of organisms and genetic analyses. Sager was able to induce mating by nitrogen deprivation. When mated, the mating types fuse together to form diploid zygospores. These develop into four meiotic products that then can be separated and grown as clones (identical reproductive products of one cell) on agar.

Large numbers of *Chlamydomonas* were grown in the presence of streptomycin (an antimicrobial drug) and other chemicals, thereby selecting streptomycin resistant strains, and other mutants. Streptomycin affects chloroplast development by inhibiting protein synthesis in chloroplast ribosomes. Crosses between contrasting parents (streptomycin-sensitive and streptomycin-resistant) gave rise to offspring showing characteristics of only one parental type (plus). Thus, these mutants did not exhibit Mendelian segregation. This pattern of 4:0 inheritance resembled maternal cytoplasmic inheritance in higher plants.

Sager was able to map nonchromosomal genes by performing crosses and identifying recombinants. She did this by treating the plus mating type parent with a dose of ultraviolet light prior to mating. Following this procedure, 50% of the zygotes showed biparental inheritance. The offspring of these zygotes contained complete sets of nonchromosomal genes from each parent. These cells, called *cytohets*, were heterozygous for cytoplasmic genes. Nobody had previously performed mating experiments of this type, because they had thought that nonchromosomal genes were transmitted only by the female. Sager showed that males also could transmit these genes. By crossing maternal and paternal nonchromosomal genes and producing hybrids and then recrossing them, Sager was able to establish patterns of nonchromosomal inheritance.

The nonchromosomal genes were shown to be very stable over a period of several years, even when exposed to heat, cold, or ultraviolet treatment. The first demonstration that these nonnuclear DNA (genes) existed in the chloroplasts was made by Sager and M. Ishida (1963). The technique utilized was cesium chloride density gradient centrifugation. DNA extracted from the whole *Chlamydomonas* cell was centrifuged to equilibrium in cesium chloride. Then the position of the banded DNA was photographed with ultraviolet light, and the pattern was traced. This pattern was then compared with one made utilizing only DNA from the chloroplast. In the original tracing the chloroplast DNA could now be identified. Sager was the first to isolate the DNA from the chloroplasts utilizing ultracentrifugation.

Cytoplasmic genes are a second genetic system that provides the organism with stability and some degree of flexibility under environmental stress conditions. Sager suggested that cytoplasmic genes may be an earlier genetic system that existed prior to the origin of chromosomes. They were found to be present as a single circular linkage group or circular cytoplasmic "chromosome." The DNA of chloroplasts was determined to be much larger than that of mitochondria, another DNA containing cytoplasmic organelle. The chloroplast DNA could potentially code for several hundred proteins.

She presented her findings at the first Gilbert Morgan Smith Memorial Lecture (1960). Subsequently, in 1963, she reported further information at The Hague in the Netherlands at the 11th International Congress of Genetics. Her theories were initially received with much skepticism by scientists. However, her investigations were soon corroborated by other scientists using different organisms.

Sager and M. G. Hamilton (1967) also characterized cytoplasmic and chloroplast ribosomes of *Chlamydomonas* by ultracentrifugation. Cytoplasmic ribosomes resembled those of animals and higher plants in most respects. However, like bacterial ribosomes, they required high levels of magnesium ions for stability.

Another related area of research that Sager and coworkers entered involved the many facets of tumor formation in cancer, viral interactions with cells, and chemotherapeutic agents. Early on, Sager proposed that cancer is a kind of hereditary disease, since cancerous cells clone themselves as such (Sager,

Cytoplasmic . . . 1972). Viral genes have also been postulated to be linked to cancer formation. Indeed, some forms of cancer are virally induced.

Sager and coworkers at the Dana-Farber Cancer Institute have been investigating many aspects of cancer relationships, such as viral interactions, and cell products such as cytokines. The long-term goals are to identify antioncogenic mechanisms. One of Sager's more recent concerns was tumor suppression. In "Tumor suppressor genes: The puzzle and the promise," Sager proposed that oncogenesis is primarily a later-life disease in humans, involved with the loss or inactivation of tumor suppressor gene function (Sager 1989). Research has focused on identifying the tumor suppressor genes that have malfunctioned in different types of cancers and, perhaps, according to Sager, utilizing these genes as anticancer therapies. Thus far, Sager's group has identified more than 30 genes that play a protective role in the cell. Sager stated that a new gene called maspin (an acronym for mammary serpin), a serine protease inhibitor, is a most promising discovery. "The maspin protein, encoded by the maspin gene, is a potent inhibitor of the process by which tumor cells eat their way out of their normal location in the breast, and spread (metastasize) to other tissues" (Personal communications 1994). Although investigations are not complete, Sager's group has high hopes for the therapeutic potential of maspin.

Among Sager's many organizational memberships were the International Society for Cell Biology, the American Society for Cell Biology, the Genetics Society of America, the American Society of Naturalists, the National Academy of Sciences, the American Academy of Arts and Sciences, Sigma Xi, the American Society of Biological Chemists, the American Association of Cancer Research, and the American Society of Human Genetics. Sager's numerous committee assignments included National Institutes of Health Genetics Study Section (1981–83); National Science Foundation Presidential Young Investigator Award Panel (1984); Scientific Advisory Board, Friedrich Miescher Institute, Basel, Switzerland (1989–95); and the National Advisory Council on Aging (1993–96).

Sager appeared to be happiest when faced with a myriad of genetic puzzles. She kept on with her research, never fearful of challenging the basic tenets of biology.

NOTE

Ruth Sager was instrumental in providing many of the materials used in the preparation of this manuscript. The assistance of Hollee Haswell, Columbiana curator, Julio L. Hernandez-Delgado, Hunter College archivist, and Renee D. Mastrocco, Rockefeller University archivist, is also gratefully acknowledged.

BIBLIOGRAPHY

Works by Ruth Sager

Scientific Works

Space does not permit the listing of the complete works of Ruth Sager. Listed here are all works by Sager except those cited in *Cytoplasmic* . . . (1972); "Nuclear and . . ." (1974); Sager (1986); "Mutation rates . . ." (1988); Sager (1991). Included here are her dissertation and all references cited in the text.

"Nutritional Status of the Tomato Seedling in Relation to Successful Transplanting." M.S. thesis, Rutgers University, 1944.

"Mutability of Waxy Locus in Maize." Ph.D. diss., Columbia University, 1948.

"On the mutability of the waxy locus in maize." *Genetics* 36 (1951): 510–540.

"Studies of inheritance with *Chlamydomonas reinhardi*." *Atti del IX Congress Internationale di Genetica, Carologia*, vol. suppl. (1954): 1107–1109.

"The architecture of the chloroplast in relation to its photosynthetic activities." *Brookhaven Symposia in Biology* 11 (1958): 101–117.

"Chromosomal and non-chromosomal determinants in *Chlamydomonas*." *Pathologie-Biologie* 9 (1961): 760–761.

(with Francis J. Ryan) *Cell Heredity*. New York: John Wiley, 1961.

(with Y. Tsubo) "Genetic analysis of streptomycin resistance and dependence in *Chlamydomonas*." *Zeitschrift fuer Vererbungslehre* 92 (1961): 430–438.

(with M. Ishida) "Chloroplast DNA in *Chlamydomonas*." *Proceedings of the National Academy of Sciences* 50 (1963): 725–730.

(with I. B.Weinstein and Y. Ashkenazi) "Coding ambiguity in cell-free extracts of *Chlamydomonas*." *Science* 140 (1963): 304–306.

"Non-chromosomal genes in *Chlamydomonas*." In *Genetics Today, Proceedings XI International Congress of Genetics*, 579–589. Oxford, England: Pergamon Press, 1964.

"Non-chromosomal heredity." *New England Journal of Medicine* 271 (1964): 352–357.

"Studies of cell heredity with *Chlamydomonas*." In *Biochemistry and Physiology of the Protozoa*, vol. 3, edited by S. H. Hunter, 297–318. New York: Academic Press, 1964.

"Genes outside the chromosome." *Scientific American* 212 (1965): 70–81.

(with M. G. Hamilton) "Cytoplasmic and chloroplast ribosomes of *Chlamydomonas*: Ultracentrifugal characterization." *Science* 157 (1967): 709–711.

(with Z. Ramanis) "A genetic map of non-Mendelian genes in *Chlamydomonas*." *Proceedings of the National Academy of Sciences* 65 (1970): 593–600.

(————)"Genetic studies of chloroplast DNA in *Chlamydomonas*." In *Control of Organelle Development, Society for Experimental Biology Symposium 23*, edited by P. L. Miller, 401–417. London: Cambridge University Press, 1970.

(————) "Methods of genetic analysis of chloroplast DNA in *Chlamydomonas*." In *Autonomy and Biogenesis of Mitochondria and Chloroplasts, Australian Academy of Science Symposium*, edited by N. K. Boardman et al., 250–259. Amsterdam, Holland: North Holland, 1970.

(with R. Wells) "Denaturation and renaturation kinetics of chloroplast DNA from *Chlamydomonas reinhardi*." *Journal of Molecular Biology* 58 (1971): 611–622.

Cytoplasmic Genes and Organelles. New York: Academic Press, 1972.

"Evolution of preferential transmission mechanisms in cytoplasmic genetic systems." *Brookhaven Symposia in Biology* 23 (1972): 495–502.

"Genetic control of biogenesis and enzyme formation in green algae." In *First International Symposium on the Genetics of Industrial Microorganisms*, 163–173. Prague: Czech Academy of Science, 1973.

(with V. R. Flechtner) "Ethidium bromide induced selective and reversible loss of chloroplast DNA." *Nature New Biology* 241 (113) (1973): 277–279.

(with Z. Ramanis) "The mechanism of maternal inheritance in *Chlamydomonas*. Biochemical and genetic studies." *Theoretical and Applied Genetics* 43 (1973): 101–108.

"Genetic and biophysical studies of chloroplast DNA in *Chlamydomonas*." *Archives de Biologie* 85 (1974): 524–525.

"Nuclear and cytoplasmic inheritance in green algae." In *Algal Physiology and Biochemistry*, edited by W.D.P. Stewart, 314–345. Berkeley: University of California Press, 1974.

(with J. Blamire and V. R. Flechtner) "Regulation of nuclear DNA replication by the chloroplast in *Chlamydomonas*." *Proceedings of the National Academy of Sciences* 71 (1974): 2867–2871.

(with Z. Ramanis) "Mutations that alter the transmission of chloroplast genes in *Chlamydomonas*." *Proceedings of the National Academy of Sciences* 71 (1974): 4698–4702.

(with G. Schlanger) "Correlation of chloroplast DNA and cytoplasmic inheritance in *Chlamydomonas* zygotes." *Journal of Cell Biology* 63 (1974): 301a.

(————)"Localization of five antibiotic resistances at the subunit level in chloroplast ribosomes." *Proceedings of the National Academy of Sciences* 71 (1974): 1715–1719.

"Patterns of inheritance of organelle genomes: Molecular basis and evolutionary significance." In *Genetics and Biogenesis of Mitochondria and Chloroplasts*, edited by C. W. Birky, Jr., 252–257. Columbus: Ohio State University Press, 1975.

(with R. Kitchin) "Selective silencing of eukaryotic DNA." *Science* 189 (1975): 426–433.

(with N. Ohta and M. Inouye) "Identification of a chloroplast ribosomal protein altered by a chloroplast gene mutation in *Chlamydomonas*." *Journal of Biological Chemistry* 250 (1975): 3655–3659.

(with Z. Ramanis) "Effects of UV irradiation on segregation and recombination of chloroplast genes in *Chlamydomonas*." *Genetics* 80 (1975): s72.

"The circular diploid model of chloroplast DNA in *Chlamydomonas*." In *Genetics and Biogenesis of Chloroplast and Mitochondria*, edited by Theodor Bucher et al., 295–303. Amsterdam, Holland: Elsevier-North, 1976.

(with Z. Ramanis) "Chloroplast genetics of *Chlamydomonas*. I. Allelic segregation ratios." *Genetics* 83 (1976): 303–321.

(————)"Chloroplast genetics of *Chlamydomonas*. II. Mapping by cosegregation frequency analysis." *Genetics* 83 (1976): 323–340.

(———— and B. Singer) "Chloroplast genetics of *Chlamydomonas*. III. Closing the circle." *Genetics* 83 (1976): 341–354.

(with G. Schlanger) "Chloroplast DNA: Physical and genetic studies." In *Handbook of Genetics*, vol. 5, edited by R. C. King, 371–424. New York: Plenum, 1976.

"Cytoplasmic inheritance." In *Cell Biology: A Comprehensive Treatise*, edited by L. Goldstein and S. Prescott, 279–317. New York: Academic Press, 1977.

(with W. G. Burton, R. J. Roberts, et al.) "A site-specific single strand endonuclease from the eukaryote *Chlamydomonas*." *Proceedings of the National Academy of Sciences* 74 (1977): 2687–2691.

(with A. N. Howell) "Selection of cytoplasmic gene mutations in cultured mouse cells." *Genetics* 86 (1977): s30.

(————)"Transmission of viral induced cell transformations in hybrid and cybrids." *Journal of Cell Biology* 75 (1977): 385a.

(————)"Non-coordinate expression of SV40 transformation in cell hybrids." *Journal of Cell Biology* 79 (1978): 3986a.

(————) "Tumorigenicity and its suppression in cybrids of mouse and Chinese hamster cell lines." *Proceedings of the National Academy of Sciences* 75 (1978): 2358–2362.

(with A. N. Howell and P. Kovac) "Chromosome reduction in selected mouse and Chinese hamster cell hybrids." *Genetics* 88 (1978): s545.

(with J. M. Wilson, A. N. Howell, et al.) "Polyethylene glycol mediated cybrid formation: High efficiency techniques and cybrid formation without enucleation." *Somatic Cell Genetics* 4 (1978): 745–752.

"Methylation and restriction of chloroplast DNA." *Journal of Supramolecular Biology* 15 (1979): 97–110.

"Methylation and restriction of chloroplast DNA: The molecular basis of maternal inheritance in *Chlamydomonas*." In *Extrachromosomal DNA*, edited by D. J. Cummings et al., 97–112. New York: Academic Press, 1979.

"Transposable elements and chromosomal rearrangements in cancer—a possible link." *Nature* 282 (1979): 447–449.

(with W. G. Burton, C. T. Grabowy, et al.) "The role of methylation in the modification and restriction of chloroplast DNA in *Chlamydomonas*." *Proceedings of the National Academy of Sciences* 76 (1979): 1390–1394.

(with P. V. Cherington) "Loss of epidermal growth factor requirement and malignant transformation." *Proceedings of the National Academy of Sciences* 76 (1979): 3937–3941.

(with N. Howell) "Cytoplasmic genetics of mammalian cells conditional sensitivity to mitochondrial inhibitors and isolation of new mutant phenotypes." *Somatic Cell Genetics* 5 (1979): 833–846.

(————) "Phenotypic expression and cytoplasmic inheritance of mutations affecting mitochondrial functions." *Journal of Cell Biology* 83 (1979): 385a.

(with P. E. Kovac) "Genetic analysis of tumorigenesis. IV. Chromosome reduction and marker segregation in progeny clones from Chinese hamster cell hybrids." *Somatic Cell Genetics* 5 (1979): 491–502.

(with C. J. Marshall) "Genetic control of anchorage in rodent cells." *Journal of Cell Biology* 83 (1979): 449A.

(with H. D. Royer) "Methylation of chloroplast DNA in the life cycle of *Chlamydomonas*." *Proceedings of the National Academy of Sciences* 76 (1979): 5794–5798.

"The application of DNA methylation studies to the analysis of chloroplast evolution." *Annals of the New York Academy of Sciences* 361 (1980): 209–218.

(with R. M. Kitchin) "Genetic analysis of tumorigenesis. V. Chromosomal analysis of tumorigenic and non-tumorigenic diploid Chinese hamster cell lines." *Somatic Cell Genetics* 6 (1980): 75–87.

(with H. Sano) "Deoxyribonucleic acid methyl transferase from the eukaryote *Chlamydomonas reinhardi.*" *European Journal of Biochemistry* 105 (1980): 471–480.

(——— and H. D. Royer) "Identification of 5-methylcytosine in DNA fragments immobilized on nitrocellulose paper." *Proceedings of the National Academy of Sciences* 77 (1980): 3581–3585.

(with J. Spudich) "Regulation of the *Chlamydomonas* cell cycle by light and dark." *Journal of Cell Biology* 85 (1980): 136–145.

(with A. Anisowicz and N. Howell) "Genomic rearrangements and tumor forming potential in an SV-40 transformed mouse cell line and its hybrid and cybrid progeny." *Cold Spring Harbor Symposium on Quantitative Biology* 45 (Parts 1, 2) (1981): 747–754.

(———) "Genomic rearrangements in a mouse cell line containing integrated SV40 DNA." *Cell* 23 (1981): 41–50.

(with C. Grabowy and H. Sano) "The mat-1 gene in *Chlamydomonas* regulates DNA methylation during gametogenesis." *Cell* 23 (1981): 41–47.

"Chromosome modifications and cancer." *Harvey Lectures* 78 (1982): 173–190.

(with R. M. Kitchin and C. J. Marshall) "Genetic control of anchorage requirement in Chinese hamster embryo fibroblast CHEF cells." *Journal of Cell Biology* 95 (1982): 448A.

(with P. Kovac) "Pre-adipocyte determination either by insulin or azacytidine." *Proceedings of the National Academy of Sciences* 79 (1982): 480–484.

(with C. J. Marshall and R. M. Kitchin) "Genetic analysis of tumorigenesis. XII. Genetic control of the anchorage requirement in CHEF cells." *Somatic Cell Genetics* 8 (1982): 709–722.

(with H. Sano) "Tissue specificity and clustering of methylated cytosines in bovine Satellite I DNA." *Proceedings of the National Academy of Sciences* 79 (1982): 3584–3588.

(with B. L. Smith) "The multistep origin of tumor-forming ability in CHEF cells." *Cancer Research* 42 (1982): 389–396.

(———, A. Anisowicz, et al.) "DNA transfer of focus-and tumor-forming ability into non-tumorigenic CHEF cells." *Proceedings of the National Academy of Sciences* 79 (1982): 1964–1968.

(———) "DNA transfer of tumor forming ability in non-tumorigenic CHEF cells." *Genetics* 100 (1982): s60.

(with D. P. Dubey, D. E. Staunton, et al.) "Lysis of Chinese hamster embryo fibroblast mutants by human natural cytotoxic (NK) cells." *Proceedings of the National Academy of Sciences* 80 (1983): 3025–3029.

(with J. J. Harrison, A. Anisowicz, et al.) "AZA cytidine induced tumorigenesis of CHEF-18 cells correlated DNA methylation and chromosome changes." *Proceedings of the National Academy of Sciences* 80 (1983): 6606–6610.

(with H. Sano and H. Noguchi) "Characterization of DNA methyltransferase from bovine thymus cells." *European Journal of Biochemistry* 135 (1983): 181–185.

(with K. Tanaka, C. C. Lau, et al.) "Resistance of human cells to tumorigenesis induced by clone transforming genes." *Proceedings of the National Academy of Sciences* 80 (1983): 7601–7605.

(with I. K. Gadi and J. J. Harrison) "Loss of chloroplast DNA methylation during dedifferentiation of *Chlamydomonas reinhardi* gametes." *Molecular Cell Biology* 4 (1984): 2103–2108.

(with H. Sano and C. T. Grabowy) "Control of maternal inheritance by DNA methylation in *Chlamydomonas.*" *Current Topics in Microbiological Immunology* 108 (1984): 157–173.

"Chloroplast genes." *Bioessays* 3 (1985): 180–183.

"The development and use of the Chinese hamster embryo fibroblast CHEF cell line." In *Molecular Cell Genetics*, edited by M. M. Gottesman, 75–94. New York: John Wiley and Sons, 1985.

"Genetic studies of transformation and tumorigenesis in Chinese hamster embryo fibroblasts." In *Molecular Cell Genetics*, edited by M. M. Gottesman, 811–828. New York: John Wiley and Sons, 1985.

"Genomic changes and the origin of neoplasia." *International Journal of Cancer* 35 (1985): 1–4.

(with J. J. Harrison) "Adipocyte differentiation in CHEF cells." In *Chemistry, Biochemistry, and Biology of DNA Methylation*, edited by G. Cantoni, 293–404. New York: Alan Liss, 1985.

(———— and E. Soudry) "Adipocyte conversion of CHEF cells in serum-free medium." *Journal of Cell Biology* 100 (1985): 429–434.

(with B. L. Smith) "Genetic analysis of tumorigenesis. XXI. Suppressor genes in CHEF cells." *Somatic Cell and Molecular Genetics* 11 (1985): 25–34.

"Genetic suppression of tumor formation: A new frontier in cancer research." *Cancer Research* 46 (1986): 1573–1580.

(with R. Sklar and D. Altman) "An endonuclease from *Chlamydomonas reinhardi* that cleaves the sequence TATA." *Journal of Biological Chemistry* 261 (1986): 6806–6810.

(with A. Anisowicz and L. Bardwell) "Constitutive over expression of a growth-regulated gene in transformed Chinese hamster and human cells." *Proceedings of the National Academy of Sciences* 84 (1987): 7188–7192.

(with G. Stenman) "Genetic analysis of tumorigenesis: A conserved region in the human and Chinese hamster genomes contains genetically identified tumor-suppression genes." *Proceedings of the National Academy of Sciences* 84 (1987): 9099–9102.

(————, E. O. Delorme, et al.) "Transfection with plasmid pSV2gptEJ induces chromosome rearrangements in CHEF cells." *Proceedings of the National Academy of Sciences* 84 (1987): 184–188.

"Genetic basis of cancer and its suppression." In *Cancer—The Outlaw Cell*, edited by R. LaFond, 75–96. New York: American Cancer Society, 1988.

"Mutation rates and mutational spectra in tumorigenic cell lines." *Cancer Surveys* 7 (1988): 325–333.

"Tumor suppression and the GRO gene." In *Gene Expression and Regulation: The Legacy of Luigi Gorini*, edited by M. Bissell et al., 343–352. Amsterdam, Holland: Elsevier, 1988.

"Tumor suppression genes." *Journal of Cellular Biology* 32 (1988): 353–357.

(with A. Anisowicz and D. Zajchowski, et al.) "Functional diversity of GRO gene expression in human fibroblast and mammary epithelial cells." *Proceedings of the National Academy of Sciences* 85 (1988): 9645–9649.

(with R. W. Craig and I. K. Gadi) "Genetic analysis of tumorigenesis. XXXI. Retention of short arm of chromosome 3 in suppressed CHEF cell hybrids containing C-HA-RAS EJ gene." *Somatic Cell and Molecular Genetics* 14 (1988): 41–54.

(with H. Sano and M. Imokawa) "Detection of heavy methylation in human repetitive DNA subsets by a monoclonal antibody against 5-methylcytosine." *Biochimica et Biophysica Acta* 951 (1988): 157–165.

(with G. Stenman and A. Anisowicz) "Genetic analysis of tumorigenesis. XXXII. Location of constitutionally amplified KRAS sequences to Chinese hamster chromosomes X and Y by *in situ* hybridization." *Somatic Cell and Molecular Genetics* 14 (1988): 639–644.

(with D. Zajchowski, V. Band, et al.) "Expression of growth factors and oncogenes in normal and tumor-derived human mammary epithelial cells." *Cancer Research* 48 (1988): 7041–7047.

"Cytokines and tumor suppression." In *Current Communications in Molecular Biology: Recessive Oncogenes and Tumor Suppression*, edited by Webster Cavenee et al., 197–202. New York: Cold Spring Harbor Laboratory, 1989.

"New approaches in breast cancer research." In *Ovarian Cancer*, edited by F. Sharp et al., 45–53. London: Chapman and Hall, 1989.

"Tumor suppressor genes: The puzzle and the promise." *Science* 246 (1989): 1406–1412.

(with V. Band, D. Zajchowski, et al.) "A newly established metastic breast tumor cell line with integrated amplified copies of ERBB2 and double minute chromosomes." *Genes, Chromosomes and Cancer* 1 (1989): 48–58.

(with D. A. Kaden, L. Bardwell, et al.) "High frequency of large spontaneous deletions of DNA in tumor-derived cells." *Proceedings of the National Academy of Sciences* 86 (1989): 2306–2310.

(with D. Kaden, I. K. Gadi, et al.) "Spontaneous mutation rates of tumorigenic and nontumorigenic Chinese hamster embryo fibroblast cell lines." *Cancer Research* 49 (1989): 3374–3379.

"Genetic strategies of tumor suppression." *Respiratory Diseases* 142 (1990): s40–s43.

"GRO as a cytokine." In *Molecular and Cellular Biology of Cytokines*, edited by M. C. Powanda et al., 327–332. New York: Alan Liss, 1990.

(with A. Anisowicz, M. C. Pike, et al.) "Structural, regulatory, and functional studies of the GRO gene and protein." In *Interleukin 8 and Related Chemotactic Cytokines*, vol. 4, edited by M. Baggiolini and C. Sorg, 96–116. Basel: Karger, 1990.

(with V. Band, D. Zajchowski, et al.) "Tumor progression in four mammary epithelial cell lines derived from the same patient." *Cancer Research* 50 (1990): 7351–7357.

(with S. Haskill, A. Peace, et al.) "Identification of three related human GRO genes encoding cytokine functions." *Proceedings of the National Academy of Sciences* 87 (1990): 7732–7736.

(with D. K. Trask, V. Band, et al.) "Keratins as markers that distinguish normal and tumor-derived mammary epithelial cells." *Proceedings of the National Academy of Sciences* 87 (1990): 2319–2323.

(with P. Yaswen, A. Smoll, et al.) "Down-regulation of a calmodulin-related gene during transformation of human mammary epithelial cells." *Proceedings of the National Academy of Sciences* 87 (1990): 7360–7364.

(with D. Zajchowski, V. Band, et al.) "Suppression of tumor forming ability and related traits in MCF7 breast cancer cells by fusion with immortal breast epithelial cells.'' *Proceedings of the National Academy of Sciences* 87 (1990): 2314–2318.

"Senescence as a mode of tumor suppression." *Environmental Health Perspectives* 93 (1991): 59–62.

(with A. Anisowicz, M. Messineo, et al.) "An NF-Kappa B-like transcription factor mediates IL-1/TNF-alpha induction of GRO in human fibroblasts." *Journal of Immunology* 147 (1991): 520–527.

(with V. Band) "Tumor progression in breast cancer." In *Neoplastic Transformation in Human Cell Systems in Vitro: Mechanisms of Carcinogenesis*, edited by J. S. Rhine and A. Dritschilo, 169–178. Clifton, NJ: Humana Press, 1991.

(———, J. A. DeCaprio, et al.) "Loss of p53 protein in human papilloma virus type 16 E6–immortalized human mammary epithelial cells." *Journal of Virology* 65 (1991): 6671–6676.

(with S. Haskill, A. Anisowicz, et al.) "GRO: A novel chemotactic cytokine." *Advances in Experimental Medical Biology* 305 (1991): 73–77.

(with S. W. Lee and C. Tomasetto) "Positive selection of candidate tumor-suppressor genes by subtractive hybridization." *Proceedings of the National Academy of Sciences* 88 (1991): 2825–2829.

(with D. A. Zajchowski) "Induction of estrogen-regulated genes differs in immortal and tumorigenic human mammary epithelial cells expressing a recombinant estrogen receptor." *Molecular Endocrinology* 5 (1991): 1613–1623.

"Tumor suppressor genes in the cell cycle." *Current Opinion in Cell Biology* 4 (1992): 155–160.

(with A. Anisowicz, M. C. Pike, et al.) "Structural, regulatory, and functional studies of the GRO gene and protein." *Cytokines* 4 (1992): 96–116.

(with B. J. Dezube, A. B. Pardee, et al.) "Cytokine dysregulation in AIDS: *In vivo* overexpression of mRNA of tumor necrosis factor-alpha and its correlation with that of the inflammatory cytokine GRO." *Journal of Acquired Immune Deficiency Syndrome* 5 (1992): 1099–1104.

(with S. W. Lee, C. Tomasetto, et al.) "Down-regulation of a member of the S100 gene family in mammary carcinoma cells and reexpression by azadeoxycytidine treatment." *Proceedings of the National Academy of Sciences* 89 (1992): 2504–2508.

(———) "Transcriptional down-regulation of gap-junction proteins blocks junctional communication in human mammary tumor cell lines." *Journal of Cell Biology* 118 (1992): 1213–1221.

(with P. Liang, L. Averboukh, et al.) "Differential display and cloning of messenger RNAs from human breast cancer versus mammary epithelial cells." *Cancer Research* 52 (1992): 6966–6968.

(with K. Swisshelm and M. Leonard) "Preferential chromosome loss in human papilloma virus DNA-immortalized mammary epithelial cells." *Genes, Chromosomes and Cancer* 5 (1992): 219–226.

(with A. Anisowicz, P. Liang, et al.) "Identification by differential display of alpha 6 integrin as a candidate tumor suppressor gene." *Federation of American Societies of Experimental Biology Journal* 7 (1993): 964–970.

(with C. Tomasetto, M. J. Neven, et al.) "Specificity of gap junction communication among human mammary cells and connexin transfectants in culture." *Journal of Cell Biology* 122 (1993): 157–167.

Other Works

Curriculum vitae, 1993.
Letter to Sam Granick, January 1949. Rockefeller Archives, Sam Granick papers, Record
    Group 450G766. Rockefeller Archive Center, North Tarrytown, NY.
Personal communications,1994.

Works about Ruth Sager

Current Biography Yearbook 1967, 367–370. New York: Wilson 1967, 1968.
Debus, A. G. (ed.). *World Who's Who in Science*. Chicago: Marquis Who's Who, 1968.
Granick, Sam. Report on supervision of Sager, February 20, 1950. Rockefeller Archives,
    Sam Granick papers, Record Group 450G766. Rockefeller Archive Center, North
    Tarrytown, NY.
Hillary, J. "Genes pictured as living on own." *New York Times* (Mar. 14, 1965): 54.
Letter to editor. *New York Times* (Mar. 24, 1977): 26.
Osmundsen, J. A. "Second system of genes traced in study by a Columbia scientist."
    New York Times (Sept. 8, 1963): 1, 84.
Pace, Eric. "Dr. Ruth Sager, 79, researcher on location of genetic material." *New York
    Times* (Apr. 4, 1997): A28.
Sapp, J. *Beyond the Gene: Cytoplasmic Inheritance and the Struggle for Authority in
    Genetics*. New York: Oxford University Press, 1987.
———. *Where the Truth Lies: Franz Moewus and the Origins of Molecular Biology*.
    Cambridge, England: Cambridge University Press, 1990.
Stadler, L. J. Letter to Herbert Gasser, April 5, 1951. Rockefeller Archives, Sam Granick
    papers, Record Group 450G766. Rockefeller Archive Center, North Tarrytown,
    NY.

Other References

Correns, Carl. "Vererbungsversuche mit blass (gelb) grunen und bluntblattrigen Sippen
    bei *Mirabilis, Urtica,* und *Lunaria.*" *Zeitschrift fuer Vererbungslehre* 1 (1909):
    291–329.
———. "Zur kenntnis der Rolle von Kern und Plasma bei der Vererbung." *Zeitschrift
    fuer Vererbungslehre* 2 (1909): 331–340.

Sources

Columbiana.
Hunter College Archives.

BERTA VOGEL SCHARRER (1906–1995)

Birgit H. Satir and Peter Satir

BIOGRAPHY

Berta Vogel was born on December 1, 1906, in Munich, Germany, the eldest of four children of Johanna (Greis) and Karl Vogel. Her father was a judge and vice president of Federal Court in Bavaria who was involved in sentencing Adolf Hitler to jail in the 1920s, for which the parents subsequently feared reprisals. As a young girl, she became interested in biology and expressed the desire to become a research scientist. This interest solidified in the course of a rigorous education at the gymnasium for girls. After passing the general gymnasium examinations, she entered the University of Munich during the years of the Weimar Republic. There she began to study honeybee behavior in the Department of Zoology under Professor Karl von Frisch, who won the Nobel Prize in Physiology or Medicine (1973). At this time in German universities, some women were able to pursue a Ph.D. degree. Under the mentorship of von Frisch this pursuit was both possible and pleasant. At the University she was also influenced by Heinrich Wieland (Nobel Prize winner in Chemistry, 1927), Karl von Goebel (botany), and Richard Hertwig (embryology), who took part in her doctoral examination. She received her Ph.D. degree in 1930, and her thesis was published in 1931.

Vogel has indicated that in the early 1930s opportunities for an academic career in Germany were generally bleak, and for a woman, virtually nonexistent. She decided to obtain a certificate for teaching in a German gymnasium. At the same time, she began to make joint plans with a fellow student of von Frisch's, Ernst Scharrer. Both found laboratory space at the Research Institute of Psychiatry in Munich, directed by Walter Spielmeyer, while Ernst Scharrer also completed an additional degree in medicine. By 1933 her research interests focused on spirochetes.

In 1934, after Ernst Scharrer's appointment at the Edinger Institute of Neurology in Frankfurt-am-Main and with a small stipend from one of her uncles to help support them, Berta Vogel and Ernst Scharrer married. This was the beginning of a remarkable collaboration that lasted 30 years, sometimes skirting nepotism rules (then strongly enforced) and ending only with Ernst Scharrer's death in a swimming accident after the American Association of Anatomists meeting in Miami Beach, Florida (1965). At the Edinger Institute, where Berta Scharrer held an unsalaried position as research associate, Ernst and Berta Scharrer established the scientific avenues of research, professional interests, and friendships with individuals like Wolfgang Bargmann, Albrecht Bethe, and Tilly Edinger that were to last their lifetimes.

In 1928 Ernst Scharrer had discovered that certain hypothalamic neurons of a fish had secretory activity similar to that of an endocrine gland cell. This was an unexpected, controversial, but potentially very significant finding. Based on this study, Ernst and Berta Scharrer decided to initiate a comparative analysis. He continued his studies of similar cells in vertebrates, while she, abandoning her spirochete work, searched for comparable phenomena in the invertebrates. The results of this analysis led to the concept of neurosecretion and to a new discipline, neuroendocrinology, for which they both became famous. Berta Scharrer's first classic paper on the presence of neurosecretory cells in invertebrates was published in 1935 ("Über das Hanströmsche . . ." 1935).

The Edinger Institute was an exciting place to work, but in the 1930s in Germany the political climate was rapidly deteriorating with the rise of Nazism. Although the Scharrers were not Jewish, they found the climate intolerable. Berta Scharrer once stated, "What the two of us were particularly opposed to was this senseless and immoral philosophy of the Nazis, the idea of racial superiority, anti-semitism, genocide . . . we decided that it was impossible for us to be part of this system any longer." In 1937, using as a pretext a one-year Rockefeller Fellowship that Ernst Scharrer obtained at the University of Chicago, the couple left for the United States. They came with two suitcases and four dollars each and, as was characteristic, with their integrity intact. At Chicago, Berta Scharrer again held the unsalaried position of research associate. Her research space was minimal, and she had no money to purchase experimental material. She had to start over again, almost from scratch, in a new language, but with some help she began to publish in English. At no cost she found her experimental animal, the cockroach, in the basement of the building.

In 1938, when the fellowship at the University of Chicago was finished, Ernst Scharrer became a visiting investigator at the Rockefeller Institute for Medical Research (now Rockefeller University) in New York. Berta Scharrer moved with him, still as research associate. Although her initial work at Chicago was with the American cockroach, at Rockefeller Berta Scharrer discovered a colony of the South American cockroaches (or woodroaches) that had arrived as stowaways in a shipment of laboratory monkeys. This sturdy creature became the model organism for the ablation experiments that were necessary to give insect neurosecretion its physiological importance.

In 1940 Ernst Scharrer joined the faculty of Western Reserve University (now Case Western Reserve University) in Cleveland. Berta Scharrer received the title of Instructor and Fellow in Anatomy, again without salary. She has recalled the environment at that time as rather inhospitable to women; originally barred, she was permitted to attend departmental seminars only on the condition that she make tea for the faculty (and only after her insistence that she would not make tea if she still could not attend the presentations). Ernst and Berta Scharrer became U.S. citizens on June 15, 1945, in the Northern District of Ohio. In 1947 Ernst Scharrer's career took him to Denver as Associate Professor at the University of Colorado Medical School. Berta Scharrer won a prestigious Guggenheim Fellowship and a U.S. Public Health Service Special Fellowship for her support.

Finally, in 1950, two decades after completing her Ph.D. degree, when she was invited to organize an international symposium in Paris and was asked to give her academic rank, Berta Scharrer received an academic title. The Dean of the University of Colorado agreed that he would grant her the listing as Assistant Professor (research), unsalaried. Philosophically, she has said, "It seems unusual, but it was what happened at the time" (Millen 1989). Her *Scientific American* article (1951) on the woodroach gives this academic title.

In 1955 Ernst Scharrer accepted the position of Chairman of Anatomy at the newly opened Albert Einstein College of Medicine of Yeshiva University in New York City. The Dean of that school, Marcus Kogel, offered Berta Scharrer, who by this time had acquired a strong, independent, international reputation for her work, the position of tenured Professor in the new department, her first regular academic appointment. Dean Kogel said, "I know about the nepotism rule, but we are an entirely new school . . . we can do something a little progressive" (Miller 1989). Berta Scharrer became the course leader for histology and an important adviser to the fledgling medical students of the first classes. The Scharrers built a house within walking distance of the school and happily collaborated in building the department. However, the progressive nature of her appointment was marred by the fact that while Ernst Scharrer was alive, she worked for half salary. After his death, Berta Scharrer's career continued for 30 years, during which she saw full acceptance of the neuron as a secretory cell, the purification of the neuropeptide hormones, and the emergence of another new discipline, neuroimmunology, in which she participated significantly in her later years. Berta Scharrer died on July 23, 1995, at her home in the Bronx, New York, at the age of 88. She was actively pursuing her research interests, preparing and editing new manuscripts, until two days before her death.

Not until Ernst Scharrer's death did Berta Scharrer begin to receive scientific honors. There were not many women members of the National Academy of Sciences of the United States when Berta Scharrer became a member in 1967. She also became a fellow of the American Academy of Arts and Sciences (1967) and of the Deutsche Akademie der Naturforscher Leopoldina (1972). In 1978–79, she was elected president of the American Association of Anatomists. She was a foreign member of the Royal Netherlands Academy of Arts and Sciences and an honorary member of a number of academic associations, including the

International Society for Neuroendocrinology and the American Society of Zoologists. She received honorary degrees in medicine, science, or law from 11 institutions, including the University of Giessen (1976), Harvard University (1982), the University of Frankfurt (1992), and her own institution, Yeshiva University (1983). In 1985, at a White House ceremony, she received the highest scientific recognition awarded by the United States, the National Medal of Science. She was also the recipient of the Kraepelin Gold Medal (1978), the Fred C. Koch Award of the Endocrine Society (1980), the Henry Gray Award of the American Association of Anatomists (1982), and the Schlieden Medal of the Leopoldina (1983). In 1994, in a special presentation at her home by the German ambassador, she received the Order of Merit from the free state of Bavaria. A species of cockroaches, the *scharrerae*, was named in her honor. She wore these honors lightly.

WORK

In assessing the independent contributions of Ernst and Berta Scharrer to the concept of neurosecretion, it is well to keep in mind that they were close professional colleagues, working in a dedicated manner toward common goals. From 1936 to 1965, Berta Scharrer went where Ernst Scharrer's academic career took them, primarily because he had the salaried position. The honors that later befell Berta Scharrer were perhaps honors they might have shared; the honors that eluded Berta Scharrer, in particular, the Nobel Prize, might not have eluded the pair. In 1963 they published *Neuroendocrinology*, a book based on their Jesup lectures given at Columbia University in 1960. This volume summarizes the theory of, and evidence for, neurosecretion and neurohormonal control across the animal kingdom, and they are coequal contributors, especially as parallels between the vertebrate and invertebrate neuroendocrine systems are central to the theme.

In the late 1920s, when Ernst Scharrer first saw neurosecretory cells in histological preparations of the brains of fishes, the concept of the chemical transmission of the nerve impulse was just beginning to be established. The idea that the neuron was a secretory cell, like a pancreatic islet cell, seemed patently absurd, especially since it was based on cytological evidence. The comparative approach was necessary to refute the suggestion that the cytological characteristics of neurosecretory cells of the mammalian hypothalamus were artifacts caused by postmortem changes or pathology. Berta Scharrer's work on invertebrates was central and critical to this refutation.

The first invertebrate ganglia in which Berta Scharrer saw neurosecretory activity, in her initial 1935 paper, were those of the opisthobranch snail *Aplysia*, but the cockroach proved more important. Not only were the parallels in organization between the brain-corpus cardiacum-corpus allatum system of insects and the hypothalamic-hypophysial system of vertebrates very significant, but the cockroach, particularly the woodroach *Leucophaea maderae*, was especially useful. Microsurgical dissection of the corpus cardiacum or corpus allatum permitted the physiological and developmental roles of the neurosecretory substances to be determined. Neurosecretory neurons of the brain could be shown to transmit their secretory granules to storage sites in the corpus cardiacum by axoplasmic transport, and the substances transmitted could be shown to affect growth and maturation in the insect. The insect experiments were arguably the most convincing early proofs of the physiological significance of neurosecretion. The maturing field was surveyed in 1953 at the First International Symposium on Neurosecretion, held at the Statione Zoologica in Naples, Italy, a pivotal event for Berta and Ernst Scharrer since it represented a wide validation of their pioneering work.

By the early 1960s the concept of neurosecretion was further clarified by the demonstration of electron dense, membrane-bounded, secretory vesicles in neurosecretory neurons by the new techniques of electron microscopy. These methods revealed specifics regarding the sites of synthesis and release of such vesicles, and Berta Scharrer was particularly involved in the demonstration of their exocytosis at the axon terminals in insect neurosecretory systems.

Berta and Ernst Scharrer's work was influential in establishing that neurosecretory neurons terminated on blood vessels and that the neuropeptides secreted by those cells acted as hormones. After Ernst Scharrer's death, Berta Scharrer expanded this concept by placing the neurosecretory neuron in a wider evolutionary and cell biological framework. She pointed out that vascular transmission was not the sole means of distribution of neurohormones; in some cases neurosecretory release sites could even be observed to be in direct contact with endocrine cells. Berta Scharrer then formulated a general evolutionary picture to replace the earlier concept that the neurosecretory cell was a specialized evolutionary development. She suggested instead that the secretory neuron was the precursor in more primitive organisms of the specialized neuronal and endocrine systems of insects and vertebrates, because such cells could subserve both systems. The remarkable conservation and wide cellular distribution of the neuropeptides gave wide credence to her views, and she remained an important spokesperson for the broadest view of the significance of the neuropeptides throughout her life.

Her work took an unexpected and highly important turn in her later years. Just as she pioneered studies of invertebrate neurosecretion in the mid-1980s, she pioneered studies of invertebrates, especially the edible mussel *Mytilus* and the woodroach *Leucophaea*, in the emerging field of comparative neuroimmunology. She collaborated with George Stefano and with the Danish scientists George and Bente Hansen in these studies, which have been influential in establishing the commonality of basic principles of communication between the nervous and immune system within invertebrates and vertebrates. In her words as senior editor of the book *Neuropeptides and Immunoregulation* (1994), the information on invertebrate neuroimmunology "opened new vistas regarding the immunomodulatory role played by opioid and other neuropeptides as signal molecules in the cell-mediated internal regulation of the immune system.'' The remarkable parallels observed between vertebrates and the complex invertebrate organism ''have established certain general principles of neuroimmunobiological integration that has basic biological as well as biomedical applications.'' Berta Scharrer herself was certainly one of the most notable contributors to the establishment of these parallels, unanticipated by most biologists, essentially repeating her earlier history in the field of neurosecretion. At the time of her death she was an associate editor of *Advances in Neuroimmunology*, which reflects her prominence in this new field, as well as of *Cell and Tissue Research*, which indicated her long-standing fundamental interests in cell biology and histology.

At the Albert Einstein College of Medicine, Berta Scharrer became Acting Chairperson of the Department of Anatomy when her husband died. With characteristic generosity, she refused to be a candidate for the permanent chair, deciding that it would be best for the school to recruit a strong outside candidate, who turned out to be Dominick Purpura. In 1976, when Purpura moved from the Department of Anatomy to his newly established chair in the Department of Neuroscience, Scharrer once again became the acting chairperson of the Department of Anatomy. She was active in convincing the authors of this article to move to Einstein, one of us as Chair, one of us as a tenured Professor, in essence repeating her own history. In 1978 she became Professor Emerita. She held the title Distinguished University Professor Emerita until her death. As part of the 25th Anniversary Celebration at the Albert Einstein College of Medicine in 1980 and in recognition of her 25 years of contribution to the school, a symposium entitled ''Neuropeptides. Retrospect and Prospect'' was held in her honor. She also became an honorary alumna of the school.

Despite her academic standing at Einstein, Scharrer was not concerned with personal or academic power. Instead, with her scientific discoveries, she built an intellectual empire of international scope around herself. She led a life of scholarship and scientific inquiry, pursuing broad and often controversial concepts with meticulous attention to experimental detail and to writing. The determination and stoicism that enabled her to work with little or no remuneration for so many years and the integrity that made her leave Germany at the time of the Nazis were never absent from her scientific endeavors. She inspired generations of medical students and colleagues by her concern for them as individuals, by her vitality even into old age, and, most of all, by her love of her work.

BIBLIOGRAPHY

Works by Berta Vogel Scharrer

Scientific Works

Space does not permit the listing of the complete works of Berta Vogel Scharrer. The editors have eliminated all works by Scharrer cited in Scharrer and Scharrer (1963);

Scharrer, "Insects as . . ." (1987); Downer and Laufer (1983, 37, 52, 178); and Scharrer, "The neuropeptide saga" (1990). Included here are her dissertation and all references cited in the text.

"Über die Beziehungen zwischen Süssgeschmack und Nährwert von Zuckern bei der Honigbiene." *Sitzungsberichte der Gesellschaft zur Morphologie und Physiologie. München* 40 (1931): 63–65 (as B. Vogel).

"Über die Beziehungen zwischen Süssgeschmack und Nährwert von Zuckern und Zuckeralkoholen bei der Honigbiene." *Zeitschrift für Vergleichende Physiologie* 14 (1931): 273–347 (as B. Vogel).

"Vergleichend-morphologische Untersuchungen an Hühner- und Recurrensspirochäten." *Sitzungsberichte der Gesellschaft zur Morphologie und Physiologie. München* (1933): 1–5 (as B. Vogel).

"Ein serumfreier Nährboden für Hühnerspirochäten." *Zentralblatt für Bakteriologie . . . I. Originale* 132 (1934): 243–244.

"Über die Verweildauer von Hühnerspirochäten im Zentralnervensystem von Hühnern und Tauben." *Zeitschrift für Hygiene* 116 (1934): 206–208.

"Über das Hanströmsche Organ X bei Opisthobranchiern." *Pubblicazioni della Stazione Zoologica di Napoli* 15 (1935): 132–142.

"Über das Verhalten von Tauben gegenüber Infektionen mit verschiedenen Hühnenspirochätenstämmen." *Zeitschrift für Hygiene* 117 (1935): 163–170.

"Über die Feststellung von Dickenunterschieden an lebenden mikroskopischen Objekten, dargelegt am Beispiel der Unterscheidung von Hühner- und Recurrensspirochäten." *Archiv für Protistenkunde* 85 (1935): 87–99.

"Über Spirochaeta (Treponema) minutum Dobell bei Amphibien." *Zoologischer Anzeiger* 111 (1935): 1–7.

(with F. Jahnel) "Ein Beitrag zur Frage der Syphilisempfindlichkeit im Tierreich, insbesondere verschiedener Mäusearten." *Dermatologische Zeitschrift* 71 (1935): 1–6.

"Über 'Drüsen-Nervenzellen' im Gehirn von *Nereis virens* Sars." *Zoologischer Anzeiger* 113 (1936): 299–302.

(with E. Hadorn) "The structure of the ring-gland (corpus allatum) in normal and lethal larvae of *Drosophila melanogaster*." *Proceedings of the National Academy of Sciences* 24 (1938): 236–242.

"The differentiation between neuroglia and connective tissue sheath in the cockroach (*Periplaneta americana*)." *Journal of Comparative Neurology* 70 (1939): 77–88.

"Neurosecretion. II. Neurosecretory cells in the central nervous system of cockroaches." *Journal of Comparative Neurology* 74 (1941): 93–108.

"Neurosecretion. III. The cerebral organ of the nemerteans." *Journal of Comparative Neurology* 74 (1941): 109–130.

(with E. Scharrer) "Endocrines in invertebrates." *Physiological Reviews* 21 (1941): 383–409.

"The influence of the corpora allata on egg development in an orthopteran (*Leucophaea maderae*)." *Anatomical Record* 87 (1943): 471.

"Experimental tumors in an insect." *Science* 102 (1945): 102.

"Fat absorption in the foregut of *Leucophaea maderae* (Orthoptera)." *Anatomical Record* 99 (1947): 638.

"Experimental studies on the prothoracic glands of *Leucophaea maderae* (Orthoptera)." *Anatomical Record* 101 (1948): 725.

"Hormones in insects." In *The Hormones, Physiology, Chemistry, and Application*, vol.

1, edited by G. Pincus and K. V. Thimann, 121–158. New York: Academic Press, 1948.

"Malignant characteristics of experimentally induced tumors in the insect, *Leucophaea maderae* (Orthoptera)." *Anatomical Record* 100 (1948): 774.

"Gastric cancer experimentally induced in insects by nerve severance." *Journal of the National Cancer Institute* 10 (1949): 375–376.

"Tumor mortality and sex in *Leucophaea maderae* (Orthoptera)." *Anatomical Record* 105 (1949): 624.

(with J. Wilson) "Fat metabolism in tumor bearing insects *Leucophaea maderae* (Orthoptera)." *Anatomical Record* 105 (1949): 625.

"Tumors in invertebrates." In *The 1950 Year Book of Pathology and Clinical Pathology*, edited by H. T. Karsner and A. H. Sanford, 57–61. Chicago: Year Book, 1951.

"The woodroach." *Scientific American* 185 (1951): 58–62.

"The effect of the interruption of the neurosecretory pathway in the insect, *Leucophaea maderae*." *Anatomical Record* 112 (1952): 386–387.

"Further studies of the intercerebralis-cardiacum-allatum system of insects." *Biological Bulletin* 103 (1952): 284.

"Hormones in insects." In *The Action of Hormones in Plants and Invertebrates*, edited by K. V. Thimann, 125–169. New York: Academic Press, 1952.

"Insect tumors induced by nerve severance: Incidence and mortality." *Cancer Research* 13 (1953): 73–76.

"Metabolism and mortality in insects with gastrointestinal tumors induced by nerve severance." *Journal of the National Cancer Institute* 13 (1953): 951–954.

(with E. Scharrer) "Symposium on Neurosecretion at Naples, Italy, May 11–16." *Science* 118 (1953): 579–580.

"Neurosecretion in invertebrates: A survey." *Pubblicazione della Stazione Zoologica di Napoli* 24 (Suppl.) (1954): 38–40.

"Castration cells in the central nervous system of an insect (*Leucophaea maderae*, Blattaria)." *Transactions of the New York Academy of Sciences* s. 2, 17 (1955): 520–525.

"Hormones in the roach." In *Scientific American Books, First Book of Animals*, 173–178. New York: Simon and Schuster, 1955.

"Corrélations endocrines dans la reproduction des insectes." *Annales des Sciences Naturelles, Zoologie* 11 s. (1956): 231–234.

(with L. Arvy and M. Gabe) "Invertebrate endocrinology." *Science* 123 (1956): 382.

(with M. von Harnack) "A study of the corpora allata of gonadectomized *Leucophaea maderae* (Blattaria)." *Anatomical Record* 125 (1956): 558.

"The corpus allatum of *Leucophaea maderae* (Blattaria)." *Proceedings of the Xth International Congress of Entomology* 2 (1958): 57.

(with E. Scharrer) "Neurosecretion." *Science* 127 (1958): 1396–1398.

"In memory of Masashi Enami." *Gunma Journal of Medical Sciences* 8 (1959): vi–vii.

"Functional analysis of the corpus allatum of the insect, *Leucophaea maderae*, with the electron microscope." *Biological Bulletin* 121 (1961): 370.

"The neurosecretory system of *Leucophaea maderae* and its role in neuroendocrine integration." *General and Comparative Endocrinology* 2 (1962): 30.

"Ultrastructure and function in the corpus allatum of an insect (*Leucophaea madera*)." *Anatomical Record* 142 (1962): 275.

(with C. Jones) "The ultrastructure of the corpus cardiacum, a neuroglandular organ of insects." *Anatomical Record* 145 (1963): 367.

(with E. Scharrer) *Neuroendocrinology*. New York: Columbia University Press, 1963.

"The fine structure of blattarian prothoracic glands." *Zeitschrift für Zellforschung und Mikroskopische Anatomie* 64 (1964): 301–326.

"Ultrastructural study of the prothoracic glands of *Leucophaea maderae* (Blattaria)." *American Zoologist* 4 (1964): 328.

"The ultrastructure of the corpus allatum of *Blaberus craniifer* (Blattaria)." *American Zoologist* 4 (1964): 327–328.

"The fine structure of an unusual hemocyte in the insect *Gromphadorhina portentosa*." *Life Sciences* 4 (1965): 1741–1744.

"Hemocytes within prothoracic glands of insects." *American Zoologist* 5 (1965): 235.

"Recent progress in the study of neuroendocrine mechanisms in insects." *Archives d'Anatomie Microscopique et de Morphologie Experimentale* 54 (1965): 331–342.

"An ultrastructural study of cellular regression as exemplified by the prothoracic gland of *Leucophaea maderae*." *Anatomical Record* 151 (1965): 411.

"An electron microscopic study of insect hemocytes." *Anatomical Record* 154 (1966): 416.

"Ultrastructural study of the regressing prothoracic glands of blattarian insects." *Zeitschrift für Zellforschung und Mikroskopische Anatomie* 69 (1966): 1–21.

"Ultrastructural specializations of neurosecretory terminals in the corpus cardiacum of cockroaches." *American Zoologist* 7 (1967): 721–722.

"Ultrastructural study of hormone release mechanism in the corpus cardiacum of insects." *General and Comparative Endocrinology* 9 (1967): 528.

(with E. Harper and S. Seifter) "Electron microscopic and biochemical characterization of collagen in blattarian insects." *Journal of Cell Biology* 33 (1967): 385–393.

(with S. Wurzelmann) "Ultrastructural study of nucleolar activity in oocytes of the lungfish, *Protopterus aethiopicus*." *Anatomical Record* 157 (1967): 316.

"Neuroendocrine factors in the control of reproduction." In *Reproduction and Sexual Behavior*, edited by M. Diamond, 145–149. Indianapolis: Indiana University Press, 1968.

"Comparative aspects of neurosecretory phenomena." In *Progress in Endocrinology. Proceedings of the III International Congress of Endocrinology, Mexico, 1968* (1969): 365–367.

"Current concepts in the field of neurochemical mediation." *Medical College of Virginia Quarterly* 5 (1969): 27–31.

"Neurohumors and neurohormones: Definitions and terminology." *Journal of Neuro-Visceral Relations* (Suppl. 9) (1969): 1–20.

(with S. B. Kater) "Ultrastructural correlates of experimentally induced neurohormone release in insects." *Anatomical Record* 163 (1969): 256.

(with S. Wurzelmann) "Ultrastructural study on nuclear-cytoplasmic relationships in oocytes of the African lungfish, *Protopterus aethiopicus*. I. Nucleolo-cytoplasmic pathways." *Zeitschrift für Zellforschung und Mikroskopische Anatomie* 96 (1969): 325–343.

(———) "Ultrastructural study on nuclear-cytoplasmic relationships in oocytes of the African lungfish, *Protopterus aethiopicus*. II. The microtubular apparatus of the nuclear envelope." *Zeitschrift für Zellforschung und Mikroskopische Anatomie* 101 (1969): 1–12.

"General principles of neuroendocrine communication." In *The Neurosciences: Second Study Program*, edited by F. O. Schmitt, 519–529. New York: Rockefeller University Press, 1970.

"Ultrastructural study of the sites of origin and release of a cellular product in the corpus allatum of insects." *Proceedings of the National Academy of Sciences* 66 (1970): 244–245.

(with W. Bargmann, eds.) *Aspects of Neuroendocrinology*. New York: Springer-Verlag, 1970.

(with M. Weitzman) "Current problems in invertebrate neurosecretion." In *Aspects of Neuroendocrinology*, edited by W. Bargmann and B. Scharrer, 1–23. New York: Springer-Verlag, 1970.

"Concepts of neurochemical mediation." *Neurocirugia* [Santiago, Chile] 29 (1971): 257–262.

"Histophysiological studies on the corpus allatum of *Leucophaea maderae*. V. Ultrastructure of sites of origin and release of a distinctive cellular product." *Zeitschrift für Zellforschung und Mikroskopische Anatomie* 120 (1971): 1–16.

"Comparative aspects of neuroendocrine communication." *General and Comparative Endocrinology* (Suppl.) 3 (1972): 515–517.

"Contribution to the cytophysiology of corpus allatum activity." *General and Comparative Endocrinology* 18 (1972): 622–623.

"Cytophysiological aspects of insect hemocytes." *Anatomical Record* 172 (1972): 465.

"Cytophysiological features of hemocytes in cockroaches." *Zeitschrift für Zellforschung und Mikroskopische Anatomie* 129 (1972): 301–319.

"Principles of neuroendocrine communication." In *Proceedings of the International Symposium on Brain-Endocrine Interaction: The Median Eminence*, 3–6. Basel: S. Karger, 1972.

"Neural control of endocrine glands in invertebrates." In *Proceedings of the Fourth International Congress of Endocrinology, Washington, D.C., 1972*, 210–214. Amsterdam: Excerpta Medica, 1973.

"The concept of neurosecretion, past and present." In *Proceedings of the Symposium on Hypothalamic Function, University of Calgary, Canada, 1973*, 1–7. Basel: S. Karger, 1974.

"The spectrum of neuroendocrine communication." In *Proceedings of the Symposium on Hypothalamic Function, University of Calgary, Canada, 1973*, 8–16. Basel: S. Karger, 1974.

(with S. Wurzelmann) "Observations on synaptoid vesicles in insect neurons." *Zoologische Jahrbücher, Abteilung für Allgemeine Zoologie und Physiologie der Tiere* 78 (1974): 387–396.

"The concept of neurosecretion and its place in neurobiology." In *The Neurosciences: Paths of Discovery*, edited by F. G. Worden, J. P. Swazey, et al., 231–243. Cambridge, MA: MIT Press, 1975.

"The role of neurons in endocrine regulation: A comparative overview." *American Zoologist* 15 (Suppl. 1) (1975): 7–11.

"Neurosecretion—comparative and evolutionary aspects." In *Progress in Brain Research*, vol. 45, edited by M. A. Corner and D. F. Swaab, 125–137. Amsterdam: Elsevier, 1976.

"Evolutionary aspects of neuroendocrine control processes." In *Reproductive Behavior and Evolution*, edited by J. S. Rosenblatt and B. R. Komisaruk, 111–124. New York: Plenum Press, 1977.

"A new look at the evolutionary history of the neurosecretory neuron." *Journal of General Biology (Moscow)* 38 (1977): 849–854 (in Russian).

(with S. Wurzelmann) "Neurosecretion. XVI. Protrusions of bounding membranes of neurosecretory granules." *Cell Tissue Research* 184 (1977): 79–85.

"Current concepts on the evolution of the neurosecretory neuron." In *Neurosecretion and Neuroendocrine Activity, Evolution, Structure, and Function*, edited by W. Bargmann, B. Scharrer, et al., 9–14. New York: Springer-Verlag, 1978.

"An evolutionary interpretation of the phenomenon of neurosecretion." *Forty-Seventh James Arthur Lecture on the Evolution of the Human Brain*. New York: American Museum of Natural History, 1978.

"Peptidergic neurons: Facts and trends." *General and Comparative Endocrinology* 34 (1978): 50–62.

(with W. Bargmann et al., eds.) *Neurosecretion and Neuroendocrine Activity, Evolution, Structure, and Function*. New York: Springer-Verlag, 1978.

"Neurosecretion and neuroendocrinology in historical perspective." In *Hormonal Proteins and Peptides*, vol. 7, edited by C. H. Li, 279–292. New York: Academic Press, 1979.

"Vergleichende Histologie der Glia." In *Handbuch der Mikroskopischen Anatomie des Menschen. Neuroglia I*, vol. 4, pt. 10, 113–141. New York: Springer-Verlag, 1980.

(with M. Weitzman) "Die Glia der wirbellosen Tiere." In *Handbuch der Mikroskopischen Anatomie des Menschen. Neuroglia I*, vol. 4, pt. 10, 157–175. New York: Springer-Verlag, 1980.

"Neuroendocrinology and histochemistry." In *Histochemistry: The Widening Horizons*, edited by J. M. Polak and P. J. Stoward, 11–20. Chichester, England: John Wiley and Sons, 1981.

"Peptidergic neurons—retrospect and prospect." *Verhandlungen der Anatomischen Gesellschaft* 75 (1981): 977–978.

"Recent results on the neuroendocrine system of *Leucophaea*." In *Current Topics in Insect Endocrinology and Nutrition*, edited by G. Bhaskaran, S. Friedman, et al., 47–52. New York: Plenum Press, 1981.

"Peptidergic neurons." *Acta Morphologica Neerlands-Scandinavica* 20 (1982): 219–223.

(with G. B. Stefano and P. Assanah) "Opioid binding sites in the midgut of the insect *Leucophaea maderae* (Blattaria)." *Life Sciences* 31 (1982): 1397–1400.

"Current views on the mechanism of release of neurosecretory products." In *Neurohemal Organs of Arthropods*, edited by A. P. Gupta, 602–607. Springfield, IL: C. C. Thomas, 1983.

"Neurosecretory cells." In *Neuroendocrine Aspects of Reproduction*, edited by R. L. Norman, 1–5. New York: Academic Press, 1983.

"A new look at the phenomenon of neurosecretion." *New York State Journal of Medicine* 83 (1983): 817–818.

"Peptiderge Neurone." *Leopoldina*, s. 3, 30 (1984): 129–130.

"Evolution of intercellular communication channels." In *Functional Morphology of Neuroendocrine Systems*, edited by B. Scharrer, H. G. Hartwig, et al., 1–8. Berlin: Springer-Verlag, 1987.

"Insects as models in neuroendocrine research." *Annual Review of Entomology* 32 (1987): 1–16.

"Opening remarks." *Annals of the New York Academy of Sciences* 493 (1987): 1–2.

(with G. N. Hansen and B. L. Hansen) "Gastrin/CCK-like immunoreactivity in the corpus cardiacum-corpus allatum complex of the cockroach *Leucophaea maderae.*" *Cell Tissue Research* 248 (1987): 595–598.

(with H. G. Hartwig and H. W. Korf, eds.) *Functional Morphology of Neuroendocrine Systems.* Berlin: Springer-Verlag, 1987.

"Neurobiology of opioids in *Leucophaea maderae.*" In *Insect Models in Research*, edited by I. Huber, 85–102. Boca Raton, FL: CRC Press, 1989.

"Neuropeptide research in historical perspective." In *Biologically Active Peptides in Insects*, edited by J. J. Menn, T. J. Kelly, et al. Washington, DC: American Chemical Society Books, 1990.

"The neuropeptide saga." *American Zoologist* 30 (1990): 887–895.

(with T. K. Hughes, Jr., E. M. Smith, et al.) "Interaction of immunoactive monokines (interleukin 1 and tumor necrosis factor) in the bivalve mollusc *Mytilus edulis.*" *Proceedings of the National Academy of Sciences* 87 (1990): 4426–4429.

(with M. A. Shipp, G. B. Stefano, et al.) "Downregulation of enkephalin-mediated inflammatory responses by CD10/neutral endopeptidase 24.11." *Nature* 347 (6291) (1990): 394–396.

(with G.B. Stefano, P. Cadet, and A. Dokun) "A neuroimmunoregulatory-like mechanism responding to stress in the marine bivalve *Mytilus edulis.*" *Brain, Behavior, and Immunity* 4 (1990): 323–329.

(with G. B. Stefano, P. Cadet, J. Sinisterra, et al.) "Neural activation of the cellular immune system involving an opioid mechanism in the mollusc *Mytilus edulis.*" *Annals of the New York Academy of Sciences* 594 (1990): 494–498.

"Neuroimmunology: The importance and role of a comparative approach." *Advances in Neuroimmunology* 1 (1991): 1–6.

(with M. A. Shipp, G. B. Stefano, et al.) "CD10 (CALLA, common acute lymphoblastic leukemia antigen)/neutral endopeptidase 24.11 (NEP, 'enkephalinase'): Molecular structure and role in regulating met-enkephalin mediated inflammatory responses." *Advances in Neuroimmunology* 1 (1991): 139–149.

(with G. B. Stefano, P. Cadet, and J. Sinisterra) "Comparative aspects of the response of human and invertebrate immunocytes to stimulation by opioid neuropeptides." In *Comparative Aspects of Neuropeptide Function*, edited by E. Florey and G. B. Stefano, 329–334. Manchester, England: Manchester University Press, 1991.

"Recent progress in comparative neuroimmunology." *Zoological Science* [Japan] 9 (1992): 1097–1100.

(with G. B. Stefano, P. Melchioni, et al.) "(D-Ala^2) deltorphin I binding and pharmacological evidence for a special subtype of delta opioid receptor on human and invertebrate immune cells." *Proceedings of the National Academy of Sciences* 89 (1992): 9316–9320.

(with G. B. Stefano, A. Digenis, et al.) "Opiatelike substances in an invertebrate: A novel opiate receptor on invertebrate and human immunocytes, and a role in immunosuppression." *Proceedings of the National Academy of Sciences* 90 (1993): 11099–11103.

(with G. B. Stefano, M. K. Leung, et al.) "Autoimmunoregulation and the importance of opioid peptides." *Annals of the New York Academy of Sciences* 712 (1993): 92–101.

(with E. M. Smith and G. B. Stefano, eds.) *Neuropeptides and Immunoregulation*. Berlin: Springer-Verlag, 1994.

(with D. Sonetti, E. Ottaviani, et al.) "Microglia in invertebrate ganglia." *Proceedings of the National Academy of Sciences* 91 (1994): 9180–9184.

(with G. B. Stefano) "Endogenous morphine and related opiates, a new class of chemical messengers." *Advances in Neuroimmunology* 4 (1994): 57–67.

(———) "Neuropeptides and autoregulation immune processes." In *Neuropeptides and Immunoregulation*, edited by B. Scharrer, E. M. Smith, et al. Berlin: Springer-Verlag, 1994.

(with G. B. Stefano) "The presence of the μ_3 opiate receptor in invertebrate neural tissue." *Comparative Biochemistry and Physiology* 113C (1996): 369–373.

Other Works

Against the Tide. Women in Cell Biology.VHS Videotape. Philadelphia: University of Pennsylvania (K. R. Porter Endowment in Cell Biology), 1987.

    Berta Scharrer discusses her early experiences as a woman in science.

Pioneers in Modern Biology, Berta Scharrer: A Partner in the Discovery of Neurohormones. VHS Videotape. Philadelphia: University of Pennsylvania (K. R. Porter Endowment in Cell Biology), 1987.

    Berta Scharrer and others discuss the origins of the concept of neurosecretion.

Works about Berta Vogel Scharrer

"Citation for the Fred Conrad Koch Award of the Endocrine Society to Berta Scharrer." *Endocrinology* 107 (1980): 364–365.

Farrell, G. (ed.). "Neuroendocrine mechanisms." *Science* 147 (1965): 1136.

"Henry Gray Award." *Anatomical News* s. 3 (8) (May 1982).

    A brief review of Berta Scharrer's career.

Kreeger, K. Y. "Pioneering neuroscientist Berta Vogel Scharrer dies." *The Scientist* (Sept. 4, 1995): 17.

Martin, D. "Roach Queen retires." *New York Times* (Feb. 9, 1995): B1–B2.

Millen, S. K. (ed.). "On journeys well traveled." In *Einstein* [Albert Einstein College of Medicine, Bronx, NY] (Spring/Summer 1989): 3–6.

    Based on an interview with Berta Scharrer.

Saxon, W. "Berta Scharrer, 88, research scientist and roach expert." *New York Times* (July 25, 1995): A13.

Siebert, C. "What the roaches told her." *New York Times Magazine* (Dec. 31, 1995): 26–27.

Szilard, G. W. (ed.). "Neuroendocrinology: Pioneering efforts." *Science* 200 (June 9, 1978): 1107.

    Letter.

Other References

Downer, R.G.H., and H. Laufer (eds.). *Endocrinology of Insects*. New York: Alan R. Liss, 1983.

FLORENCE BARBARA SEIBERT (1897–1991)

Paris Svoronos

BIOGRAPHY

Florence Barbara Seibert was born in Easton, Pennsylvania, on October 6, 1897, the daughter of George Peter and Barbara (Memmert) Seibert. At the age of three she and her older brother, who eventually became an architect, were stricken by polio, and they were both forced to wear braces. Seibert had to focus her attention on academics because she "could not go out and dance and play like other children." She remembered living in a comfortable house, playing store in imitation of her father, pretending to be a doctor, like her uncle, and imagining herself a schoolteacher (Yost 1943, 180).

Upon graduating from Easton High School at the top of her class, she won a scholarship to Goucher College in Towson, Maryland. Three more scholarships followed, making her undergraduate studies possible. She hoped to attend medical school, possibly Johns Hopkins University, to become a physician. Faculty and friends, however, advised her against it because they felt that the life of a woman doctor would be difficult. Despite their objections, Seibert took all the required premedical courses.

In her senior year (1918), Dr. Jessie Minor of the Chemistry Department suggested that she work with her at the laboratories of the Hammersley Paper Mills in Garfield, New Jersey. Seibert, who enjoyed chemistry, accepted the offer, hoping that the opportunity would provide her with experience in scientific work and earn her money for future graduate work. After graduating with an A.B. degree in biology (1918), she worked as a chemist at the mill (1918–20), taking advantage of the post–World War I era, which gave women the opportunity to get involved in many laboratory projects. This was the turning point in Seibert's career. She realized that she would be more productive as a biochemical researcher than as a medical practitioner. She realized that "she had to use her head in order to save her feet" (Yost 1943, 182).

In 1920 she won the Dean Van Meter Scholarship at Yale University, where she earned her Ph.D. degree in biochemistry (1923). She also received the competitive Porter Fellowship of the American Philosophical Society (1923–24) at the University of Chicago. In 1924 she became an instructor in pathology and assistant to Dr. Esmond R. Long at the Otho S. A. Sprague Memorial Institute in Chicago. During her first year she perfected an improved and efficient way of getting bacteria-free water to be used in intravenous injections.

At the end of her first year, Long asked Seibert to join him in a tuberculosis research project. She devised a method for isolating the tuberculosis protein molecules, later using the material to develop a skin test reaction for the tuberculosis infection. From its inception this project was financed through funds raised by the sale of the tuberculosis Christmas seals.

In 1932 Seibert followed Long to Philadelphia, where he became director of the Henry Phipps Institute of the University of Pennsylvania. There she was promoted through the ranks to Professor (1955) and held that position until her retirement (1959).

In 1934 Lafayette B. Mendel, Seibert's former thesis adviser, who was dying of diabetes, asked what last favor he could do for her. Knowing that he was on a fellowship committee, she asked for a Guggenheim Fellowship. Unfortunately, he died before the committee met. A few years later Florence Sabin* of the Rockefeller Institute joined the committee. She strongly supported Seibert's application to learn the latest protein-separation techniques in Sweden. Sabin stressed the impact that the trip would make on the future of tuberculosis research. The Guggenheim Fellowship gave Seibert the chance to go to Uppsala University (1937–38) to join Professor Theodor Svedberg (Nobel laureate, 1926) and to establish herself as a leading scientist.

Seibert was a petite woman who weighed approximately 100 pounds. A fellow worker teased her once that her weight was less than the weight of all the medals and awards she had earned during her lifetime. She has been described as a person with "a happy, eager expression in her eyes that makes you know she has found life very good and expects to continue to find it so." She was an adamant believer in hard work, and she has been quoted as saying that "science has a lot of big men in it. And big men are quick to give opportunities to women as well as to men if they see the kind of ability a scientific problem calls for and a willingness to put into it the kind of work it needs. But science is not a lazy man's job—or a lazy woman's either" (Yost 1943, 185).

Seibert was the recipient of the Trudeau Medal of the National Tuberculosis Association in 1938. Four years later she was presented the Garvan Medal of the American Chemical Society (1942). She also won the first Achievement Award of the American Association of University Women (1943), the National Achievement Award of Chi Omega (1944) presented by First Lady Eleanor Roosevelt, the Gimbel Award (1945), the John Scott Award (1947), and the John Elliott Award of the American Association of Blood Banks (1962). She earned honorary doctorates from five different institutions, including the University of Chicago at its 50th anniversary celebration. In 1990 she was inducted into the National Women's Hall of Fame, the ceremony for which she could not attend because of ill health.

Seibert died on August 23, 1991, at the Palm Springs Nursing Home in St. Petersburg, Florida. She was survived by a younger sister, Mabel, of St. Petersburg, who had assisted Seibert in her research over the years and lived with her almost until the end of her life. Mabel was the only sibling who was not affected by infantile paralysis, and she was one of the secretaries at the Phipps Institute who contributed to several of her publications. The cause of death was not disclosed, but Teresa Hicks, an aide who helped care for Seibert, said she had been in declining health in her last two years, suffering from complications of childhood polio. In 1968 Seibert wrote her autobiography, entitled *Pebbles on the Hill of a Scientist*. In 1993 Mayor Thomas F. Goldsmith of her native Easton dedicated a marker in her memory that was erected in her birthplace.

WORK

Seibert's two-year collaboration with Minor at the Hammersley Paper Mills in Garfield provided her with her first three papers on the chemistry of pulps and papermaking and firmly established her belief that biochemistry and research were her fields. Her work in chemistry and immunology was initiated at the Sprague Institute of Chicago. There she met Long, the man who was to have the greatest influence on her scientific life. Long supported her research, using a grant of his own from the National Tuberculosis Association. She discovered that some bacteria managed to survive already existing distillation methods, including triple distillations. Seibert observed that such contaminated water would cause fever in patients. She devised a simple spray-catching trap for the distillation apparatus that eliminated the fever-producing chemical, thereby providing a safer water distillate. As a result she won the Ricketts Prize of the University of Chicago (1924).

Robert Koch (Nobel laureate, 1905), the discoverer of tuberculosis, had isolated tuberculin (a protein) by permitting tubercle bacilli to grow in beef broth, filtering off the bacilli, and then evaporating the culture liquid on a steam bath to one-tenth of its original volume (Brock 1975). This product, if injected beneath the skin of a person suffering from the disease, would cause a reaction indicating the presence of the disease. His tuberculin, however, which is known as "the old tuberculin," did not give accurate results because of the impurities incorporated during its preparation. The evaporated solution included all the unused medium constituents along with the active component, plus numerous other substances made by the tubercle bacillus during its growth. The tuberculin principle itself was difficult to purify. As a result, in some cases, people who were severely infected with tuberculosis failed to give a positive test.

Long had improved Koch's substance by using a synthetic culture broth, but the tuberculin was still not pure enough, and the test was not always dependable.

His project dealt with ascertaining the chemical nature of the specific substance in raw tuberculin and then perfecting the skin test. Pure tuberculin was believed to be a protein. Proteins are traditionally one of the most difficult groups of compounds to purify and crystallize. After ten years of hard work, Seibert isolated pure tuberculin by filtration, using a guncotton membrane of a specific thickness. The precipitated creamy white powder was the purified protein derivative of the tubercle bacillus, known today as "PPD." Her skill in chemical procedures enabled her to recrystallize the tuberculin protein 14 times.

A positive PPD test does not necessarily indicate that a person has tuberculosis. The tubercle bacilli can always exist in a human being in a mild form that can easily be fought off by the immune system. If the test is positive, the patient is required to undergo the more expensive X-ray examination to determine if the disease is in an active state. However, a negative test is a guarantee that the person is not affected by tuberculosis.

Seibert furnished the National Tuberculosis Association with a large quantity of pure tuberculin for commercial use. It is said that she, in her own laboratory, made enough PPD to serve as the international standard for testing all the 2 billion inhabitants of the world. She never patented her discoveries, which would have made her wealthy. Instead, she depended on scholarships and grants.

At the Henry Phipps Institute in Philadelphia, Seibert received the much overdue recognition and promotion she deserved. The Institute was famous for its tuberculosis clinic and offered excellent research facilities. There Seibert received many grants from the National Tuberculosis Association.

The unfortunate death of Mendel diminished her hopes of working in Sweden with Svedberg, but Sabin's efforts earned her the Guggenheim Fellowship. At Uppsala she dealt with Professor Arne Tiselius. He had developed the electrophoresis apparatus, which aids the separation of molecules by passing them through an electric field. At the end of her stay in Sweden, she received a letter from Long asking her whether he could provide her with an apparatus that could be used in her research when she returned to the United States. She quickly responded and asked for an electrophoresis apparatus.

At a symposium sponsored by the American Association for the Advancement of Science (AAAS), Seibert announced her finding concerning the presence of the tuberculosis bacillus in the body, which leads to changes in the proportions of various blood proteins, with the quantity of albumin being drastically reduced.

Seibert's efforts also included experimental work on cancer in 1954. Unfortunately, she did not spend much time on this research since she retired in 1959.

She was a fellow of the AAAS and the New York Academy of Sciences. She was also a member of the American Chemical Society, the American Association of the Organization of Women, and the American Society of Biological Chemists, as well as an honorary member of the Trudeau Society. She served as an alumnae trustee of Goucher College (1943–46) and became the board corporator of the Woman's Medical College (1946–49).

NOTE

The author would like to express his feelings of appreciation for the efforts of Dr. LeLeng To, chair, Department of Biological Sciences, Goucher College, for her great efforts in providing all possible information pertaining to Dr. Seibert.

BIBLIOGRAPHY

Works by Florence Barbara Seibert

Scientific Works

Space does not permit the listing of the complete works of Florence Barbara Seibert. The complete bibliography of Seibert is included in her autobiography (Pebbles 1968). This list includes all works cited in the text as well as her dissertation.

"Febrile Reactions Following the Injection of Non-specific Agents into Rabbits." Ph.D. diss., Yale University, 1923.

Pebbles on the Hill of a Scientist. St. Petersburg, FL: n.p., 1968.

Works about Florence Barbara Seibert

"Dr. Florence B. Seibert, inventor of standard TB test, dies at 93." *New York Times* (Aug. 31, 1991).

"Dr. Florence Seibert." *Newsday* (Sept. 2, 1991): 21.

Hall, D. L. "Academics, bluestockings, and biologists: Women at the University of Chicago, 1892–1932." In *Annals of the New York Academy of Sciences*, vol. 323, edited by A. M. Briscoe and S. M. Pfafflin, 300–320. New York: New York Academy of Sciences, 1979.

Hollinshead, Ariel. "Florence Barbara Seibert (1897–1991)." In *Women in Chemistry and Physics*, edited by L. S. Grinstein et al., 526–529. Westport, CT: Greenwood Press, 1993.

Yost, Edna. *American Women of Science*. Philadelphia: J. B. Lippincott, 1943.

Other References

Brock, T. D. (ed.). *Milestones in Microbiology*, 109–118. Washington, DC: American Society for Microbiology, 1975.

LYDIA WHITE SHATTUCK (1822–1889)

Philip Duhan Segal

BIOGRAPHY

Lydia White Shattuck was born on June 10, 1822, in East Landaff (now Easton), New Hampshire. The family, prominent in early New England days, traces its origin to William Shattuck, born in England in 1621 or 1622. His name appears in a 1642 list of the proprietors of Watertown, Massachusetts. In 1798, together with other members of the family, her grandfather Shattuck moved from Massachusetts to New Hampshire and settled in Landaff. Timothy Shattuck, her father, a cousin of the eminent physician George C. Shattuck of Boston, married Betsey Fletcher of Acton, Massachusetts, on January 28, 1812.

Her father was a farmer whose land near the Franconia Mountains was not very productive. Intellectual, kindly, and religious, he was a man of deep and firm convictions. Although strict in all religious observances, he did not require his children to adopt either his practices or his opinions. He often visited the sick and dying and was active in good works. Her mother was more reticent, but she displayed a love of the beautiful. She often found companionship in nature, rather than among the neighbors of their sparsely settled community.

Lydia, their fifth child, was the first who lived. William L., the only son, survived his sister, and for most of his life resided in Wing Road, New Hampshire, a few miles from the place of their birth. The daughter resembled her father physically. Full of life and strength, she climbed the hills with her brother or wandered up and down the streams that feed the Ammonoosuc River.

The young girl absorbed all the influences about her. She remembered her father's words and ways, tempered by her mother's feelings and love of nature. A longtime friend observed that nature "took her into loving confidence; wandering about in meadow, marsh, and forest, in the valleys and over the hills; delighting in the whisper of the breeze, the notes of insects, the songs of birds, and the ever-varying phases of the vegetable world." The future botanist "gained a knowledge of flying and creeping things, of green and beautiful

growths, far beyond that of most of her own age, and indeed of her own time"
(Stow 1890, 7).

After her early schooling was completed, she taught in district schools (1837–
48), with short periods of study in various academies in Vermont and New
Hampshire. In 1848 she entered Mount Holyoke Seminary in South Hadley,
Massachusetts, an institution founded by Mary Lyon in 1837 to train women to
be teachers. Mary Lyon regarded the sciences as a primary element of the curriculum, and she taught the first chemistry courses until her death in 1849.
Shattuck, who studied briefly with Lyon, paid her own way through the Seminary, largely by working extra hours as a domestic.

After graduating with honors from the Seminary in 1851, she was engaged
at once to teach there, and she remained as a faculty member for the rest of her
life. She won recognition as a botanist of "unquestioned standing in the world
of science" (Cole 1940, 156). She was an early proponent of evolution, which
she reconciled with her strong religious beliefs. Her scientific liberalism may
have been encouraged by her friendship with a Harvard University botanist, Asa
Gray, whom she invited to teach as one of many visiting lecturers.

Chemistry was her second love, and she was among the founders of the
American Chemical Society (ACS). She said that she liked chemistry best in
winter and botany best in summer. At various times during her tenure, she taught
algebra and geometry, physiology and natural philosophy, astronomy and physics.

She was raised in a religious atmosphere, and her father's house was always
open to welcome the itinerant Methodist preacher for his fortnightly lectures.
For 11 years before her death in 1889, she was a member of the Methodist
Church in Springfield. After several months in poor health, due to problems
with heart, lungs, and kidneys, she died on November 2, 1889, and was buried
in Evergreen Cemetery, South Hadley, Massachusetts. In 1893 a new building
to house the chemistry and physics departments was named Shattuck Hall in
her honor. In 1954, the name was transferred to a later building for physics,
after the original Shattuck Hall was demolished.

WORK

Lydia Shattuck's importance in science and science education is well established. She carried on the tradition of excellence in science teaching begun by
Mary Lyon, expanded Mount Holyoke's influence in the world of science, and
helped define its role as a college. "As a teacher and as an inspiration for
younger teachers, she had a profound influence on the teaching of biology in
this country and elsewhere in the world" (Shmurak and Handler 1991, 130).
She never published scientific papers, but she did active research on the classification of plants, taught a generation of women who became scientists and
science teachers, and was an outstanding leader. Her experimental notebooks,
her lecture notes, and her correspondence with leading scientists create a sub-

stantial record. Recollections by her students, her faculty colleagues, and fellow scientists give us a balanced picture of her work.

Mount Holyoke archives hold Shattuck's lecture notes for 1860 and 1866. They include the standard chemistry of her day (acids and bases, specific gravity, spectral analysis, descriptions of the elements, and boiling under pressure), household tips such as how to improve a "soft and miserable bread," and astronomical references to the formation of the solar system according to the nebular hypothesis. Her lecture notes refer to religion, such as scientific law, as the "human way of looking at Divine actions" or prayer as a way to "change even a law." This combination of science and religion was not unusual at the time (Warner 1978) and also appears in the chemistry lectures of visiting professors such as Edward Lasell of Williams College.

More than 100 experiments, through which students could learn about various elements, are described in her notebooks. Her notebooks reveal that she used experiments from the *Manual of Inorganic Chemistry* (Eliot and Storer 1867). In their preface, the authors indicated that their text is "intended to enable the student to see, smell and touch for himself." Evidently, the laboratory method of teaching, which was pioneered by Amos Eaton and adopted by Mary Lyon, was used by many enlightened science teachers in the 1860s.

A contemporary noted that "[h]er interest in all branches of science was remarkable" (Stow 1890, 10). One of her students, who became a distinguished teacher of science, wrote, "In these days of specialties, such scattering work would not be considered advantageous as a preparation for science teaching." However, she noted that Shattuck "regretted, truly for her pupils' sake, the amount of territory she was thus obliged to cover." Yet she was able to convert this regrettable necessity into an "opportunity for broadening her own vision," and "with the ability to sift and assimilate" from wide reading, she was able to achieve a level of "general information and broad culture which every one who talked with her recognized" (Hooker 1890, 30–31).

A most important pedagogical experience for her occurred in 1873, when she was among the first 15 women to attend the Anderson School of Natural History on Penikese Island off the coast of Massachusetts. She was carefully selected from hundreds of applicants by Louis Agassiz, Swiss-American zoologist and geologist, who had studied at several European universities before accepting a professorship at Harvard. He was helped in the selection by his wife, Elizabeth Agassiz,* who later became the first president of Radcliffe College. Since the Anderson School offered women a rare opportunity for advanced education in biology at a time when such training was otherwise closed to women, it played a critical role in improving science teaching in the United States (Rossiter 1982).

In 1874 she was one of a handful of women chemists who attended the Priestley Centennial in Philadelphia, which led to the formation of the ACS. However, it is noteworthy that she and her female colleagues were not allowed in the official photograph (Rossiter 1982).

During the 1880s she was a corresponding member of the Torrey Botanical

Club of New York City. Expert on the Connecticut Valley region of Massachusetts, she was, at one time, president of the Connecticut Valley Botanical Association. She made every effort to keep current in her two primary fields,
botany and chemistry. In 1878 she went to a teachers' convention in Montreal,
followed by a field trip to the Saguenay River. She also attended an exposition
in New Orleans (1884).

Shattuck traveled widely in the United States, Canada, Europe, and Hawaii,
expanding her own knowledge, collecting botanical and mineral specimens, and
establishing contacts with many active scientists. In the summer and fall of 1869
she toured Europe. In 1886 and 1887, while studying the flora of the Hawaiian
Islands, she took a week's detour to observe a lava flow from the Mauna Loa
volcano.

Her active mind was always intellectually curious. For example, in the summer of 1868, on an excursion to the Great Lakes, a party of teachers visited an
island in Lake Superior where they hoped to find the fragrant *Aspidum*, rare
elsewhere in the United States. However, chlorastrolites were also on the island.
Mineralogy won over botany. While she sought the green star-stones, others
became the first to locate the rare fern with its violet fragrance. Fascinated by
geologic changes, she read the "testimony of the rocks" (Stow 1890, 10).

Her enthusiasm inspired students and colleagues. Many of her students, including those who became missionaries in foreign lands, sent specimens from
their new lands. In this way, the Mount Holyoke Seminary herbarium came to
house more than 7,000 carefully analyzed plants, seeds, and woods from all
over the world. Her field of research—the classification of plants and the setting
up of collections—was "typical of her generation," and "she excelled in both"
(Haywood 1971, 274). She established a botanical garden, collecting and transplanting many of its wild plants herself.

While educating her students in science, she also helped them appreciate the
beauty of nature. One biographer noted that "it was to her great credit as a
scientist that she accepted the controversial theory of evolution and promoted it
at Mount Holyoke, where conservative religious tradition was strong" (Haywood 1971, 274). A contemporary of hers reports that she said, "I would rather
be a descendant of a good monkey than of a wicked man" (Stow 1890, 11).

A prominent student of Shattuck's was Cornelia Clapp, who became a noted
biologist and the first woman given a research post at the Woods Hole Biological
Laboratory. Although Penikese Island supplied Shattuck's only advanced education, Clapp went on to earn two doctorates soon after graduate schools opened
their doors to women in the 1880s. Mount Holyoke produced more women who
later earned doctorates in the physical sciences from 1910 to 1969 than any
other undergraduate institution (Tidball and Kistiakowsky 1976).

Shattuck spoke to alumnae at the semicentennial celebration of Mount Holyoke Seminary (1887). She observed, "There is an earnest spirit of study here
favorable to scientific research, and science does not tolerate any half-way

work." She noted that "the Seminary has had a scientific trend from the first, without tendency to convert us into agnostics or infidels" (Carr 1918, 160).

Her last work was raising funds for the new chemistry and physics building that was to be named for her.

NOTE

I should like to thank Dr. Elizabeth Kennan, President of Mount Holyoke College, and Patricia J. Albright, archivist of the Mount Holyoke College Library, for their gracious assistance in supplying source material for this article.

BIBLIOGRAPHY

Works by Lydia White Shattuck

Scientific Works

"Notebooks and other papers of Lydia Shattuck." Hadley, MA: Mount Holyoke College Library Archives.

Works about Lydia White Shattuck

Carr, E. "The department of chemistry: Historical sketch." *Mount Holyoke Alumnae Quarterly* 2 (1918): 159–162.

Clapp, C. "Oral interview in Cornelia Clapp Papers." Hadley, MA: Mount Holyoke College Library Archives.

Cole, A. *A Hundred Years of Mount Holyoke College.* New Haven, CT: Yale University Press, 1940.

Haywood, C. "Shattuck, Lydia White." In *Notable American Women 1607–1950: A Biographical Dictionary*, edited by E. T. James et al., 273–274. Cambridge, MA: Belknap Press of Harvard University Press, 1971.

Hooker, H. "Miss Shattuck as a student and teacher of science." In *Memorial of Lydia W. Shattuck*, 25–31. Boston: Beacon Press, 1890.

Rossiter, M. *Women Scientists in America.* Baltimore: Johns Hopkins University Press, 1982.

Shmurak, C., and B. Handler. "Lydia Shattuck: 'A streak of the modern.' " *Teaching Education* 3 (1991): 127–131.

Stow, S. "Biographical sketch." In *Memorial of Lydia W. Shattuck*, 5–23. Boston: Beacon Press, 1890.

Other References

Barker, G. *A Text-Book of Elementary Chemistry, Theoretical and Inorganic.* Louisville, KY: Morton, 1870.

Eliot, C., and F. Storer. *A Manual of Inorganic Chemistry.* Boston: privately published, 1867.

Kohlstedt, S. "In from the periphery: American women in science, 1830–1880." *Signs* 4 (1978): 81–96.

McLean, M. "Chemical lectures by Professor Lasell." Hadley, MA: Mount Holyoke College Library Archives.
    Notebook.

Shattuck, Lemuel. *Memorials of the Descendants of William Shattuck*. Boston: Dutton and Wentworth, 1855.

Tidball, M., and V. Kistiakowsky. "Baccalaureate origins of American scientists and scholars." *Science* 193 (Aug. 20, 1976): 646–652.

Warner, D. "Science education for women in antebellum America." *Isis* 69 (1978): 58–67.

MAUD CAROLINE SLYE (1869–1954)

Jeanie Strobert Payne

BIOGRAPHY

The American medical researcher and pathologist Maud Caroline Slye was born on February 8, 1869, in Minneapolis, Minnesota. Some records suggest that she may have been born in 1879. She was the second of three children and the younger daughter of Florence Alden (Wheeler) and James Alvin Slye. The family was poor but well educated, and her ancestors could be traced back to colonial times. Her father was a lawyer and author. Her mother enjoyed poetry and had aspirations that her daughter would become an artist or writer. Instead, Maud Slye turned toward a career in biology (Sicherman et al. 1980, 651).

Slye's family moved to Iowa. There she attended public schools in Des Moines and in Marshalltown. Maud Slye graduated from Marshalltown High School in 1886.

Despite financial obstacles, Slye was determined to go to college, and she wanted to go to the University of Chicago. She enrolled in September 1895, with very limited funds at her disposal. As a freshman in 1895, Slye worked as a clerk for William Rainy Harper, President of the University. It was difficult for her to meet tuition and living expenses, but for three years she carried a full academic load in addition to her clerical work. She really had no choice, since at that time there were few, if any, scholarships available to women. In time, this workload became unbearable, and in the spring of 1898, she suffered what people politely called a "nervous breakdown" (McCoy 1977, 14). She was asked to leave the University. On the advice of her physician she took some time off to restore her strength while staying in Massachusetts with close relatives.

Slye rested for about two months and then decided to take courses at the Woods Hole Laboratory. By the fall of 1898, Slye was able to enter Brown University, where she successfully completed her bachelor's degree in 1899.

After graduating from Brown University, Slye accepted a professional position at the Rhode Island Normal School, which was a two-year teacher-training institution. There Slye developed an interest in genetics and psychiatry, but her chief interest was in genetics and human heredity. In 1908 Slye was appointed a graduate assistant to Professor Charles Whitman, who was head of the Biology Department at the University of Chicago. She was given a small fellowship and a corner of the basement in the zoology building as her laboratory (McCoy 1977, 15), where she worked and studied the Japanese waltzing mice, using limited funds from her fellowship.

Otho S. A. Sprague, an extremely wealthy man, had died and left millions of dollars for medical research. The Sprague Memorial Institute was established at the University of Chicago, and an eminent pathologist, H. Gideon Wells, was appointed its director. In 1911 Slye was named to the Institute staff and was permitted to go ahead with her cancer research. Now, for the first time, she was given a salary and a laboratory large enough for her to work in comfort and ease (McCoy 1977, 25).

Slye was promoted from Instructor to Assistant Professor of Pathology (1922), then to Associate Professor of Pathology (1926), and she remained in that rank until 1944 (Sicherman 1980, 651). In 1936 she became director of the Cancer Laboratory of the Sprague Memorial Institute. Slye was named Professor Emeritus in 1944 after her retirement. She died on September 17, 1954, of heart failure.

During her lifetime, Slye received several awards, including the American Medical Association's Gold Medal (1914), the Ricketts Prize from the University of Chicago (1915), and the Gold Medal from the Radiological Society (1922). In 1937 Brown University awarded her an honorary D.Sc. degree.

Slye's success as a dedicated research scientist did not come easily. She was ably assisted by Harriet Holmes and Wells, who were also well-known pathologists. They helped Slye in microscopic analyses of tissues from the mice that had died due to various cancers. Slye's work was constantly under attack by some of her peers, especially Dr. Clarence Cook Little. Little was one of the country's leading researchers who had earned his D.Sc. degree in zoology from Harvard University in 1914. His main interest was in the field of genetics, and he had conducted various inheritance experiments while at Harvard (McCoy 1977, 139).

In addition to being a research scientist, Slye was a poet. She enjoyed writing and expressing herself in verse, some of which was accepted for publication. She took pleasure in music, gardening, and the enjoyment of nature after so many long days and hours spent in the laboratory. Although Slye never married, she did have a very close relationship with an artist in her early years, but the artist died, and she dedicated the rest of her life to science. ''Nevertheless, beyond the wall of the scientific enclave, away from the unrelenting routine of research, she felt and reacted to life: to people, animals, sounds, smells, noises; to the joy of listening to great music; the satisfaction of painting a landscape; the peace that came in her garden'' (McCoy 1977, 114).

WORK

Maud Slye was named the "mouse lady" because of her research with Japanese waltzing mice (apparently afflicted with a nervous disorder). Her work with these mice helped her study the possibility of the inheritance of cancer. During this time in England, members of the Royal Society of Medicine also considered the possibility that heredity might be a factor in cancer. But the society was by no means unified on the theory. Some members believed in it, but nobody had ever conducted a lengthy study of the problem. Therefore, no reliable data existed (McCoy 1977, 19).

What was needed was a carefully controlled mouse-breeding program to establish whether heredity was or was not a factor in cancer. As yet, there was no such program, but the idea intrigued Maud Slye. In her mouse laboratory were the basic ingredients for such a research project, and her research interest changed from studying the inheritance of nervous disorders to the inheritance of cancer (McCoy 1977, 19).

Slye continued to breed her mice, and soon a mouse developed cancer of the breast. Then another mouse developed breast tumors, and after that more spontaneous tumors appeared, which had not been grafted onto the mice but had developed naturally. Subsequently, Slye had the first litter of baby mice born to a cancerous mother. She looked at the tiny squirming babies, pink and hairless, and knew that now she had the material to determine whether or not cancer was inherited (McCoy 1977, 22).

From 1911 to 1913 Slye performed over 5,000 autopsies with the assistance of Holmes and Wells. Out of the 5,000 mice that were autopsied, 298 had cancer. She kept careful records and charts of all her research findings, and on May 5, 1913, she presented her findings to the American Society for Cancer Research. Slye explained that a mouse could be called purebreeding noncancerous when it fulfilled three conditions: it died without cancer as shown by autopsy; when mated with a cancerous mate, only cancer-free young were produced; and it was possible to derive an extracted line of wholly cancer-free descendants (McCoy 1977, 32). Slye later proved that susceptible mice developed carcinoma, and in nonsusceptible strains the result was never malignancy, but only inflammatory or septic changes in the tissues (McCoy 1977, 175). As a result of her research, Slye wanted to compare her studies to human cancer inheritance, since most of her peers did not believe that there was any relationship between the inheritability of human cancer and mouse cancer. Based on her studies, Slye persisted in believing that cancers were inherited. She repeatedly suggested that a cancer statistics collection bureau be established to trace the inheritability of cancer in humans. Unfortunately, this bureau was never established. Only a code system was developed by Slye for the Cook County Hospital in Chicago.

Slye's dedication to the field of science did not go unnoticed, and her membership in several professional organizations helped support her work. Some of these were the Sprague Memorial Institute of Chicago, the Association of Cancer Research, the Chicago Institute of Medicine, and an honorary membership in the Seattle Academy of Surgery.

Slye was definitely ahead of her time, because her research involved the inheritability of cancer long before the discovery of deoxyribonucleic acid (DNA). She was tenacious in her belief that cancer was inherited, despite criticism from many of her peers. Consequently, after the death of Slye, doctors began to question patients about any history of cancer in their families (McCoy 1977, 181).

BIBLIOGRAPHY

Works by Maud Caroline Slye

Scientific Works

"The incidence and inheritability of spontaneous cancer in mice. Preliminary report." *Zeitschrift für Krebsforschung* 13 (1913): 500–504.
"The incidence and inheritability of spontaneous tumors in mice. Second report." *Journal of Medical Research* 30 (3) (1914): 281–298.
(with H. F. Holmes and H. G. Wells) "The primary spontaneous tumors of the lungs in mice." *Journal of Medical Research* 30 (3) (1914): 417–442.
(———) "Spontaneous primary tumors of the liver in mice." *Journal of Medical Research* 33 (2) (1915): 171–182.
"The inheritability of spontaneous tumors of specific organs and of specific types in mice." *Journal of Cancer Research* 1 (1916): 479–502.
"The inheritability of spontaneous tumors of the liver in mice." *Journal of Cancer Research* 1 (1916): 503–522.
"The inheritance behavior of infections common to mice." *Journal of Cancer Research* 2 (1917): 213–238.
(with H. F. Holmes and H. G. Wells) "Comparative pathology of cancer of the stomach with particular reference to the primary spontaneous malignant tumors of the alimentary canal in mice." *Journal of Cancer Research* 2 (1917): 401–425.
(———) "Primary spontaneous sarcoma in mice." *Journal of Cancer Research* 2 (1917): 1–37.
(———) "Primary spontaneous tumors of the testicle and seminal vesicle in mice and other animals." *Journal of Cancer Research* 4 (1919): 207–228.
"Relation of inbreeding to tumor production." *Journal of Cancer Research* 5 (1920): 53–79.
"Relation of pregnancy and reproduction to tumor growth." *Journal of Cancer Research* 5 (1920): 25–52.
(with H. F. Holmes and H. G. Wells) "Primary spontaneous tumors of the ovary in mice." *Journal of Cancer Research* 5 (1920): 205–226.
"The influence of heredity in determining tumor metastases." *Journal of Cancer Research* 6 (1921): 139–173.
(with H. F. Holmes and H. G. Wells) "Primary spontaneous squamous cell carcinomas in mice." *Journal of Cancer Research* 6 (1921): 57–86.

(————) "Primary spontaneous tumors in the kidney and adrenal of mice." *Journal of Cancer Research* 6 (1921): 305–336.

"Biological evidence for the inheritability of cancer in man." *Journal of Cancer Research* 7 (1922): 107–147.

"The fundamental harmonies and the fundamental differences between spontaneous neoplasms and all experimentally produced tumors." *Journal of Cancer Research* 8 (1924): 240–273.

(with H. F. Holmes and H. G. Wells) "Primary spontaneous tumors of the uterus in mice." *Journal of Cancer Research* 8 (1924): 96–118.

"Heredity in relation to cancer." In *Our Present Knowledge of Heredity*, 101–156. Phildadelphia: W. B. Saunders, 1925.

"The inheritance behavior of cancer as a simple Mendelian recessive." *Journal of Cancer Research* 10 (1926): 15–49.

"Some misconceptions regarding the relation of heredity to cancer and other diseases." *Journal of the American Medical Association* 86 (1926): 1599–1605.

"The relation of inbreeding to tumor production." *Medical Woman's Journal* 33 (1926): 87–99.

(with H. F. Holmes and H. G. Wells) "The comparative pathology of cancer of the thyroid, with report of primary spontaneous tumors of the thyroid in mice and in a rat." *Journal of Cancer Research* 10 (1926): 175–193.

"Some observations in the nature of cancer. Preliminary report." *Journal of Cancer Research* 11 (1927): 135–151.

"The relation of heredity to spontaneous thyroid tumors in mice." *Journal of Cancer Research* 11 (1927): 54–71.

"Cancer and heredity." *Annals of Internal Medicine* 1 (1928): 951–976.

"The relation of heredity to cancer." *Journal of Cancer Research* 12 (1928): 83–133.

Cancer and Heredity. Baltimore: Williams and Wilkins, 1929.

"The interrelation between hereditary predisposition and external factors in the causation of cancer." *Annals of Surgery* 93 (1931): 40–49.

"The relation of heredity to cancer occurrence as shown in strain 621." *American Journal of Cancer* 15 (1931): 2675–2726.

"The relation of heredity to the occurrence of spontaneous leukemia, pseudoleukemia, lymphosarcoma and allied diseases in mice: Preliminary report." *American Journal of Cancer* 15 (1931): 1361–1386.

(with H. F. Holmes and H. G. Wells) "Intracranial neoplasms in lower animals." *American Journal of Cancer* 15 (1931): 1387–1400.

"The relation of heredity to cancer occurrence as shown in strain 73." *American Journal of Cancer* 18 (1933): 535–582.

(with H. F. Holmes and H. G. Wells) "The comparative pathology of carcinoma of the pancreas, with report of two cases in mice." *American Journal of Cancer* 23 (1935): 81–86.

(with H. G. Wells) "Tumors of islets tissue with hyperinsulinism in a dog." *Archives of Pathology* 19 (1935): 537–542.

"Genetics of cancer and its localization." *Occasional Publications of the American Association for the Advancement of Science* (4) (June 1937): 3–16.

"The relation of heredity to the occurrence of cancer." *Radiology* 29 (1937): 406–433.

(with H. G. Wells and H. F. Holmes) "Comparative pathology of cancer of the alimentary canal, with report of cases in mice.'' *American Journal of Cancer* 33 (1938): 223–238.

''Cancer and heredity.'' In *Medical Genetics and Eugenics*, 109–127. Philadelphia: Woman's Medical College of Pennsylvania, 1940.

(with H. G. Wells and H. F. Holmes) ''The occurrence and pathology of spontaneous carcinoma of the lung in mice.'' *Cancer Research* 1 (4) (1941): 259–261.

Nonscientific Works

Songs and Solaces. Boston: Stratford, 1934.

I in the Wind, Symphony no. 1 and Minor Songs. Boston: Stratford, 1936.

Works about Maud Caroline Slye

Jaffe, B. *Outposts of Science: A Journey to the Workshops of our Leading Men of Research*. New York: Simon and Schuster, 1935.

McCoy, J. J. *The Cancer Lady: Maud Slye and Her Heredity Studies*. Nashville, TN: Thomas Nelson, 1977.

Sicherman, B., H. Green, et al. *Notable American Women: The Modern Period*. Cambridge, MA: Belknap Press of Harvard University Press, 1980.

Strong, L. C. ''Genetic studies on the nature of cancer.'' *American Naturalist* 60 (668) (1926): 201–226.

MARJORY STEPHENSON (1885–1948)

Rebecca Meyer Monhardt

BIOGRAPHY

Marjory Stephenson was born on January 24, 1885, near Cambridge, England, in the village of Burwell. Her family, which had a long history in this area, was well-to-do and public-minded. Her father was Robert Stephenson, and her mother was Elizabeth Rogers of Newmarket. She was the youngest of the Stephensons' four children. Her great-grandfather trained racehorses in the area during the eighteenth century, and records indicate that he was a well-educated man. Little is known of her grandfather other than that he was a successful and prosperous farmer.

Robert Stephenson made his living as a farmer and horse breeder in the fenland. Although her father had never been to a university himself, he did have an excellent secondary school background and valued a good education. He believed in Charles Darwin's theory of evolution and also had an interest in the ideas of Gregor Mendel. Mr. Stephenson applied scientific theory to farming and introduced fruit orchards to Cambridgeshire. Marjory Stephenson recollected that when she was a young child, her father explained the facts of symbiotic nitrogen fixation to her while the two walked together through a clover field on their farm.

Her mother saw to it that both her daughters received a higher education. Mrs. Stephenson developed in her younger daughter an interest in literature and art. As an adult, she appreciated the great Dutch painters, as well as those of the nineteenth-century French school. In the late 1930s, Marjory Stephenson took up painting as a hobby for a few years and gained considerable pleasure from it.

Because Stephenson was a great deal younger than her three siblings, she felt somewhat isolated in the family and at times thought that her arrival was resented, even by her mother. During her childhood she was greatly influenced by her governess, a woman of great character and intellect who recommended that Stephenson be sent to Berkhampsted High School for Girls. Stephenson won a scholarship and after entering the school in 1897, spent six years there. Although she considered the science education poor, she did receive one year of adequate training in physiology, which proved to be a good foundation for her later studies in biology.

In 1903 she entered Newnham College and took a Part I Natural Science Tripos in chemistry, physiology, and zoology. At this time, physiology was the only laboratory science taught in the university, and women were excluded from classes in the two other subjects. She received a Class II in the Part I examination in 1906. She maintained a close link with Newnham College throughout her life and became a member of the governing body in 1931 and of the council in 1944.

Stephenson wanted to attend medical school after leaving Newnham but had to earn her living by teaching. She spent five years teaching at the Gloucester County Training College in Domestic Science and the Kings College of Household Science in London. These were not especially happy years for her. It was common for women scientists of this time to study nutrition and dietetics, which offered little chance to do science. Stephenson was not content to be a passive learner, as evidenced by her later scientific work, but because the more elite academic disciplines were virtually closed to women at this time, she had to be content with teaching.

It is interesting to note that later in her career she found teaching at Cambridge quite rewarding. She was never a good lecturer and took no pleasure in simply presenting the facts to her students. What she did enjoy was the interaction with students during the laboratory part of the class. She was known as a generous and inspiring mentor to her students. Stephenson had many recruits to her research, and perhaps through her guidance she had her greatest influence. She allowed her students to pursue their own ideas but insisted on persistence and thoroughness, which were characteristic of her own work. She believed that research and teaching were complementary endeavors and that it was also important for an instructor to be an active researcher.

Stephenson's career was characterized by being in the right place at the right time. Nutrition, the ''woman's discipline,'' provided an entry for her into the developing profession of physiological chemistry. In 1911 her talents attracted the attention of Dr. R.H.A. Plimmer, a nutritional biochemist. He offered her a job at University College, London, where she assisted him in advanced teaching and research on lipid metabolism and metabolic disease. She worked successively on the lactase of intestinal mucosa, on the synthesis of palmitic acid esters, and finally on experimental diabetes. In 1913 Stephenson was awarded a Beit Memorial Fellowship for medical research, which gave her the opportunity to explore her own interests, but the outbreak of World War I in 1914 interrupted her career. She served in the British Red Cross from 1914 to 1918, where she put her knowledge of nutrition and diet to practical use in France and then in Salonika. She was made an associate of the Royal Red Cross and was also awarded the Member of the Order of the British Empire (MBE) for her service during the war.

In 1919 she returned to Cambridge to join the enthusiastic group of scientists whom Professor Frederick G. Hopkins was assembling. Under his leadership, this group made Cambridge the center of biochemical thought in Britain for a generation. A paper on essential amino acids written by Hopkins before the war had made him well known as a researcher on "vitamins" (Hopkins 1912). Stephenson joined Hopkins's group, hoping to move from the metabolism of fats into the emerging field of fat-soluble vitamins. Her research on the possible role of vitamin A deficiency in keraomalacia got nowhere, and Hopkins, unable to compete with better organized and better funded research teams in London, lost interest in vitamins and immersed himself in the chemistry of biological oxidation. Under these circumstances Stephenson changed her focus to the less fashionable field of bacterial biochemistry.

Hopkins had a long interest in the biochemistry of microorganisms. He envisioned general biochemistry as encompassing every kind of organism, from bacteria to man. During the war, he arranged for a young chemist, Harold Raistrick, to begin a line of work on the chemistry of microbes. However, in 1921, Raistrick left to take a job in industry just as a large endowment and a new laboratory provided Hopkins with the means to do further research in comparative biochemistry. Thus, an important position became vacant just as Stephenson's work on vitamins had reached a dead end. Once again, she found herself in the right place at the right time. In this position she had free rein to develop bacterial biochemistry according to her own vision. The field was yet to be defined, and the intellectual space created between bacteriology and chemistry was waiting for someone to provide a certified body of knowledge to define it.

Stephenson was interested in microbes as organisms with distinctive physiologies. Her early work built upon the classical research of Louis Pasteur, Max Rubner, and other biologists that focused on metabolic balance. She was always fascinated by the diversity of microbial metabolism. At the same time Stephenson was deeply influenced by the mainstream lines of the biochemical work in enzymology and oxidation that dominated Hopkins's expanding school. Women research scientists were very rare at Cambridge at this time but were welcomed by Hopkins, who was the chair of biochemistry. Stephenson found this stimulating environment much to her liking, and she became a leading personality in the department. She credits Sir Frederick Hopkins's influence alone in giving her the incentive to pursue biochemical research. Stephenson worked on fat-soluble vitamins, but by 1922 she was publishing papers on bacterial metabolism, the subject that she developed and that became her life's work.

After the Beit Fellowship expired, she worked on annual grants from the Medical Research Council for nearly ten years, but finally in 1929 she was appointed to the permanent staff at Cambridge. In 1945 she was elected fellow of the Royal Society, being the first woman in biological science to receive that honor and recognition. In 1936 she was also one of the first women to receive the Sc.D. degree of Cambridge, then titular only. Stephenson faced the uphill struggle for women in the scientific world with her characteristic sense of humor. It amused her greatly to observe the reactions of some of the older males when women were finally admitted to full rights of the degree and appeared in full regalia.

In her early years Stephenson worked for women's suffrage. Later she lost interest in women's issues, except that she remained an active supporter of education for women. Perhaps she herself did not feel oppressed as a woman, or perhaps she was simply too busy with other interests to concern herself with feminist issues. Her generous nature could not abide the oppression of scientific workers, especially of Jews. She was quite active in providing assistance to refugee scientists. She thus found an outlet for her personal generosity.

Marjory Stephenson was a true Englishwoman. Her judgment of people was not always infallible, but she was quite willing to reconsider her thinking and admit that she could have been wrong. Her politics leaned to the intellectual Left. She had many friends internationally and traveled to Russia and Hungary. She even traveled to America, where she spent a holiday at a dude ranch, an experience she recalled happily in the grim period of the 1940s.

In 1935 she built a house where she lived for the last 14 years of her life. She never married. Stephenson shared her father's interest in fruit trees and planted many of them on her property. She became very knowledgeable on this subject and drew comfort and happiness from watching things grow. The building of her house gave her the opportunity to entertain various people. In her home she created a pleasant atmosphere where colleagues and friends could converse and discuss work. During the last years of her life, she became ill. Few of her friends realized the seriousness of her illness until the end, because she carried on with her usual gaiety and vigor. She was able to continue with her work almost until her death. The last stages of her illness progressed rapidly, and she died on December 12, 1948.

The teaching of advanced biochemistry at Cambridge was shared by the leading workers in the field, whether they were university staff members or not. Marjory Stephenson shouldered her full responsibility in this area from 1925 onward. The kind of teaching she enjoyed involved active learning. She was not especially interested in the number of publications that her students produced but rather insisted that they obtain the maximum advantage from their first years of work. She was an early believer in the constructivist philosophy toward learning and allowed students to figure things out for themselves, even though she sometimes had to intervene. She believed that "infection not instruction is the secret of education" (Woods 1950). A steady stream of students worked on her research team. Through informal talks during laboratory sessions she truly got to know her students. As development and interest in chemical and general microbiology increased, she worked hard for the establishment of a special Part II Biochemistry at Cambridge. This was achieved in 1947, and that same year she was recognized by the University for her long service to teaching by making her its first Reader in Chemical Microbiology.

Stephenson had a strong sense of duty and was active in college and professional matters. She was intolerant of pretentiousness, professional or personal. Although she worked in the field for nearly 30 years, she did not publish extensively. Her papers, however, provided major contributions to the subject and gave insight into the way she worked. She refused to publish until the work had been thoroughly established and had reached a certain level of completeness. Her name never appeared on a paper unless she had been responsible for a full share of the research. For this reason, it is difficult to use her papers alone in assessing her influence on the development of the subject. Much of the work published independently by her students was suggested by her, and its successful completion was made possible through her help.

WORK

Marjory Stephenson worked in the field of chemical microbiology from 1920 to 1948. When she joined Hopkins's laboratory, she first worked on fat-soluble vitamins. But a desire to know more about fat metabolism led her to her first work with bacteria. Her first papers on the metabolism of bacteria dealt with fat metabolism of the Timothy grass bacillus (*Mycobacterium phlei*). She demonstrated through "balance sheet" experiments that growth could go on even after glucose had been used up. She found that there was a difference in the relative amounts of nitrogen and lipoid material depending on whether the carbonaceous food in a synthetic inorganic medium was supplied as lactic acid or as glucose. The nitrogen/lipoid ratio was higher in the glucose medium. The lipoid content was used and burned while the protein content remained stable (Robertson 1948).

Her work in this area focused her attention on the anaerobic way of life. This interest is reflected not only in her own work but also in the work of other members of her group. She joined forces with Juda Quastel, a young organic chemist whose research had led him to similar considerations. Quastel and Stephenson published a significant paper in 1926 that dealt with the relationship of *B. sporogenes* to oxygen. They demonstrated that the organism could tolerate a range of exposures to oxygen and to hydrogen peroxide. Subsequent growth of the organism was delayed, which was indicated to be dependent on the time required for the nonproliferating bacteria to produce various reducing bodies from the medium. This work was of considerable importance and blazed a new trail in the study of anaerobes. Other scientists built on this knowledge to develop the conception of the reduction of oxygen potential in studies of tetanus.

Stephenson made an important contribution to microbiology through her development of the washed suspension technique. The technique was first used by A. Harden and S. S. Zilva in 1915 and was useful in the study of bacterial reactions isolated from the complexities of growth. This technique permitted the use of whole cells as sources of enzyme systems for the detailed and quantitative study of cell reactions, especially of a catabolic type. It was important to the progress of bacterial biochemistry to establish whether the nature and activity of bacterial enzymes were similar to those of eucaryotes. Stephenson focused on the separation of enzymes in a cell-free state. She accomplished this separation for lactic dehydrogenase of *E. coli*, which became the first bacterial enzyme to be obtained in a cell-free state. This was the first of many studies by Stephenson concerning hydrogen transfer.

In 1930 Stephenson's book, *Bacterial Metabolism*, was published. This work helped establish the study of bacterial biochemistry. It provided a valuable body of new knowledge and was a useful guide to laboratory methods for further research. It has gone into three editions and is still considered a standard. Research and revision of this book developed her own knowledge, leading her to new avenues of thought.

The period from 1930 to 1937 was a very productive time for Stephenson's research career. At this time she was established as a world authority in her field. Her ingenuity allowed her to get past doors that might otherwise have been closed to her.

In 1930 Stephenson and L. H. Stickland began a series of investigations on the transfer of hydrogen. The two found an enzyme that activates molecular hydrogen in a mixed bacterial culture obtained from the river Ouse in the fenland. They called it hydrogenase. Their research showed that it had a wide distribution among coliform and other organisms. In the presence of this enzyme system, formate is reduced by some organisms to methane, and sulfate is reduced to sulfide. Stephenson and Stickland showed that three enzymes are involved in hydrogen transfer in relation to formate: formic dehydrogenase, formic hydrogenlyase, and hydrogenase. These studies led to investigation of the growth conditions that gave rise to the formation of formic hydrogenlyase within the cell. It was discovered that the enzyme was formed only when growth occurred on a medium containing the substrate or substances from which the substrate was probably formed, as an intermediate during degradation. Enzyme formation could also occur without significant cell division. During this research Stephenson became increasingly interested in the factors that control the formation of enzymes in bacteria (Robertson 1948).

In 1936 Stephenson and John Yudkin, her student, published a paper on galactozymase. They showed that galactozymase is an adaptive enzyme and that adaptation can occur as a direct response to the environment. Although this investigation was not confirmed for almost ten years, eventually it was accepted. In 1945 Sol Spiegelmann reported that the diploid cultures, but not the haploid cultures of the organism, adapt under the conditions described by Stephenson and Yudkin (Robertson 1948). The investigations with adaptive enzymes made a deep impression on Stephenson and changed her general technique of investigation. In all her work from that point on, a section was included on the effect of the age of the culture used.

From 1936 to 1939 Stephenson worked with E. F. Gale on the factors that control the formation of enzymes involved in the deamination of amino acids. Published just before the outbreak of World War II, their work discussed the malic dehydrogenase and codehydrogenase of *B. coli*. In 1938 Stephenson also worked with A. R. Trim on the metabolism of adenine compounds in *E. coli*. This began her interest in nucleic acids and their derivatives in cell metabolism, but this area of study was delayed by the war, and Stephenson did not return to it until the last years of her life.

The war took Stephenson on an interesting side trip from the main line of her research. She and R. Davies studied acetone-butyl alcohol fermentation (1941). The fermentation of *Cl. acetobutylicum* had never been studied before, other than in fermenting liquors. Davies and Stephenson worked out a method of preparing active cell suspensions. Important contributions to the intermediate chemistry of fermentation resulted, which led to further research.

During the war years Stephenson organized an informal conference of researchers experienced with pathogenic spore-bearing anaerobes. At this meeting B.C.J.G. Knight and M. Macfarlane announced their work on the alpha toxin of *Cl. welchii* as a lethicinase. This marked an advance in the understanding of the toxic action of these anaerobic products. As a result many investigators again turned to the study of these pathogenic anaerobes (Robertson 1948).

Following the war Stephenson became interested in reports of acetylcholine in sauerkraut and thought that the enzymes involved in the production of this substance were worth investigating. From her own brews of sauerkraut she isolated strains of organisms responsible for this substance. She found the presence of pantothenic acid in the medium was necessary for the formation of the choline acetylase in the organism (Stephenson and Rowatt 1947).

During the last years of her life Stephenson returned to the studies of nucleic acids of bacteria and of their breakdown by enzymes within cells. Unfortunately, her illness prevented her from progressing past the preparation of these enzymes in a cell-free condition and the preliminary investigation of their kinetics.

Stephenson was largely responsible for establishing bacterial chemistry as a separate discipline within the field of biochemistry. Her work on adaptive enzymes became one of the growth points for molecular genetics. Among biochemists, she is best remembered for her discovery of a new kind of enzyme. It is, however, the biological side of bacterial chemistry where her contributions are most distinctive. While most enzymologists viewed cells as convenient bags of enzymes and preferred to work in cell-free systems, Stephenson always returned to the cell to do her work. With Margaret Whetham, she perfected the "resting cell" technique for studying the chemical activities of heavy suspensions of nongrowing cells in minimal media (Stephenson, Quastel, and Whetham 1925).

A striking feature of Stephenson's research was the fact that she used a wide variety of experimental methods, both biochemical and microbiological. She was always willing to try new methods, however difficult they were, even if the purpose was to establish some relatively minor point. She took great pains to learn the techniques of microbiologists in isolating pure cultures and counting viable cells. She made it a standard part of every experiment to determine the proportions of viable and dead cells. This was mainly to quell critics in the field of biology who argued that her resting cells were not resting but dead and that their chemical activities were not normal physiology but a chemical artifact.

Marjory Stephenson was president of the Society for General Microbiology at the time of her death. She was a vigorous and active member of the Medical Research Council's Research Unit in Chemical Microbiology at Cambridge. She organized a summer school in 1946 to meet the growing demand for researchers in bacterial biochemistry.

Stephenson began her career at a time when there were few opportunities for women in science. She was excluded from university classes in chemistry and zoology in her youth. She never became cynical or resentful. She took each opportunity that came her way and made the most of it. Stephenson believed in the value of hard work and expected the same effort from those with whom she worked. She faced life with a lively sense of humor, and the expression of her feelings about people and situations was always positive. Stephenson was an independent thinker who matured slowly in her professional life. She left her mark on the scientific world and laid the foundation for generations of researchers to build upon.

BIBLIOGRAPHY

Works by Marjory Stephenson

Scientific Works

Space does not permit the listing of the complete works of Marjory Stephenson. Listed here are all works by Stephenson except those cited in Robertson (1948). Included here are all references cited in the text.

(with M. D. Whetham) "Studies in the fat metabolism of the Timothy Grass *Bacillus.*" *Proceedings of the Royal Society* 93 (1922): 262–280.

(———) "Studies in the fat metabolism of the Timothy Grass *Bacillus.* II. The carbon balance sheet and respiratory quotient." *Proceedings of the Royal Society B* 95 (1923): 200–206.

(with J. H. Quastel and M. D. Whetham) "Some reactions of resting bacteria in relation to anaerobic growth." *Biochemical Journal* 19 (1925): 304–317.

(with J. H. Quastel) "Experiments on 'strict' anaerobes. I. The relationship of *B. sporogenes* to oxygen." *Biochemical Journal* 20 (1926): 1125–1137.

Bacterial Metabolism. Monographs on Biochemistry. London: Longmans, Green, 1930; 2d ed., 1939; 3d ed., 1949.

(with L. H. Stickland) "Hydrogenase: A bacterial enzyme activating molecular hydrogen. I. The properties of the enzyme." *Biochemical Journal* 25 (1931): 205–214.

(————) "Hydrogenase. II. The reduction of sulphate to sulphide by molecular hydrogen." *Biochemical Journal* 25 (1931): 215–220.

(————) "Hydrogenlyases. Bacterial enzymes liberating molecular hydrogen." *Biochemical Journal* 26 (1932): 712–724.

(————) "Hydrogenase. III. The bacteria formation of methane by the reduction of one-carbon compounds by molecular hydrogen." *Biochemical Journal* 27 (1933): 1517–1527.

(————) "Hydrogenlyases. III. Further experiments on the formation of formic hydrogenlyase by *B. coli.*" *Biochemical Journal* 27 (1933): 1528–1532.

(with J. Yudkin) "Galactozymase considered as an adaptive enzyme." *Biochemical Journal* 30 (1936): 506–514.

(with E. F. Gale) "The adaptability of glucozymase and galactozymase in *B. coli.*" *Biochemical Journal* 31 (1937): 1311–1315.

(————) "Factors influencing bacterial deamination. I. The deamination of glycine, *d l*-alanine and *l*-glutamic acid by *B. coli.*" *Biochemical Journal* 31 (1937): 1316–1322.

(————) "Factors influencing bacterial deamination. II. Factors influencing the activity of *d l*-serine deaminase in *B. coli.*" *Biochemical Journal* 32 (1938): 392–404.

(with A. R. Trim) "The metabolism of adenine compounds by *B. coli.*" *Biochemical Journal* 32 (1938): 1740–1751.

(with E. F. Gale) "*l*-Malic dehydrogenase and codehydrogenase of *B. coli.*" *Biochemical Journal* 33 (1939): 1245–1256.

(with R. Davies) "Studies on the acetone-butyl alcohol fermentation. I. Nutritional and other factors involved in the preparation of active suspension of *Cl. acetobutylicum* (Weizmann)." *Biochemical Journal* 35 (1941): 1320–1331.

(with E. Rowatt) "The bacterial production of acetylcholine." *Journal of General Microbiology* 1 (1947): 279–298.

(with J. Moyle) "Nucleic acid metabolism in *Escherichia coli.*" Biochemical Journal 45 (1949): iv.

Other Works

"Obituary notice of Dorothy Jordan Lloyd." *Newnham Letter* (1947).

"Obituary notice of Frederick Gowland Hopkins, 1891–1947." *Biochemical Journal* 42 (1948): 161–169.

Works about Marjory Stephenson

Needham, Dorothy. "Women in Cambridge biochemistry." In *Women Scientists: The Road to Liberation*, edited by Derek Richter, 161–163. London: Macmillan Press, 1982.

Robertson, Muriel. "Marjory Stephenson, 1885–1948." *Obituary Notices of the Fellows of the Royal Society* 6 (1948): 563–577.

Woods, D. D. "Obituary notice of Marjory Stephenson, 1885–1948." *Biochemical Journal* 46 (1950): 377–383.

Other References

Harden, A., and S. S. Zilva. "The reducing enzyme of *Bacillus coli communis*." *Biochemical Journal* 9 (1915): 379–384.

Hopkins, F. G. "Feeding experiments illustrating the importance of accessory factors in normal dietaries." *Journal of Physiology* 44 (1912): 425–460.

NETTIE MARIA STEVENS (1861–1912)

Marilyn Bailey Ogilvie

BIOGRAPHY

Nettie Maria Stevens was born on July 7, 1861, in Cavendish, Vermont. Her father, Ephraim Stevens, was of an old New England family. In 1635 his ancestor, Henry Stevens, came to Boston from Chelmsford, England. Her mother, Julia (Adams) Stevens, also had a New England background. Nettie Stevens was the third of their four children. Her two brothers, Charles Merrill and William Porter, died when they were very young, leaving only one sister, Emma Julia, born on January 14, 1863. The mother died about six months after Emma was born, and Stevens's father remarried on February 3, 1865. Her stepmother was Ellen C. (Thompson) Stevens of West Haven, Vermont. After the second marriage, the family moved to Westford, Massachusetts, where the father worked as a carpenter, joiner, and general handyman. By 1875 he had accumulated a comfortable amount of property, at least enough to guarantee his family freedom from poverty.

Nettie Stevens received her early education in the public schools of Westford, where she displayed considerable ability at an early age. She then attended Westford Academy, from which she graduated in 1880, one of 11 college preparatory students to graduate from the academy between 1872 and 1883. Of these students, three were women, including her sister, Emma. After graduation she spent three terms as a Latin, English, mathematics, physiology, and zoology teacher at the high school in Lebanon, New Hampshire. She then entered the Westfield (Massachusetts) Normal School, where she concentrated on the sciences. She studied physics, astronomy, chemistry, physiology, botany, zoology, mineralogy, geology, and geography, graduating in 1883 with the highest scores in her class of 30. She demonstrated exceptional ability in geometry, chemistry, and algebra.

During the years 1883–96 Stevens earned her living as a schoolteacher and librarian in the towns of Westford, Chelmsford, and Billerica, Massachusetts. By 1897 she had saved enough money to resign from her teaching job at Billerica and to attend Stanford University in California. President David Starr Jordan had attracted gifted professors to Stanford, including Dr. Oliver Peebles Jenkins, Stevens's first major professor, and Frank Mace MacFarland, under whom she later worked. Stevens matriculated in September 1896 as a special student. She was awarded regular freshman standing in January 1897 and three months later was admitted to advanced standing. Her father and sister followed her to California in 1899. Although she first intended to major in physiology under Jenkins, she later became more interested in histology and became MacFarland's student. While she was at Stanford, Stevens spent four summer vacations at the Hopkins Seaside Laboratory at Pacific Grove, California, pursuing research in histology and cytology. During the summers of 1898 and 1899, MacFarland was Instructor, while Jacques Loeb, a physician and Associate Professor of Physiology at the University of Chicago, held the Investigator's Chair in 1898. She received the bachelor's degree from Stanford in 1899 and the master's degree in 1900. Her M.A. thesis, "Studies on Ciliate Infusoria," was published in 1901.

Stevens returned to the East in 1900 to pursue a Ph.D. degree at Bryn Mawr College. This college was a fine choice for a potential cytologist or histologist because two well-known biologists, Edmund Beecher Wilson and Thomas Hunt Morgan (Nobel laureate, 1933), were on the faculty. Wilson left before Stevens arrived, but he still was influential at the college and continued to correspond with Morgan, who became one of Stevens's teachers. During her first six months at Bryn Mawr, Stevens's superior abilities earned her a fellowship to study abroad. This European interlude (1901–2) greatly expanded her research experience, for she had the opportunity to study at the Naples Zoological Station in Naples, Italy, and the Zoological Institute of the University of Würzburg, Germany, under Theodor Boveri. Stevens was awarded the Ph.D. degree by Bryn Mawr in 1903. Her dissertation, "Further Studies on the Ciliate Infusoria, *Licnophora* and *Boveria*," was published the same year.

Stevens remained affiliated with Bryn Mawr for the rest of her life. During the years 1903–5 her research was funded by a grant from the Carnegie Institution of Washington. In 1905 she was awarded the Ellen Richards Prize by the association that maintained the American Women's Table at the Naples Zoological Station, for her paper, "A study of the germ cells of *Aphis rosae* and *Aphis oenotherae*" (1905). The highest academic rank she attained was that of associate in experimental morphology (1905–12). In 1908–9 she again studied at Würzburg with Boveri. The trustees of Bryn Mawr finally created a research professorship for Stevens, but it was too late for her to benefit from it. She died of breast cancer at the Johns Hopkins Hospital in Baltimore on May 4, 1912.

From her letters, Stevens appears to be a shy, diffident scholar, with little time for extracurricular pursuits. Since she had to support herself without help from her family, she was constantly concerned with finances. She took time out from school to teach and work as a librarian so that she could pursue further education. A lively correspondence between Stevens and various officials of the Carnegie Institution of Washington, regarding her application for support, also indicates her financial worries. Stevens never married. All of her considerable energy was funneled into her research work. Unhappy with conditions at Bryn Mawr, she considered leaving in 1910 for a job at Cold Spring Harbor. However, once Professor Charles Benedict Davenport's actual job offer arrived, she decided that the situation at Bryn Mawr had improved and declined the offer.

In a eulogy to Stevens, a student characterized her as a shy and unassuming person. The student indicated that because of this diffidence, few people (including her students) were aware of her reputation as a biologist. Stevens worked quietly and unostentatiously and published in journals with which the majority of the undergraduates were unfamiliar. However, the students who worked with her praised both her excellence as a biologist and the personal interest that she took in them. She was never too busy to answer their questions.

WORK

Stevens published approximately 38 papers, most of which were in cytology, although she was also concerned with experimental physiology. Initially, she dealt with the morphology and taxonomy of ciliate protozoa. An early work begun at Stanford was based on research conducted at the station at Pacific Grove, California, where she studied one-celled parasitic ciliate infusoria. After she described two new species of ciliates of the genera *Licnophora* and *Boveria*, she began to work on regeneration in these forms. She expanded her interest to include other species and varieties belonging to these genera. For her dissertation she isolated processes that occurred during ciliate conjugation and did some experimental work to determine patterns of ciliate regeneration. This dissertation demonstrated that Stevens was a solid, thoughtful scholar. Her creativity, thus far, lay in her ability to choose and carefully execute appropriate experiments, implying general theoretical conclusions rather than stating them. Stevens augmented these early ciliate studies by working on regeneration in other forms, such as hydroids and planarians.

Yet it is not for these carefully conceived and superbly executed studies that Nettie Stevens is remembered. The importance of her greatest contribution to science, the demonstration that sex is determined by a particular chromosome, must be understood within the general context of the history of genetics. During the period of Stevens's research, investigators were exploring the relationship between the chromosomes and heredity. Although the behavior of the chromosomes had been described and explained, speculations about their relation to Mendelian heredity had not been experimentally confirmed. No trait had been traced from the chromosomes of the parent to those of the offspring, nor had a specific chromosome been linked with a specific characteristic. Hints that the inheritance of sex might be related to a morphologically distinct chromosome suggested the possibility of connecting a particular trait to a specific chromosome. If sex were shown to be inherited in a Mendelian fashion, then a chromosomal basis for heredity would be demonstrated. This would represent an elegant linking of two heretofore parallel strands of information, one from breeding data and the other from cytological observations.

Stevens's interest in the problem of sex determination and its relationship to the chromosomes was definitely in her mind by 1903. When she applied for the Carnegie Institution of Washington grant, she indicated an interest in histological problems of heredity and their relationship to Mendelian genetics. The grant that she received was for investigating problems relating to sex determination. At approximately the same time, Wilson was doing research on the same problem (1905). Wilson is often credited with this discovery, although the question of priority has been recently raised (Brush 1978; Ogilvie and Choquette 1981). It is apparent, however, from studying the dates of the work, that the two arrived at their conclusions quite independently.

In 1905 Stevens published *Studies in Spermatogenesis with Especial Reference to the "Accessory Chromosome."* This study resulted from her research on the common mealworm, *Tenebrio molitor.* She observed that in this species the egg pronuclei always contained 10 large chromosomes but that there were two possibilities for the nuclei of the spermatocytes: they could have either 10 large ones or 9 large ones and 1 small one. The unreduced somatic cells of the female contained 20 large chromosomes, while the unreduced somatic cells of the male contained 19 large ones and 1 small one. Stevens concluded that the spermatozoa containing the small chromosome would determine maleness, while those with the 10 large chromosomes of equal size would determine femaleness. She cautiously concluded that sex may, in some cases, be determined by a difference in the amount and quality of the chromatin.

Since the chromosome number was so variable in different species of insects, Stevens was hesitant about proclaiming that she had found the answer to sex determination. She continued to examine different organisms, hoping to find confirmation. In 1908 she published a paper on the germ cells of Diptera, which laid the foundation for future *Drosophila* cytogenetics. Her work on this group still did not answer the basic question that was causing her uncertainty: whether the dimorphic chromosomes themselves determined sex, or whether sex was a character borne by the heterochromosomes and segregated in the maturation of the germ cells of each sex. She concluded that the test of the Mendelian nature of sex determination would be found in breeding experiments with forms that had been shown to be favorable by cytological studies.

Stevens continued to add new information to her theory throughout her later work. In 1909 she described two new coleopterans with heterochromosomes, and in 1910 she studied the mosquito, *Culex,* but was disappointed by not finding the expected heterochromosomes in this group. In 1910 she found that the earwig, *Forficula,* had an unequal pair of heterochromosomes.

Morgan praised Stevens's work for its precision and empirical productivity. However, he indicated that she was extremely careful, to an extent that hindered her creativity. Morgan failed to note that, amid her observations and experimental results, was a very important theory. Her theory of chromosomal sex determination did not automatically emerge from the data collected on *Tenebrio*. It was the product of a fertile mind that was capable of processing the facts to generalize a theory. Her later work was an impressive effort to test her conclusions. Insofar as theories of sex determination are concerned, Stevens was certainly the theoretical equal of Wilson, the master cytologist of his time, and she was far ahead of the critical and skeptical Morgan. Stevens certainly made a substantial contribution to theoretical genetics. She also contributed a large amount of factual information to the body of scientific knowledge, information that has been used to confirm or deny other theoretical proposals. The facts she presented and the experiments she chose to perform were carefully selected for their relevance to impending theoretical questions.

BIBLIOGRAPHY

Works by Nettie Maria Stevens

Scientific Works

"Studies on Ciliate Infusoria." Master's thesis, Stanford University, 1900.
"On the force of contraction of the frog's gastrocnemius in rigor, and on the influence of 'chloretone' on that process." *American Journal of Physiology* 5 (1901): 374–386.
"Notes on regeneration in *Planaria lugubris*." *Archiv für Entwicklungsmechanik der Organismen* 13 (1901): 396–409.
"Regeneration in *Tubularia mesembryantheum*." *Archiv für Entwicklungsmechanik der Organismen* 13 (1901): 410–415.
"Studies on ciliate infusoria." *Proceedings of the California Academy of Sciences.* 3d s. *Zoology* 3 (1901): 1–42.
    Based on master's thesis.
"Experimental studies on eggs of *Echinus microtuberculatus*." *Archiv für Entwicklungsmechanik der Organismen* 15 (1902): 421–428.
"Regeneration in *Antennularis ramosa*." *Archiv für Entwicklungsmechanik der Organismen* 15 (1902): 429–447.
"Regeneration in *Tubularia mesembryanthemum*." *Archiv für Entwicklungsmechanik der Organismen* 15 (1902): 317–326.
"Further Studies on the Ciliate Infusoria, *Licnophora*, and *Boveria*." Ph.D. diss., Bryn Mawr College, 1903.
"Notes on regeneration in *Stentor coeruleus*." *Archiv für Entwicklungsmechanik der Organismen* 16 (1903): 461–471.
"On the ovogenesis and spermatogenesis of *Sagitta bipunctata*." *Zoologische Jahrbücher. Abteilung für Anatomie und Ontogenie der Tiere* 18 (1903): 227–240.
"Further studies on the ciliate infusoria, *Licnophora* and *Boveria*." *Archiv für Protistenkunde* 3 (1904): 1–43.
    Based on Ph.D. dissertation.
"On the germ cells and the embryology of *Planaria simplissima*." *Proceedings of the Academy of Natural Sciences of Philadelphia* 56 (1904): 208–220.

(with Theodor Boveri) "Über die Entwicklung dispermer *Ascaris*-Eier." *Zoologischer Anzeiger* 2 (1904): 406–417.

(with T. H. Morgan) "Experiments on polarity in *Tubularia*." *Journal of Experimental Zoology* 1 (1904): 559–585.

"Further studies on the ovogenesis of *Sagitta*." *Zoologische Jahrbücher. Abteilung für Anatomie und Ontogenie der Tiere* 21 (1905): 243–252.

Studies in Spermatogenesis with Especial Reference to the "Accessory Chromosome." Publication no. 36, pt. 1. Washington, DC: Carnegie Institution, 1905.

"A study of the germ cells of *Aphis rosae* and *Aphis oenotherae*." *Journal of Experimental Zoology* 2 (1905): 313–333.

Studies in Spermatogenesis. A Comparative Study of the Heterochromosomes in Certain Species of Coleoptera, Hemiptera and Lepidoptera, with Especial Reference to Sex Determination. Publication no. 36, pt. 2. Washington, DC: Carnegie Institution, 1906.

Studies on the Germ Cells of Aphids. Publication no. 51. Washington, DC: Carnegie Institution, 1906.

(with A. M. Boring) "*Planaria morgani* n. sp." *Proceedings of the Academy of Natural Sciences of Philadelphia* 58 (1906): 7–9.

(———) "Regeneration in *Polychoerus caudatus*." Part I. "Observations on living material," by N. M. Stevens. Part II. "Histology," by A. M. Boring. *Journal of Experimental Zoology* 2 (1906): 335–346.

"Color inheritance and sex inheritance in certain aphids." *Science* n.s. 26 (1907): 216–218.

"A histological study of regeneration in *Planaria simplicissima, Planaria maculata* and *Planaria morgani*." *Archiv für Entwicklungsmechanik der Organismen* 24 (1907): 350–373.

"The chromosomes in *Diabrotica soror* and *Diabrotica 12-punctata*. A contribution to the literature on heterochromosomes and sex determination." *Journal of Experimental Zoology* 5 (1908): 453–470.

"A study of the germ cells of certain Diptera, with reference to the heterochromosomes and the phenomena of synapsis." *Journal of Experimental Zoology* 5 (1908): 359–374.

"The effect of ultraviolet light upon the developing eggs of *Ascaris megalocephala*." *Archiv für Entwicklungsmechanik der Organismen* 27 (1909): 622–639.

"Further studies on the chromosomes of the Coleoptera." *Journal of Experimental Zoology* 6 (1909): 101–113.

"Notes on regeneration in *Planaria simplicissima* and *Planaria morgani*." *Archiv für Entwicklungsmechanik der Organismen* 27 (1909): 610–621.

"An unpaired heterochromosome in the aphids." *Journal of Experimental Zoology* 6 (1909): 115–123.

"The chromosomes and conjugation in *Boveria subcylindrica*, var. *concharum*." *Archiv für Protistenkunde* 20 (1910): 126–131.

"Chromosomes in *Drosophila ampelophila*." *Proceedings of the Seventh Zoological Congress* (1910).

"The chromosomes in the germ cells of *Culex*." *Journal of Experimental Zoology* 8 (1910): 207–225.

"Further studies on reproduction in *Sagitta*." *Journal of Morphology* 21 (1910): 179–319.

"A note on reduction in the maturation of male eggs in *Aphis*." *Biological Bulletin of the Marine Biological Laboratory, (Woods Hole, Mass.)* 18 (1910): 72–75.

"An unequal pair of heterochromosomes in *Forficula*." *Journal of Experimental Zoology* 8 (1910): 227–241.

"Various types of heterochromosomes in the Coleoptera." *Proceedings of the Seventh International Zoological Congress* (1910).

"Further studies on heterochromosomes in mosquitoes." *Biological Bulletin of the Marine Biological Laboratory, (Woods Hole, Mass.)* 20 (1911): 109–120.

"Heterochromosomes in the guinea-pig." *Biological Bulletin of the Marine Biological Laboratory, (Woods Hole, Mass.)* 21 (1911): 155–167.

"Preliminary note on heterochromosomes in the guinea pig." *Biological Bulletin of the Marine Biological Laboratory, (Woods Hole, Mass.)* 20 (1911): 121–122.

"Further observations on the supernumerary chromosomes and sex ratios in *Diabrotica soror*." *Biological Bulletin of the Marine Biological Laboratory, (Woods Hole, Mass.)* 22 (1912): 231–238.

"Supernumerary chromosomes and synapsis in *Ceuthophilus*." *Biological Bulletin of the Marine Biological Laboratory, (Woods Hole, Mass.)* 22 (1912): 219–230.

Works about Nettie Maria Stevens

Brush, Stephen G. "Nettie M. Stevens and the discovery of sex determination by chromosomes." *Isis* 69 (June 1978): 163–172.

Morgan, Thomas Hunt. "The scientific work of Miss N. M. Stevens." *Science* n.s. 36 (928) (1912): 468–470.

Ogilvie, Marilyn Bailey, and Clifford J. Choquette. "Nettie Maria Stevens (1861–1912): Her life and contributions to cytogenetics." *Proceedings of the American Philosophical Society* 125 (Aug. 21, 1981): 292–311.

Other References

Wilson, E. B. "The chromosomes in relation to the determination of sex in insects." *Science* 22 (1905): 500–502.

HELEN BROOKE TAUSSIG (1898–1986)

Laura Gray Malloy

BIOGRAPHY

With dogged determination and the foundations provided by a supportive and privileged academic family, Helen Brooke Taussig overcame learning disabilities, deafness, and a medical culture that was not ready for women in order to devote her life to the health of other people's children. She was the first pediatric cardiologist and the designer of the first surgical treatment for "blue babies." The path that led her to a distinguished career began with an appreciation for service and intellectual activity that she learned as a child.

Helen Taussig was born in Cambridge, Massachusetts, on May 24, 1898, the youngest child of Edith (Guild) and Frank William Taussig. She had two sisters, Catherine and Mary, and a brother, William. Her paternal grandfather, William Taussig, was a medical specialist who treated children with impaired eyesight. He emigrated to the United States from Germany and settled in Missouri, where he was highly respected for his work and helped establish the William Taussig School for Handicapped Children in St. Louis.

Helen Taussig's mother was one of the earliest students enrolled in Radcliffe College, before it became part of Harvard University. Her primary academic interests were in education and the natural sciences, particularly botany and zoology. She encouraged her children's interests in these areas as well. Edith Taussig died in 1909 after a two-year illness that was first diagnosed as Hodgkin's disease and later recognized as tuberculosis. While prolonged illness and early death limited their interactions, her mother left Helen Taussig with clear and happy memories. Taussig retained a sense of her mother's confidence in her future. She had learned from her mother an appreciation for the natural world, which contributed to her career and her lifelong fondness for gardening and the outdoors.

Helen Taussig's father was the Henry Lee Professor of Economics at Harvard University and the cofounder of the Harvard School of Business Administration. Highly respected as a teacher and scholar, he wrote textbooks in economics while at his summer home on Cape Cod, pursued his scholarship at the University of California at Berkeley, and was a member of the National Tariff Commission in Washington, D.C. His dedication to academic life extended to his relationships with his children. As a child Helen Taussig benefited from hours of patient tutoring from her father. His pattern of preserving summer mornings exclusively for writing also provided her with a model for routine and discipline that she successfully imitated as an adult. When she pursued her scholarship at their summer home on Cape Cod, the older Taussig urged the family to respect her writing time without interruption, as they had respected his. He advised his daughter in her academic and career decisions and supported her choices.

Helen Taussig's early education was complicated by both illness and learning disability. She was a frail baby and contracted a mild mediastinal tuberculosis as a child. This restricted her to attending school only half-time for two or three years. She first attended a small private school in Waverly, Massachusetts, then Buckingham School, and ultimately she graduated from the Cambridge School for Girls (1917). Her efforts in school were complicated by difficulties in reading, spelling, number recognition, and languages. Though her father privately wondered if she would even finish grammar school, he tutored her with extreme patience and encouragement. As an adult, Taussig concluded that her early academic problems were the outgrowth of dyslexia, a learning disability that went undiagnosed until the 1920s. She compensated for these difficulties by taking more time to read and review her work. Her academic performance was exemplary by the time she reached college. Even as an adult, reading and number recognition did not come easily, but the methodical approaches to study that she developed as survival strategies in childhood served her well.

Taussig followed her mother's example and enrolled in Radcliffe College (1917), where she continued to improve as a student and also became a tennis champion. At the end of her first year at Radcliffe, shortly after her father moved to Washington, D.C., Taussig considered a transfer to the University of California at Berkeley. She was impressed with Berkeley, which she visited with her father; she felt a need to become more self-reliant (at Radcliffe she was known as her father's daughter); and she had a friend at Berkeley, while many of her local friends in Cambridge had moved on. At her father's recommendation she remained at Radcliffe for a second year, improved her grades, and then moved to Berkeley in 1919. She was highly successful there, won membership in Phi Beta Kappa, and graduated with a B.A. degree in 1921.

After undergraduate school, Taussig wanted to pursue a career in medicine. Her father recommended that she consider a degree in public health, since it might be a more hospitable career for a woman. Harvard University, whose medical school did not admit women, had just opened its School of Public Health. Taussig met the Dean of the new school, Dr. Milton J. Rosenau, to inquire about the program. In a 1978 interview with W. Proctor Harvey, editor of the *Medical Times*, she recalled that meeting. She was told that women were permitted to take courses in the public health program and that the first two years of the program were the study of medicine. However, women could not be admitted as degree candidates. When Taussig inquired, "Who is going to be such a fool as to spend four years studying and not get a degree?", the Dean replied, "No one, I hope." Taussig answered, "I'll not be the first to disappoint you." She dismissed the idea of the public health program and decided to study medicine however she could.

By special permission Taussig was able to enroll in bacteriology and histology at Harvard Medical School, where she was seated in an isolated corner of the lecture hall for classes and placed in a room alone to look at slides. Reflecting upon this sometime later, she conjectured that this was to prevent her from "contaminating" the male students. The histology professor, Dr. John Lewis Bremmer, was supportive, however. He met with her daily, respected her work, and finally recommended that she switch to Boston University. There she would be permitted to take anatomy and get credit for a full year's work, instead of slipping into courses here and there. She took his advice and, after traveling in Europe with her family for eight months, returned to her studies in the fall of 1922 as a research student with Dr. Alexander Begg, Dean of Boston University Medical School and Professor of Anatomy. Dr. Begg sparked her interest in cardiology when he asked her to work on the anatomy of the muscle bundles of the heart. She then became interested in physiological studies of the rhythmic contractions of heart muscle and published her first paper on this topic (1925).

After completing courses in anatomy, physiology, and pharmacology, Taussig was encouraged by Dr. Begg to apply to the medical school at Johns Hopkins University, which admitted women. Taussig obtained strong support for her application from Dr. Walter Cannon, an eminent Professor of Physiology at Harvard and a family friend. Cannon appreciated her work and wrote that if women were admitted to Harvard, he would support her admission. She was admitted to the Hopkins program, and in 1924 she began a lifetime career affiliation with this institution.

Taussig pursued research throughout her medical school training. She was disappointed that the Physiology Department was not enthusiastic about her interests but established a working relationship with Dr. Edward P. Carter at the hospital heart station, where she spent all of her elective course time. She completed her M.D. degree in 1927, having published three scholarly articles as a medical student. She was also elected to the medical honor society, Alpha Omega Alpha. However, she was denied a medical internship at Hopkins because there was only one place for a woman. The grades of the other woman candidate were two-tenths of a point higher than Taussig's.

Subsequently, Dr. Carter, whom she viewed as one of her earliest mentors, invited her to continue her research and take the Archibald Fellowship at the Hopkins Heart Station for one year. She accepted this position and followed it with a 15-month internship in pediatrics under Dr. Edwards A. Park, the new Chief of Pediatrics. Dr. Park had just established several new clinics to treat chronically ill children, including a pediatric heart clinic, where Taussig concentrated her efforts. During this time Park also became an important mentor for Taussig. Both he and Carter have been credited for encouraging her to pursue the study of cardiac birth defects, which led to her most important contributions to medicine. Certainly, their support and overlapping interests in cardiology and pediatrics paved the way for Taussig's efforts in establishing this area as a full-fledged medical specialty. After her internship in pediatrics at the Children's Heart Clinic at Harriet Lane Home of Johns Hopkins University Hospital, Taussig returned to Boston. She was offered a permanent professional position as head of the same clinic in September 1930.

Taussig remained physician-in-charge of the Harriet Lane Home Cardiac Clinic at Hopkins from 1930 until her retirement in 1963. Her hospital appointment was accompanied by academic appointments in the medical school as Instructor in Pediatrics (1930–46), Associate Professor of Pediatrics (1946–59), Professor of Pediatrics (1959–63), and Professor Emeritus (1963–86). As an instructor in pediatrics she conducted basic research on the anatomical patterns of congenital malformations of the heart and their relationships to clinical symptoms. Her methodically collected observations, functional understanding of birth defects, and insights regarding therapy laid the foundation for her life's work.

Helen Taussig was the first woman in the history of Johns Hopkins University School of Medicine to achieve the rank of full Professor. She did not choose to end her career when she retired. Instead, she continued as Professor Emeritus and Thomas M. Rivers Research Fellow. She continued her research follow-up on patients who had been treated surgically for malformations of the heart and began a new project on the evolutionary implications of cardiac malformations.

Helen Taussig was awarded 20 honorary degrees, including D.Sc. degrees from Boston University School of Medicine (1948), Columbia University (1951), the University of Athens, Greece (1956), Harvard University (1959), Göttingen University, Germany (1960), the University of Vienna, Austria (1965), and Duke University (1968). She particularly appreciated her honorary degree from Boston University, where she had begun her research career, and relished the recognition from Harvard University, where she had been turned away as a degree candidate. Between 1947 and 1987 she was honored 48 times with awards, citations, prizes, medals, fellowships, symposia, and leadership positions. She was the first woman to receive many of these acknowledgments, as well as the first woman president of the American Heart Association. She was named Chevalier Legion d'Honneur, France (1947) and was awarded the Feltrinelli Prize, Rome, Italy (1954), and the Albert Lasker Award (1954). Two cardiac clinics were named in her honor, at the University of Göttingen, Germany (1964), and at her own alma mater, Johns Hopkins University (1970). She received the Presidential Medal from the Republic of Peru, presented by President Fernando Belaunde Terry (1968), and the single highest honor awarded to a civilian in the United States, the Medal of Freedom, which was awarded by

President Lyndon B. Johnson on September 14, 1964. She was awarded the National Medal of Science in 1977.

Taussig achieved this level of professional accomplishment in the face of an even greater challenge: she began to lose her hearing at age 31. She adapted to this loss by using a hearing aid and a specially amplified stethoscope. She was not shy about asking for help when she needed it. She became expert at using her sense of touch to detect the characteristics of a heartbeat, and she encouraged her students to learn and use these skills as well. When, at age 65, she had surgery that significantly improved her hearing, Dr. Park, her mentor and life-long friend, accompanied her into the operating room to hold her hand. After the surgery, Taussig was pleased to use an unamplified stethoscope but continued to use the other diagnostic strategies that she had refined during her career.

Taussig's sense of social responsibility extended well beyond the limits of her medical specialty. Her political affiliation was with the Democratic Party, and she was an active participant in the political process. In 1967 she was a member of the U.S. deputation to the World Health Assembly in Geneva. In the same year she helped express concerns for war-injured Vietnamese children as the honorary chair of the Physician's Committee for Social Responsibility. As an outgrowth of her professional experience with birth defects and her personal concern for the welfare of patients and families, she was an outspoken proponent for a patient's right to choice regarding abortion. In her retirement years she also wrote about women in science, the history of medicine, and adapting to hearing loss.

Helen Taussig never married or had children, although she encouraged her women students to do both. She preserved close ties with her immediate family throughout her life and maintained a home on Cape Cod as both a refuge and a link with her childhood. She acknowledged the encouragement she received from her parents and academic mentors, Drs. Bremmer, Begg, Carter, and especially Park, and considered it a privilege to have been raised in an academic family. She viewed it her responsibility, having had such advantage, to make a contribution, extend the limits of knowledge, and share the knowledge with others. She often spoke of her patients and research fellows as her extended family and their children as her grandchildren. In later life Taussig mourned the losses of several of her research fellows whom she survived. On May 20, 1986, just four days before her 88th birthday, she was involved in a car accident close to her retirement home in Crosslands, Kennett Square, Pennsylvania. She apparently failed to see an oncoming car as she drove out of a parking area while driving a friend to a polling place. She died an hour later in Chester County Hospital in Pennsylvania.

The effects of her scholarship, mentoring, and approach to medicine continue to have an impact on medical practice. In May 1987 Helen Taussig was awarded, posthumously, the American Association of Physicians' Kober Award, and her final scholarly paper, entitled "Evolutionary origins of cardiac malformations," was published in 1988.

WORK

Taussig founded the medical discipline of pediatric cardiology, pioneered one of the earliest surgical procedures to treat malformation of the heart, and investigated the birth defects caused by thalidomide, thereby helping to prevent its use in the United States.

The early years at the Harriet Lane Home Cardiac Clinic at Hopkins were difficult for Taussig. In 1930 little was known about congenital heart defects, and they generated little interest from other physicians. Dr. Maude Abbott had written the first article on congenital heart disease (1915) and later published an atlas (1936) that documented 1,000 cases of congenital malformations, dividing them into functional classifications. However, it was very difficult to recognize these specific malformations in a living patient, and the prevailing viewpoint was that nothing could be done about them.

Taussig took these difficulties as a challenge: treating patients where she could and learning as much as possible about them. There are many different ways in which the heart or blood vessels around it can be misformed. Different malformations can cause the same lack of oxygen that gives a child the characteristic blue color of so-called "blue babies." By looking at the heart from several angles, Taussig began to deduce which structures were enlarged or malformed. She called these deductions the "crossword puzzles" of the cardiac clinic. She would predict the anatomical variations, track the patients through their lifetime, and if the children died, correlate her predictions with autopsy findings. With these data she was able to recognize distinct, repeating patterns among patients born with complex anatomical malformations. Motivated by an academic sense of duty, she spent ten years assembling these findings into a monograph, *Congenital Malformations of the Heart* (1947). In simple, fluid language, with comprehensive illustrations, it established the clinical criteria for diagnosis of heart malformations and became the standard reference text for pediatric cardiology.

Her most influential and celebrated contribution to pediatrics was the insight that the negative effects of complex cardiac malformations might still be treated with simple surgical steps. Though it may not be possible to rebuild the circulation exactly as in a normal child, she believed ways could be found to compensate for the physiological problems caused by the malformations. Taussig suspected that the root cause of death in most "blue babies" was not heart failure, as was currently thought, but restricted blood flow to the lungs. If differences in blood flow distribution could explain the clinical symptoms of the various "blue babies," then the possibility existed to redirect blood flow to correct the problem.

One cardiac birth defect that Taussig saw frequently was a condition known as *patent*, or open, *ductus arteriosus*. This blood vessel, the *ductus*, shunts blood around the lungs in a normal fetus but closes off after birth. If this vessel fails to close off, it prevents adequate blood flow to the lungs, making some babies look blue from lack of oxygen. In 1939 Dr. Robert Gross at Harvard University successfully developed a surgical procedure to correct this problem.

The critical inference that Taussig made about other "blue babies" was that in children with more than one defect, a *ductus arteriosus* that remains open can help the child. She realized this after observing two children who died shortly after the *ductus* closed off. She reasoned that when multiple defects prevent adequate blood flow to the lungs, blood flow through the *ductus* might reverse and direct blood toward the lungs. This would improve the oxygenation of the blood and help the children survive. Her goal then became to find some other way to redirect blood to the lungs, one that would not close off with time.

In 1939, when Dr. Gross's procedure to close the *ductus* was published, Taussig traveled to Harvard to seek his collaboration in trying to build a *ductus* for patients in whom this would be a corrective procedure. Dr. Gross was quite skeptical and, flush with recent success, rebuffed her proposal.

Alfred Blalock, a vascular surgeon, was appointed Chairman of Surgery at Hopkins two years later. He was intrigued by Taussig's ideas. He took the project into the research laboratory to work out the surgical steps in animals, and, in 1944 performed the first surgery in humans to redirect blood flow from an artery in the arm to the lungs. Blalock and Taussig reported the results of their first three operations in a landmark paper (1945). Together they assembled a clinical team of physicians, nurses, and technicians that performed hundreds of successful "blue baby" operations. Furthermore, the striking success of this treatment inspired the development of surgical techniques to treat other cardiac defects and the growth of pediatric heart clinics all over the United States.

In the 20 years following the first Blalock-Taussig surgery, Taussig trained more than 60 pediatric cardiologists. Her evaluation and reassessment of patients treated with the Blalock-Taussig surgery, which spanned more than 30 years, provided her students with exposure to an analytical approach that was a model for a systematic and meticulous scientific method. Annually, Taussig hosted a two-day, continuing education, scientific symposium to bring former fellows in contact with current students. She applied for funding from the National Institutes of Health and the Children's Bureau to start a formal academic training program in pediatric cardiology, with both clinical and research components. Subsequently, the American Academy of Pediatrics established a cardiology section, and Taussig helped define the national standards for certification in Pediatric Cardiology as a founding member of its Board of Pediatric Cardiology.

Late in 1961, when Taussig was nearing retirement age, she was visited by a former fellow from Germany, Dr. Alois Beuren, who told her about some striking birth defects he had seen: children born with missing limbs or flipperlike hands without arms. Taussig was struck by the urgency of understanding this problem. In February 1962 she went to Germany on a personal fact-finding tour. She learned that the sleeping capsule thalidomide was suspected as the cause of these impairments. Taussig saw evidence that she considered conclusive. Thalidomide had been taken by the mothers of all the children showing the birth defects, and these impairments were absent within the U.S. Army population in Germany, where thalidomide was not available. It appeared that even a small quantity of the drug taken once in early pregnancy could have devastating results.

Taussig returned to the United States with pictures and clinical statistics to report her observations throughout the medical community. Her information was used by the Food and Drug Administration to prevent the release of thalidomide in the United States. Not only did Taussig's efforts help prevent an outbreak of birth defects in the United States, but they also highlighted the need for greater control over the evaluation and marketing of new drugs. Taussig testified before a Senate committee to draft legislation to address this issue (the Kefauver committee), and she called for specific language in the law to address birth defects.

Frequently in Taussig's career, her insight into an urgent medical problem had both immediate and far-reaching effects. Although the thalidomide affair was her last widely publicized contribution to medicine, she spent the remaining 24 years of her life conducting research and working toward a better understanding of pediatric heart disease.

BIBLIOGRAPHY

Works by Helen Brooke Taussig

Scientific Works

Space does not permit the listing of the complete works of Helen Brooke Taussig. Included here are all references by Taussig cited in the text and those not cited in McNamara et al. (1987).

(with F. L. Meserve) "Rhythmic contractions in isolated strips of mammalian ventricle." *American Journal of Physiology* 72 (1925): 89–98.

"The anatomy of the heart in two cases of *situs inversus.*" *Bulletin of the Johns Hopkins Hospital* 39 (1926) : 199–202.

"Septic endocarditis in an infant seven weeks of age." *American Journal of the Diseases of Children* 48 (1926): 355–358.

(with A. Blalock) "The surgical treatment of malformations of the heart in which there is pulmonary stenosis or pulmonary atresia." *Journal of the American Medical Association* 128 (1945): 189–202.

Congenital Malformations of the Heart. New York: Commonwealth Fund, 1947. Rev. ed., 2 vols. Published for the Commonwealth Fund. Cambridge, MA: Harvard University Press, 1960.

"Congenital malformations of the heart." *Medical Times* 94 (4) (1966): 455–473.

(with C. E. Anagnostopoulos, P. G. Coleman, et al.) "Single ventricle and pulmonary stenosis. Surgical management in a patient over a period of 25 years." *American Journal of Cardiology* 32 (6) (1973): 855–859.

(with N. J. Fortuin) "Late occurrence of subacute bacterial endocarditis with closed atrial septal defect." *Annals of Internal Medicine* 78 (4) (1973): 609.

(with R. Keinonin, N. Momberger, et al.) "Long-time observations on the Blalock-

Taussig operation. VII. Transpositions of the great vessels and pulmonary stenosis." *Johns Hopkins Medical Journal* 135 (1974): 161–170.

(with A. Blalock) "Landmark article May 19, 1945: The surgical treatment of malformations of the heart in which there is pulmonary stenosis or pulmonary atresia." *Journal of the American Medical Association* 251 (16) (1984): 2123–2138.

"Evolutionary origin of cardiac malformations." *Journal of the American College of Cardiology* 12 (4) (1988) : 1079–1086.

Works about Helen Brooke Taussig

Baldwin, J. *To Heal the Heart of a Child: Helen Taussig, M.D.* New York: Walker, 1992.

————. "A troubling tribute: Jamie Wyeth's painting of Dr. Helen Taussig." *American Medical News* 32 (44) (1989): 27.

Bing, R. J. "The Johns Hopkins: The Blalock-Taussig era." *Perspectives in Biology and Medicine* 32 (1) (1988): 85–90.

Dietrich, H. J., Jr. "Helen Brooke Taussig 1898–1986." *Transactions and Studies of the College of Physicians of Philadelphia* 8 (4) (1986) : 265–271.

Engle, M. A. "Dr. Helen Brooke Taussig, living legend in cardiology." *Clinical Cardiology* 8 (6) (1985) : 372–374.

Harvey, A. M. "Helen Brooke Taussig, May 24, 1898–May 20, 1986." *1986 Yearbook of the American Philosophical Society of Philadelphia* (1987): 180–185.

Harvey, W. P. "A conversation with Helen Taussig." *Medical Times* 106 (11) (1978): 28–44.

————. "A warm thank you to Helen Taussig." *Medical Times* 106 (11) (1978): 17–18.

McNamara, D. G. "Helen B. Taussig, the original pediatric cardiologist." *Medical Times* 106 (11) (1978): 23–27.

————. "Helen Brooke Taussig: 1898–1986." *Pediatric Cardiology* 7 (1) (1986): 1–2.

————, et al. "Helen Brooke Taussig: 1898 to 1986." *Journal of the American College of Cardiology* 10 (3) (1987) : 662–671.

Neill, C. A. "Dr. Helen Brooke Taussig, May 24, 1898–May 21, 1986. International cardiologist." *International Journal of Cardiology* 14 (2) (1987): 255–261.

————. "A tribute to Helen Brooke Taussig, 1898–1986." *Circulation* 74 (1986): 1180.

Ross, R. S. "Presentation of the George M. Kober Medal (posthumously) to Helen B. Taussig." *Transactions of the Association of American Physicians* 100 (1987): cxii–cxxv.

"Taussig, Helen B(rooke)." *Current Biography* (1966): 401–403.

"Taussig, Helen B(rooke)." *New York Times* (May 22, 1986): B16.

Other References

Abbott, M. E. *Atlas of Congenital Cardiac Disease.* New York: American Heart Association, 1936.

————. "Congenital cardiac disease." In *Osler and McCrae's Modern Medicine*, 2d ed., vol. 4, edited by W. Osler and T. McCrae. Philadelphia: Lea and Febiger, 1915.

Gross, R. E., and J. P. Hubbard. "Surgical ligation of a *patent ductus arteriosus*; report of first successful case." *Journal of the American Medical Association* 112 (1939): 729–731.

BIRGIT VENNESLAND (1913–)

Paris Svoronos

BIOGRAPHY

Birgit Vennesland, the daughter of Gunnuf Olaf and Sigrid Kristine (Bandsborg) Vennesland, was born on November 17, 1913, in Kristiansand, Norway. Both her parents were Lutheran and met while attending normal school in Kristiansand. Soon after graduating from high school, her father emigrated to the United States, where he pursued a variety of activities. He returned to visit Norway several times until he finally convinced her mother, a schoolteacher, to marry him in 1913. He listed his profession on his marriage license as timber merchant. After honeymooning in Germany, the Venneslands went to Calgary, Canada. When her mother found out she was pregnant, she decided to give birth in Norway, so she sailed for Kristiansand. The father crossed the U.S. border and enrolled in dental school in Chicago. Birgit Vennesland and her only other sibling, her twin sister Kirsten, lived in Norway until May 1917, when her mother decided to emigrate. The ship went by way of Iceland to avoid the German submarines.

The U.S. entry record listed Vennesland and her sister, then four years old, as ''illiterate,'' although both could read simple Norwegian fairly well. In Chicago both girls had English-speaking playmates, and so the transition to English was swift and painless. Norwegian, however, was still the language spoken at home. Vennesland became a naturalized American citizen in 1925. Her father voted for Warren G. Harding and Herbert C. Hoover but switched parties and voted for Franklin D. Roosevelt in 1932. Her mother had always been free-spirited and considered herself a liberal. Despite her extensive teaching in Norway, she never taught after she emigrated to the United States.

Vennesland liked school and still remembers many of her primary and high school teachers with affection. Her career interest at the time was medicine. Upon graduation from Roosevelt High School, she earned a scholarship for the University of Chicago, which was awarded to her on the basis of a competitive examination in physics. She liked all science courses but chose biochemistry because, in her words, "it combined the physical and biological sciences," both of which seemed very attractive to her (Personal communication 1994).

While Vennesland was studying at the University of Chicago, her sister, Kirsten, was working toward her medical degree and contracted tuberculosis. Kirsten eventually recovered, went into chest research work, and during 1967–78 served as the tuberculosis control officer for Hawaii.

Fred Koch was the Chairman of the Department of Biochemistry at the University of Chicago when Vennesland enrolled as a graduate student after earning her B.S. degree in 1934. She decided to work with Martin Hanke as her thesis adviser and earned her Ph.D. degree in 1938. After serving as an assistant at the University of Chicago for one year, she decided to work as a postdoctoral fellow under Albert Baird Hastings in the Department of Biochemistry at the Harvard Medical School. Two years later (1941) she joined the faculty of the University of Chicago as an instructor in biochemistry. She was promoted through the ranks to full professor (1957) and remained there till 1968.

Her long research association with Otto W. Warburg (1883–1970) made her a frequent visitor at the Max Planck Institute for Cell Physiology in Berlin. As a result, she was elected a member of the Max Planck Society (1967) and served as a Director of the Planck Institute (1968–70). She eventually headed the Forschungsstelle Vennesland of the Max Planck Society (1970–81) and became an emeritus following her retirement in 1981. Vennesland moved to Hawaii in 1987 to live with her sister and served for several years as an adjunct professor of biochemistry and biophysics at the University of Hawaii.

For her great contributions to science she earned the Stephen Hales Award (presented by the American Society of Plant Physiologists) in 1950 and the Garvan Medal of the American Chemical Society in 1964. Mount Holyoke College awarded her an honorary D.Sc. degree (1960).

Vennesland has always been an avid reader and a persistent research scientist who thoroughly enjoyed teaching. At an early age she appreciated literature and poetry but gradually switched to history and biography.

WORK

Vennesland has always praised her doctoral mentor, Hanke, for his tolerance and patience. Her doctoral thesis was entitled "The Oxidation-Reduction Potential Requirements of a Non-Spore-Forming Obligate Anaerobe." In her words "there was nothing remarkable about [it]," except the fact that she selected the topic herself, without advice, since "[she] didn't know any better" (Personal communication 1994). One thing that emerged from her thesis studies, however, had an important effect on her later work as a postdoctoral fellow at Harvard. This was the totally unexpected finding that the bacteria she was working with would not grow in the absence of a trace of carbon dioxide. Several years later, in collaboration with other scientists, she showed that the light-catalyzed oxidation-reduction reactions in the green leaf require the participation of carbon dioxide.

As a postdoctoral fellow under Hastings, Vennesland became involved with the early use of radioactive carbon as the tracer in the understanding of the basic steps of metabolic pathways. Using the very unstable ^{11}C isotope (half-life 20 minutes), the Harvard group showed that the carbon of carbon dioxide was converted into the carbon of glycogen in the rat liver. At about the same time, E. A. Evans, Jr., and Louis Slotin (1941) at the University of Chicago showed that the carbon of carbon dioxide entered the Krebs cycle for the degradation of carbohydrates. Vennesland returned to Chicago as an instructor and joined Evans and Slotin in a search for the enzyme (or enzymes) responsible for the entry of carbon dioxide into metabolic pathways in animal tissues. An important point of entry was eventually identified as the Wood-Weikman reaction, a process first shown to occur in bacteria.

During World War II there was little or no opportunity to do basic research. After the war the situation improved gradually when graduate students came back to school, and many eminent scientists emigrated from Europe. Vennesland was now back in Chicago as an instructor and was finally established as a faculty member. Eric Conn was her first graduate student, who was officially working with the chairman of the department but actually worked with her. Later, Harvey Fisher asked to do a joint thesis involving both chemistry and biochemistry. Frank Westheimer, who was on the chemistry staff at that time, suggested the use of deuterium to study the action mechanism of the pyridine nucleotide dehydrogenase enzymes. Conn, who had already made an active preparation of alcohol dehydrogenases, also became involved in the problem. This was the beginning of Vennesland's work with the hydrogen transfer reactions by the pyridine nucleotide dehydrogenases. Both Conn (of the University of California at Davis) and Westheimer (of Harvard University) are now retired but keep in touch with Vennesland.

Her research career dealt primarily with intermediary metabolism, a process that involves determining the chemical reaction pathways by which all small organic molecules essential to the cell are synthesized, degraded, and interconverted. Her particular achievements dealt with the enzymology of the dark metabolism of plant tissues. It was demonstrated that plants that synthesize carbohydrates during photosynthesis using light, degrade these carbohydrates in the dark by reaction paths that are similar to those occurring in animal tissues. Such a degradation starts with hydrogen and electron transfers to a pyridine nucleotide in a process catalyzed by dehydrogenase enzymes. Vennesland's contributions established that the hydrogen transfers were direct and stereospecific. In 1950–51 she showed that the enzyme acted by holding together its substrates, the alcohol and the pyridine nucleotide, in such an orientation that a hydrogen atom could be transferred from the alcohol to the pyridine nucleotide in a specific direction. These studies not only showed conclusively how a dehydrogenase enzyme worked but also enhanced the understanding of asymmetry in organic molecules.

In 1969 she and her colleagues started studying the reduction of nitrate by chlorophyll-containing cells. This process is sensitive to hydrogen cyanide, a respiratory poison. They found that hydrogen cyanide, which is catalytically formed from the aromatic amino acids, complexes with the enzyme nitrate reductase to inactivate it and, therefore, is a normal metabolic product with a regulatory role.

Vennesland is a member of the American Society of Biological Chemists and the American Chemical Society. She is also a fellow of the American Association for the Advancement of Science, the New York Academy of Sciences, and the American Society of Plant Physiologists.

NOTE

The author would like to express his feelings of deep appreciation to Dr. B. Vennesland for her prompt, courteous, and cooperative efforts in compiling this biography.

BIBLIOGRAPHY

Works by Birgit Vennesland

Scientific Works

Space does not permit the listing of the complete works of Birgit Vennesland. Listed here are all works by Vennesland except those cited in Vennesland ("Carbohydrate . . ." 1948); Vennesland and Conn (1953); Vennesland, (1956); and Vennesland, Pistorius and Gewitz (1981). Included are her dissertation and all references cited in the text.

(with H. Holt and R. W. Keeton) "The effect of gonadectomy on body structure and body weight in albino rats." *American Journal of Physiology* 114 (1936): 515–535.

(with M. B. Blauch and F. Saunders) "Studies in proteins. V. Acrystalline globulin from the paradise nut, *Lecythis zabucayo.*" *Journal of the American Chemical Society* 59 (1937): 174.

"The Oxidation-Reduction Potential Requirements of a Non-Spore-Forming Obligate Anaerobe." Ph.D. diss., University of Chicago, 1939.

(with E. A. Evans, Jr., and J. J. Schneider) "Effect of spermine on tissue oxidations." *Proceedings of the Society for Experimental Biology and Medicine* 41 (1939): 467–470.

(with M. E. Hanke) "The oxidation-reduction potential requirements of a non-spore-forming obligate anaerobe." *Journal of Bacteriology* 39 (1940): 139–169.

(with E. G. Ball, H. F. Tucker, et al.) "The source of pancreatic juice bicarbonate." *Journal of Biological Chemistry* 140 (1941): 119–129.

(with J. B. Conant, R. D. Cramer, et al.) "Metabolism of lactic acid containing radioactive carboxyl carbon." *Journal of Biological Chemistry* 137 (1941): 557–566.

(with D. E. Green, W. W. Westerfeld, et al.) "Pyruvic and α-ketoglutaric carboxylases in animal tissues." *Journal of Biological Chemistry* 140 (1941): 683–684.

(with A. K. Solomon, F. W. Klemperer, et al.) "The participation of carbon dioxide in the carbohydrate cycle." *Journal of Biological Chemistry* 140 (1941): 171–182.

(with E. A. Evans, Jr., and L. Slotin) "Carbon dioxide assimilation in cell-free liver extracts." *Journal of Biological Chemistry* 143 (1942): 565.

(with D. E. Green, W. W. Westerfeld, et al.) "Carboxylases of animal tissues." *Journal of Biological Chemistry* 145 (1942): 69–84.

(with A. K. Solomon, J. M. Buchanan, R. D. Cramer, et al.) "Metabolism of lactic acid containing radioactive carbon in the α or β position." *Journal of Biological Chemistry* 142 (1942): 371–377.

(with A. K. Solomon, J. M. Buchanan, and A. B. Hastings) "Glycogen formation from glucose in the presence of radioactive carbon dioxide." *Journal of Biological Chemistry* 142 (1942): 379–386.

(with E. A. Evans, Jr., and L. Slotin) "The mechanism of carbon dioxide fixation in cell-free extracts of pigeon liver." *Journal of Biological Chemistry* 147 (1943): 771–784.

(with E. A. Evans, Jr.) "The formation of malonic acid from oxalacetic acid by pig heart preparations." *Journal of Biological Chemistry* 156 (1944): 783–784.

(with J. W. Moulder and E. A. Evans, Jr.) "A study of enzymic reactions catalyzed by pigeon liver extracts." *Journal of Biological Chemistry* 160 (1945): 305–325.

(with H. G. Wood and E. A. Evans, Jr.) "The mechanism of carbon dioxide fixation by cell-free extracts of pigeon-liver: Distribution of labelled carbon dioxide in the products." *Journal of Biological Chemistry* 159 (1945): 153–158.

(with E. A. Evans, Jr., and A. M. Francis) "The action of metmyoglobin, oxygen and manganese on oxalacetic acid." *Journal of Biological Chemistry* 163 (1946): 573–574.

"Carbohydrate metabolism." *Annual Review of Biochemistry* 17 (1948): 227– 252.

"Nitrogen and carbon isotopes: Their application *in vivo* to the study of the animal organism." In *Advances in Biological and Medical Physics*, vol 1. edited by J. H. Lawrence and J. G. Hamilton, 45–116. New York: Academic Press, 1948.

"The β-carboxylases of plants. II. The distribution of oxalacetic carboxylase in plant tissues." *Journal of Biological Chemistry* 178 (1949): 591–597.

(with J. J. Ceithaml) "The synthesis of tricarboxylic acids by carbon dioxide fixation in parsley root preparations." *Journal of Biological Chemistry* 178 (1949): 133–134.

(with E. E. Conn and L. M. Kraemer) "Distribution of a triphosphopyridine nucleotide-specific enzyme catalyzing the reversible oxidative decarboxylation of malic acid in higher plants." *Archives in Biochemistry* 23 (1949): 179–197.

(with M. B. Mathews) "Enzymatic oxidation of formic acid." *Journal of Biological Chemistry* 186 (1950): 667–682.

(with E. E. Conn) "The enzymatic reduction of glutathione by triphosphopyridine nucleotide." *Nature* 167 (1951): 976–977.

(with F. H. Westheimer, H. F. Fisher, et al.) "The enzymatic transfer of hydrogen from alcohol to diphosphopyridine nucleotide." *Journal of the American Chemical Society* 73 (1951): 2403.

(with D. G. Anderson, H. A. Stafford, et al.) "The distribution in higher plants of triphosphopyridine nucleotide-linked enzyme systems capable of reducing glutathione." *Plant Physiology* 27 (1952): 675–684.

(with E. E. Conn, L. M. Kraemer, et al.) "The aerobic oxidation of reduced triphosphopyridine nucleotide by a wheat germ enzyme system." *Journal of Biological Chemistry* 194 (1952): 143–151.

(with R. C. Barnett, H. A. Stafford, et al.) "Phosphogluconic dehydrogenase in higher plants." *Plant Physiology* 28 (1953): 115–122.

(with E. E. Conn) "Carboxylating enzymes in plants." *Annual Review of Plant Physiology* 4 (1953): 307–332.

(with H. A. Stafford) "Alcohol dehydrogenase of wheat germ." *Archives of Biochemistry and Biophysics* 44 (1953): 404–419.

(with D. G. Anderson) "The occurrence of di- and tri-phosphopyridine nucleotides in green leaves." *Journal of Biological Chemistry* 207 (1954): 613–620.

(with E. E. Conn) "The enzymatic oxidation and reduction of glutathione." In *Glutathione: A Symposium*, edited by S. P. Colowick et al., 105–127. New York: Academic Press, 1954.

(with H. A. Stafford and A. Magaldi) "Enzymatic oxidation of DPNH by dioxosuccinate and dihydroxyfumarate." *Science* 120 (1954): 265–266.

(———) "The enzymatic reduction of hydroxypyruvic acid to D-glyceric acid in higher plants." *Journal of Biological Chemistry* 207 (1954): 621–629.

(with T. T. Tchen and F. A. Loewus) "Mechanism of enzymatic carbon dioxide fixation into oxaloacetate." *Journal of the American Chemical Society* 76 (1954): 3358–3359.

"Glutathione reductase (plant)." In *Methods in Enzymology*, vol. 2, edited by S. P. Colowick and N. O. Kaplan, 719–722. New York: Academic Press, 1955.

"Some applications of deuterium to the study of enzyme mechanisms." *Discussions of the Faraday Society* 20 (1955): 240.

(with T. T. Tchen) "Enzymatic carbon dioxide fixation into oxaloacetate in wheat germs." *Journal of Biological Chemistry* 213 (1955): 533–546.

(——— and F. A. Loewus) "The mechanism of enzymatic carbon dioxide fixation into oxaloacetate." *Journal of Biological Chemistry* 213 (1955): 547–555.

"Steric specificity and hydrogen transfer in pyridine nucleotide dehydrogenase reactions." *Journal of Cellular and Comparative Physiology* 47 (Suppl. 1) (1956): 201–216.

(with J. L. Graves, M. F. Utter, et al.) "The mechanism of the reversible carboxylation of phosphoenolpyruvate." *Journal of Biological Chemistry* 223 (1956): 551–557.

(with H. R. Levy and F. A. Loewus) "The enzymatic transfer of hydrogen. V. The reaction catalyzed by glucose dehydrogenase." *Journal of Biological Chemistry* 222 (1956): 685–693.

(with F. A. Loewus and H. R. Levy) "The enzymatic transfer of hydrogen. VI. The reaction catalyzed by D-glyceraldehyde-3-phosphate dehydrogenase." *Journal of Biological Chemistry* 223 (1956): 589–597.

(with J. L. Graves) "The stereospecific hydrogen exchange in the dihydroorotic dehydrogenase reaction." *Journal of Biological Chemistry* 226 (1957): 307–316.

(with H. R. Levy) "The stereospecificity of enzymatic hydrogen transfer from diphosphopyridine nucleotide." *Journal of Biological Chemistry* 228 (1957): 85–96.

(——— and F. A. Loewus) "The optical rotation and configuration of a pure enantiomorph of ethanol 1–d.'' *Journal of the American Chemical Society* 79 (1957): 2949–2953.

(with M. Mazelis) ''Carbon dioxide fixation into oxaloacetate in higher plants.'' *Plant Physiology* 32 (1957): 591–600.

(with R. H. Nemon) ''Cytochrome *c* photooxidase of spinach chloroplasts.'' *Science* 125 (1957): 353–354.

''Stereospecificity of hydrogen transfer in pyridine nucleotide dehydrogenase reactions.'' *Federation Proceedings* 17 (1958): 1150–1157.

(with C. T. Chow) ''The nonenzymatic decarboxylation of diketosuccinate and oxaloglycolate (dihydroxyfumarate).'' *Journal of Biological Chemistry* 233 (1958): 997–1002.

(with H. C. Friedmann) ''Purification and properties of dihydroorotic dehydrogenase.'' *Journal of Biological Chemistry* 233 (1958): 1398–1406.

(with A. Marcus) ''Formation of keto-pyruvate in the dehydrogenation catalyzed by yeast lactic oxidase.'' *Journal of the American Chemical Society* 80 (1958): 1123–1125.

(———) ''Studies with acetyl coenzyme A and condensing enzyme in D$_2$O.'' *Journal of Biological Chemistry* 233 (1958): 727–730.

(——— and J. R. Stern) ''The enzymatic transfer of hydrogen. VII. The reaction catalyzed by β-hydroxybutyryl dehydrogenase.'' *Journal of Biological Chemistry* 233 (1958): 722–726.

(with N. I. Bishop, H. Nakamura, et al.) ''Kinetics and properties of cytochrome *c* photooxidase of spinach.'' *Plant Physiology* 34 (1959): 551–557.

(with J. Koukol and C. T. Chow) ''Photophosphorylation by digitonin-fragmented spinach chloroplasts.'' *Journal of Biological Chemistry* 234 (1959): 2196–2201.

(with D. W. Krogmann) ''Oxidative photosynthetic phosphorylation by spinach chloroplasts.'' *Journal of Biological Chemistry* 234 (1959): 2205–2210.

(with T. Nakamoto and D. W. Krogmann) ''The effect of oxygen on riboflavin phosphate dependent photosynthetic phosphorylation by spinach chloroplasts.'' *Journal of Biological Chemistry* 234 (1959): 2783–2788.

(with H. Nakamura and C. T. Chow) ''A note on the distribution of phosphorus in photophosphorylating spinach chloroplast preparations.'' *Journal of Biological Chemistry* 234 (1959): 2202–2204.

(with R. H. Nieman) ''Photoreduction and photooxidation of cytochrome *c* by spinach chloroplast preparations.'' *Plant Physiology* 34 (1959): 255–262.

(——— and H. Nakamura) ''Fractionation and purification of cytochrome *c* photooxidase of spinach.'' *Plant Physiology* 34 (1959): 262–267.

(with S. Udaka and J. Koukol) ''Lactic oxidase of *Pneumococcus*.'' *Journal of Bacteriology* 78 (1959): 714–725.

''Decarboxylases in plants.'' In *Encyclopedia of Plant Physiology*, vol. 12, pt. 1, edited by W. Ruhland, 478–525. Berlin: Springer-Verlag, 1960.

(with C. Ardao) ''Chlorophyllase activity of spinach chloroplastin.'' *Plant Physiology* 35 (1960): 368–371.

(with H. C. Friedmann) ''Crystalline dihydroorotic dehydrogenase.'' *Journal of Biological Chemistry* 235 (1960): 1526–1532.

(with H. Kondo and H. C. Friedmann) ''Flavin changes accompanying adaptation of *Zymobacterium oroticum* to orotate.'' *Journal of Biological Chemistry* 235 (1960): 1533–1535.

(with T. Nakamoto) ''The enzymatic transfer of hydrogen. VII. The reactions catalyzed by glutamic and isocitric dehydrogenases." *Journal of Biological Chemistry* 235 (1960): 202–204.

(with B. K. Stern) "The carbon dioxide requirement for the photoevolution of oxygen by chloroplast preparations." *Journal of Biological Chemistry* 235 (1960): PC51–53.

(———) "The enzymatic transfer of hydrogen. IX. The reactions catalyzed by glucose-6-phosphate dehydrogenase and 6-phosphogluconic dehydrogenase." *Journal of Biological Chemistry* 235 (1960): 205–208.

(———) "The stereospecificity of glutathione reductase for pyridine nucleotides." *Journal of Biological Chemistry* 235 (1960): 209–212.

"Conversion of P-pyruvate to oxalacetate (plant)." In *Methods in Enzymology*, vol. 5, edited by S. P. Colowick and N. O. Kaplan, 617–622. New York: Academic Press, 1961.

"The mechanism of the Hill reaction and its relationship to photophosphorylation." In *Proceedings of the First IUB/IUBS International Symposium*, vol. 2, edited by T. W. Goodwin and O. Lindberg, 411–429. New York: Academic Press, 1961.

(with G. Krakow) "The equilibrium constant of the dihydroorotic dehydrogenase reaction." *Journal of Biological Chemistry* 236 (1961): 142–144.

(with T. Nakamoto and B. Stern) "The effect of oxygen on photophosphorylation and the effect of carbonate on the Hill reaction." In *Light and Life*, edited by W. D. McElroy and B. Glass, 609–614. Baltimore: Johns Hopkins University Press, 1961.

(with M. Stiller) "Photophosphorylation and the Hill reaction." *Nature* 191 (1961): 677–678.

(with G. Krakow and S. Udaka) "The stereochemistry of the reaction catalyzed by D-glyceric-3-dehydrogenase." *Biochemistry* 1 (1962): 254–258.

(with H. R. Levy and P. Talalay) "The steric course of enzymatic reactions at meso carbon atoms: Application of hydrogen isotopes." In *Progress in Stereochemistry*, vol. 3, edited by P.B.D. de La Mare and W. Klyne, 299–349. Washington, DC: Butterworth, 1962.

(with B. K. Stern) "The effect of carbon dioxide on the Hill reaction." *Journal of Biological Chemistry* 237 (1962): 596–602.

(with M. Stiller) "Photophosphorylation accompanying the Hill reaction with ferricyanide." *Biochimica et Biophysica Acta* 60 (1962): 562–579.

(with S. Udaka) "Properties of triphosphopyridine nucleotide-linked dihydroorotic dehydrogenase." *Journal of Biological Chemistry* 237 (1962): 2018–2024.

"Some flavin interactions with grana." In *Photosynthetic Mechanisms of Green Plants*, pub. no. 1145, 421–435. Washington, DC: National Academy of Sciences, National Research Council, 1963.

(with W. W. Gattung and E. Birkicht) "Fluorometric measurement of the photoreduction of flavin by illuminated chloroplasts." *Biochimica et Biophysica Acta* 66 (1963): 285–291.

(with G. Krakow) "The stereospecificity of glyoxylate reduction in leaves." *Biochemische Zeitschrift* 338 (1963): 31.

(———, J. Ludowieg, et al.) "Some stereospecificity studies with tritiated pyridine nucleotides." *Biochemistry* 2 (1963): 1009.

"The energy conversion reactions of photosynthesis." *Record of Chemical Progress* 25 (1964): 211

(with N. K. Gupta) "Glyoxylate carboligase of *Escherichia coli*: A flavoprotein." *Journal of Biological Chemistry* 239 (1964): 3787–3789.

(with R. N. Ammeraal and G. Krakow) "The stereospecificity of the Hill reaction with diphosphopyridine nucleotide." *Journal of Biological Chemistry* 240 (1965): 1824–1828.

(with G. Krakow and R. N. Ammeraal) "The stereospecificity of the Hill reaction with triphosphopyridine nucleotide." *Journal of Biological Chemistry* 240 (1965): 1820–1823.

(with E. Olson and R. N. Ammeraal) "Role of carbon dioxide in the Hill reaction." *Federation Proceedings* 24 (1965): 873–880.

"Involvement of CO_2 in the Hill reaction." *Federation Proceedings* 25 (1966): 893–898.

(with N. K. Gupta) "Glyoxylate carboligase of *Escherichia coli*: Some properties of the enzyme." *Archives of Biochemistry and Biophysics* 113 (1966): 255–264.

(with E. O. Turkington) "The relationship of the Hill reaction to photosynthesis: Studies with fluoride-poisoned blue-green algae." *Archives of Biochemistry and Biophysics* 116 (1966): 153–161.

(with R. L. Hall and F. J. Kezdy) "Glyoxylate carboligase of *Escherichia coli*: Identification of carbon dioxide as the primary reaction product." *Journal of Biological Chemistry* 244 (1969): 3991–3998.

(with C. Jetschmann) "The nitrate dependence of the inhibition of photosynthesis by carbon monoxide in *Chlorella*." *Archives of Biochemistry and Biophysics* 144 (1971): 428–437.

(———) "The nitrate reductase of *Chlorella pyrenoidosa*." *Biochimica et Biophysica Acta* 229 (1971): 554–564.

(——— and L. P. Solomonson) "Activation of nitrate reductase by oxidation." *Biochimica et Biophysica Acta* 275 (1972): 276–278.

(with L. P. Solomonson) "Nitrate reductase and chlorate toxicity in *Chlorella vulgaris* Beijerinck." *Plant Physiology* 50 (1972): 421– 424.

(———) "The nitrate reductase of *Chlorella* (species or strain differences)." *Plant Physiology* 49 (1972): 1029–1031.

(———) "Properties of a nitrate reductase of *Chlorella*." *Biochimica et Biophysica Acta* 276 (1972): 544–557.

(with C. Jetschmann and L. P. Solomonson) "Die Regulierung der Nitratreduktaseaktivität in *Chlorella vulgaris*." *Hoppe-Seyler's Zeitschrift für Physiologische Chemie* 353 (1973): 1530–1531.

(with L. P. Solomonson and C. Jetschmann) "Reversible inactivation of nitrate reductase of *Chlorella vulgaris* Beijerinck." *Biochimica et Biophysica Acta* 309 (1973): 32–43.

"Stereospecificity in biology." *Forschungsstelle Vennesland der Max-Planck-Gesellschaft, Berlin Dahlem* 48 (1974): 39–65.

(with R. Gerster and G. H. Lorimer) "The extra oxygen evolved during nitrate utilization by *Chlorella*." *Plant Science Letters* 5 (1975): 255–260.

(with M. G. Guerrero) "Stereospecificity of hydrogen removal from pyridine nucleotide: The reaction catalyzed by nitrate reductase and by xanthine oxidase." *Federation of European Biochemical Societies (FEBS) Letters* 51 (1975): 284–286.

"Recollections of small confessions." *Annual Review of Plant Physiology* 32 (1981): 1–20.

(with E. E. Conn, C. J. Knowles, et al., eds.) *Cyanide in Biology*. London: Academic Press, 1981.

(with H.-S. Gewitz and J. Piefke) "Purification and characterization of demolybdonitrate reductase (NADH-Cytochrome *c* Oxidoreductase) of *Chlorella vulgaris.*" *Journal of Biological Chemistry* 256 (1981): 11527–11531.

(with E. K. Pistorius and H.-S. Gewitz) "Hydrogen cyanide production by microalgae." In *Cyanide in Biology*, edited by B. Vennesland, E. E. Conn, et al., 349–361. London: Academic Press, 1981.

(with C. S. Ramadoss and T.-C. Shen) "Molybdenum insertion *in vitro* in demolybdonitrate reductase of *Chlorella vulgaris.*" *Journal of Biological Chemistry* 256 (1981): 11532–11537.

(with P. A. Castric, E. E. Conn, et al.) "Cyanide metabolism." *Federation Proceedings* 41 (1982): 2639–2648.

(with T.-C. Shen and C. S. Ramadoss) "Effect of reduced pyridinenucleotide and tungstate on the *in vitro* insertion of molybdenum into demolybdonitrate reductase of *Chlorella vulgaris.*" *Biochimica et Biophysica Acta* 704 (1982): 227–234.

(with E. A. Funkhouser and H.-S. Gewitz) "Accumulation of demolybdo-nitrate reductase during induction and its separation from active nitrate reductase in *Chlorella.*" *Plant Cell Physiology* 24 (1983): 1565–1568.

"Isotope time—50 years ago." *Federation of American Societies for Experimental Biology and Medicine* 5 (1991): 2868–2869.

Other Works

Personal communication to the author, 1994.

Works about Birgit Vennesland

"Hastings, Albert Baird." In *McGraw-Hill Modern Scientists and Engineers*, vol. 2, 27–28. New York: McGraw-Hill, 1980.

"Warburg, Otto Heinrich." In *McGraw-Hill Modern Scientists and Engineers*, vol. 3, 274–275. New York: McGraw-Hill, 1980.

Other References

Evans, E. A., and L. Slotin. "Carbon dioxide utilization by pigeon liver." *Journal of Biological Chemistry* 141 (1941): 439–450.

SALOME GLUECKSOHN SCHOENHEIMER WAELSCH (1907–)

Paris Svoronos

BIOGRAPHY

Salome Gluecksohn was born on October 6, 1907, in Danzig, Germany, the daughter of Ilya and Nadia (Pomeranz) Gluecksohn. Her father passed away in the flu epidemic of 1918, and her mother lost practically all of the family savings during the inflation that followed World War I. After completing her high school education, Gluecksohn enrolled at the University of Königsberg in East Prussia and later at the University of Berlin.

While growing up in the post–World War I era, Gluecksohn was exposed to the internal struggle of a defeated country. She recalls a childhood that was marred with taunts from her peers, anti-Semitic songs and rhymes. As a teenager she became an ardent Zionist who, like many other Jews, hoped and dreamed of eventually settling in Palestine.

Initially, she expressed a great interest in teaching classical languages but changed her mind as soon as she took her first biology class upon the recommendation of a fellow student. As a graduate student at the University of Freiburg, she joined the group of Dr. Hans Spemann, the 1935 winner of the Physiology or Medicine Nobel Prize, with whom she earned her doctorate in 1932. She then moved to the University of Berlin, where she worked as a research assistant in cell biology for one year.

In 1932 she married Rudolf Schoenheimer, one of Germany's most prominent biochemists. Feeling the pressure of the upcoming Nazi takeover in Germany, Schoenheimer and her husband left for the United States in 1933. Upon their settlement in New York City, Rudolf Schoenheimer received an appointment at the College of Physicians and Surgeons at Columbia University, but she could not obtain a post for three years. In 1936 she was offered a position at Columbia University in the laboratory of Leslie C. Dunn, a renowned geneticist, where she remained as a research associate in genetics for 17 years.

In 1941 Rudolf Schoenheimer died, and on January 8, 1943, she married the Columbia University biochemist Heinrich Waelsch. For two years (1953–55) she served as a research associate in obstetrics and gynecology (genetics) at the College of Physicians and Surgeons of Columbia. Her salary at the time was only $1,500 a year, but she was happy because she was actively involved with her research. For years she was deprived of any chance for a career at Columbia. She was advised by her husband to try something new. Every time she attempted to discuss her promotion from the title of research associate, she was told that there was no chance for advancement. Finally, in 1955, she moved to the newly founded Albert Einstein College of Medicine, where she has been working ever since. In 1988 she became a distinguished Professor Emeritus in Genetics.

Waelsch's second marriage was blessed with two children, Naomi Barbara and Peter Benedict, now both dedicated teachers. Peter works at an educational institution in Cambridge, Massachusetts, teaching high school dropouts and other adults. Naomi has a teaching job in a school in Westchester, New York. Waelsch is eager to point out that her children did not adversely affect her research activities, because she maintained a very strict routine and a high degree of organization. She gives a lot of credit to her husband, who was instrumental in helping her raise their two children. She always refers to her devoted baby-sitter, Ellen, who served as the second mother at home. Once during her career, Waelsch was away from her children for an extended period of time (for about three months in Scotland), but Ellen took care of the family. Waelsch attributes her success to her happy life with her second husband. She recalls that they rarely went out together, and their lives revolved around the children and work. Heinrich Waelsch died in 1966.

Waelsch overcame obstacles to earn acknowledgment for her contributions. She ardently believes that women nowadays do not experience the same kind of restricted options that she was subjected to in her field. Waelsch feels that women were given much less challenging projects in the past. She believes that there is "a greater variability in work styles within sex than between the sexes." Nowadays, Waelsch observes that there are salary differences not necessarily between men and women but between people. She points out that in small group meetings she has never seen any discrimination against women. In her laboratory she always enjoyed working with a small number of people. She tries to arrive at work before anybody else and leaves after everybody else has gone (Zuckerman et al. 1991, 79, 89). In September 1993, she was selected as one of eight scientists to receive the National Medal of Science, a prestigious honor awarded by the President of the United States. Columbia University awarded her an honorary D.Sc. degree in 1995.

WORK

Gluecksohn came onto the scene of experimental embryology, as it was called at the time, in 1928. By then she had already completed several semesters of zoology and chemistry in Königsberg and Berlin. In an embryology course she became acquainted with Spemann's work and decided to become his graduate student. Despite a rather unsettling introductory first meeting, Spemann accepted her and assigned her to a "rather boring" descriptive study of limb development for her Ph.D. dissertation. The work was to be done with two species of salamanders having different types of limb development. She was asked to describe the development of each species, while a male fellow student was supposed to transplant tissues between them. Nevertheless, the work in Spemann's laboratory was both instructive and stimulating. She had to share the laboratory with six predoctoral and postdoctoral international students. She found that the work in the laboratory was particularly strenuous during the "season" (i.e., the period in which amphibians laid their eggs), since the students had to observe the embryonic development day and night under the constant, strict supervision of Spemann.

During that time Gluecksohn met several eminent scientists, such as Paul Weiss, Richard Goldschmidt, and Walter Vogt. She then realized the limitations in the scientific thinking of Spemann. She considers him to be a strong German nationalist, full of mistrust toward other nationalities, and feels that he was prejudiced toward women's ability. However, she still considers him one of the most influential people in twentieth-century embryology and a scientist to whom the work ethic was the basis of judging people.

While in Spemann's laboratory, Gluecksohn became acquainted with one of his assistants, Viktor Hamburger, who supervised her experimental results. Her associations with him also provided her with stimulating theoretical discussions. Hamburger escaped Germany in 1933 and eventually became one of the founders and leaders of modern neuroembryology.

Upon her arrival in the United States, Gluecksohn Schoenheimer met with Ross G. Harrison, whose dynamic approach to solving problems of development and differentiation appealed to her much more than Spemann's frame of thinking. She was particularly impressed with Harrison's early experiment in which he proved that the nerve fiber arose as an outgrowth of a single neuronal cell. His result led to the development of a method of tissue culture (Harrison 1910).

For her scientific progress Gluecksohn Schoenheimer gives great credit to her association with Dunn of Columbia University, who had earlier recognized the significance of genes in the processes of development and differentiation. She joined his progressive laboratory in 1935 and enjoyed several years of scientific success. This was the time of the depression, and she was glad to be able to work in his laboratory, even without a salary at first. At the time there were no outside agencies, such as the American Cancer Society and the National Institutes of Health, where she might obtain grants. Only in the late 1930s did the National Research Council come into existence, and it awarded Dunn money that he used to pay her (Zuckerman et al. 1991, 81–82). Gluecksohn Schoenheimer also taught a developmental genetics course, substituting for Dunn, whose interest in teaching was limited.

During the late 1930s and 1940s she became an active adherent to the idea of overlap between the fields of embryology and genetics. She joined the ranks of famous pioneers, such as E. B. Wilson, Richard Goldschmidt, Edgar Zwilling, Conrad H. Waddington, and Boris Ephrussi.

Upon moving to Albert Einstein, Gluecksohn Waelsch earned the long overdue respect and acknowledgment of her peers. She was invited to present her own research papers at prestigious conferences. She worked in New York City, where she wanted to be, and in two years she became full Professor. In 1963 she became Chair of the Department of Genetics. She credits her recognition to Ernst Scharrer.

In a speech delivered at a conference in Dubrovnik, Yugoslavia (1985), Gluecksohn Waelsch expressed her personal view of immediate and future concerns of causal developmental analysis. She pointed out that the ultimate problem remains finding out how a zygote (which contains one set of maternal and one set of paternal genes) manages to differentiate itself into a multicellular organism, with its multitude of cells derived by mitosis.

Waelsch does not travel a great deal. She performs genetic and developmental work using mice, while her collaborators do molecular and biochemical work. Waelsch obtains grants that pay both her own salary and the salaries of her coworkers. For more than 36 years she has received grants from the American Cancer Society and the National Institutes of Health.

She is a fellow of the American Association for the Advancement of Science (since 1988) and the American Academy of Sciences (elected in 1980). Waelsch is a member of the National Academy of Sciences (elected in 1979) and a member of the New York Academy of Sciences (honorary life member since 1979). She is also a member of the American Society for Zoologists, the American Association for Anatomists, the Genetics Society, the Society of Developmental Biology, the American Society of Naturalists, the American Society of Human Genetics, and Sigma Xi.

NOTE

The author is indebted to Dr. Waelsch for her valuable help in completing this biographical entry.

BIBLIOGRAPHY

Works by Salome Gluecksohn Schoenheimer Waelsch

Scientific Works

Space does not permit the listing of the complete works of Salome Gluecksohn Schoenheimer Waelsch. Listed here are all works by Waelsch except those cited in Waelsch (1951); Waelsch and Erickson (1970); Waelsch (1977); and Waelsch (1979). Included here are all references cited in the text.

"Aussere Entwicklung der Extremitäten und Stadieneinteilung der Larvenperiode von

Triton taeniatus Leyd. und *Triton cristatus* Laur." *Archiv für Entwicklungsmechanik der Organismen* 125 (1931): 341–405 (as S. Gluecksohn).

(with L. C. Dunn) "A dominant short-tail mutation in the house mouse with recessive lethal effects." *Genetics* 23 (1938): 146–147 (as S. Schoenheimer).

(———) "The inheritance of taillessness (Anury) in the house mouse. II. Taillessness is a second balanced lethal line." *Genetics* 24 (1939): 587–609.

(——— and V. Bryson) "A new mutation in the mouse affecting spinal column and urogenital system." *Journal of Heredity* 31 (1940): 343–348.

"The development of early mouse embryos in the extraembryonic coelom of the chick." *Science* 93 (1941): 502–503.

(with L. C. Dunn) "Stub, a new mutation in the mouse with marked effects on the spinal column." *Journal of Heredity* 33 (1942): 235–239.

(———) "A specific abnormality associated with a variety of genotypes." *Proceedings of the National Academy of Sciences* 30 (1944): 173–176.

(———) "Dominance modification of physiological effect of genes." *Proceedings of the National Academy of Sciences* 31 (1945): 82–84.

"Genetically determined duplications and twin formation in the house mouse." *Anatomical Record* 94 (1946): 462.

(with L. C. Dunn) "A new type of hereditary harelip in the house mouse." *Anatomical Record* 102 (1948): 279–287.

"Mutations and the causal analysis of mouse development." *Hereditas* 35 (1949): 583–584.

"Physiological genetics of the mouse." *Advances in Genetics* 4 (1951): 1–51.

(with L. C. Dunn) "On the origin and genetic behavior of a new mutation (t^3) at a mutable locus in the mouse." *Genetics* 36 (1951): 4–12.

"Lethal factors in development." *Quarterly Review of Biology* 28 (1953): 115–135.

"Genetic control of embryonic growth and differentiation." *Journal of the National Cancer Institute* 15 (1954): 629–634.

"Abnormalities of the developing nervous system in embryos from mothers immunized against adult brain tissue." *Anatomical Record* 122 (1955): 472.

"Genetic factors and the development of a nervous system." In *Biochemistry of the Developing Nervous System*, edited by H. Waelsch, 375–396. New York: Academic Press, 1955.

(with S. A. Kammell) "Physiological investigations of a mutation in mice pleiotropic effects." *Physiological Zoology* 28 (1955): 68–73.

(with Helene M. Ranney) "Filter-paper electrophoresis of mouse haemoglobin: Preliminary note." *Annals of Human Genetics* 19 (1955): 269–272.

(with D. Hagadorn and B. F. Sisken) "Genetics and morphology of a recessive mutation in the house mouse affecting head and limb skeleton." *Journal of Morphology* 99 (1956): 465–479.

"Discussion of embryonic tissue antigens and their relation to maternal-fetal immunologic interaction." In *Etiological Factors in Mental Retardation*, edited by S. J. Onesti, Jr., 75–76. Columbus, OH: Ross Laboratories, 1957.

"The effect of maternal immunization against organ tissues on embryonic differentiation in the mouse." *Journal of Embryology and Experimental Morphology* 5 (1957): 83–92.

"Trends in mammalian teratology." *Pediatrics* 19 (1957): 777–781.

(with H. M. Ranney and B. F. Sisken) "The hereditary transmission of hemoglobin differences in mice." *Journal of Clinical Investigation* 36 (1957): 753–756.

"Abnormalities of the nervous system in the mouse." In *Progress in Neurobiology*, vol. 4, *The Biology of Myelin*, edited by S. R. Korey, 108–121. New York: Harper and Brothers, 1959.

(with B. F. Sisken) "A developmental study of the mutation (Phocomelia) in the mouse." *Journal of Experimental Zoology* 142 (1959): 623–641.

"The inheritance of hemoglobin types and other biochemical traits in mammals." *Journal of Cellular and Comparative Physiology* 56 (Suppl. 1) (1960): 89–101.

(with H. M. Ranney and G. M. Smith) "Haemoglobin differences in inbred strains of mice." *Nature* 188 (1960): 212–214.

"Developmental genetics of mammals." *American Journal of Human Genetics* 13 (1961): 113–121.

"Developmental aspects of kidney form and function. Section II." In *Hereditary, Developmental and Immunological Aspects of Kidney Disease*, edited by J. Metcoff, 39–64. Evanston, IL: Northwestern University Press, 1962.

Discussion of "Organic culture analysis of inherited abnormalities." *National Cancer Institute, Monograph* 11 (1962): 244–246.

"Mammalian genetics in medicine." In *Progress in Medical Genetics*, vol. 2, edited by A. G. Steinberg and A. G. Bearn, 295–330. New York: Grune and Stratton, 1962.

(with T. R. Rota) "Development in organ tissue culture of kidney rudiments from mutant mouse embryos." *Developmental Biology* 7 (1963): 432–444.

(with N. S. Kosower) "Hemoglobin polymorphism and susceptibility to *Plasmodium* infection in mice." *Genetics* 50 (1964): 841–845.

"Genetic control in mammalian differentiation." In *Genetics Today*, vol. 2, edited by S. J. Geerts, 209–219. Oxford: Pergamon Press, 1965.

(with A. S. Pai) "Developmental genetics of a lethal mutation, muscular dysgenesis (mdg), in the mouse. I. Genetic analysis and gross morphology." *Developmental Biology* 11 (1965): 82–92.

(————) "Developmental genetics of a lethal mutation, muscular dysgenesis (mdg), in the mouse. II. Developmental analysis." *Developmental Biology* 11 (1965): 93–109.

(with P. Greengard, G. P. Aunin, et al.) "Genetic variations of an oxidase in mammals." *Journal of Biological Chemistry* 242 (1967): 1271–1273.

(with G. C. Moser) "Electrophoretic patterns of lens proteins from genetically caused cataract in the mouse." *Experimental Eye Research* 6 (1967): 297–298.

"Gene regulation in mammalian cells." *Science* 167 (1970): 1524–1526.

(with R. P. Erickson) " The T-locus on the mouse: Implications for mechanisms of development." *Current Topics in Developmental Biology* 5 (1970): 281–316.

(with A. C. Platzer) "Fine structure of mutant (muscular dysgenesis) embryonic mouse muscle." *Developmental Biology* 28 (1972): 242–252.

(with M. B. Schiffman, M. L. Santorineou, et al.) "Lipid deficiencies, leukocytosis, brittle skin—a lethal syndrome caused by a recessive mutation, edematous (oed) in the mouse." *Genetics* 81 (1975): 525–536.

(with K. C. Atwood, M. T. Yu, et al.) "Does the T-locus in the mouse include ribosomal DNA?" *Cytogenetics and Cell Genetics* 17 (1976): 9–17.

(with M. Vojtiskova, V. Viklicky, et al.) "The effects of a t-allele (t-^{AE5}) in the mouse on the lymphoid system and reproduction." *Journal of Embryology and Experimental Morphology* 36 (1976): 443–451.

"Developmental genetics." In *Handbook of Teratology*, vol. 2, edited by J. G. Wilson and F. C. Fraser, 19–40. New York: Plenum Press, 1977.

(with S. E. Lewis) "Developmental analysis of lethal effects of homozygosity for the c^{25H} deletion in the mouse." *Developmental Biology* 65 (1978): 553–557.

———, H. A. Turchin, et al.) "Fertility studies of complementing genotypes at the albino locus of the mouse." *Journal of Reproduction and Fertility* 53 (1978): 197–202.

"Genetic control of morphogenetic and biochemical differentiation: Lethal albino deletions in the mouse." *Cell* 16 (1979): 225–237.

(with M. S. Deol) "The role of inner hair cells in hearing." *Nature* 278 (1979): 250–252.

(with L. S. Teicher, L. Pick, et al.) "Genetic rescue of lethal genotypes in the mouse." *Developmental Genetics* 1 (1980): 219–228.

"Dynamic concepts of developmental genetics." In *Hamburger Festschrift*, 44–52. London: Oxford University Press, 1981.

"The genetic basis of pattern formation: Systems of genetically caused abnormal morphogenesis." In *Morphogenesis and Pattern Formation: Implications for Normal and Abnormal Development*, edited by L. Brinkley, 205–213. New York: Raven Press, 1981.

(with C. F. Cori, H. P. Klinger, et al.) "Complementation of gene deletions by cell hybridization." *Proceedings of the National Academy of Sciences* 78 (1981): 479–483.

(with A. E. Goldfeld, C. S. Rubin, et al.) "Genetic control of insulin receptors." *Proceedings of the National Academy of Sciences* 78 (1981): 6359–6361.

(with L. Pick, S. G. Schiffer, et al.) "Tyrosine aminotransferase (TAT) induced in cells genetically and epigenetically deficient for this enzyme. I." *Developmental Biology* 92 (1982): 275–278.

"Developmental genetics of specific locus mutations." In *Environmental Science Research*, edited by F. J. deSerres, 259–266. New York: Plenum Press, 1983.

"50 years of developmental genetics." In *Festschrift for Sarah Ratner*, edited by Maynard E. Pullman, 243–251. New York: New York Academy of Sciences, 1983.

"Genetic control of differentiation." In *Cold Spring Harbor Conferences on Cell Proliferation*, vol. 10, edited by Lee M. Silver, et al., 3–13. New York: Cold Spring Harbor, 1983.

(with C. F. Cori, P. S. Shaw, et al.) "Correction of a genetically caused enzyme defect by somatic cell hybridization." *Proceedings of the National Academy of Sciences* 80 (1983): 6111–6114.

(with A. E. Goldfeld, G. L. Firestone, et al.) "A recessive lethal deletion on mouse chromosome 7 affects glucocorticoid receptor binding activities." *Proceedings of the National Academy of Sciences* 80 (1983): 1431–1434.

(with P. A. Shaw) "Epidermal growth factor and glucagon receptors in mice homozygous for a lethal chromosomal deletion." *Proceedings of the National Academy of Sciences* 80 (1983): 5379–5382.

"Deletions near the albino locus in chromosome 7 of the mouse affect the level of tyrosine aminotransferase mRNA." *Proceedings of the National Academy of Sciences* 82 (1985): 2866–2869.

(with J. M. SalaTrepat, M. Poiret, et al.) "A lethal deletion in chromosome 7 of the mouse affects the regulation of liver cell specific functions: Posttranscriptional control of serum protein and transcriptional control of aldolase B synthesis." *Proceedings of the National Academy of Sciences* 82 (1985): 2442–2446.

"Trans regulation of the phosphoenolpyruvate carboxykinase (GTP) gene identified by deletions in chromosome 7 of the mouse." *Proceedings of the National Academy of Sciences* 83 (1986): 5184–5188.

(with D. S. Loose, P. A. Shaw, et al.) "Developmental genetics of hepatic gluconeogenic enzymes." In *Annals of the New York Academy of Sciences* 478 (1986): 101–108.

(with P. A. Shaw and A. M. Adamany) "Glycosylation in livers of newborn mice homozygous for a lethal deletion." *Proceedings of the Society for Experimental Biology and Medicine* 183 (1986): 118–124.

(with A. W. Wolcoff and R. J. Stockert) "Selective expression of hepatocellular membrane proteins in mice homozygous for a lethal chromosome deletion." *Proceedings of the Society for Experimental Biology and Medicine* 181 (1986): 270–274.

"Regulatory genes in development." *Trends in Genetics* 3 (1987): 123–127.

(with D. DeFranco, S. M. Morris, Jr., et al.) "Metallothionine mRNA expression in mice homozygous for chromosomal deletions around the albino locus." *Proceedings of the National Academy of Sciences* 85 (1988): 1161–1164.

(with M. E. Donner and C. M. Leonard) "Developmental regulation of constitutive and inducible expression of hepatocyte-specific genes in the mouse." *Proceedings of the National Academy of Sciences* 85 (1988): 3049–3051.

"In praise of complexity." *Genetics* 122 (1989): 721–725.

(with S. L. McKnight and M. D. Lane) "Is CCAAT/enhancer binding protein a central regulator of energy metabolism?" *Genes and Development* 3 (1989): 2021–2024.

(with H. Nitowski) "The genesis of teaching human genetics at medical schools." *American Journal of Human Genetics* 46 (1990): 1222.

(with D. DeFranco) "Lethal chromosomal deletions in the mouse, a model system for the study of development and regulation of postnatal gene expression." *BioEssays* 13 (1991): 557–561.

(————, D. Bali, et al.) "The glucocorticoid hormone signal transduction pathway in mice homozygous for chromosomal deletions causing failure of cell type specific inducible gene expression." *Proceedings of the National Academy of Sciences* 88 (1991): 5607–5610.

"An overview of the developmental genetics in mammals." *Current Opinion in Genetics and Development* 2 (1992): 498–503.

(with J. C. Collins, D. N. Buchanan, et al.) "Metabolic studies in a mouse model of hepatorenal tyrosinemia: Absence of perinatal abnormalities." *Biochemistry and Biophysics Research Communications* 187 (1992): 340–346.

(with C. S. Giometti, M. A. Gemmell, et al.) "Evidence for regulatory genes on mouse chromosome 7 that affect the quantitative expression of proteins in the fetal and newborn liver." *Proceedings of the National Academy of Sciences* 89 (1992): 2448–2452.

(with M. Lia and D. Bali) "Regulatory genes linked to the albino locus in the mouse confer competence for inducible expression on the structural gene encoding serine

dehydratase." *Proceedings of the National Academy of Sciences* 89 (1992): 2453–2455.
(with K. S. Zaret, P. Milos, et al.) "Selective loss of a DNAse I hypersensitive site upstream of the tyrosine aminotransferase gene in mice homozygous for lethal albino deletions." *Proceedings of the National Academy of Sciences* 89 (1992): 6540–6544.

Other Works

Personal communication to the author, 1994.

Works about Salome Gluecksohn Schoenheimer Waelsch

Beck, Phil. "Medal of science winners: Eight pioneers of research." *The Scientist* 7 (Nov. 15, 1993): 7–8.
Osmundsen, J. A. "Muscle research yields new clue." *New York Times* (March 6, 1964): 33.
Zuckerman, Harriet, et al. *The Outer Circle: Women in the Scientific Community*. New York: W. W. Norton, 1991.

Other References

Harrison, Ross G. "The outgrowth of the nerve fiber as a mode of protoplasmic movement." *Journal of Experimental Biology* 9 (1910): 781–846.

ROSALYN SUSSMAN YALOW (1921–   )

Brina Nathanson

BIOGRAPHY

Rosalyn Sussman was born in the Bronx, New York, on July 19, 1921, to Clara (Zipper) and Simon Sussman. She had a brother, Alexander, who was five years older. Both parents were of Jewish heritage. Her father was born on the Lower East Side of New York, graduated from the eighth grade, went to work as a conductor on the Third Avenue Line, and then went into business for himself, selling paper and twine. Her mother came from Germany with her family. They first went to Ohio, where her grandfather was a tailor. When they left Ohio, they moved to New York City, where her mother finished the sixth grade and started working at the age of 12 in a department store.

Sussman was able to read even before she attended kindergarten. Her older brother was responsible for a weekly trip to the library to exchange books. She credits her father for encouraging the view that girls could do anything that boys could do. By the seventh grade she was earning excellent grades and was particularly good in mathematics. She was encouraged to study chemistry while at Walton High School (which was an all-girls' high school at the time) by a chemistry teacher, Morris Mondzak. When she was in high school, she felt that women could have a career as well as a husband and a family. This idea was in contrast to that held by many of her contemporaries at that time.

After graduating from high school, Sussman entered Hunter College (which was then an all-girls' school and is now part of the City University of New York). Hunter College was tuition-free, which was important because her parents could not afford to send her anywhere else. She majored in physics and chemistry and was encouraged by Professors Herbert Otis and Duane Roller. When she was a junior, she heard a talk by Enrico Fermi on nuclear fission. This experience, in addition to reading the biography of Marie Curie by Curie's daughter, encouraged Sussman to pursue a career in nuclear physics.

Sussman graduated Phi Beta Kappa and magna cum laude from Hunter College in 1941, at the age of 19, as its first physics major. Since it was the depression, she felt that it would be difficult for her as a Jewish woman to obtain a job in physics. She wanted to go to graduate school, even though her parents thought that she should become an elementary school teacher. Because of the prejudice at the time, her physics professors felt that the most likely way to enter graduate school was through the back door, as a secretary to a good scientist who would then help her enter graduate school. Since she was an excellent typist, in the last half of her senior year she was offered a part-time position as secretary to Dr. Rudolf Schoenheimer, a prominent biochemist at Columbia University's College of Physicians and Surgeons. She obtained the job on condition that when she graduated from Hunter, she would take a stenography course. In January 1941 she entered business school. Then, in February 1941 she was offered a teaching assistantship at the University of Illinois at Champaign-Urbana (which was the best of the physics schools to which she had applied). Since she did not have a full complement of physics courses, she went to City College of New York (now part of the City University of New York) at night and took a physics course while working as Schoenheimer's secretary until June of that year. Over the summer, she took two tuition-free courses (under government sponsorship) at New York University.

In the fall she enrolled at the University of Illinois in the physics department of the College of Engineering. She was the only woman in the school of 400 and the first graduate woman in physics since World War I. Because of World War II, young men were drafted, and this made it possible for her to be admitted to graduate school. At the school, she was assigned to teach premedical students. Since she did not have the same physics courses as other students, she had to take extra classes to catch up, while she was also a half-time teaching assistant. Two years later, while she was still a graduate student, she was made an instructor of engineering physics. Even though she received straight As in two of her courses, A in the lecture half of the course in optics, and A− in its laboratory, the chairman of the Physics Department said of her record, "This A− confirms that women do not do well at laboratory work" (Yalow, "Rosalyn . . ." 1978, 238). This was the type of discrimination against women that was common at that time. She spent long hours in the laboratory working on an experimental thesis under the direction of Drs. Maurice and Gertrude Goldhaber. Since her research dealt with nuclear physics, she became skilled at making and using apparatus for measurement of radioactive substances.

In her first day of graduate school, she met Aaron Yalow, a rabbi's son from upstate New York and a fellow graduate student who would become her husband on June 6, 1943. In 1942 she received an M.S. degree from the University of Illinois and was awarded a Ph.D. degree in nuclear physics from the University of Illinois (January 1945). Because her husband's degree was delayed, and it was felt that the best place for two physicists to find work was New York City, she went there and accepted a position as an assistant engineer at the Federal Telecommunications Laboratory, which was a research laboratory for International Telephone and Telegraph (ITT). Yalow was the only woman engineer there. The laboratory was run by Jewish engineers from France who had escaped before German forces took over their country. In 1946 the ITT laboratory was disbanded.

Yalow obtained a full-time teaching position at Hunter College, where she taught physics to returning veterans who were preengineering students. Since there were no research facilities at that time, Yalow looked for a place to do research. Her husband, who had come to New York in 1945, had a position in Medical Physics at Montefiore Hospital in the Bronx. He told her about Dr. Edith Quimby,* a leading medical physicist at Columbia University's College of Physicians and Surgeons. Yalow volunteered to work part-time in her laboratory to gain experience in the medical application of radioisotopes. The development of nuclear reactors made large amounts of radioisotopes available to researchers. Quimby introduced Yalow to Dr. Gioacchino Failla, "dean" of American medical physicists. Failla recommended Yalow to Dr. Bernard Roswit, Chief of the Radiotherapy Service at the Bronx Veteran's Administration (VA) Hospital. Because they needed a physicist, Yalow was hired as a part-time consultant in December 1947. She was given a large janitor's closet as an office. While she was still teaching full-time at Hunter College, she equipped and developed the hospital's Radioisotope Service and began doing research projects with Roswit and other physicians.

Yalow states that although both the doctors and the medical technicians were men, she never felt any hostility toward herself, because they needed her skills (Personal communication 1995).

In January 1950, Yalow left Hunter College and joined the VA full-time, where she was appointed physicist and assistant chief of the hospital's Radioisotope Service. The early work that she did in collaboration with Roswit and other physicians resulted in eight publications in different areas of clinical investigation. From the beginning, the laboratory has been supported solely by the VA Medical Research Program. In 1950 Yalow requested that the use of radioisotopes be extended beyond radiotherapy. In July 1950 Dr. Solomon A. Berson, an internist, joined the radioisotope unit as head of the Radioisotope Service. Yalow and Berson collaborated on research. Although neither Berson nor Yalow had postdoctoral training, they learned from each other: Yalow learning medicine from Berson, and Berson learning physics, chemistry, and mathematics from Yalow. The research they initiated on diabetes led them to develop the techniques of radioimmunoassay (RIA).

At the time she was working with Berson, Yalow and her husband lived in the Riverdale section of the Bronx, about one mile from the VA hospital, where she put in an 80–100-hour workweek. She and her husband helped each other in physics. At this time her husband was a professor at Cooper Union in New York City.

Yalow keeps a kosher home and observes Jewish traditions that were important to her husband. In 1952 she gave birth to a son, Benjamin. She took a week off and then went back to work. In 1954 she gave birth to a daughter, Elanna.

She was kept in the hospital for eight days. After she was released, she went to Washington, D.C., to give a lecture. Until Benjamin was 12, the Yalows had live-in help; after that, they had part-time help. On Sunday afternoons Yalow would return to the hospital to work, taking her children with her. Both children attended public schools, including the Bronx High School of Science. Benjamin became a computer analyst for City University of New York (CUNY). Her daughter obtained a Ph.D. degree in educational psychology at Stanford University. She is currently Director of Corporate Development at the Children's Discovery Centers of America in San Raphael, California. Yalow is pleased that her daughter has kept Yalow as her professional name, even though she is married (Stone 1978).

In 1972 Berson (who was a heavy smoker) died of a heart attack at the age of 54, while attending a medical meeting in Atlantic City, New Jersey. He had been the head of the Department of Medicine at Mt. Sinai Hospital since 1968. Although previously no survivor of a research team had won the Nobel Prize alone, Yalow decided to carry on the research that she and Berson had been involved in for 22 years. She felt that since he had become the head of Medicine in 1968 and did not have the time to spend more than one night a week at the VA, she had essentially run the laboratory independently for the past four years. In honor of her collaborator she named the laboratory the Solomon A. Berson Research Laboratory. She then worked to honor the commitments made by her and Berson, as well as continuing their joint research. One example was the book that they had been editing, *Methods in Investigative and Diagnostic Endocrinology*, vol. 2 (1973).

From 1968 to 1974 Yalow was Research Professor at the Department of Medicine of Mount Sinai School of Medicine and was later appointed Distinguished Service Professor (1974–79). She was appointed Director of the Solomon A. Berson Research Laboratory at the VA Hospital in 1973. The Bronx VA Hospital is now associated with Mt. Sinai School of Medicine. She has served as Physicist and Assistant Chief, Radioisotope Service, VA Hospital, Bronx (1950–70); Acting Chief, Radioisotope Service (1968–70); Senior Medical Investigator, VA Hospital from 1972 on; Chief, RIA Reference Laboratory from 1969 on; Chief, Nuclear Medical Service (1970–80); and Solomon A. Berson Distinguished Professor-at-Large, Mt. Sinai School of Medicine, CUNY, from 1986 on. Yalow served as consultant with the Radioisotope Unit of the VA Hospital of the Bronx (1947–50) and Lenox Hill Hospital (1952–62).

At Montefiore Hospital and Medical Center, Bronx, she served as Chairman, Department of Clinical Science (1980–85); at Albert Einstein Medical College, Yeshiva University, she was Distinguished Professor-at-Large (1979–85) and Professor Emeritus from 1985 to the present. She is currently Emeritus Distinguished Professor-at-Large at Albert Einstein Medical College. In 1986 the Solomon A. Berson Research Laboratory was designated a Nuclear Historical Landmark by the American Nuclear Society.

After the death of Berson, a number of different researchers joined her laboratory. One of them was Dr. Eugene Straus, who joined as a fellow in 1972, then became Research Associate, and finally Clinical Investigator. Another investigator was John Eng at Mount Sinai School of Medicine, who extended the earlier studies of vasoactive intestinal peptide (VIP) and cholecystokinin by analyzing the amino acid sequence of these polypeptides (Yalow, Eng, Yu, et al. 1992).

Yalow considers the young investigators associated with the laboratory her professional children. She has trained many researchers, and they are scattered throughout the world. A symposium was held October 11, 1991, at the Mount Sinai Medical Center in New York City ("Festschrift . . ." 1992) on the occasion of Yalow's 70th birthday. By all accounts, Yalow is not only a remarkable woman but also a wonderful human being. This was attested to by the participants in the symposium who related how she had aided them not only on a professional basis but also with their personal lives. Her husband attended the symposium but died a year later, on August 8, 1992. In 1991 she was made Professor Emeritus by the VA, and while she is not paid by the VA, she has a secretary there and interacts with, and gives help to, the physicians, even though her name does not appear on their papers.

Yalow has received over 50 awards and prizes. In 1976 she became the first woman to win the Albert Lasker Prize for Basic Medical Research. Her other awards include the Lilly Award, the Banting Medal of the American Diabetes Association, the Koch Award of the Endocrine Society, and the Georg von Hevesy Nuclear Medicine Pioneer Medal. She has rejected women's awards, such as the Ladies' Home Journal Woman of the Year (in 1978), because she does not believe in accepting gender-related awards.

In 1977 she was awarded the Nobel Prize in Physiology or Medicine for the development of radioimmunoassay. She received half of the $145,000 prize money; the other half was shared by Andrew V. Schally and Roger Guillemin, who used the technique of RIA to investigate brain hormones.

Yalow has received more than 50 honorary doctorate degrees from universities in the United States and abroad. Among them are Yeshiva University, New York Medical College, Tel-Aviv University, Bar-Ilan University, University of Ghent, and Technion-Israel Institute of Technology.

WORK

In 1947 Yalow set up the Radioisotope laboratory at the VA Hospital. This was to provide diagnostic and therapeutic services for the returning veterans. Once Berson joined the Radioisotope Service, Yalow concentrated on their collaboration rather than working with the other physicians in the service. They initially developed the isotope dilution technique to study the use of radioactive iodine in the diagnosis and treatment of thyroid disease. They derived the thyroid iodine clearance test, which is still in use today. According to J. E. Rall ("Festschrift . . ." 1992), the results obtained by measuring the blood cleared of iodine by the thyroid per unit of time provide more valid physiological information than just measuring uptake rate as a function of time. They extended the techniques that they used to the study of the distribution of globulin and other serum proteins. They measured blood volume by tagging red blood cells or plasma proteins with radioactive phosphorus or potassium.

Their work with RIA was initiated when Berson and Yalow investigated the metabolism of ^{131}iodine insulin in diabetic and nondiabetic patients after intravenous administration. They observed a slower rate of disappearance of ^{131}I-insulin from the plasma of the patients who had been treated with insulin, whether or not they were diabetic. This result was contrary to what would have been expected if adult diabetes were the result of rapid degradation of insulin by insulinase (as proposed by Dr. I. A. Mirsky 1952). They postulated that the slower disappearance was the result of binding to antibodies that had been produced in response to the foreign insulin that had been administered. They also found that these adult diabetics had more than enough insulin present in the blood. Before the development of RIA, the concentration of peptide hormones (like insulin) could not be determined in blood by classical techniques, because they are in too low a concentration. Yalow and Berson used different techniques, including paper electrophoresis, to demonstrate that the binding of radioisotopically labeled insulin to antibody is an inverse function of the amount of unlabeled insulin present in the plasma. When they submitted their findings to journals, *Science* refused to publish the article, and the *Journal of Clinical Investigation* published it only after the words "insulin antibody" were deleted from the title.

The RIA technique that they developed is *not* an isotope dilution technique but rather a competitive radioimmunoassay. The technique that they used is as follows: ^{131}I-tagged insulin is added to antiserum and incubated, and the amount of insulin bound to antibody is measured, using paper electrophoresis. A peak at the origin indicates unbound insulin, while a radioactive peak migrating with the gamma globulin fraction of the serum indicates insulin that is bound to antibody. A series of known concentrations of unlabeled insulin is prepared and is then added to the tubes of radioactively tagged insulin and antiserum. After a suitable incubation time, samples are subjected to paper electrophoresis, and the height of the radioactive insulin peak migrating with the serum gamma globulin is determined. A standard curve can then be prepared by plotting the decrease of the radioactive insulin that is bound, as a function of concentration of the unlabeled insulin standards. The higher the concentration of the standard, the greater is the decrease in the peak of the radioactively bound insulin. The concentration of insulin in an unknown sample of blood can, therefore, be determined by measuring the height of the peak of the bound radioactive insulin incubated with the untagged sample and comparing it with the standard curve that had been previously prepared. The procedure that they used is very sensitive and can be performed in a test tube.

The sensitivity of RIA is such that, as little as 0.1 picogram (pg=1x10^{-12}g) per mL of gastrin (a peptide hormone) can be measured. Peptide hormones in the unstimulated state are in concentrations as low as 10^{-12} to 10^{-10} molar. Thyroid and steroid hormones are in the range of 10^{-6} molar. The concentration of drugs in the blood may be even higher. One concrete example may be given to illustrate the usefulness of RIA. Thompson et al. (1986) compared the detection of Staphylococcal enterotoxin by enzyme-linked immunoassay (ELISA) and RIA to the "classical" microslide double-gel immunoassay in which the sensitivity of detection is 0.1 microgram of toxin per mL of food extract. When RIA is used, the sensitivity is equal to one nanogram ($ng=1x10^{-9}g$) per mL, and therefore, investigators can use the food extract without the concentration procedures used in the double-gel immunoassay method, which take up to three days to complete.

After acceptance of Berson and Yalow's paper, as well as subsequent articles on insulin-binding antibodies, insulin-resistance, insulin antagonists in plasma, and immunoassay of endogenous plasma insulin in man, a number of investigators wanted to learn their RIA techniques. Berson and Yalow organized a series of workshops on RIA in the years 1960, 1963, and 1965 (Yalow, "Remembrance . . ." 1991), and by the late 1960s RIA had become widely used in endocrinology laboratories throughout the world. Yalow and Berson showed that RIA could be used to measure almost any substance as long as that substance could elicit the production of antibodies.

RIA measures the presence and titer of hepatitis virus in a drop of blood. This is utilized by the Red Cross and other blood banks to prevent the spread of hepatitis. RIA is also used to measure purified protein derivative (PPD), which allows the more rapid diagnosis of the growth of *Mycobacterium tuberculosis*. RIA has served to measure peptide hormones such as thyroid-stimulating hormone (TSH), follicle-stimulating hormone (FSH), luteinizing hormone (LH), vasopressin (ADH), oxytocin, insulin, glucagon, and parathyroid hormone (PTH). It can be used to measure nonpeptide hormones such as thyroidal hormones, prostaglandin, and steroid hormones. RIA can also determine concentrations of nonhormonal substances, such as drugs (cardiac glycosides, morphine, LSD), enzymes, viruses, tumor antigens, rheumatoid factors, and Staphylococcal enterotoxins. One particularly important use is in the radioallergosorbant test (RAST) and the modified RAST test (MRT) developed in other laboratories (Fadel 1992). These are *in vitro* tests for the measurement of total and allergenspecific immunoglobin E (IgE) reactivity (IgE is the agent for immediate hypersensitivity reactions). Only patients with serum titers of allergen-specific IgE can be considered suitable candidates for immunotherapy (Fadel 1992).

RIA has been utilized to measure hundreds of substances of biological interest. Now 70 different RIA kits are used throughout the world. Yalow and Berson did not patent RIA, because they wanted it made available to anyone who needed it. Yalow has chosen not to become a consultant to the drug companies, because she feels that she would not be free to speak her mind about the uses of RIA.

Yalow served as coeditor of *Hormones and Metabolic Research* (1973–79). She was on the editorial boards of *Endocrinology* (1967–72), *Mount Sinai Journal of Medicine* (1976–79), and *Diabetes* (1976–79) and on the Board of Directors of the New York Diabetes Association (1974–77). Yalow has participated in several national committees and study groups. She was on the Task Force on Immunological Diseases, National Institutes of Allergy and Infectious Disease (1972–73); World Health Organization (WHO) consultant, Radiation Medical Center, Bombay (1978); and a consultant for the New York City Department of Health. She belongs to many professional organizations and honorary societies, including the Endocrine Society (president-elect, 1977–78 and president, 1978–79), National Academy of Sciences, fellow of the New York Academy of Sciences (chairman, Biophysics Division, 1964–65), Radiation Research Society, American Association of Physicists in Medicine, and the Society of Nuclear Medicine. She holds honorary membership in several scientific and medical organizations, including the American College of Nuclear Physicians and the New York Academy of Medicine.

Yalow has primarily concentrated on applying RIA methods to endocrinology. Recent papers by Yalow and her associates have dealt with the evolutionary relationships between the marsupial and placental mammals of the Old and New Worlds; and the use of amino acid sequencing of VIP, cholecystokinin, and gastrin to determine the relationships between these organisms.

Because Yalow feels that students in junior high school and high school should be encouraged to set future goals for themselves, she frequently gives talks at these schools. She believes that all high school students should be required to have a yearlong science course that includes aspects of physics, biology, and chemistry. Exposure to science in this way will give them sufficient knowledge to evaluate matters relating to science when they are old enough to vote on political issues dealing with science. Finally, Yalow has been an outspoken opponent of the manner in which the Environmental Protection Agency (EPA) has dealt with the subject of exposure to low-level ionizing radiation delivered at low doses. In a recent paper (Yalow, "Concerns . . ." 1990) she criticizes the reports of the EPA and other governing agencies. Predictions of radiation injury in these reports are based upon faulty data. Yalow believes that it is important to consider dose-rate effects for injury from low-energy radiation and to differentiate between fractionated doses at high dose rates and continuous irradiation at low dose rates. She feels that it is her duty to enlighten the general public on the usefulness of nuclear energy, especially with respect to medical usage.

BIBLIOGRAPHY

Works by Rosalyn Sussman Yalow

Scientific Works

Space does not permit the listing of the complete works of Rosalyn Sussman Yalow. This list includes all works by Yalow with the exception of those cited in Biermann and Biermann (1993), Rall (1990), and Yalow (*Science* 1978). Also included are all references cited in the text.

(with H. Conway, B. Roswit, et al) "Radioactive sodium clearance as a test of circulatory efficiency of tube pedicles and flaps." *Proceedings of the Society for Experimental Biology and Medicine* 77 (1951): 348–351.

(with R. S. Feder) "Uniformity of cell counts in smears of bone marrow particles." *American Journal of Clinical Pathology* 21 (1951): 541–545.

(with B. Roswit, J. A. Rosenkrantz, et al.) "Evaluation of diagnostic methods in diseases of thyroid function with particular reference to radioiodine tracer tests." *American Journal of Medical Science* 223 (1952): 229–238.

(with O. N. Vasquez, K. Newerly, et al.) "Determination of trapped plasma in the centrifuged erythrocyte volume of normal human blood with radio-iodinated (I^{131}) human serum albumin and radiosodium (Na^{24})." *Journal of Laboratory and Clinical Medicine* 39 (1952): 595–604.

(with B. B. Cohen) "Apparatus for intracavitary administration of colloidal gold." *Nucleonics* 11 (1953): 65–67.

(with J. Freund and L. H. Wishaw) "The effect of priscoline on the clearance of radiosodium from muscle and skin in normal and diseased limbs." *Circulation* 8 (1953): 89–97.

(with M. A. Rothschild, A. Bauman, et al.) "Effect of splenomegaly on blood volume." *Journal of Applied Physiology* 6 (1954): 701–706.

(with O. N. Vasquez, K. Newerly, et al.) "Estimation of trapped plasma with I^{131} albumin." *Journal of Applied Physiology* 5 (1954): 437–440.

(with K. R. Paley and E. S. Sobel) "A comparison of the thyroidal plasma I^{131} clearance and the plasma protein-bound I^{131} tests for the diagnosis of hyperthyroidism." *Journal of Clinical Endocrinology and Metabolism* 15 (1955): 995–1009.

(with S. A. Berson) "Kinetics of reaction between insulin and insulin-binding antibody." *Journal of Clinical Investigation* 36 (1957): 873.

(with K. R. Paley and E. S. Sobel) "Effect of oral and intravenous cobaltous chloride on thyroid function." *Journal of Clinical Endocrinology and Metabolism* 18 (1958): 850–859.

(————) "Some aspects of thyroidal iodine metabolism in a case of iodine-induced hypothyroidism." *Journal of Clinical Endocrinology and Metabolism* 18 (1958): 79–90.

"The effects of alpha-particle irradiation on I^{131}-labeled iodotyrosines." *Radiation Research* 11 (1959): 30–37.

(with S. A. Berson) "Application of light scattering to biological systems: Deoxyribonucleic acid and the muscle proteins." In *Advances in Biological and Medical*

Physics, vol. 6, edited by C. A. Tobias and J. H. Lawrence, 349–430. New York: Academic Press, 1959.

(———) "Correction of sample absorption of radioactivity." *Science* 131 (1960): 606.

(———) "Insulin inhibitors and insulin resistance." *New York State Journal of Medicine* 60 (1960): 3658–3665.

(———) "Immunoassay of insulin content of crystalline glucagon preparations." *Proceedings of the Society for Experimental Biology and Medicine* 107 (1961): 148–151.

(———) "Plasma insulin in health and disease." *American Journal of Medicine* 31 (1962): 1044–1045.

(———) "Immunoassay of plasma insulin (methods and techniques)." *Methods in Biochemical Analysis* 12 (1964): 69–96.

(———) "Reaction of fish insulin with human antibodies to beef, pork insulin: Potential in the treatment of insulin resistance." *Physics in Medicine and Biology* 9 (1964): 101–102.

(———) "Radioimmunoassay of peptide hormones." *Folia Endocrinologica Japonica* 45 (1969): 545–554.

(———) "Dynamics of insulin secretion in early diabetes in humans." In *Advances in Metabolic Disorders*, suppl. 1, edited by R. A. Camerini-Davalos and H. S. Cole, 95–103. New York: Academic Press, 1970.

(———) "Various problems in radioimmunoassay theory of radioimmunoassay and methodological observation" (in Japanese). *Recent Medicine* 25 (1970): 724–735.

(———) "Applications of radioimmunoassay to problems of clinical interest. Sampling of some physiological applications and diagnostic uses of radioimmunoassay indicates potentialities of method." *Rhode Island Medical Journal* 54 (1971): 501–508.

(———) "Clinical applications of radioimmunoassay of plasma parathyroid hormone." *American Journal of Medicine* 50 (1971): 623–629.

"Radioimmunoassay methodology: Applications to problems of heterogeneity of peptide hormones." *Pharmacological Reviews* 25 (1973): 161–178.

"Radioimmunoassay, practices and pitfalls." *Circulation Research* 32 (Suppl. 1) (1973): 116–128.

"Solomon A. Berson: The VA years." *Mount Sinai Journal of Medicine* 40 (1973): 281–283.

(with S.A. Berson) "General radioimmunoassay." In *Methods in Investigative and Diagnostic Endocrinology*, vol. 1, edited by S. A. Berson and R. S. Yalow, 84–120. Amsterdam: North Holland, 1973.

(———) "Radioimmunoassay of insulin." In *Methods in Investigative and Diagnostic Endocrinology*, vol. 3, edited by S. A. Berson and R. S. Yalow, 864–870. Amsterdam: North Holland, 1973.

(———, eds.) *Methods in Investigative and Diagnostic Endocrinology*, 3 vols. Amsterdam: North Holland, 1973.

"Heterogeneity of peptide hormones." *Recent Progress in Hormone Research* 30 (1974): 597–633.

(with G. Gewirtz, B. Schneider, et al.) "Big ACTH: Conversion to biologically active ACTH by trypsin." *Journal of Clinical Endocrinology and Metabolism* 38 (1974): 227–230.

(with E. Straus and C. D. Gerson) "Hypersecretion of gastrin associated with the short bowel syndrome." *Gastroenterology* 66 (1974): 175–180.

(with K. Hall and R. Luft) "Radioimmunoassay of somatomedin B." *Advances in Metabolic Disorders* 8 (1975): 73–83.

"Multiple forms of corticotropin (adrenocorticotropic hormone, ACTH) and their significance." *Ciba Foundation Symposium* 41 (1976): 159–181.

(with E. Straus) "Heterogeneity of gastrointestinal hormones." In *Hormonal Receptors in Digestive Tract Physiology*, edited by S. Bonfils et al., 79–93. Amsterdam: North Holland, 1977.

"Radioimmunoassay: A probe for the fine structure of biological systems." *Medical Physics* 5 (1978): 247–257.

"Radioimmunoassay: A probe for the fine structure of biological systems." *Science* 200 (1978): 1236–1245.

"Significance of the heterogeneity of parathyroid hormone." In *Endocrinology of Calcium Metabolism*, edited by D. H. Cope and R. V. Talmage, 308–312. Amsterdam: Excerpta Medica, 1978.

"Radioimmunoassay: Past and potential role in clinical medicine." *Proceedings of the Rudolf Virchow Medical Society in the City of New York* 33 (1979): 16–21.

"Clinical significance of the heterogeneity of parathyroid hormone." *Contributions to Nephrology* 20 (1980): 15–20.

"Presidential address: Reflections of a non-establishmentarian." *Endocrinology* 106 (1980): 412–414.

(with E. Straus) "Radioimmunoassay of some peptides in the brain." In *Ferring Symposium on Brain and Pituitary Peptides*, Munich, 1979, edited by W. Wuttke et al., 89–96. Basel: S. Karger, 1980.

(with A. A. Yalow) "The physics of radioimmunoassay." *Transactions of the New York Academy of Sciences* 40 (1980): 253–266.

"Diverse applications of radioimmunoassay." *Radioisotopes* 30 (1981): 340–347.

(with E. Straus) "Immunochemical studies relating to cholecystokinin in brain and gut." *Recent Progress in Hormone Research* 87 (1981): 447–475.

"Radioimmunoassay and tumor diagnosis." *Progress in Clinical Cancer* 8 (1982): 1–7.

(with J. Eng) "Insulin in the central nervous system." *Advances in Metabolic Disorders* 10 (1983): 341–354.

(——— and E. Straus) "The role of CCK-like peptides in appetite and regulation." *Advances in Metabolic Disorders* 10 (1983): 435–456.

"Radioactivity in the service of man." *Pharos of Alpha Omega Alpha Honor Medical Society* 47 (1984): 40–43.

"Radioimmunoassay in laboratory medicine." *Japanese Journal of Clinical Pathology* 32 (1984): 490–505.

(with B. H. Du and J. Eng) "Methods for concentration of urinary immunoreactive insulin." *Journal of Korean Medical Science* 1 (1984): 1–4.

"Practices and pitfalls in immunologic methodology." *Advances in Prostaglandin, Thromboxane, and Leukotriene Research* 16 (1986): 327–338.

(with J. H. Yu, J. Eng, et al.) "Opossum insulin, glucagon and pancreatic poly-peptide: Amino acid sequences." *Peptides* 10 (1989): 1195–1197.

"Concerns with low level ionizing radiation: Rational or phobic?" *Journal of Nuclear Medicine* 31 (1990): 17A-18A, 26A.

"The contributions of ^{131}I to the understanding of radiation carcinogenesis." *Endocri-*

nology 126 (1990): 1787–1789.
    Editorial.

(with J. Eng and H. R. Li) "Purification of bovine cholecystokinin-5B and sequencing of its N-terminus." *Regulatory Peptides* 30 (1990): 15–19.

(with Y. Shinomura, J. Eng, et al.) "Opossum (*Didelphis virginiana*) 'little' and 'big' gastrins." *Comparative Biochemistry and Physiology—B: Comparative Biochemistry* 96 (1990): 239–242.

(with J. H. Yu and J. Eng) "Isolation and amino acid sequences of squirrel monkey (*Saimiri sclurea*) insulin and glucagon." *Proceedings of the National Academy of Sciences* 87 (1990): 9766–9768.

"Remembrance project: Origins of RIA." *Endocrinology* 129 (1991): 1694–1695.

(with J. H. Yu, Y. Xin, et al.) "Rhesus monkey gastroenteropancreatic hormones: Relationship to human sequences." *Regulatory Peptides* 32 (1991): 39–45.

"The Nobel lectures in immunology. The Nobel Prize for Physiology or Medicine, 1977 awarded to Rosalyn S. Yalow (classical article)." *Scandanavian Journal of Immunology* 35 (1992): 1–23.

(with J. Eng, J. Yu, et al.) "Isolation and sequences of opossum vasoactive intestinal polypeptide and cholecystokinin octapeptide." *Proceedings of the National Academy of Sciences* 89 (1992): 1809–1811.

"Concerns with low level ionizing radiation." *Mayo Clinic Proceedings* 69 (1994): 436–440.

"Radiation effects." In *Nuclear Test Results: A Woman's Perspective*, 63–70. New York: City University of New York Academy for the Humanities and Sciences, 1995.

"Effects in society of ionizing radiation." *Mount Sinai Journal of Medicine* (submitted).

"A reasonable appraisal of danger from low-level radiation." In *Radiation Risk: A Primer*, edited by Murray Janower et al. American College of Radiology, Commission on Physics and Radiation Safety (submitted).

Other Works

Personal communications to the author, 1995.

"Rosalyn S. Yalow." In *Les Prix Nobel* (1977), 237–242. Stockholm: Nobel Foundation, 1978.

Works about Rosalyn Sussman Yalow

Altman, L. K. "Nobel winner's test has variety of uses." *New York Times* (Oct. 14, 1977): A-18.

Biermann, C. A., and L. Biermann. "Rosalyn Sussman Yalow." In *Women in Chemistry and Physics: A Biobibliographic Sourcebook*, edited by L. S. Grinstein, R. K. Rose, and M. H. Rafailovich, 626–639. Westport, CT: Greenwood Press, 1993.

Reed, R. "Feminist tone colors speech by a Nobel laureate." *New York Times* (Oct. 14, 1977): A-18.

Stone, E. "A Mme. Curie from the Bronx." *New York Times Magazine* (Apr. 19, 1978): 29–103.

Other References

Fadel, R. G. "Experience with RAST-based immunotherapy." *Otolaryngologic Clinics of North America* 25 (1992): 43–60.o
    Review.
"Festschrift for Rosalyn S. Yalow: Hormones, metabolism and society." *Mount Sinai Journal of Medicine* 59 (1992): 95–100.
Mirsky, I. A. "The etiology of diabetes mellitus in man." *Recent Progress in Hormone Research* 7 (1952): 436–465.
Rall, J. E. "Solomon A. Berson." *Biographical Memoirs of the National Academy of Sciences*, vol. 59, 55–70. Washington, DC: National Academy Press, 1990.
Thompson, N. E., M. Razdan, et al. "Detection of Staphylococcal enterotoxin by enzyme-linked immunosorbant assays and radioimmunoassays. Comparison of mono-clonal and polyclonal antibody systems." *Applied and Environmental Microbiology* 5 (1986): 885–890.

APPENDIX A: CHRONOLOGICAL LIST OF BIOGRAPHEES

Hildegard of Bingen (1098–1179)
Manzolini (1716–1774)
Pinckney (1722–1793)
Colden (1724–1766)
Phelps (1793–1884)
Agassiz (1822–1907)
Shattuck (1822–1889)
Bodley (1831–1888)
Hyde (1857–1945)
Britton (1858–1934)
DeWitt (1859–1928)
Eastwood (1859–1953)
Stevens (1861–1912)
Bailey (1863–1948)
Doubleday (1865–1918)
Potter (1866–1943)
Eckerson (ca. 1867–1954)
Chase (1869–1963)
Hamilton (1869–1970)
Slye (1869–1954)
Morgan, Lilian (1870–1952)
Lepeshinskaia (1871–1963)
Sabin (1871–1953)
Arber (1879–1960)
Evans (1881–1975)
Morgan, Ann (1882–1966)
Nice (1883–1974)
Harvey (1885–1965)
Pearce (1885–1959)
Stephenson (1885–1948)
Hyman (1888–1969)
Braun (1889–1971)

Quimby (1891–1982)
Klieneberger-Nobel (1892–1985)
Lancefield (1895–1981)
Cori (1896–1957)
Seibert (1897–1991)
Esau (1898–)
Mangold (1898–1924)
Taussig (1898–1986)
Auerbach (1899–1994)
Fell (1900–1986)
Alexander (1901–1968)
McClintock (1902–1992)
Scharrer (1906–1995)
Carson (1907–1964)
Hamerstrom (1907–)
Patrick (1907–)
Waelsch (1907–)
Levi-Montalcini (1909–)
Colwin (1911–)
Russell (1911–1967)
Vennesland (1913–)
Ray (1914–1994)
Elion (1918–)
Sager (1918–1997)
Franklin (1920–1958)
Mintz (1921–)
Yalow (192l–)
Clark (1922–)
Cobb (1924–)
Lyon (1925–)
Rowley (1925–)
Neufeld (1928–)
Fossey (1932–1985)

APPENDIX B: BIOGRAPHEES BY PLACE OF BIRTH, PLACE OF WORK, AND FIELD OF SCIENTIFIC INTEREST

Name	Place of Birth	Place of Work	Field of Interest
Agassiz	USA	USA	Biology, natural history
Alexander	USA	USA	Bacteriology, medical research, microbiology, poliovirus infections research, influenzal meningitis research, serum development, treatment of *Haemophilus influenzae* infections
Arber	England	England	Biology, botany, natural philosophy
Auerbach	Germany	Scotland, USA	Biology, mutation research, genetics, mutagenic agents research
Bailey	USA	USA	Natural history, ornithology, ornithological education
Bodley	USA	USA	Chemistry, botany
Braun	USA	USA	Botany, conservation, geographical botany, plant succession, botany education, plant ecology
Britton	USA	USA	Botany, horticulture, bryology
Carson	USA	USA	Biology, conservation, ecology, marine biology, natural history, DDT studies, environmental protection, biology education
Chase	USA	USA	Botany, systematic agrostology (grasses)

Clark	USA	USA	Ichthyology, marine biology, marine zoology, zoology, behavior and ecology of fish, shark behavior and research, blowfish research, fish taxonomy, deepwater fish and tropical bony fish research
Cobb	USA	USA	Biology, cell physiology, biology education, chemotherapeutic agents, neoplastic disease research, melanoma research
Colden	USA	USA	Botany, plant taxonomy, plant identification
Colwin	USA	USA	Cytology, embryology, zoology, sperm and egg interactions, invertebrate reproduction, fertilization research
Cori	Czech Republic	Austria, USA	Carbohydrate metabolism, glycogen, glucose, enzymes, biochemistry
DeWitt	USA	USA	Anatomy, pathology, bacteriology, histology, medical research
Doubleday	USA	USA	Natural history, ornithological education, biology education, horticulture
Eastwood	Canada	USA	Field botany, conservation, natural history, plant taxonomy, plant collections, fern studies
Eckerson	USA	USA	Botany, microchemistry, botany education, physiology, germination, plant mineral nutrition, nitrate reduction in plants, cell walls, cellulose, starch grains, tomato mosaic virus
Elion	USA	USA	Organic synthesis, biochemistry, medical research, chemotherapeutic agents, pharmacology, immunology
Esau	Ukraine	USA	Botany, virology, phloem research, photomicrography, sugar beet infections, developmental studies of plants, plant pathology
Evans	USA	USA	Bacteriology, medical research, microbiology, human brucellosis

Fell	England	England, Scotland	Skeletal system biology, bone histology, cytology, organ culture research, physiology, embryology, wound-healing research, study of retinoids, biochemistry, vitamin A research
Fossey	USA	USA, Africa	Biology, ethology, conservation, zoology, gorilla behavior, gorilla research, gorilla ecology, primatology
Franklin	England	England, France	X-ray diffraction, DNA crystallography, biophysics, genetics, tobacco mosaic virus (TMV), virus crystallography
Hamerstrom	USA	USA	Ornithology, natural history, wildlife field biology, raptor research, bird breeding, harrier population research, bird banding, chickadee research, wildlife management, prairie chicken research
Hamilton	USA	USA	Bacteriology, pathology, industrial toxicology
Harvey	USA	USA, Italy, Bermuda	Biology, cytology, embryology, zoology, hydra research
Hildegard of Bingen	Germany	Germany	Biology, botany, folk medicine, natural history, zoology
Hyde	USA	USA	Biology, cytology, physiology, invertebrate zoology
Hyman	USA	USA, Italy	Biology, invertebrate zoology, biology education, invertebrate taxonomy
Klieneberger-Nobel	Germany	England	Bacteriology, mycoplasma research, pleuropneumonia-like organisms (PPLOs), "rolling disease"
Lancefield	USA	USA	Bacteriology, microbiology, immunology, medical research, streptococcal research, type-specificity of streptococci
Lepeshinskaia	Russia	Russia	Biology, agronomy, histology, Lysenkoism
Levi-Montalcini	Italy	Italy, USA	Biology, embryology, neurology, nerve growth factor
Lyon	England	England	Biology, genetics, Lyon hypothesis, radiobiology, X-chromosome inactivation

Mangold	Germany	Germany	Experimental embryology, Spemann research
Manzolini	Italy	Italy	Anatomy
McClintock	USA	USA	Biology, botany, cytology, maize genetics, chromosome research, transposons and transposition, molecular genetics
Mintz	USA	USA	Medical and developmental genetics, embryology, cancer research, allophenic mice
Morgan, Ann	USA	USA	Biology of mayflies, conservation, ecology, entomology, zoology, respiration of aquatic insects
Morgan, Lilian	USA	USA	Fruit fly genetics, X-chromosome, genetics, anatomy, embryology
Neufeld	France	USA	Biochemistry, mucopolysaccaridoses, genetics, pathology, lysosome studies, enzymology
Nice	USA	USA	Biology, conservation, ethology, ornithology, behavioral ornithology
Patrick	USA	USA	Botany, ecology, limnology, diatom research, riverine communities, plankton research, taxonomy, biodynamics of rivers
Pearce	USA	USA, China	Medical research, pathology, parasitology, syphilis research, cancer research
Phelps	USA	USA	Botany, plant taxonomy, botany education
Pinckney	West Indies	USA	Botany, horticulture, agronomy
Potter	England	England	Botany, mycology, natural history
Quimby	USA	USA	Biophysics, radioactive isotopes, radiation therapy
Ray	USA	USA	Biology, marine biology, zoology, science education, atomic energy, environmental protection
Rowley	USA	USA	Genetics, cancer research, chromosomes, Philadelphia chromosome, translocation, leukemia, lymphoma, proto-oncogenes, oncogenes

Russell	USA	USA	Biochemistry, endocrinology, medical research, carbohydrate metabolism, growth hormone, pituitary research
Sabin	USA	USA	Anatomy, biology, immunology, medical research, cardiovascular research, tuberculosis pathology
Sager	USA	USA	Biology, genetics, physiology, cytoplasmic genes, cancer research, tumor suppressor genes
Scharrer	Germany	USA	Anatomy, biology, neuroimmunology, neuroendocrinology, roach research
Seibert	USA	USA	Biochemistry, bacterial pyrogens, standardized tuberculin, pathology
Shattuck	USA	USA	Botany, horticulture, natural history, science education, chemistry
Slye	USA	USA	Biology, genetics, medical research, pathology, cancer research, mouse-breeding programs
Stephenson	England	England	Bacteriology, vitamin A studies, chemical microbiology, bacterial biochemistry, enzymology
Stevens	USA	USA	Cytology, experimental physiology, parasitic ciliates, Mendelian genetics, studies of sex determination, invertebrate zoology, X-chromosome discovery
Taussig	USA	USA	Thalidomide research, cardiac malformations, heart muscle physiology, treatment of "blue babies," pediatric cardiology founder and educator
Vennesland	Norway	USA	Biochemistry, metabolic pathway studies, metabolism of plant tissues, enzymology, photosynthesis research
Waelsch	Germany	Germany, USA	Experimental embryology, developmental genetics, mutation research, development of nervous system
Yalow	USA	USA	Insulin studies, radioimmunoassay (RIA), biochemistry, biophysics, biology education

APPENDIX C: REFERENCES IN BIOGRAPHICAL DICTIONARIES AND OTHER COLLECTIONS

In order to avoid duplication of entries, the following conventions were adopted. Some of the sources cited list other references. Those primary sources are Fruton2, Grinstein, Herz, Ireland70, Ireland88, Pell, and Siegel. If a biographee is included in several editions of a source such as AmM&WS, only one edition is cited. A key to the source codes appears in Appendix D.

Agassiz	Abir, AWSC, DAB, Elliott, Herz, Hinding, Hoyrup, Ireland70, Ireland88, Mchenry, Mozans, NCAB, Norw, NotAmW, Ogilvie, Oneil, Pell, Ross1, Searing, Sweeney, Uglow, Weiser, WhoWasWho, WWWST
Alexander	AmM&WS(12th), Bail, Herman, Herz, Ireland88, Kass, NCAB, NotAmW80, NTCS, NWLS, Oneil, Pell, Ross2, Siegel
Arber	Abir, DSB, Fruton2, Fruton3, Golemba, Herz, Hinding, Hoyrup, Ireland88, NTCS, Pell, Searing, Uglow
Auerbach	Fruton2, Herz, Hoyrup, Morton, NTCS, Pell, Uglow
Bailey	Abir, AmM&WS(8th), AWSC, Bonta, DAB, Debus, Golemba, Herz, Hinding, Ireland70, Ireland88, Mchenry, Mozans, NCAB, Norw, NotAmW, NotAmW80, NTCS, NWLS, Ogilvie, Oneil, Pell, Read, Ross1, Stille
Bodley	Abir, AWSC, Chaff, Elliott, Herz, Hinding, Hoyrup, Ireland88, Lovejoy, NotAmW, NWLS, Ogilvie, Pell, Rudolph, Siegel, Uglow
Braun	Abir, AmM&WS(12th), AWSC, Bail, Bonta, Herz, Hinding, Ireland88, Kahle, NotAmW80, Pell, Porter, Read, Ross2, Searing, Siegel, Torrey
Britton	Abir, AmM&WS(6th), AWSC, Bonta, Golemba, Herz, Hinding, Ireland88, Mchenry, Mozans, NCAB, NotAmW, NWLS, Pell, Read, Ross1, Ross2, Rudolph, Searing, Siegel, Torrey, Visher, WWWST

Carson	AWSC, Bail, Bonta, Debus, DSB, Herman, Herz, Hinding, Hoyrup, Ireland70, Ireland88, Kahle, Mchenry, NCAB, Norw, NotAmW, NTCS, NWLS, Oneil, Pell, Porter, Ross1, Ross2, Searing, Siegel, Stille, Sweeney, Uglow, Vare, Vernoff, Weiser, WSD, WWWST
Chase	Abir, AWSC, Bail, Bonta, DAB, Herz, Hinding, Ireland88, Norw, NotAmW80, NWLS, Ogilvie, Pell, Porter, Rudolph, Siegel, Torrey, Visher, WWWST
Clark	AmM&WS(19th), AWSC, Debus, Herz, Hoyrup, Ireland70, Ireland88, Norw, Oneil, Pell, Ross2, Searing
Cobb	AmM&WS(19th), Herz, Ireland88, NTCS, NWLS, Oneil, Ross2, Stille, WWAW
Colden	Abir, AWSC, Bonta, DAB, Elliott, Golemba, Herz, Hoyrup, Ireland70, Ireland88, Kass, Norw, NotAmW, Ogilvie, Pell, Read, Ross1, Rudolph, Searing, Siegel, Vare, WhoWasWho
Colwin	AmM&WS(19th), Ross2
Cori	AWSC, Bail, Grinstein, NTCS, Pell, Porter, Ross2, WSD
DeWitt	AmM&WS(5th), AWSC, Herz, Ireland88, NCAB, NotAmW, Ogilvie, Pell, Ross1, Siegel, Visher, WhoWasWho
Doubleday	AmM&WS(13th), AWSC, DAB, Elliott, Herz, Ireland88, NCAB, NotAmW, Pell, Siegel, Torrey
Eastwood	Abir, AmM&WS(1st), AWSC, Bail, Bonta, Debus, Herz, Hinding, Hoyrup, Ireland70, Ireland88, Norw, NotAmW80, NWLS, Ogilvie, Pell, Porter, Ross1, Rudolph, Searing, Siegel, Stille, Torrey, Visher, WWWST
Eckerson	AmM&WS(6th), AWSC, Herz, NWLS, Pell, Ross1, Rudolph, Siegel, Torrey, Visher
Elion	Grinstein, NTCS, Ross2, Stille, WSD
Esau	Abir, AmM&WS(19th), AWSC, Debus, Fruton2, Golemba, Herz, Hinding, Hoyrup, Ireland88, NTCS, NWLS, Oneil, Pell, Ross1, Ross2, Torrey, Uglow, Vare, Vernoff, Weiser, WWAW
Evans	AmM&WS(11th), AWSC, Bail, Chaff, Fruton2, Herman, Herz, Hinding, Hoyrup, Ireland70, Ireland88, Kass, Lovejoy, NotAmW80, NWLS, Oneil, Pell, Porter, Read, Ross1, Siegel, Vare, WhoWasWho, WSD, WWAW
Fell	Fruton2, Herz, Hoyrup, Morton, NTCS, Pell, Uglow
Fossey	Herz, Ireland88, Norw, NTCS, Oneil, Pell, Read, Stille, Uglow
Franklin	Grinstein, NTCS, NWLS, Pell, Stille, WSD

Hamerstrom	AmM&WS(19th), AWSC, Ross2
Hamilton	AmM&WS(10th), AWSC, Bail, Chaff, Debus, Golemba, Herz, Hinding, Hoyrup, Ireland70, Ireland88, Kass, Lovejoy, Morton, NCAB, NotAmW80, NTCS, NWLS, Oneil, Pell, Porter, Read, Ross1, Searing, Siegel, Sweeney, Uglow, Vernoff, Weiser
Harvey	AmM&WS(10th), AWSC, Bail, Herman, Herz, Ireland88, Kass, NotAmW, NWLS, Pell, Porter, Ross1, Ross2, Siegel, Visher, WhoWasWho, WWWST
Hildegard	Chaff, DSB, Herz, Hoyrup, Ireland70, Ireland88, Kass, Mozans, NWLS, Ogilvie, Pell, Searing, Uglow, Vare, Weiser
Hyde	AmM&WS(8th), AWSC, Bail, Herz, Hinding, Ireland88, Kahle, Kass, Mozans, NCAB, NotAmW, NTCS, NWLS, Ogilvie, Oneil, Pell, Read, Searing, Siegel, Stille
Hyman	AmM&WS(12th), AWSC, Bail, Debus, DSB, Fruton2, Golemba, Herman, Herz, Hinding, Ireland70, Ireland88, Kahle, Kass, Mchenry, NotAmW80, NWLS, Oneil, Pell, Porter, Ross1, Ross2, Siegel, Vernoff, Visher, Weiser, WWWST
Klieneberger-Nobel	Fruton2, Hoyrup, Morton, Searing
Lancefield	AmM&WS(12th), AWSC, Bail, Fruton2, Fruton3, Herz, Hoyrup, Ireland88, Morton, NTCS, NWLS, Oneil, Pell, Ross1, Ross2, Vernoff, Weiser
Lepeshinskaia	Debus, Fruton2, Herz, Hoyrup, Ireland70
Levi-Montalcini	AmM&WS(19th), AWSC, Fruton2, Herz, Hoyrup, Ireland88, Kass, Mcgrayne, Morton, NTCS, NWLS, Oneil, Pell, Porter, Ross2, Stille, Vare, Weiser, WSD
Lyon	Herz, Kass, Ross1, Weiser
Mangold	Mcgrayne
Manzolini	Chaff, Golemba, Herz, Hoyrup, Ireland70, Lovejoy, Mozans, Ogilvie, Pell, Uglow, Weiser
McClintock	Abir, AmM&WS(18th), AWSC, Bail, Fruton2, Fruton3, Golemba, Herz, Hinding, Hoyrup, Ireland88, Kahle, Kass, Mcgrayne, NTCS, NWLS, Oneil, Pell, Porter, Ross1, Ross2, Stille, Sweeney, Torrey, Uglow, Vare, Vernoff, Visher, Weiser, WSD
Mintz	AmM&WS(19th), AWSC, Herz, Ireland88, NTCS, Oneil, Ross2, Weiser
Morgan, A. H.	AmM&WS(8th), AWSC, Bail, Bonta, Fruton3, Herman, Herz, Ireland88, NotAmW80, NWLS, Pell, Porter, Ross1, Ross2, Siegel, Visher, WhoWasWho, WWWST
Morgan, L. V.	Fruton3, Herz, Ross1, Searing

Neufeld   AmM&WS(19th), Herz, Ireland88, NTCS, NWLS, Oneil, Ross2, Weiser, WWAW

Nice   Abir, AmM&WS(12th), AWSC, Bail, Bonta, Herz, Hinding, Hoyrup, Ireland88, NotAmW80, NTCS, NWLS, Pell, Porter, Ross1, Ross2, Siegel

Patrick   AmM&WS(19th), AWSC, Herz, Hinding, Ireland88, NTCS, NWLS, Oneil, Pell, Ross2, Stille, Torrey, Vare, Weiser

Pearce   AmM&WS(10th), AWSC, Chaff, Debus, Fruton2, Herman, Herz, Ireland70, Ireland88, Lovejoy, NotAmW80, Pell, Ross1, Siegel, Vare, WWWST

Phelps   Abir, AWSC, DAB, Elliott, Herz, Hinding, Hoyrup, Ireland70, Ireland88, Mchenry, Mozans, NCAB, Norw, NotAmW, Ogilvie, Pell, Ross1, Rudolph, Searing, Siegel, Sweeney, WhoWasWho

Pinckney   Abir, AWSC, DAB, Elliott, Golemba, Hoyrup, Norw, NotAmW, NWLS, Ross1, Sweeney, Uglow, Vare, WhoWasWho

Potter   Herz, Hinding, Hoyrup, Ireland70, Ireland88, Oneil, Searing, Vare, Vernoff

Quimby   AmM&WS(12th), AWSC, Debus, Golemba, Herman, Herz, Ireland70, Ireland88, Kahle, NCAB, NTCS, NWLS, Pell, Ross1, Ross2, Weiser, WWAW

Ray   AmM&WS(10th), AWSC, Herman, Herz, Ireland70, Ireland88, Mchenry, NTCS, Oneil, Pell, Ross2, Weiser

Rowley   AmM&WS(19th), Herz, NTCS

Russell   AmM&WS(10th), AWSC, Bail, Fruton3, Herman, Herz, Ireland88, Pell, Ross2, Searing, Siegel, WWWST

Sabin   AmM&WS(18th), AWSC, Bail, Chaff, DAB, Debus, DSB, Fruton2, Golemba, Herz, Hinding, Hoyrup, Ireland70, Ireland88, Kass, Lovejoy, Mchenry, Morton, NCAB, NotAmW, NTCS, NWLS, Ogilvie, Oneil, Pell, Read, Ross1, Ross2, Searing, Siegel, Stille, Sweeney, Uglow, Vare, Vernoff, Visher, Weiser, WSD, WWWST

Sager   AmM&WS(19th), AWSC, Debus, Fruton2, Fruton3, Golemba, Herz, Ireland70, Ireland88, Kass, Mchenry, NTCS, Oneil, Pell, Ross2, Vernoff, Weiser

Scharrer   AmM&WS(19th), AWSC, Debus, Fruton2, Herz, Ireland88, Kass, NTCS, Oneil, Pell, Ross2, Weiser

Seibert   AWSC, Bail, Grinstein, Mozans, NTCS, Oneil, Pell, Ross1, Ross2, WSD, WWAW

Shattuck   Abir, AWSC, Elliott, Herz, Ireland88, NotAmW, NWLS, Pell, Ross1, Rudolph, Siegel

Slye	AWSC, Bail, Debus, Fruton2, Herz, Hinding, Hoyrup, Ireland70, Ireland88, Kass, Lovejoy, Morton, NCAB, NotAmW80, NTCS, NWLS, Oneil, Pell, Ross1, Ross2, Searing, Siegel, Sweeney, WWWST
Stephenson	DSB, Fruton2, Herz, Hoyrup, Kahle, Kass, Morton, Pell, Searing
Stevens	AmM&WS(3rd), AWSC, Bail, DSB, Elliott, Fruton2, Herman, Herz, Hoyrup, Ireland88, Kahle, Kass, Mchenry, NotAmW, NTCS, NWLS, Ogilvie, Pell, Porter, Read, Ross1, Searing, Siegel, Stille, Uglow, Vare, Visher, WSD
Taussig	AWSC, NWLS, Pell, Ross2, WSD
Vennesland	AmM&WS(19th), AWSC, Debus, Fruton3, Herz, Oneil, Pell, Ross2, WWAW
Waelsch	AmM&WS(19th), AWSC, Debus, Herz, NTCS, Pell, Ross2, Weiser
Yalow	AmM&WS(19th), AWSC, Grinstein, Hoyrup, Ireland88, Mcgrayne, Mchenry, NTCS, NWLS, Pell, Porter, Ross2, Stille, Uglow, Vare

APPENDIX D: INDEX OF SOURCES

Some of the title codes are from *Biography and Genealogy Master Index*, 2d ed., and *1983 Supplement*, edited by M. C. Herbert and B. McNeil. Detroit: Gale Research, 1980, 1983.

Abir	Abir-Am, P. G., and D. Outram. *Uneasy Careers and Intimate Lives: Women in Science 1789–1979.* New Brunswick, NJ: Rutgers University Press, 1987.
AmM&WS	*American Men and Women of Science* (formerly *American Men of Science*). Edited by J. Cattell. New York: Bowker, 1910– .
AWSC	Bailey, M. J. *American Women in Science.* Santa Barbara, CA: ABC-CLIO, Inc., 1994.
Bail	Bailey, B. *The Remarkable Lives of 100 Women Healers and Scientists.* Holbrook, MA: Bob Adams, 1994.
Bonta	Bonta, M. M. *Women in the Field. America's Pioneering Women Naturalists.* College Station, TX: Texas A&M University Press, 1991.
Chaff	Chaff, S. L., et al. *Women in Medicine. A Bibliography of the Literature on Women Physicians.* Metuchen, NJ: Scarecrow Press, 1977.
DAB	*Dictionary of American Biography.* New York: Scribner's 1928–80.
Debus	Debus, A. G. (ed.). *World Who's Who in Science.* Chicago: Marquis Who's Who, 1968.
DSB	*Dictionary of Scientific Biography and Supplements.* Edited by C. C. Gillespie. New York: Scribner's, 1970–78.
Elliott	Elliott, C. A. *Biographical Index to American Science. The Seventeenth Century to 1920.* Westport, CT: Greenwood Press, 1990.

Fruton2   Fruton, J. S. *A Bio-Bibliography for the History of the Biochemical Sciences since 1800*. Philadelphia: American Philosophical Society, 1982.

Fruton3   Fruton, J. S. *A Supplement to a Bio-Bibliography for the History of the Biochemical Sciences since 1800*. Philadelphia: American Philosophical Society, 1985.

Golemba   Golemba, B. E. *Lesser-Known Women: A Biographical Dictionary*. Boulder, CO: Lynne Rienner, 1992.

Grinstein   Grinstein, L. S., R. K. Rose, and M. H. Rafailovich. *Women in Chemistry and Physics*. Westport, CT: Greenwood Press, 1993.

Herman   Herman, K. *Women in Particular: Index to American Women*. Phoenix, AZ: Oryx Press, 1984.

Herz   Herzenberg, C. L. *Women Scientists from Antiquity to the Present: An Index*. West Cornwall, CT: Locust Hill Press, 1986.

Hinding   Hinding, A., and A. S. Bower (eds.). *Women's History Sources*. New York: Bowker, 1979.

Hoyrup   Hoyrup, E. *Women of Science, Technology, and Medicine: A Bibliography*. Roskilde, Denmark: Roskilde University Library, 1987.

Ireland70   Ireland, N. O. *Index to Women of the World from Ancient to Modern Times*. Westwood, MA: Faxon, 1970.

Ireland88   Ireland, N. O. *Index to Women of the World from Ancient to Modern Times, A Supplement*. Metuchen, NJ: Scarecrow Press, 1988.

Kahle   Kahle, J. B. *Women in Science: A Report from the Field*. Philadelphia: Falmer Press, 1985.

Kass   Kass-Simon, G., and P. Farnes (eds.). *Women of Science: Righting the Record*. Bloomington, IN: Indiana University Press, 1990.

Lovejoy   Lovejoy, E. P. *Women Doctors of the World*. New York: Macmillan, 1957.

Mcgrayne   McGrayne, S. B. *Nobel Prize Women in Science: Their Lives, Struggles and Momentous Discoveries*. New York: Carol, 1993.

Mchenry   McHenry, R. *Famous American Women*. New York: Dover, 1983.

Morton   Morton, L. T., and R. J. Moore. *A Bibliography of Medical and Biomedical Biography*. Hants, England: Scolar Press, 1989.

Mozans	Mozans, H. J. (pseudonym for J. A. Zahm). *Women in Science*. New York: Appleton, 1913. Reprint. Cambridge, MA: MIT Press, 1974.
NCAB	*National Cyclopaedia of American Biography*. New York and Clifton, NJ: White, 1892– . Reprint. Vols. 1–50. Ann Arbor, MI: University Microfilms, 1967–71.
Norw	Norwood, V. *Made from this Earth: American Women and Nature*. Chapel Hill: University of North Carolina Press, 1993.
NotAmW	James, E. T., et al. (eds.). *Notable American Women, 1607–1950: A Biographical Dictionary*. Cambridge, MA: Belknap Press of Harvard University Press, 1971.
NotAmW80	Sicherman, B., et al. (eds.). *Notable American Women: The Modern Period*. Cambridge, MA: Belknap Press of Harvard University Press, 1980.
NTCS	McMurray, E. J. (ed.). *Notable Twentieth-Century Scientists*. Detroit: Gale Research, 1995.
NWLS	Shearer, B. F., and B. S. Shearer (eds.). *Notable Women in the Life Sciences*. Westport, CT: Greenwood Press, 1996.
Ogilvie	Ogilvie, M. B. *Women in Science: Antiquity Through the Nineteenth Century*. Cambridge, MA: MIT Press, 1991.
Oneil	O'Neil, L. D., ed. *The Women's Book of World Records and Achievements*. Garden City, NY: Doubleday, 1979.
Pell	Pelletier, P. A. *Prominent Scientists. An Index to Collective Biographies*. 3d ed. New York: Neal-Schuman, 1994.
Porter	Porter, R. (ed.). *The Biographical Dictionary of Scientists*. 2d ed. New York: Oxford University Press, 1994.
Read	Read, P. J., and B. L. Witlieb. *The Book of Women's Firsts*. New York: Random House, 1992.
Ross1	Rossiter, M. W. *Women Scientists in America: Struggles and Strategies to 1940*. Baltimore: Johns Hopkins University Press, 1982.
Ross2	Rossiter, M. W. *Women Scientists in America: Before Affirmative Action*. Baltimore: Johns Hopkins University Press, 1995.
Rudolph	Rudolph, E. D. "Women who studied plants in the pre-20th century United States and Canada." *Taxon* 39 (2) (1990): 151–205.

Searing
Searing, S. E. *The History of Women and Science, Health, and Technology: A Bibliographic Guide to the Professions and the Disciplines.* Madison: University of Wisconsin System Women's Studies Librarian, 1988.

Siegel
Siegel, P. J., and K. T. Finley. *Women in the Scientific Search: An American Bio-Bibliography, 1724–1979.* Metuchen, NJ: Scarecrow Press, 1985.

Stille
Stille, D. R. *Extraordinary Women Scientists.* Chicago: Children's Press, 1995.

Sweeney
Sweeney, P. E. *Biographies of American Women.* Santa Barbara, CA: ABC-CLIO, 1990.

Torrey
Torrey Botanical Club. *Index to American Botanical Literature, 1886–1966.* Boston: G. K. Hall, 1969.

Uglow
Uglow, J. S. *The Continuum Dictionary of Women's Biography.* New York: Continuum, 1989.

Vare
Vare, E. A., and G. Ptacek. *Mothers of Invention. From the Bra to the Bomb: Forgotten Women and Their Unforgettable Ideas.* New York: William Morrow, 1987.

Vernoff
Vernoff, E., and R. Shore. *The International Dictionary of Twentieth Century Biography.* New York: New American Library, 1987.

Visher
Visher, S. S. *Scientists Starred 1903–1943 in "American Men of Science."* Baltimore: Johns Hopkins University Press, 1947.

Weiser
Weiser, M.P.K., and J. S. Arbeiter. *Womanlist.* New York: Atheneum, 1981.

WhoWasWho
Who Was Who in America. Chicago: Marquis Who's Who, 1943–85.

WSD
Travers, B. (ed.). *World of Scientific Discovery.* Detroit: Gale Research, 1994.

WWAW
Who's Who of American Women. Chicago: Marquis Who's Who, 1958.

WWWST
Who Was Who in American History—Science and Technology. Chicago: Marquis Who's Who, 1976.

INDEX

Page numbers in *italics* refer to main entries in the sourcebook.

ABOUT THE CONTRIBUTORS

I. EDWARD ALCAMO, Ph.D., is Professor of Microbiology at the State University of New York (SUNY) at Farmingdale. He has taught at the college level for 30 years and is the author of *Fundamentals of Microbiology* (1994), *AIDS: The Biological Basis* (1996), and *DNA Technology: The Awesome Skill* (1995). Dr. Alcamo has been honored for excellence in teaching by the State University of New York (1986, 1990) and by the National Association of Biology Teachers (NABT) (1990). He is a member of numerous scientific societies and a fellow of the American Academy of Microbiology.

TERESA T. ANTONY came to the United States as a Fulbright Scholar and earned her doctorate in biochemistry from Fordham University. She did part of her postdoctoral work on muscle proteins at the Institute for Muscle Diseases. Her work later at Columbia University was on lens proteins, an evolutionary pattern from fishes to humans. Returning to India, she worked as a Senior Research Fellow at Delhi University on aflatoxins. She has been a faculty member at New York City Technical College of the City University of New York (CUNY) since 1969. She is a member of Sigma Xi and several scientific and educational professional organizations. She has authored and coauthored several publications, and her papers have appeared in various journals, including *Nature*, *Archives of Biochemistry and Biophysics*, and *Biochimica et Biophysica Acta*.

CAROL A. BIERMANN is Professor of Biological Sciences at Kingsborough Community College of CUNY, where she has taught since 1966. She received her M.A. degree from Brooklyn College (now part of CUNY), and her Ed.D. degree in science education from Rutgers University. She has authored and coauthored many papers for various journals, including *Journal for Research in Science Teaching*, *Journal of College Science Teaching*, *American Biology Teacher*, and *Science Education*. She has written several essays on women in the biological sciences, including Rosalyn Yalow and Ruth Sager. Dr. Biermann has presented papers at various scientific meetings. She is an active member of the community college subsection of the NABT and has served as editor of its newsletter, the *Biology Observer*. Her interests include science education as well as the role and status of women in the biological sciences.

JANET NEWLAN BOWER has an M.A. degree in history from the University of California at Los Angeles and an M.A. degree in education from the United States International University. She has taught U.S. history and Western civilization as a member of the adjunct faculty at San Diego Mesa College before moving to South Carolina, where she teaches at Midlands Technical College. She has authored articles on history and natural history.

VIRGINIA L. BUCKNER received her M.S. degree from the University of Missouri-Kansas City. She has taught biology at Johnson County Community College since 1970. Among her publications are several articles and a biology textbook, *A Manual for Life Science* (1985). She is the coordinator for region IV of the NABT.

MARTHA OAKLEY CHISCON earned her Ph.D. degree at Purdue University and has published in immunobiology, plant tissue culture, and problems affecting women in science. Proud of being a Purdue teacher of undergraduates, Dr. Chiscon also is Assistant Dean of Science. Recipient of numerous teaching and service awards, she recently was named Indiana Professor of the Year by the National Council for Advancement and Support of Education. An undergraduate teaching award at Purdue has been named in honor of Dr. Chiscon and her husband.

BEATRIZ CHU CLEWELL is Principal Research Associate at the Urban Institute in Washington, D.C. She received her Ph.D. degree in educational policy, planning, and analysis from Florida State University. Her main research activities have focused on factors that encourage or impede equal access to educational opportunity for members of racial/ethnic minority groups and women. Clewell has coauthored such publications as *Women of Color in Mathematics, Science, and Engineering: A Review of the Literature* (1991), *Building the Nation's Work Force from the Inside Out: Educating Minorities for the Twenty-first Century* (1991), and *Breaking the Barriers: Helping Female and Minority Students Succeed in Mathematics and Science* (1992). Recently she authored *Asian Americans in Mathematics and Science* (1995).

MARJORIE McCANN COLLIER received a Ph.D. degree in biology from CUNY. Her research interests are oogenesis and early development of the marine gastropod *Ilyanassa obsoleta*. She is Professor of Biology at Saint Peter's College, Jersey City, New Jersey, and served as Chairperson of the Biology Department from 1987 to 1995.

DIANA M. COLON is a biology faculty member at Northwest State Community College in Archbold, Ohio. She serves as the biology laboratory coordinator and program chair of General Studies. She has a master's degree in biology from Bowling Green State University, concentrating in physiology and endocrinology. Currently she is completing a Ph.D. degree in higher education, with a minor in biology, at the University of Toledo.

ALAN CONTRERAS is a graduate of the University of Oregon and serves as Senior Associate at the Oregon Community College Association. He is coauthor of *Birds of Oregon: Status and Distribution* (1994), a former editor of *Oregon Birds*, and past president of Oregon Field Ornithologists. His articles on birds have appeared in *American Birds*, *Western Birds*, *Birding*, *Oregon Birds*, *Missouri Conservationist*, *British Columbia Birds*, and *The Bluebird*.

LESTA J. COOPER-FREYTAG is Professor of Biology at Raymond Walters College, University of Cincinnati. Her graduate work was done at the University of Cincinnati and the University of Pennsylvania. She has been teaching at the University of Cincinnati since 1967 and has received the University's Cohen Award for Outstanding Teaching. Her professional interests are primarily in the fields of human genetics and the history of women in science. She is a member of Sigma Xi, Phi Beta Kappa, Iota Sigma Pi, and the NABT.

MARY R. S. CREESE is an Associate at the Hall Center for the Humanities, University of Kansas (formerly Research Associate in the Department of Medicinal Chemistry). Her present interest is nineteenth-century women research scientists. She has recently published papers on British women chemists and British women geologists in the *British Journal for the History of Science* and has contributed articles to the *Dictionary of National Biography* (1993 Supplement), to *American Chemists and Chemical Engineers* (1994), and to *Women in Chemistry and Physics* (1993).

JUDITH A. DILTS is Professor and Chair of Biology at William Jewell College. She received her Ph.D. degree in genetics from Indiana University. Her research interest is the molecular biology of bacterial endosymbionts in *Paramecium*. She was the 1995 Project Kaleidoscope Scientist-in-Residence. Dr. Dilts is a Biology Councilor for the Council on Undergraduate Research. She was principal investigator for two successful NSF-ILI grants and a collaborating investigator for an NIH-AREA grant.

EDNAMAY DUFFY is Associate Professor in the Division of Educational Studies at Lesley College and serves as Director of Academic Services to Graduate School students and faculty in off-campus sites in 15 states. She has a Ph.D. degree in higher education administration and science education from Boston College and has been a faculty member and Academic Administrator in higher education posts for over 23 years. Trained as a clinical microbiologist, she has taught in both two-year and four-year colleges, as well as at the graduate level in areas of microbiology and teacher education. She has been an active member and held various offices in both the NABT and the New England Chapter of the Association for Women in Science.

SUSAN E. EICHHORN received a B.S. degree in mathematics with a minor in physiology and genetics from the University of Illinois, Urbana. She worked on the academic staff of the University of Wisconsin in the Zoology and Botany Departments from 1961 to 1967 and taught mathematics at Yeshiva High School in Washington, D.C., for a year. She returned to the University of Wisconsin in 1969 and, studying under Ray Evert, was awarded an M.S. degree. She now works as a Research Program Manager with Evert, assisting him with his research and teaching programs at the University of Wisconsin.

RAY F. EVERT is Katherine Esau Professor of Botany and Plant Pathology at the University of Wisconsin, Madison. He received his Ph.D. degree in botany from the University of California at Davis after completing his undergraduate work at Pennsylvania State University. Dr. Evert has served as president of the Botanical Society of America and has won a number of honors, among them the Alexander von Humboldt Award and election to the American Academy of Arts and Sciences. His teaching accolades include the prestigious Emil H. Steiger Award for Excellence in Teaching from the University of Wisconsin. He spent an academic year (1965–66) in Germany as a Guggenheim fellow. His publications include 200 scientific papers and contributions to 24 books, including the classic reference *Plant Anatomy*, edited by Professor Emeritus Katherine Esau of the University of California at Santa Barbara.

THOMAS A. FIRAK has a Ph.D. degree in microbiology and immunology from the University of Illinois. He did postdoctoral work with Dr. Janet Rowley at the University of Chicago. The work entailed mapping genes on chromosome 5 related to acute myeloid leukemia arising from cytotoxic therapy. At present he is Associate Professor of Biology at Oakton Community College in Des Plaines, Illinois. He is a member of the NABT, the Illinois Association of Community College Biologists, the Illinois Society of Microbiologists, and the Woodstock Institute for Science and the Humanities. Dr. Firak writes a monthly science column for the *Woodstock Independent*.

MAURA C. FLANNERY, Ph.D., is Associate Dean and Professor of Biology at St. Vincent's College of St. John's University, Jamaica, New York. Her research interests include the aesthetic of biology and teaching nonscience majors. She is author of the 1991 book *Bitten by the Biology Bug*; writes a monthly column, ''Biology Today,'' for the *American Biology Teacher*; and has contributed to such journals as *Perspectives in Biology and Medicine, Science Education, Art and Academe*, and the *Journal of College Science Teaching*.

PAULA FORD teaches writing at Penn State's Altoona Campus and writes magazine articles and books, such as her recent *Birder's Guide to Pennsylvania*. She has been compiling a census of birds for over two years at one of her favorite places, a reclaimed strip mine in western Pennsylvania.

LOUISE S. GRINSTEIN received a Ph.D. degree from Columbia University in mathematics education. She has worked in industry as a computer programmer and systems analyst. She is currently Professor Emeritus of mathematics and computer science at Kingsborough Community College of CUNY. She is the coeditor of *Calculus: Readings from the Mathematics Teacher* (1977), *Women of Mathematics* (1987), *Mathematics Education in Secondary Schools and Two-year Colleges: A Sourcebook* (1988), and *Women in Chemistry and Physics* (1993). She is also the author of *Mathematical Book Review Index, 1800–1940* (1992).

ROBIN M. HALLER, coauthor of *The Physician and Sexuality in Victorian America* (1995), is currently a freelance indexer and writer in Carbondale, Illinois.

KATALIN HARKÁNYI is the Chemistry and Engineering Bibliographer and a Science Reference Librarian at the San Diego State University Library in San Diego, California. In addition to academic degrees in chemistry and library science, she has academic, special, and public library experience. She is a member of Special Libraries Association (SLA), History of Technology Society, and Beta Phi Mu. She has published a number of chemistry and engineering bibliographies and a full-length book, *The Natural Sciences and American Scientists in the Revolutionary Era* and is a coauthor of an expert system, ChemRAS. She is a contributor to the "Current Bibliography in the History of Technology," an annual supplement to *Technology and Culture*.

ROBERT M. HENDRICK, Ph.D., is Professor of History at St. Vincent's College of St. John's University, Jamaica, New York. His main area of research interest is the popularization of science in nineteenth-century America and France. He has published articles on this topic in *Science and Education, American Biology Teacher, History of Education, History Today*, and *Nineteenth-Century Contexts*.

MAUREEN M. JULIAN, Adjunct Professor and Senior Research Scientist, Department of Geological Sciences, Virginia Polytechnic Institute and State University at Blacksburg, Virginia, is currently doing theoretical calculations in lattice dynamics and applied group theory and is completing a biography of

Dame Kathleen Lonsdale. She attended Hunter College, received her Ph.D. degree from Cornell University, and held a research position with Professor Lonsdale at University College, London. Her latest paper is "Comparison of stretching force constants in symmetry coordinates between T_d and C_{3v} point groups."

JANE BUTLER KAHLE is presently Condit Professor of Science Education at Miami University. She works extensively in the assessment of the special needs and issues surrounding women and minorities in science. Her current research involves the development and testing of a theoretical model to guide research concerning the recruitment, retention, and success of girls and women in nontraditional courses and careers. Professor Kahle has received national awards and several international fellowships for her research. She has served as president of the NABT and section chairperson for the American Association for the Advancement of Science (AAAS). Professor Kahle is the author of more than 80 articles, 4 monographs, 13 book chapters, and 6 books. She has served on the editorial review boards of more than 10 scholarly journals.

LEE B. KASS is Associate Professor of Botany and Curator of the Elmira College Herbarium, as well as Adjunct Associate Professor at the L. H. Bailey Hortorium, Cornell University. Her research interests are in the flora of the Bahamas and the history of botany. She has published on the flora of San Salvador Island and has written biographies of American botanists. She is the recipient of a 1995–96 Fulbright Scholar Award to teach at the College of the Bahamas and to initiate a National Herbarium for the Bahamas.

KATHERINE KEENAN received a Ph.D. degree in biology from CUNY. She studied the physiology and ecology of both marine and freshwater microorganisms. Her interest in Lilian Morgan was sparked in 1970, while working in the Biology Department at Columbia University. She heard stories even then from colleagues about "Mrs. Morgan and the graduate students in the fly room." She did most of her research on Lilian Morgan later, while an Assistant Professor of Biology at Yeshiva University. In addition, she has written about the life of Ellen Swallow Richards. She also studied the history of marine laboratories during summers at the Marine Biological Laboratory in Woods Hole, Massachusetts. She is currently Senior Training Associate, Department of Clinical Research and New Drug Development, Hoechst Roussel Pharmaceuticals, Inc., in Somerville, New Jersey.

LINDA H. KELLER, Ph.D., is Research Associate Professor of Avian Medicine and Pathology at the University of Pennsylvania School of Veterinary Medicine at New Bolton Center. Her research and teaching interests focus on the mechanisms of bacterial pathogenesis. Current projects on *Salmonella enteritidis* infections of chickens involve bacterial gene manipulation and the study of chicken genetic parameters that affect the immune response to the pathogen. Results are published in both avian and microbiological research journals.

HARRIET KOFALK has published the only biography to date of Florence Merriam Bailey, *No Woman Tenderfoot* (1989), and recently completed a biography of Bailey's naturalist husband, Vernon. Her research now encompasses other interests, including vegetarian and solar cooking, which resulted in the publication of *The Peaceful Cook* (1993) and *Solar Cooking-Primer/Cookbook* (1995). She is currently working on a book about sea vegetables.

BONNIE KONOPAK is Professor and Chair of the Department of Instructional Leadership and Academic Curriculum at the University of Oklahoma. Her primary area of research interest is literacy development and instruction, with an emphasis on science education. In a series of investigations she has focused on conceptual learning from science text in middle school, the role of text within upper elementary and middle school science classrooms, and the beliefs of science teachers regarding reading and learning from text in middle and secondary schools.

JOHN KONOPAK is Assistant Professor of Instructional Leadership and Academic Curriculum at the University of Oklahoma. A curriculum theorist/generalist, his research interests focus on the practices of reflective inquiry in relation to teacher education. He has published articles on historical trends in reading research, writing critically, and the phenomenology of cultural assimilation through textual mediation. Professor Konopak has presented papers at professional conferences. He was formerly a journeyman carpenter, newspaper writer/editor, and broadcaster.

LAURA GRAY MALLOY is Associate Professor and Chair of the Department of Biology at Bates College, in Lewiston, Maine. She holds a B.A. in English and philosophy from SUNY Binghamton and a Ph.D. in physiology from the University of Virginia. She conducted postdoctoral research in pharmacology at the University of Vermont. Currently, she pursues research in cardiovascular physiology, physiology education, and the history of physiology. She has various publications in these areas.

BARBARA MANDULA received a Ph.D. degree in biochemistry, but has always been interested in ecological issues. In her current position with the Environmental Protection Agency in Washington, D.C., she is developing ways to share ecosystem information with the public. In previous positions she analyzed data on academic research and development for the National Science Foundation and led projects on human health-risk assessment. She is particularly concerned with issues that affect women scientists and has been active with the Association for Women in Science (AWIS).

MARY CLARKE MIKSIC received a Ph.D. degree from Columbia University. Her career has included histological and embryological research as well as teaching basic sciences at several institutions. Currently, she is at Queensborough Community College of CUNY. She is particularly interested in the interrelationship between structure and function in biological organ systems. She is working on a book for the general reader that attempts to highlight the significance of this interdependent relationship to nervous system development and potential.

VERONICA REARDON MONDRINOS received her Ed.D. degree from Rutgers University. She is retired from the Hillside, New Jersey, public school system, where she taught secondary school science for 20 years. Her research interests lie in undergraduate biology education. She is concerned with identifying the factors that determine women's success in the undergraduate biological sciences.

REBECCA MEYER MONHARDT is a doctoral candidate in science education at the University of Iowa. She received an M.Ed. degree in curriculum and instruction from the Texas A & M University. She spent 12 years as an elementary science teacher, teaching gifted and talented students, as well as students labeled at-risk. Her current research interests include issues dealing with equity and writing-to-learn in science.

RANDY MOORE is Professor of Biology and Dean of the College of Arts and Sciences at the University of Akron, Ohio. He studies botany and scientific writing and has published numerous articles in journals such as *Science*, *Scientific American*, and *Natural History*. He has also written several successful textbooks, including *Botany, Biology*, and *Writing to Learn Biology*. He has received many research grants and awards, including the Excellence in Educational Journalism Award and the Teacher Exemplar Award. He has been designated the ''Most Outstanding Faculty Member'' at both Baylor University and Wright State University. Professor Moore is editor of the *American Biology Teacher*.

SUZANNE E. MOSHIER, Ph.D., is currently Professor of Biology at the University of Nebraska at Omaha. Her scholarly research interests include the biochemistry of naturally occurring pigments (especially carotenoids); the culture and development of marine sponges; the relationship of marine sponges and bacteria with regard to carotenoid pigmentation; and the cellular immunology of protozoan parasitic diseases, particularly babesiosis. She is coauthor of two books, *Science in Cinema* and *Fantastic Voyages*, and several articles that explore the use of science fiction films to aid in science teaching.

BRINA NATHANSON received her Ph.D. degree from CUNY. Her doctoral dissertation was on the mode of action of extracts from the leaves of *Ricinus communis* (Castor bean plant). She belongs to the American Society of Plant Physiologists and the New York Academy of Sciences. She presently teaches microbiology and is engaged in research at CUNY dealing with algal physiology.

CONNIE H. NOBLES, Ph.D., is Assistant Professor in the Department of Teacher Education at Southeastern Louisiana University. She instructs elementary preservice teachers in science/social studies methodologies. Her research interests include gender equity issues, women in science, the archaeology of gender, and wetlands ecology. Among her publications are "Gender equity and the library" (*School Library Journal*); "Teaming up to teach archaeology" (*Underwater Archaeological Proceedings*); "Alice Eastwood" (in *Noted Women Scientists*, vol. 1); "Leona Woods Marshall Libby" (in *Noted Women Scientists*, vol. 2); and "Engaging learning through student-authored historical vignettes" (*Third International History, Philosophy and Science Teaching Conference Proceedings*).

MARILYN BAILEY OGILVIE is currently Curator of the History of Science Collections, Professor of Bibliography, and Adjunct Professor of the History of Science at the University of Oklahoma. Her research interests include women in science, history of biology, and bibliography. Among her publications are *Women in Science: Antiquity through the Nineteenth Century*, as well as chapters in *Uneasy Careers and Intimate Lives: Women in Science, 1789–1979* (1987) and *The Expansion of American Biology: The Interwar Years* (1991).

MARGARET (BETSY) G. OTT has an M.S. degree in biology from the University of Alabama at Tuscaloosa. She has been a member of the biology faculty at Tyler Junior College in Tyler, Texas, since 1982. Recently she has been elected Director-at-Large of the NABT. She currently serves on the NABT's Grants Publication Subcommittee. Professor Ott has been secretary and chair of the NABT two-year college subsection.

JEANIE STROBERT PAYNE is Associate Professor at Bergen Community College, where she teaches microbiology. She received an Ed.D. degree in microbiology and science education from Teachers College, Columbia University. Her professional interests include education and research in clinical microbiology. She has contributed to *Microbiological Applications: A Laboratory Manual in General Microbiology* by H. Benson, and she has coauthored *Microbiology Study Guide*.

VIRGINIA PEZALLA is a member of the faculty of Robert Morris College in Chicago, where she teaches biology and environmental science. Her research interests include prairie ecology and restoration, as well as dragonfly behavior.

She has published a paper on the ecology and behavior of the dragonfly, *Libellula pulchella (American Midland Naturalist)*.

GARY E. RICE is Assistant Professor in the College of Education at Louisiana State University. He received his Ph.D. degree from Syracuse University. His research interests are in the area of politics of literacy and technology, with particular interest in constructivist approaches to learning. He is a frequent contributor to the professional literature.

LINDA E. ROACH has a Ph.D. degree in science education from Louisiana State University. She is currently teaching at Northwestern State University of Louisiana. Using histories of science to enhance science teaching is one area of her research interest. Her publications include *I Have a Story about That: Historical Vignettes to Enhance the Teaching of the Nature of Science*; ''Short story science'' (*The Science Teacher*); and ''Putting the people back into science: Using historical vignettes'' (*School Science and Mathematics*).

SCOTT S. ROACH is Assistant Professor in the Department of Business at Northwestern State University of Louisiana. While his Ph.D. degree is not in science, he has had a long-term interest in this area.

ROSE K. ROSE received a Ph.D. degree in chemistry from CUNY. She is Professor of Physical Sciences at Kingsborough Community College of CUNY. She has been a contributing editor for six medical publications, including *Medical Tribune* and *Oncology News*. Her areas of interest include chemical pharmacology, medicinal chemistry, organic synthesis, chelates of palladium, liquid crystals, spectroscopy, stereoselective reactions, heterogeneous catalysis, and education in chemistry and physics, as well as women in science. She has authored or coauthored several books, including *Laboratory Experiments in Chemistry and Physics for the Allied Health Sciences* (1980, 1983, 1990), *Selected Topics in Physics for the Health Sciences* (1981, 1988), and *Women in Chemistry and Physics* (1993).

BIRGIT H. SATIR, Ph.D., received her training in biochemistry and cell biology in Denmark at the University of Copenhagen and the Carlsberg Biological Institute, with Hans Ussing and Erik Zeuthen. She came to the United States in 1961. She has held positions as a Research Associate at the University of Chicago and the University of California at Berkeley, where she became Adjunct Associate Professor in 1976. Like Berta Scharrer, she received a permanent tenured faculty position only when she moved to the Albert Einstein College of Medicine of Yeshiva University in 1977, as Professor. She continues to hold this appointment. Her awards include recognition as an ''Outstanding Scientist'' by AWIS in 1990 for her work on membrane structure and function. She is a fellow of the AAAS.

PETER SATIR, Ph.D., is currently Professor and Chairman of the Department of Anatomy and Structural Biology at the Albert Einstein College of Medicine. He has held this position since 1977, when Berta Scharrer became Professor Emeritus. Previously, he held positions at the University of California at Berkeley and at the University of Chicago. He received his graduate training in cell motility at the Rockefeller University, under the guidance of Keith R. Porter, in the early days of cell biology. His awards include a Guggenheim Fellowship for his work on cilia. He is a fellow of the AAAS.

ANNE-MARIE SCHOLER, Ed.D., has worked at Pine Manor College in Massachusetts. Currently she is Assistant Professor of Anatomy and Physiology at Endicott College in Beverly, Massachusetts. Her background is in the medical sciences, especially developmental biology, and in science education. Her research interests include gender equity in science education (especially higher education), as well as the history of women in the sciences.

JOEL S. SCHWARTZ is Associate Professor of Biology at the College of Staten Island of CUNY. He has degrees from the University of Rochester and New York University. His special area of interest is in the history of biology, particularly in the development of the theory of evolution, as well as Victorian naturalists, such as Charles Darwin, Alfred Russel Wallace, Thomas Henry Huxley, George John Romanes, and Robert Chambers. He has published numerous papers and reviews on these subjects in such journals as the *Journal of the History of Biology*, *Isis*, *Annals of Science*, *Archives Internationales d'Histoire des Sciences*, *Journal of the History of Behavioral Sciences*, and *Victorian Periodicals Review*. He is currently working on the letters of Romanes at the American Philosophical Society Library in Philadelphia, with the aid of a Mellon Research Fellowship.

PHILIP DUHAN SEGAL has an A.B. degree in chemistry and an M.A. degree in journalism, both from the University of Missouri, as well as a Ph.D. degree from Yeshiva University. He is Professor of English and has taught at CUNY since 1960. As Visiting Professor of American Literature at Hiroshima University in Japan, he did research on the response of imaginative writers to the atomic bomb. For 40 years he has been a member of the National Association of Science Writers. He has contributed over 200 articles and chapters to books and periodicals.

KEIR B. STERLING is Ordnance Branch Historian, U.S. Army, at Fort Lee, Virginia. He received a Ph.D. degree in history from Columbia University. Since 1966 he has taught at various colleges and universities, including the University of Wisconsin and Pace University in New York City. A biographer and historian of natural history, he is completing a biographical dictionary of North American environmentalists with several coauthors and is editing a multivolume international history of mammalogy.

DAVID R. STRONCK is Professor of Science Education at the California State University in Hayward. He is currently directing a grant from the National Science Foundation (NSF) to train teachers in biotechnology. In the past he directed six other grants of the NSF and five grants from the U.S. Department of Education. He is the sole author of six books, 20 major statistical research articles, and more than 200 other articles; he is also the editor of two books. From 1989 to 1991 he served as the Director of Research for the National Science Teachers Association.

RONALD L. STUCKEY earned a Ph.D. degree in botany from the University of Michigan at Ann Arbor. As Professor of Botany at Ohio State University (1965–91), he taught courses in general botany, local flora, aquatic flowering plants, and botanical nomenclature. He is a recognized international authority on geographical distribution of aquatic and wetlands plants in North America. He has also written extensively on nineteenth-century history of botanical exploration in eastern North America. Since 1991 he has been Professor Emeritus of Botany and continues his research and writing at the University Herbarium of Ohio State University.

PARIS SVORONOS earned his B.S. degree in chemistry and physics from the American University in Cairo, and his Ph.D. degree in organic chemistry from Georgetown University. After spending two years as Assistant Professor at Trinity College, he moved to Queensborough Community College of CUNY, where he is now Professor. He also has summer appointments at Georgetown University. His research interests are in the synthesis and electrochemistry of organosulfur compounds. He has published several books, including an organic laboratory manual as well as an organic problems workbook.

SORAYA GHAYOURMANESH-SVORONOS has earned B.S. and M.S. degrees at Mashad University, as well as M.S. and Ph.D. degrees in biochemistry at Georgetown University. She has done postdoctoral work at National Institutes of Health (NIH) and became a research assistant at the Cancer Institute of Columbia University. She has since worked as Adjunct Assistant Professor of Chemistry and Biochemistry at different institutions, including St. John's University, SUNY at Old Westbury, and Georgetown University. She has also contributed biographical and encyclopedia articles to various publications.

MILTON BERNHARD TRAUTMAN, a renowned ornithologist and ichthyologist, died in 1991. Having received little formal training beyond the first two years of high school, he learned much of value during apprentice years from teachers at Ohio State University and the University of Michigan. He also collaborated extensively with his wife, Mary Auten Trautman, an entomologist. He published over 150 papers, the majority about birds. His two monumental monographs were entitled *The Birds of Buckeye Lake, Ohio* (1940) and *The Fishes of Ohio* (1957). He worked as fish biologist at the Ohio Department of Conservation as well as at the Museum of Zoology of the University of Michigan. By 1955 he transferred to the Columbus Campus of Ohio State University, where he was advanced to Professor in 1969 and was named Professor Emeritus in 1972. He was awarded an honorary D.Sc. degree by Ohio State University in 1978.

SUSAN J. WURTZBURG is Lecturer in the Department of Geography, University of Canterbury, Christchurch, New Zealand. Her research interests include Maya archaeology, contributions of women to the social sciences, and the roles of women in Pacific Island societies. Recent publications include "Down in the field in Louisiana: An historical perspective on the role of women in Louisiana archaeology" in *Women in Archaeology* (1994) and "Gender relations and roles" and other contributions to *Magill's Ready Reference: American Indians* (1995), as well as articles in *Mexicon, Boletín de la Escuela de Ciencias Antropológicas de la Universidad de Yucatán* and the *International Journal of American Linguistics*.

NATIONAL UNIVERSITY
LIBRARY   SAN DIEGO

NON CIRCULATING

NATIONAL UNIVERSITY
LIBRARY   SAN DIEGO

NON CIRCULATING

ISBN 0-313-29180-2

90000>

EAN

9 780313 291807

HARDCOVER BAR CODE